# Orthogonal Waveforms and Filter Banks for Future Communication Systems

# Orthogonal Waveforms and Filter Banks for Future Communication Systems

Edited by

**Markku Renfors**
**Xavier Mestre**
**Eleftherios Kofidis**
**Faouzi Bader**

ACADEMIC PRESS
An imprint of Elsevier

Academic Press is an imprint of Elsevier
125 London Wall, London EC2Y 5AS, United Kingdom
525 B Street, Suite 1800, San Diego, CA 92101-4495, United States
50 Hampshire Street, 5th Floor, Cambridge, MA 02139, United States
The Boulevard, Langford Lane, Kidlington, Oxford OX5 1GB, United Kingdom

**Library of Congress Cataloging-in-Publication Data**
A catalog record for this book is available from the Library of Congress

**British Library Cataloguing-in-Publication Data**
A catalogue record for this book is available from the British Library

ISBN: 978-0-12-810384-5

For information on all Academic Press publications
visit our website at https://www.elsevier.com/books-and-journals

*Publisher:* Mara Conner
*Acquisition Editor:* Tim Pitts
*Editorial Project Manager:* Charlotte Kent
*Production Project Manager:* Julie-Ann Stansfield
*Designer:* Mark Rogers

Typeset by VTeX

*In the memory of my beloved father and supportive friend Georgios E. Kofidis.*

*Eleftherios Kofidis*

# Contents

## PART I  APPLICATION DRIVERS

### CHAPTER 1  New Waveforms for New Services in 5G ......... 3

Gerhard Wunder, Thorsten Wild, Frank Schaich,
Dimitri Kténas, Jean Baptiste Doré, Ivan Gaspar and
Gerhard Fettweis

### CHAPTER 2  TVWS as an Emerging Application of Cognitive Radio . . . . . . . . . . . . . . . . . . . . . . . . . . . . . . . . . . . 35

Dominique Noguet, Ranveer Chandra and Markus Mueck

**CHAPTER 4    Application of FBMC in Optical Communications.    73**
Jessica Fickers, Trung-Hien Nguyen,
Jérôme Louveaux, Simon-Pierre Gorza and
François Horlin

## PART II FILTERBANK SYSTEMS FOR COMMUNICATIONS: THEORY AND DESIGN

**CHAPTER 5    Multirate Signal Processing and Filterbanks. . . . .    89**
Markku Renfors and Juha Yli-Kaakinen

# PART III FBMC SIGNAL PROCESSING

## PART V  IMPLEMENTATION ASPECTS

**CHAPTER 18   Power Amplifier Effects and Peak-to-Average
                Power Mitigation**.............................. **461**
                Krishna Bulusu, Hmaied Shaiek, Daniel Roviras,
                Rafik Zayani, Markku Renfors, Lauri Anttila and
                Mahmoud Abdelaziz

**CHAPTER 19   FBMC Implementation for a TVWS System** ....... **491**
                Dominique Noguet, Vincent Berg, Jean-Baptiste Doré
                and Dimitri Kténas

# Contributors

**Mahmoud Abdelaziz**
Tampere University of Technology, Tampere, Finland

**Lauri Anttila**
Tampere University of Technology, Tampere, Finland

**Faouzi Bader**
CentraleSupélec, Rennes, France

**Leonardo Gomes Baltar**
Intel Deutschland GmbH, Neubiberg, Germany

**Jean Baptiste Doré**
CEA-Leti, Grenoble, France

**Nikolaos Bartzoudis**
Centre Tecnològic de Telecomunicacions de Catalunya (CTTC/CERCA), Barcelona, Spain

**Vincent Berg**
CEA-Leti, Grenoble, France

**Krishna Bulusu**
CEDRIC Laboratory, CNAM, Paris, France

**Màrius Caus**
Centre Tecnològic de Telecomunicacions de Catalunya (CTTC/CERCA), Barcelona, Spain

**Ranveer Chandra**
Microsoft Corporation, Redmond, WA, United States

**Xiaofei Chen**
San Diego State University, San Diego, CA, United States

**Yao Cheng**
Ilmenau University of Technology, Ilmenau, Germany

**Pascal Chevalier**
CEDRIC Laboratory, CNAM, Paris, France

**Jean-Baptiste Doré**
CEA-Leti, Grenoble, France

**Gerhard Fettweis**
Vodafone Chair Mobile Communications Systems, Technische Universität Dresden, Dresden, Germany

**Jessica Fickers**
Université libre de Bruxelles, Brussels, Belgium

**Oriol Font-Bach**
Centre Tecnològic de Telecomunicacions de Catalunya (CTTC/CERCA), Barcelona, Spain

**Martin Fuhrwerk**
Leibniz Universität Hannover, Hannover, Germany

**Ivan Gaspar**
Vodafone Chair Mobile Communications Systems, Technische Universität Dresden, Dresden, Germany

**Simon-Pierre Gorza**
Université libre de Bruxelles, Brussels, Belgium

**David Gregoratti**
Centre Tecnològic de Telecomunicacions de Catalunya (CTTC/CERCA), Barcelona, Spain

**Christophe Gruet**
AIRBUS Defence and Space, Elancourt, France

**Martin Haardt**
Ilmenau University of Technology, Ilmenau, Germany

**Fred Harris**
San Diego State University, San Diego, CA, United States

**François Horlin**
Université libre de Bruxelles, Brussels, Belgium

**Eleftherios Kofidis**
University of Piraeus, Piraeus, Greece;
Computer Technology Institute & Press "Diophantus" (CTI), Patras, Greece

**Dimitri Kténas**
CEA-Leti, Grenoble, France

**Didier Le Ruyet**
CEDRIC Laboratory, CNAM, Paris, France

**David Lopez Bueno**
Centre Tecnològic de Telecomunicacions de Catalunya (CTTC/CERCA), Barcelona, Spain

**Jérôme Louveaux**
Université catholique de Louvain, Louvain-la-Neuve, Belgium

**Laurent Martinod**
AIRBUS Defence and Space, Elancourt, France

**Davide Mattera**
Università degli Studi di Napoli Federico II, Napoli, Italy

**Philippe Mège**
AIRBUS Defence and Space, Elancourt, France

**Xavier Mestre**
Centre Tecnològic de Telecomunicacions de Catalunya (CTTC/CERCA), Barcelona, Spain

**Markus Mueck**
INTEL Deutschland GmbH, Munich, Germany

**Tor André Myrvoll**
SINTEF Digital, Trondheim, Norway

**Trung-Hien Nguyen**
Université libre de Bruxelles, Brussels, Belgium

**Dominique Noguet**
CEA-Leti, Grenoble, France

**Nikos Passas**
University of Athens, Athens, Greece

**Ana I. Pérez-Neira**
Centre Tecnològic de Telecomunicacions de Catalunya (CTTC/CERCA), Barcelona, Spain;
Universitat Politècnica de Catalunya, Barcelona, Spain

**Dmitry Petrov**
Magister Solutions Ltd., Jyväskylä, Finland

**Mylène Pischella**
CEDRIC Laboratory, CNAM, Paris, France

**Xavier Pons-Masbernat**
AIRBUS Defence and Space, Elancourt, France

**Markku Renfors**
Tampere University of Technology, Tampere, Finland

**Vidar Ringset**
SINTEF Digital, Trondheim, Norway

**François Rottenberg**
Université catholique de Louvain, Louvain-la-Neuve, Belgium

**Daniel Roviras**
CEDRIC Laboratory, CNAM, Paris, France

**Vincent Savaux**
b<>com, Cesson-Sévigné, France

**Frank Schaich**
NOKIA Bell Labs, Stuttgart, Germany

**Hmaied Shaiek**
CEDRIC Laboratory, CNAM, Paris, France

**Pierre Siohan**
Independent Researcher, Rennes, France

**Mario Tanda**
Università degli Studi di Napoli Federico II, Napoli, Italy

**Dimitris Tsolkas**
University of Athens, Athens, Greece

**Elettra Venosa**
Space Micro, San Diego, CA, United States

**Thorsten Wild**
NOKIA Bell Labs, Stuttgart, Germany

**Gerhard Wunder**
Heisenberg Communications and Information Theory Group, FU, Berlin, Germany

**Juha Yli-Kaakinen**
Tampere University of Technology, Tampere, Finland

**Rostom Zakaria**
CEDRIC Laboratory, CNAM, Paris, France

**Rafik Zayani**
CEDRIC Laboratory, CNAM, Paris, France

# About the Editors

**Markku Renfors** received MSc and Dr. Tech. degrees from Tampere University of Technology (TUT), Tampere, Finland, in 1978 and 1982, respectively. After research and teaching positions at TUT, he was a Design Manager with Nokia Research Center and Nokia Consumer Electronics from 1988 to 1991. Since 1992, he has been a Professor with the Department of Electronics and Communications Engineering, TUT, where he was Department Head from 1992 to 2010. Dr. Renfors is a Fellow of IEEE and recipient of the 1987 IEEE Circuits and Systems Society's Guillemin–Cauer Award (together with Tapio Saramäki). He has authored 80 papers in refereed journals, about 330 papers in conferences with review practice, and two patents. Dr. Renfors was an Associate Editor of IEEE Signal Processing Letters in 2006–2010. Currently, he is a Senior Area Editor of IEEE Transactions on Signal Processing and a member of the Editorial Board of EURASIP Signal Processing journal. He has served in the conference organization committees of ISCAS 1988, EUSIPCO 2000, ICC 2001, PIMRC 2006, SPAWC 2007, and SiPS 2009. He has supervised 18 doctoral dissertations. His research interests include multirate filtering and filter banks, especially with applications in advanced multicarrier and single-carrier waveforms, software defined radio, and algorithms for flexible communications receivers and transmitters. He has actively participated in the EU FP7 projects PHYDYAS and EMPhAtiC developing FBMC techniques especially for cognitive radio and heterogeneous fragmented spectrum use scenarios.

**Xavier Mestre** received the MSc and PhD in Electrical Engineering from the Technical University of Catalonia (UPC) in 1997 and 2002, respectively, and the Licentiate Degree in Mathematics in 2011. During the pursuit of his PhD, he was recipient of a 1998–2001 PhD scholarship (granted by the Catalan Government) and was awarded the 2002 Rosina Ribalta second prize for the best doctoral thesis project within areas of Information Technologies and Communications by the Epson Iberica foundation. From January 1998 to December 2002, he was with UPC's Communications Signal Processing Group, where he worked as a Research Assistant and participated actively in several European-funded projects. In January 2003, he joined the Telecommunications Technological Center of Catalonia (CTTC), where he currently holds a position as a Senior Research Associate and head of the Advanced Signal and Information Processing Department. During this time, he has actively participated in multiple European projects (including the coordination of the EMPhAtiC project, in 2012–2015) and in several contracts with the European Space Agency and the local industry. He is IEEE Senior member and elected member of the IEEE Sensor Array and Multi-Channel Signal Processing technical committee (2013–present). He has been an associate editor of the IEEE Transactions on Signal Processing (2008–2011, 2015–present), technical chair of the European Signal Processing Conference EUSIPCO 2011, and a general co-chair of the European Wireless Conference 2014.

**Eleftherios Kofidis** received the Diploma and PhD degrees in 1990 and 1996, respectively, both from the Department of Computer Engineering and Informatics, University of Patras, Patras, Greece. From 1996 to 1998, he served in the Hellenic Army. In the period 1998 to 2000, he was a postdoctoral fellow at the Institut National des Télécommunications (INT), Évry, France (now Télécom SudParis). From 2001 to 2004, he was a research associate at the University of Athens, and an adjunct professor at the Universities of Peloponnese and Piraeus, Greece. In 2004, he joined the Department of Statistics and Insurance Science, University of Piraeus, Greece, where he is now an Assistant Professor. He is also affiliated with the Computer Technology Institute & Press "Diophantus" (CTI), Greece. His research interests are in signal processing for communications (with emphasis on multicarrier systems) and medical imaging. Dr. Kofidis has served as technical program co-chair in two international conferences (CIP-2008 and DSP-2009) and as a technical program committee member and reviewer in a number of conferences and journals. He has served as an Associate Editor in the EURASIP Journal on Advances in Signal Processing (JASP) and the IET Signal Processing journal. He (co-)organized three special sessions on filter bank-based multicarrier systems (ISWCS-2012, EW-2014, and SPAWC-2015) and was a lead guest editor for a JASP special issue on this subject. He has actively participated in a number of national and European projects, including the EU FP7 ICT projects PHYDYAS and EMPhAtiC. Dr. Kofidis is currently serving as an Associate Editor in the IEEE Transactions on Signal Processing journal.

**Faouzi Bader** received his BSc degree in Electronic Engineering (communication speciality) in 1996 from the Mentouri University of Constantine (UMC) in Algeria and his PhD degree (with Honours) in Telecommunications in 2002 from Universidad Politécnica de Madrid (UPM), Madrid, Spain. He joined the Centre Technologic de Telecomunicacions de Catalunya (CTTC) in Barcelona (Spain) as a Research Associate in 2002 and from 2006 to 2013 as a Senior Research Associate. Since June 2013, he is an Associate Professor at the Centrale Supélec campus of Rennes (France), developing his research at the Signal, Communication & Embedded Electronics (SCEE) research group. Since January 2017, he is a honorary Adjunct Professor at the University of Technology Sydney (UTS), Australia. His research interests are in signal processing for communications (with emphasis on multicarrier waveforms and resource management). He has published over 120 papers in peer-reviewed journals and international conferences, more than 13 book chapters, and 3 books. He served as a Technical Program Committee (TPC) member and session chair in major IEEE ComSoc and VTS conferences (ICC, PIMRC, VTC spring/fall, WCNC, ISWCS, GLOBECOM, ICT). He was the general chair of the eleventh edition (2014) of the ISWCS conference, the general co-chair of ISWCS'2015, and the TPC-co-chair of the ISWCS'2017 conference.

He has served as an associate editor in Emerging Telecommunications Technologies (ETT) and as a guest editor at EURASIP Journal on Advances in Signal Processing (JASP). He is an IEEE Senior Member since 2007.

# Preface

The development of the future generations of communication systems poses an important demand for waveforms and related signal processing solutions with improved spectral characteristics and greatly increased flexibility. This is needed to effectively support a heterogeneous mixture of applications and services, with diverse key performance indicators (KPIs), operating simultaneously in the same communication framework. Filterbank-based signal processing has been recognized as an important tool in this direction, providing attractive alternatives to current solutions, which are commonly based on the Orthogonal Frequency Division Multiplexing (OFDM) scheme. Due to their excellent spectral characteristics, FilterBank Multi-Carrier (FBMC) schemes are strong waveform candidates in demanding spectrum use scenarios, such as fragmented opportunistic spectrum use cases. Intensive worldwide research activities in recent years have made FBMC a mature and viable alternative for future wireless and wired communication systems development.

The aim of this book is to provide a comprehensive and up-to-date compilation of theoretical, algorithmic, design, implementation, and application aspects of FBMC systems. While recognizing the importance of various on-going developments for effective nonorthogonal communication schemes, the scope here is limited to orthogonal (or, in practice, nearly orthogonal) multicarrier waveforms, including general aspects of filterbank-based communication signal processing. The potential application areas include multiservice cellular wireless systems beyond 5G, professional mobile radio for safety organizations (PMR/PPDR), military communications, power line communications (PLC), and optical communications.

This book was elaborated by experts from both the academic and industrial sectors and summarizes expertise acquired in important recent European research projects of the FP7 ICT programme. We hope that it will serve as a useful reference to researchers, engineering practitioners, and graduate students working on physical (PHY) and media access control (MAC) layer aspects of modern and future wireless and wired communication systems. The wide range of topics covered gives the instructors using the book the flexibility to focus on particular themes of their interest. The readers are expected to have basic working knowledge of digital signal processing (DSP), communication theory, and wireless communication systems.

There are twenty chapters in total, organized in five parts:

Part I, Chapters 1–4, motivates the study of FBMC through a number of diverse applications in future communication systems, which are expected to greatly benefit from the adoption of new flexible multicarrier waveforms. Chapter 1 is concerned with the air interfaces in 5th generation (5G) networks and the main candidate waveforms for meeting the stringent requirements of these systems. Chapter 2 presents the regulatory context of the TV white space (TVWS) and discusses the use of FBMC as an option for TVWS technologies. Existing related products and current worldwide trials are also outlined. Chapter 3 discusses the PMR/PPDR emergency communication scenarios and systems in the 400 MHz band. The different modes of PMR communications, the waveforms standardized by ETSI, and the waveform candidates for the coexistence of narrowband and broadband PMR systems are presented. Chapter 4 is about the application of FBMC in optical communications, emphasizing the DSP-based coherent receivers. It explains the challenges in the implementation of multicarrier offset quadrature amplitude modulation (OQAM) in these systems, with an emphasis on the gains in equalization that stem from the adoption of this modulation.

Part II, Chapters 5–8, starts by introducing important tools for later developments, including complex (I/Q) signal and system models, sampling of bandpass signals, basics of multirate signal processing, the Nyquist pulse shaping principle, the polyphase structure, Fast Fourier Transform (FFT), and uniform modulation-based filterbanks. Effective polyphase filterbank-based channelization filtering schemes for DSP-intensive multichannel radio transmitters and receivers are discussed in Chapter 6. Chapter 7 presents the theory and evolution of orthogonal multicarrier waveforms, leading to the time-frequency orthogonality principles and laws named after Gabor, Weyl–Heisenberg, and Balian–Low. Then the characteristics of various orthogonal multicarrier waveforms are discussed in the mentioned theoretical framework. This discussion is extended to also include certain important nonorthogonal waveform candidates and filterbank-based single-carrier waveforms. After this chapter, the book focuses on the FBMC scheme with Offset-QAM subcarrier modulation (FBMC/OQAM, also known as OFDM/OQAM), which attains maximal spectral efficiency among orthogonal FBMC waveforms. Chapter 8 reviews various alternative filterbank design/optimization methods for FBMC and introduces alternative filterbank structures for efficient implementation. The latter part of this chapter presents another alternative for effective generation and processing of FBMC/OQAM. This realization scheme is based on the fast-convolution (FC) idea, that is, realizing linear convolution in the FFT domain through bin-wise multiplication. This leads to the concept of FC-based waveform processing engine, which is able to support simultaneous processing of different and/or differently parameterized waveforms within the used signal band.

Part III, Chapters 9–12, is concerned with the challenges implied by FBMC/OQAM modulation in the implementation of single-antenna receiver tasks, namely synchronization, and estimation and equalization of the channel distortion. First, Chapter 9 offers some insight into the effect of the channel response on the received signal of any FBMC/OQAM link. Building on the polyphase implementation, a system model that explicitly takes the channel frequency selectivity into account is developed, and it is shown how the dynamics of the channel frequency response (i.e., its derivatives) play a fundamental role in the signal distortion. Synchronization in an FBMC/OQAM system is the subject of Chapter 10. The sensitivity of this FBMC modulation to a residual symbol timing offset (STO) or carrier frequency offset (CFO) is studied first, followed by an overview of different algorithms for STO and CFO estimation and compensation. Both training-based (relying on a preamble or scattered pilots) and blind approaches are considered, which operate in the time or frequency domain. The asynchronous multiuser scenario, where FBMC/OQAM would be mostly influential, is also considered through synchronization processing embedded into the FC-based filterbank (FC-FB) structure of Chapter 8. Chapter 11 is devoted to channel estimation and its FBMC/OQAM-induced challenges. Both training (preamble- or pilot-)based and (semi-)blind techniques are considered. Emphasis is given to recent advances, addressing the more demanding scenario of strong channel frequency selectivity. Channel equalization is discussed in Chapter 12, with the presentation including both single-tap and multitap equalizers, for channels of low and high frequency selectivity, respectively. In addition to the classical per-subcarrier zero forcing (ZF) and minimum mean squared error (MMSE) linear equalizers, solutions benefiting from interference cancellation or widely linear processing are also considered. Particularly for channels of strong frequency selectivity, equalizers relying on the system model developed in Chapter 9 or on per-subcarrier frequency sampling within the FC-FB structure are given special attention. Recent developments in blind FBMC/OQAM equalization are also included in this chapter.

Part IV, Chapters 13–17, is devoted to multiantenna FBMC/OQAM-based processing in a single- or multiuser environment. Chapter 13 addresses the end-to-end design of FBMC/OQAM transceivers, with special emphasis on precoding design for single- and multiuser communications. The main focus of this chapter is on linear signal processing architectures that allow one to implement spatial multiplexing capabilities under nonideal channel conditions. Chapter 14 deals more specifically with the channel effect in a multiple-input-multiple-output (MIMO) FBMC/OQAM setup and discusses channel estimation and equalization techniques that are specially designed for such systems. Both low and high frequency selectivity scenarios are considered, and different signal processing architectures are studied for each case. Chapter 15 focuses on the generation of spatial diversity on the transmit side in the absence of channel state information. In this chapter, different ways of constructing space-time codes with FBMC/OQAM modulation are presented, and a special emphasis is given to the challenging problem of realizing Alamouti space-time coding in this context. All the previous chapters are concerned with collocated MIMO architectures, where all the transmit/receive antennas are physically mounted on the same platform. Chapter 16, on the contrary, considers the special aspects of distributed and cooperative MIMO configurations. Here, one of the main targets is the study of (multicell) cooperative multipoint signal processing architectures in combination with FBMC/OQAM. Linear processing and channel state information acquisition methods in FBMC-based distributed systems are presented in detail. Part IV is concluded with Chapter 17 on radio resource management for FBMC/OQAM systems with special emphasis on multiuser scenarios. This chapter also analyzes system-level simulations for FBMC/QOAM-based systems, including physical layer abstraction mechanisms.

Part V, Chapters 18–20, is about the implementation aspects of FBMC-based systems. Chapter 18 is concerned with the nonlinearities present in practical FBMC/OQAM systems employing high-power amplifiers (HPA). Methods for reducing the corresponding peak to average power ratio (PAPR) and for digital predistortion (DPD)-based power amplifier linearization are presented and analyzed. FBMC for TVWS is detailed in Chapter 19. Dynamic spectrum access in a fragmented spectrum use scenario is at the focus of this chapter. The applicability of the frequency spreading FBMC (FS-FBMC) structure in such a setup is discussed in detail. Chapter 20 presents the design and implementation of the real-time demonstrator developed within the EMPhAtiC project, by which the developed FBMC concepts and flexible waveform processing solutions for broadband PPDR/PMR with coexistence capability with narrowband TETRA systems were successfully demonstrated in real hardware.

Supplementary materials can be accessed at the book's companion website https://www.elsevier.com/books-and-journals/book-companion/9780128103845. These materials include Matlab codes, with explanations, for selected central algorithms/techniques covered in the book.

# Acknowledgments

We are indebted to all the authors for their valuable contributions and time and effort spent for carefully preparing the texts and checking the proofs. Special thanks go to Maurice G. Bellanger for his comments and support in preparing this book, especially Chapter 8, as well as for having been the initiator and inspiring coordinator of the PHYDYAS project, a pioneering effort in developing FBMC as a viable candidate waveform for wireless communication applications. We would also like to thank the editorial staff at Elsevier, for guidance, encouragement, and patience during the preparation of this book. A major part of the technical knowledge presented in this book was conceived under the cooperative European Union FP7-ICT projects PHYDYAS (grant agreement no. 211887, http://www.ict-phydyas.org) and EMPhAtiC (grant agreement no. 318362, http://www.ict-emphatic.eu). We acknowledge the hard work and enthusiasm of all the researchers who participated in these efforts.

# Application Drivers

# I

## PART

## Application Drivers

# New Waveforms for New Services in 5G

1

**Gerhard Wunder\*, Thorsten Wild†, Frank Schaich†, Dimitri Kténas‡,
Jean Baptiste Doré‡, Ivan Gaspar§, Gerhard Fettweis§**

*Heisenberg Communications and Information Theory Group, FU, Berlin, Germany\**
*NOKIA Bell Labs, Stuttgart, Germany†*
*CEA-Leti, Grenoble, France‡*
*Vodafone Chair Mobile Communications Systems, Technische Universität Dresden, Dresden,
Germany§*

## CONTENTS

Orthogonal Waveforms and Filter Banks for Future Communication Systems. DOI: 10.1016/B978-0-12-810384-5.00001-3

## 1.1 KEY COMMUNICATION SCENARIOS

Fundamental research for 5G is well under way, and mobile communication networks are on the brink toward a new innovation cycle [1–3]. The main drivers are:

- *Gigabit wireless connectivity* is required, for example, in large crowd gatherings with possibly interactively connected devices using angle-controlled 3D video streaming, augmented reality, etc.
- The *Internet of Things* (IoT) will connect billions of devices, i.e., the things of our everyday life, which is far more than 4G can technically and economically accommodate. This will then open up new ways to monitor, assist, secure, control, e.g., in the telemedicine area, smart homes, smart factory, etc.
- Moreover, the *Tactile Internet* comprises a vast amount of real-time applications with extremely low latency requirements including *industrial wireless applications*. Motivated by the human tactile sense, which requires round-trip times in the order of 1 ms, 5G can then be applied for steering and control scenarios implying a disruptive change from today's content-driven communications. This is far shorter than current 4G cellular systems allow for, missing the target by nearly two orders of magnitude.

From a technical perspective it seems to be utmost challenging to provide uniform service experience to users under the premises of heterogeneous networking or future small-cell scenarios. Not only must the network operators be well prepared to take on the challenge of a much higher per-user rate and increasing overall required bandwidth but also to realize service differentiation with very different (virtually contradicting) application requirements. Consequently, the radio access has to be flexible, scalable, content aware, robust, reliable, and efficient in terms of energy and spectrum. In fact, with the limitations of current 4G system outlined further, the requirements will put further pressure on the common value chains on which the operators rely in order to compensate for investment costs for future user services. Hence, there is a clear motivation for an innovative and in part disruptive redesign of the physical (PHY) layer.

Let us discuss several intriguing examples.

### 1.1.1 SPORADIC TRAFFIC

Devices that generate sporadic traffic (for example, machine-type communication (MTC) devices in the IoT) should not be forced to be integrated into the bulky synchronization procedure of the LTE-A PHY layer random access [2,3], which has been deliberately designed to meet orthogonal constraints. Instead, ideally, they awake occasionally and then should transmit their messages right away and only be coarsely synchronized. By doing so, MTC traffic would be removed from standard uplink data pipes, with drastically reduced signaling overhead. Therefore, alleviating the synchronism requirement can significantly improve operational capabilities, network performance, user experience, and the lifetimes of autonomous MTC nodes, which are typically heavily resource-(energy-, computation-, memory-)constrained.

## 1.1.2 SPECTRAL AND TEMPORAL FRAGMENTATION

The LTE-A waveform imposes generous guard bands on other legacy networks to satisfy spectral-mask requirements, which either severely deteriorate spectral efficiency or even prevent band usage at all, which is again an artifact of strict orthogonality and synchronism constraints within the PHY layer [2,3]. Moreover, in a scenario with uncoordinated interference from pico- or femtocells and highly overlapping coverage, it seems illusive to provide the degree of coordination to maintain synchronism and orthogonality in the network while calling for new waveforms as well. In addition to spectral fragmentation, temporal fragmentation is another key issue. This occurs, for example, due to sporadic access in the asynchronous uplink physical layer random access channel (PRACH). Notably, asynchronous signaling also matters in the downlink in the context of cooperative multipoint (CoMP). In conclusion, such 5G scenarios where multiple users are allocated a pool of frequencies with relaxed (or even no) synchronization in time must be addressed by new waveforms. Such waveforms must implement sharp frequency notches and tight spectral masks in order not to interfere with other legacy systems and must be robust to asynchronous signaling and handle uncoordinated interference. Traditional OFDM schemes are not suited due to the inflexible handling of guard intervals (GIs), cyclic prefixes (CPs), or cyclic suffixes, as well as poor spectral localization. In a later section, we discuss waveforms achieving 100 times better localization. New waveforms thus make a real difference in fragmented spectrum and CoMP scenarios.

## 1.1.3 REAL-TIME CONSTRAINTS

Fourth-generation systems offer latencies of several tens of milliseconds between terminal and base station, which originate from resource scheduling, frame processing, retransmission procedures, and so on [2,3]. However, future application scenarios such as the tactile Internet scenario require ultralow latency matched with the human tactile sense. In such an environment, a massive number of distributed sensors and actuators will be connected to enable real-time tactile interaction in an augmented way. Sharing the medium becomes an additional challenge and imposes short wake-up cycles on the nodes and the use of burst transmission. Instead of consuming spectrum and power resources by introducing sophisticated algorithms to reach synchronism, an asynchronous approach appears promising. In order to achieve ultralow latency, each and every element of the communication and control chain must be optimized.

## 1.2 5G NEW AIR INTERFACE CORE ELEMENTS
### 1.2.1 WAVEFORMS

The ability to explore time and frequency dimensions is one core element of a flexible waveform. To better understand how these domains can be engineered, consider a signal $s$. Its time-domain representation $s(t)$ provides exact information about the

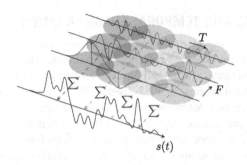

**FIGURE 1.1**

Illustration of Gabor expansion. The expanded signal is the sum of scaled time-frequency shifts of a prototype window. The scaling factors are given by the Gabor expansion coefficients.

behavior at any time instant. However, no information about frequency components at these positions is available. Instead, we can look at the Fourier transform (FT) of the signal, which provides exact information about frequency components, but no information on time-domain behavior is obtained. It is possible to gather information about frequency components of a signal at certain positions in time by looking at the FT of the multiplication of the signal with a window function, which leads to the short-time Fourier transform (STFT). But the output of the STFT can be highly redundant if the time and frequency parameters are kept independent.

In 1947, Dennis Gabor proposed to represent a signal as a linear combination of Gaussian functions that are shifted in time and frequency to positions in a regular grid; see Fig. 1.1. He chose the Gaussian function because it has the best localization in time and frequency simultaneously, so that local behavior of the signal is most accurately described. Gabor concluded that the original signal is fully characterized by the coefficients multiplying the Gaussian functions, establishing the foundation of time-frequency analysis. Later it was shown that the uniqueness and existence of such an expansion critically depends on the density of the grid of time-frequency shifts, which is defined as the product of spacing in time $T$ and frequency $F$. Densities larger than 1 imply nonunique expansions, whereas with densities smaller than 1, expansion coefficients only exist for certain signals. Nowadays, the linear combination of time-frequency shifted windows is known as a Gabor expansion, and the calculation of the STFT with a certain window at a regular grid is known as a Gabor transform. Expansion and transform windows are in a dual relation, i.e., the coefficients that are used to expand to a certain signal with a given window are provided by the Gabor transform of that signal with the dual window. In case the window and its dual are equal, the window is said to be orthogonal, and expansion and transform reduce to well-known orthogonal expansion series.

A prominent example is OFDM, which performs a Gabor expansion using a finite discrete set of rectangular window functions with length $T$ in time and shifts of $1/T$ in the frequency grid. In the discrete Gabor expansion and transform, which in the

OFDM case is the discrete Fourier transform (DFT), all signals are assumed to be periodic in time and frequency. However, nonperiodic time-continuous scenarios can be approximated by choosing long frames and appropriate sampling frequencies.

In this book chapter, the following waveform approaches will be covered:

- GFDM can be seen as a more generic block-oriented filtered multicarrier system that follows the Gabor principles. Basically, the parameterization of the waveform directly influences i) transmitter window, ii) time-frequency grid structure, and iii) transform length and can hence provide means to emulate a multitude of conventional multicarrier systems.
- FBMC/OQAM belongs to the family of filterbank-based waveforms. The principles revolve around filtering the subcarriers in the system while retaining orthogonality. As the name suggests, the essence of this candidate waveform is offset modulation, which allows avoiding interference between real and imaginary signal components.
- BFDM directly relates to the theory of Gabor frames. Signal generation can be considered a Gabor expansion, whereas the biorthogonal receive filter constitutes a Gabor transform.
- In UFMC, a pulse shaping filter is applied to a group of conventional OFDM subcarriers. This approach can be also represented in the context of the Gabor frame.

These waveforms have been thoroughly investigated, each particularly related to certain scenarios as described in detail in the next section.

## 1.2.2 UNIFIED FRAME STRUCTURE, ONE-SHOT TRANSMISSION AND AUTONOMOUS TIMING ADVANCE

The unified frame structure concept, depicted by Fig. 1.2, provides a flexible multi-service supporting solution in one single integrated 5G air interface. Each square in Fig. 1.2 represents a single resource element, a single subcarrier of a single multicarrier symbol.

For supporting the heterogeneous 5G system requirements, the frame is divided into four different areas:

- The type I area carries classical "bit pipe" traffic. High-volume data transmissions are served here. High spectral efficiency is the key performance indicator to be pursued. A high degree of orthogonality and strict synchronism is kept in this service type.
- Type II traffic is rather similar to type I traffic. Basically, the same service and device classes are supported. In contrast to type I, users being confronted with a higher degree of interference from adjacent cells are assembled here. Key building block for efficient multiuser separation is vertical layering (see the next section). Synchronization and orthogonality requirements are not as tight as with type I traffic.

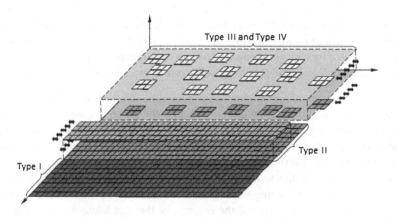

**FIGURE 1.2**

Unified frame structure.

- Type III traffic includes sporadic sensor/actor messages requiring low latencies. As outlined above, closed-loop synchronization is less suited here. Instead transmissions are only loosely synchronized (open-loop), and a contention-based access technique is used.
- Type IV traffic includes sporadic sensor/actor messages tolerating high latencies. Multiple signal layers are used, either using spreading or other multiple access schemes (e.g., interleaved division multiple access (IDMA)).

The unified frame concept shall be the main part of the standardization processes to be initiated in the future. In that context, exemplary scheduling/multiple access schemes are defined as follows:

- Dynamic, channel adaptive resource scheduling for traffic Type I using standard or advanced resource scheduling mechanisms.
- Semistatic/persistent scheduling for traffic Type II. From MAC point of view, it is necessary to decide on the amount of resources allocated for this type of traffic since schedulers will not adapt to specific parts of the frequency (may also be used for high-speed terminals).
- "One-shot transmission" using contention-like-based approaches for random access based on "sparse" signal processing methodology [4–6] respectively waveform design, which enables payload (Type III and IV) transmission in PRACH, in short "data" PRACH (D-PRACH). Clearly, by doing so, sporadic traffic is removed from standard uplink data pipes resulting in drastically reduced signaling overhead. Another issue that is closely related to the signaling overhead is the complexity and power consumption of the devices. Notably, waveform design in such a setting is necessary since the OFDM waveform used in LTE cannot handle the highly asynchronous access of different devices with possible negative delays or delays beyond the CPs. Clearly, guards could be introduced between the

individual (small) data sections, however, which make the approach again very inefficient.

• Notably, traffic types II and III rely on open-loop synchronization. The device listens to the downlink and synchronizes itself coarsely, based on synchronization channel and/or reference symbols, similar to 4G systems. Furthermore, the devices may apply some autonomously derived timing advance, which we call autonomous timing advance (ATA) relevant also particularly for MTC. ATA effectively shifts the receiver's "reception" window, which is aligned with the LTE broadcast signal, toward both positive and negative time delays. In such reception window, the multiuser interference is symmetric around the zero. This can indeed significantly lower the distortion for the new waveforms and, in particular, drastically with respect to OFDM.

## 1.3 5G WAVEFORM CANDIDATES

### 1.3.1 UFMC

#### 1.3.1.1 UFMC and UF-OFDM Overview

Universal Filtered Multicarrier modulation (UFMC) [7] is a subband-wise filtered variant of OFDM and thus also is denoted Universal Filtered (UF-) OFDM [8]. In practical OFDM systems, typically entire groups of subcarriers are allocated, organized in subbands, e.g., one or more physical resource blocks (PRB) in the terminology of LTE. In a multiuser setting, different subbands are reserved for different users. The driver for improved spectral localization is thus typically associated to subband-wise resource usage, which motivates the proposed subband-wise filtering.

With applying subband-wise filtering, improved in-carrier spectral localization is achieved, removing a weak point of OFDM. This improved spectral localization is an enabler for the multiservice air interface: It enables the system to multiplex signals with different multicarrier numerologies and allows tolerating time-frequency misalignments between different uplink allocations. An adaptation of numerologies and the capability to frequency-multiplex them is motivated by the very diverse nature of use cases and propagation environments which 5G has to support [9,10]. Robustness to time-frequency misalignments is helpful when frequency-multiplexing of broadband traffic and open-loop synchronized contention-based access traffic is carried out [11]. This allows light-weighted protocols [12] for efficient small packet transportation with low overhead and energy consumption.

A benefit of subband-wise filtering compared to subcarrier-wise filtering is that the filter is applied to an entire allocation (e.g., 12 subcarriers or more). Thus, the filters are broader in frequency domain than with subcarrier-wise filtering, and hence the filter impulse response is comparatively short in time domain, making UFMC well suited for short bursts [7], supporting low latency and MTC services well. A typical design choice so far is to use filter lengths that are in the order of typical cyclic prefix lengths in OFDM. Note that this is much shorter than FBMC, where typical

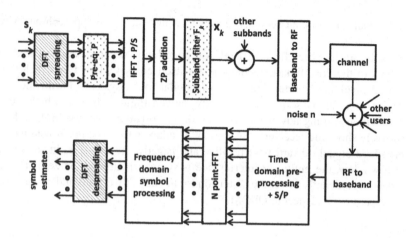

**FIGURE 1.3**

Block diagram of UFMC processing chain with optional DFT-precoding and filter precompensation.

filter lengths are in the order of 2–4 multicarrier symbols. UFMC targets QAM modulation and thus is well compatible to MIMO and coordinated multipoint (CoMP) techniques. In [13], it has been applied to uplink CoMP for improved robustness against carrier frequency offsets.

A reasonable option for UFMC is to operate with symbol durations larger than the inverse subcarrier spacing, enabling the use of a guard interval between multicarrier symbol bodies. Although most publications have set the filter length equal to the guard interval in time, this is not a must. Instead, it is a degree of freedom to balance temporal and spectral localization and thus to trade robustness in multipath propagation environments against protection against any sources of intercarrier interference [14,15]. Any variant of OFDM can use a cyclic prefix (CP) or a zero postfix (ZP) for delay spread protection [16], which are in most cases equivalent. UFMC is typically associated with the ZP-variant. The UF-OFDM concept in general both allows for CP and ZP, as discussed in [14].

### 1.3.1.2 *UFMC Basic Description*

Fig. 1.3 depicts the text-book variant of the transceiver. Variants with reduced complexity are given later. The transmitted sample vector $\mathbf{x}_k$ for the $k$th subband for one UFMC multicarrier symbol can be written as

$$\mathbf{x}_k = \mathbf{F}_k \mathbf{V}_k \mathbf{P}_k \mathbf{D}_k \mathbf{s}_k,$$

where the modulation symbol vector $\mathbf{s}_k \in \mathbb{C}^{Q \times 1}$ for a subband of $Q$ subcarriers is transformed into time domain by the tall IDFT-matrix $\mathbf{V}_k \in \mathbb{C}^{N \times Q}$ and filtered by the Toeplitz matrix $\mathbf{F}_k \in \mathbb{C}^{(N+L-1) \times N}$, $N$ is the FFT size, carrying out the linear

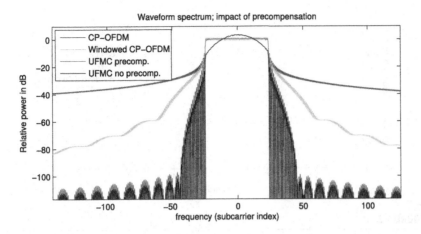

**FIGURE 1.4**

Spectrum for 4 PRBs; UFMC using Dolph–Chebychev filters with 80 dB side lobe attenuation; with and without transmit filter preequalization, compared to CP-OFDM and windowed OFDM.

convolution of a length $L$ subband-FIR filter. It is recommended to preequalize or precompensate the pass-band frequency response of the subband filter by the diagonal matrix $\mathbf{D}_k \in \mathbb{C}^{Q \times Q}$ to make the allocated band flat with equal transmitted power per subcarrier. In case low peak-to-average power ratios (PAPR) are of high importance (e.g., for the uplink), a single-carrier variant of UFMC (similar to single-carrier (SC)-FDMA as used in the LTE uplink) can be applied by precoding the modulation symbol vector with the DFT-precoding matrix $\mathbf{P}_k \in \mathbb{C}^{Q \times Q}$. In case additional protection against sources of inter-symbol interference, such as multipath delay spread, is required, the edge symbols of input vector $\mathbf{s}_k$ can be set to zero generating a low-power tail, as done in [17] for OFDM. This zero-tail-DFT-spreading technique for UFMC allows for additional adaptive delay spread protection if required. Fig. 1.4 depicts the UFMC spectrum.

For UFMC, low complex transmitter implementations exist, e.g., based on frequency domain filtering such as presented in [18]. For transmissions with relatively small allocation bandwidths (e.g., 12 or less subcarriers, often used by low-end MTC), even less complex implementations based on look-up tables exist [14]. The baseline receiver uses the $N$-point FFT [14], so the receiver complexity is almost identical to CP-OFDM. Note that subcarrier outputs after the FFT are fully orthogonal in case of a flat channel. Additional options for windowing and filtering [11,9,16] are beneficial for additional protection against intercarrier interference sources, such as mixed numerology.

Since UFMC is very close to OFDM, all the existing OFDM signal processing techniques can be reused, such as efficient frequency domain channel estimation techniques [19].

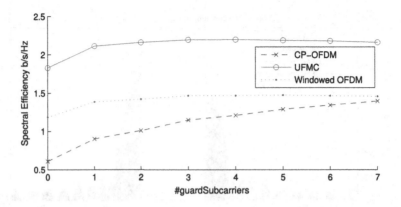

**FIGURE 1.5**

Spectral efficiency in mixed numerology uplink scenario using the parameters of [15]; 64-QAM, rate 1/2, ETU channel 3 km/h, SNR = 30 dB, 4-GHz carrier frequency.

### 1.3.1.3 *UFMC Results*

The UFMC time-frequency resource efficiency in short bursts has been compared against OFDM and FBMC in [7]. It has been shown that guard bands compared to conventional OFDM can be reduced, increasing spectral efficiency over OFDM by roughly 10%. Furthermore, the short filter lengths bring a significant spectral efficiency advantage over FBMC in case the transmission lasts only a few multicarrier symbols, when accounting for the time overhead of the filter. Filter optimization for UFMC has been addressed in [20,21], showing that SIR can be improved over OFDM by about 10 dB with carrier frequency offsets being equally distributed between ±0.1 relative to the subcarrier spacing. It is notable that the resulting filters from those leakage-based optimization are in most cases very close to the Dolph–Chebychev filters. This is not surprising as the Dolph–Chebychev design for a given main lobe width provides the lowest maximum side lobe level. Another nice property of the Dolph–Chebychev filters is that they can be tuned via the side lobe attenuation parameter.

Current 5G discussions incorporate the introduction of new nonorthogonal superposition-based multiple access schemes, such as IDMA. In [22], the compatibility of UFMC with IDMA was shown; both approaches complement each other well for efficient multiuser detection and contention-based access with relaxed time-frequency alignment.

Coexistence of UFMC and OFDM within the same carrier has shown to work well in [23], which opens up the option for a smooth migration from 4G to 5G and shows a wide range of parameterization options a 5G system can make use of when applying UFMC. UFMC results for MIMO including a practical hardware implementation in [24] have demonstrated a very good MIMO capability of the new waveform. As an interesting example result, Fig. 1.5 shows the uplink spectral efficiency performance in a mixed numerology setting when comparing UFMC against OFDM and windowed OFDM. The results include channel estimation with 1-D MMSE interpolation using an LTE-like pilot structure. The allocation of interest is 48 subcarriers wide with 15-kHz subcarrier spacing. Neighbor allocations are on each side with 48 subcarriers using 30-kHz subcarrier spacing. UFMC uses Dolph–Chebychev filters across 48 subcarriers with side lobe attenuation of 75 dB. Windowed OFDM uses an overlapping raised cosine filter where the overlap spans the cyclic prefix length. Note that maximum performance for UFMC is almost reached with a single guard subcarrier and that spectral efficiency is significantly higher than basic OFDM or windowed OFDM.

### 1.3.2 **FBMC**

With FBMC, a set of parallel data symbols are transmitted through a bank of modulated filters. The choice of the prototype filter controls the localization in frequency of the generated pulse and can provide better adjacent channel leakage performance in comparison to OFDM. OQAM, combined with Nyquist constraints on the prototype filter, is used to guarantee orthogonality between adjacent symbols and adjacent carriers while providing maximum spectral efficiency [25]. The duration $L_g$ of the prototype filter is a multiple of the size of the FFT $M$, so that $L_g = KM$, where $K$ is an integer and usually referred to as the overlapping factor. Frequency sampling technique is often considered to design the prototype filter. The technique has been proven simple and yet very efficient to build an almost optimal filter as a function of $K$. The FBMC transmitter–receiver structure can be efficiently implemented using IFFTs or FFTs combined with a polyphase network (PPN) [25]. Frequency Spreading approach (FS-FBMC) has been recently proposed in [26] and [27] as an alternative to PPN-FBMC. This technique is inspired by the frequency sampling technique used to design the prototype filter. With this approach, the number of nonzero samples in the frequency response is given by $P = 2K - 1$. For $K = 2$, the frequency domain pulse response coefficients are equal to [25]

$$G_0 = 1, \ G_1 = G_{-1} = \tfrac{\sqrt{2}}{2}. \tag{1.1}$$

For $K = 3$,

$$G_0 = 1, \ G_1 = G_{-1} = 0.911438,$$
$$G_2 = G_{-2} = \sqrt{1 - G_1^2}, \tag{1.2}$$

and for $K = 4$,

$$G_0 = 1, \ G_1 = G_{-1} = 0.971960,$$
$$G_2 = G_{-2} = \tfrac{\sqrt{2}}{2}, \ G_3 = G_{-3} = \sqrt{1 - G_1^2}. \tag{1.3}$$

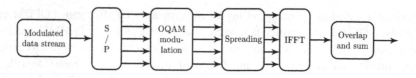

**FIGURE 1.6**

FBMC/OQAM transmitter block diagram.

The impulse response of the prototype filter is then given by

$$h(t) = G_0 + 2 \sum_{k=1}^{K-1} (-1)^k G_k \cos\left(\frac{2\pi k}{KM}(t+1)\right). \tag{1.4}$$

With the frequency spreading technique, symbols are spread over $P$ carriers by filtering the output of the OQAM process by the frequency domain pulse response. The output is then processed through an IFFT of size $KM$. OQAM precoding imposes that real and pure imaginary symbol values alternate on successive carrier frequencies and on successive transmitted symbols for any given carrier. This precoding guarantees orthogonality between adjacent carriers since the coefficients of the prototype filter are real. The output of the IFFT is converted through a parallel-to-serial (P/S) conversion and is accumulated with the following IFFT output data block stream delayed by $M/2$. This P/S conversion is called overlap-and-sum. Once the transient period is over, $2K$ of the $KM$-IFFT output samples are added together at any given time. Therefore, the time domain signal can be expressed as [28]

$$\mathbf{y}_k = \mathbf{F}_M^H \mathbf{G} \mathbf{X}_k \tag{1.5}$$
$$+ \sum_{p=1}^{2K-1} \mathbf{Q}_{-\frac{pM}{2}} \mathbf{F}_M^H \mathbf{G} \mathbf{X}_{k-p} + \mathbf{Q}_{\frac{pM}{2}} \mathbf{F}_M^H \mathbf{G} \mathbf{X}_{k+p},$$

where matrix $\mathbf{G}$ is the prototype filter matrix. $\mathbf{Q}_x$ introduces $x$ samples delay between the block of samples. Eq. (1.5) represents the sum of the $2 \times (2K - 1) + 1$ filtered FBMC/OQAM symbols that overlap over time. The corresponding transceiver is depicted in Fig. 1.6. By using the property of the prototype filter and the property of the spreading process (in frequency domain) it is possible to derive a closed-form expression of the time domain signal illustrating the use of the Fourier transform of size $M$ combined with a PPN filter. This formulation is interesting for practical implementation as demonstrated in [25].

At the receiver side, the dual operation of the overlap-and-sum operation of the transmitter is a sliding window in the time domain that selects $KM$ points every $M/2$ samples. The FFT is then applied on every block of $KM$ selected points. Equalization is applied using a single-tap equalizer and is followed by the filtering with the prototype matched filter (Fig. 1.7A). Data at the output of the matched filter is then

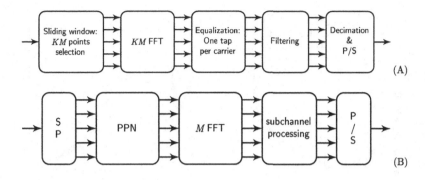

(A)

(B)

**FIGURE 1.7**

FBMC/OQAM receiver block diagram. (A) Using frequency spreading approach. (B) Using classical PPN approach.

demapped to compute a log-likelihood ratio for the input of the inner decoder. Because the size of the FFT is $K$-times larger than the multicarrier symbol time period, the signal at the output of the FFT is oversampled by a factor $K$ compared to the carrier spacing. This property gives a significant advantage to FS-FBMC receiver when the channel is exhibiting large delay spread or in case of synchronization mismatch. Of course, the classical PPN receiver scheme can also be applied [25] (Fig. 1.7B).

As reported in [27], the well-localized frequency response of FBMC/OQAM entitles the use of fragmented spectrum with minor interference on adjacent bands. Very good performance of FS-FBMC is also demonstrated in nonsynchronous uplink access due to the high stop-band attenuation of the prototype filter combined with the asynchronous frequency domain processing of the receiver [27]. Since the structure of the proposed FS-FBMC receiver is robust to channel exhibiting very large delay spread [27], this implies that carrier spacing can be significantly increased when FBMC is considered, giving the waveform a significant advantage for resilience to Doppler shift, CFO, and phase noise. It also gives advantages to the support of small data packet as the duration of the pulse can be reduced. Finally, if the carrier spacing is increased, then the number of active carriers is decreased; consequently, we could expect a better power efficiency of a transmitter if the PAPR could be reduced.

We depict in Fig. 1.8 the PSD for OFDM and FBMC/OQAM for several values of $K$. The overlapping factor directly controls the location in the frequency domain, and the spectral containment can be heavily relaxed by decreasing $K$. For $K = 3$, the spectral location is still very good compared to OFDM. For $K = 2$, the Out-Of-Band (OOB) leakage increases and is only 10 dB lower than OFDM. Using a small overlapping factor gives a lower pulse duration making this configuration interesting for short burst transmission or robustness against small coherence time of the channel. It is also noticeable that even for $K = 2$, the FBMC prototype filter shows good performance if a small frequency spacing is inserted between two adjacent users (here one RB).

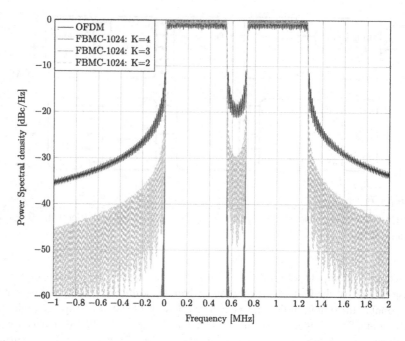

**FIGURE 1.8**

Power spectral density of OFDM and FBMC/QAM with several values of $K$.

The PAPR metric is assessed in Fig. 1.9 for 64-QAM. SC-FDMA, due to its (quasi) single carrier property, offers the lowest PAPR. Multicarrier waveforms (except SC-FDMA) have a comparable PAPR, 2 dB higher than SC-FDMA. It is well known that the higher the FFT size, the higher the PAPR. As a consequence, for FBMC/OQAM, PAPR is reduced when switching from a 1024-FFT size to a 256-FFT size, leading to almost the same performance as we get for SC-FDMA. This result is interesting for the definition of a waveform dedicated to UL, where a low PAPR is a stringent requirement.

Eventually the Bit Error Rate performance of the proposed waveform has been assessed for Extended Typical Urban model (ETU). This delay profile represents a high delay spread environment. Fig. 1.10 first shows the BER for a 64-QAM modulation over ETU channel. Here $M = 1024$ for all waveforms. Two DFT sizes are assessed for SC-FDMA, 12 and 24. Two different implementations of FBMC with $K = 4$ are tested: FBMC/OQAM with PPN receiver and FBMC/OQAM with FS receiver. SC-FDMA, despite the use of a CP, presents poor performance: the DFT spreading at the receiver indeed spreads the high fading occurrences over several carriers. This is all the more true when the DFT is realized on a larger number of carriers. Thanks to the 72 samples long CP, CP-OFDM has good performance. Both implementations of FBMC/OQAM nevertheless outperform CP-OFDM, due to the high length of the filter in the time domain, compared to the channel spreading. The FS receiver is the

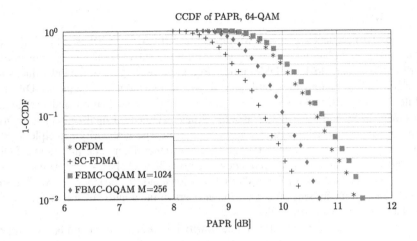

**FIGURE 1.9**

Comparison of PAPR for different waveforms using 64-QAM.

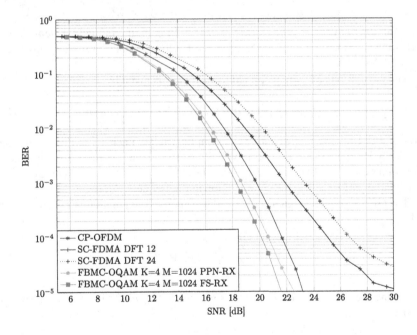

**FIGURE 1.10**

Performance of FBMC/OQAM and OFDM over ETU channel for 64-QAM.

best FBMC/OQAM receiver; it benefits from its high frequency granularity in such a frequency selective channel.

### 1.3.3 GFDM

Generalized Frequency-Division Multiplexing (GFDM) [29,30] is a block-filtered multicarrier modulation scheme, in which multiple subsymbols can be transmitted per subcarrier in a block. GFDM applies circular pulse shaping of the individual sub-carriers. GFDM can also resemble the well-known Orthogonal Frequency-Division Multiplexing (OFDM) system with the choice of a single subsymbol and a rectangular pulse shape, and the Single-Carrier Frequency-Domain Equalization (SC-FDE), using a single subcarrier with several subsymbols. The circularity principle allows GFDM to explore cyclic prefix (CP) and use Frequency-Domain Equalization (FDE) to handle multipath effects in frequency selective channel (FSC). GFDM can be designed to reduce the out-of-band (OOB) emissions and, for instance, to avoid severe interference in incumbent services or other users in the context of fragmented spectrum and dynamic spectrum allocation.

Consider a binary data vector source $\mathbf{b}$, which is encoded to obtain $\mathbf{b_c}$. A mapper, e.g., QAM, maps the encoded bits to symbols from a $2^\mu$-valued complex constellation, where $\mu$ is the modulation order. The resulting vector $\mathbf{d}$ denotes a data block that contains $N$ elements, which can be decomposed into $M$ subcarriers with $Q$ subsymbols each, according to $\mathbf{d} = \left(\mathbf{d}_0^T, \ldots, \mathbf{d}_{Q-1}^T\right)^T$ and $\mathbf{d}_q = \left(d_{0,q}, \ldots, d_{M-1,q}\right)^T$. The total number of symbols follows as $N = MQ$. Therein, the individual element $d_{m,q}$ corresponds to the data transmitted on the $m$th subcarrier and in the $q$th subsymbol of a GFDM block.

The circular pulse shaping applied to each $d_{m,q}$ is given by

$$g_{m,q}[n] = g\left[(n - qM) \mod N\right] \cdot \exp j2\pi \frac{m}{M}n, \tag{1.6}$$

with $n = 0, 1, \ldots, N - 1$ denoting the sampling index. Each $g_{m,q}[n]$ is a time- and frequency-shifted version of a prototype filter $g[n]$, where the remainder modulo $N$ operation makes $g_{m,q}[n]$ a circularly time-shifted version of $g_{m,0}[n]$ and the complex exponential performs the shifting operation in the frequency domain.

The GFDM transmit signal is obtained by superposition of all modulated subsymbols

$$x[n] = \sum_{m=0}^{M-1} \sum_{q=0}^{Q-1} g_{m,q}[n]d_{m,q}. \tag{1.7}$$

Collecting the filter samples in a vector $\mathbf{g}_{m,q} = \left(g_{m,q}[n]\right)^T$ allows us to formulate (1.7) as

$$\mathbf{x} = \mathbf{Ad}, \tag{1.8}$$

where $\mathbf{x} = (x[n])^T$, and $\mathbf{A}$ is a $MQ \times MQ$ transmitter matrix [31] with a structure according to

$$\mathbf{A} = \left(\mathbf{g}_{0,0} \quad \cdots \quad \mathbf{g}_{M-1,0} \quad \mathbf{g}_{0,1} \quad \cdots \quad \mathbf{g}_{M-1,1} \quad \cdots \quad \mathbf{g}_{0,Q-1} \quad \cdots \quad \mathbf{g}_{M-1,Q-1}\right). \tag{1.9}$$

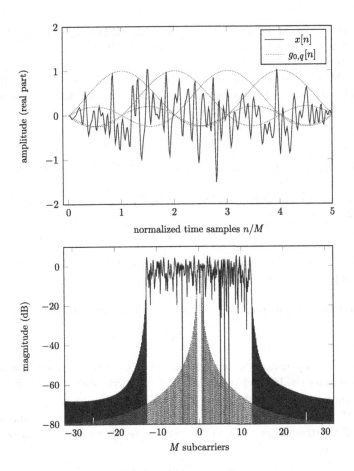

**FIGURE 1.11**

GFDM symbol in the time and frequency domains. Dotted lines show a Raised Cosine (RC) pulse shape, with roll-off $\alpha = 0.1$, cyclically shifted according to the subsymbol positions and the corresponding subcarrier spectrum when the first subsymbol is null. $M = 64$, $M_{on} = 24$ (amount of active subcarriers), $Q = 5$, $Q_{on} = 4$ (amount of active subsymbols).

At this point, **x** contains the transmit samples that correspond to the GFDM data block **d**.

The use of CP can provide robustness against interference between blocks and allow for FDE, and windowing [32] can be considered to reduce OOB, e.g., pinching the block boundaries. Inserting guard symbol (GS) is another solution that allows the GFDM block boundaries to naturally present a fade-in and fade-out behavior. Fig. 1.11 depicts a configuration where only four active subsymbols in GFDM are placed in a block with five subsymbols. The filter tails of the subsymbols are cyclically confined within the duration of the GFDM block. The use of one guard subsymbol, nulling the data at the first time slot, allows the GFDM signal to present smooth transitions at the block boundaries and improved OOB emission at the extremes of the spectrum.

The GFDM transceiver can be modeled as $\mathbf{y} = \mathbf{HAd} + \mathbf{w}$. Considering $\mathbf{z} = \mathbf{H}^{-1}\mathbf{HAd} + \mathbf{H}^{-1}\mathbf{w} = \mathbf{Ad} + \bar{\mathbf{w}}$ as the received signal after channel equalization, linear demodulation of the signal can be expressed as

$$\hat{\mathbf{d}} = \mathbf{Bz}, \tag{1.10}$$

where $\mathbf{B}$ is a $MQ \times MQ$ receiver matrix, and $\hat{\mathbf{d}}$ corresponds to the estimated received data before a slicer operation.

The Matched Filter (MF) receiver maximizes the signal-to-noise ratio (SNR) per subcarrier and is given by

$$\mathbf{B} = \mathbf{A}^H, \tag{1.11}$$

but with the effect of introducing self-interference when a nonorthogonal transmit pulse is applied, i.e., the scalar product $\langle g_{0,0}, g_{m,q} \rangle_{\mathbb{C}^N} \neq \delta_{0,m}\delta_{0,q}$ with Kronecker delta $\delta_{i,j}$.

The zero-forcing (ZF) receiver completely removes any self-interference with

$$\mathbf{B} = \mathbf{A}^{-1}, \tag{1.12}$$

but at the cost of potentially enhancing the noise. Over flat channels, the noise enhancement factor (NEF) $\xi$ determines the SNR reduction when using the ZF receiver. It is defined as $\xi = \sum_{i=0}^{N-1} \left| [\mathbf{B}]_{k,i} \right|^2 \geq 1$, which is equal for every $k$.

The Minimum Mean-Squared Error (MMSE) receiver makes a compromise between self-interference and noise enhancement considering a joint equalization, i.e., $\hat{\mathbf{d}} = \mathbf{By}$, where

$$\mathbf{B} = (\mathbf{R}_w + \mathbf{A}^H\mathbf{H}^H\mathbf{HA})^{-1}\mathbf{A}^H\mathbf{H}^H, \tag{1.13}$$

requires the knowledge of the covariance matrix of the noise, denoted as $\mathbf{R}_w$.

Finally, the received symbols $\hat{\mathbf{d}}$ are sliced and demapped to produce a sequence of bits $\hat{\mathbf{b}}_c$ at the receiver, which are then passed to a decoder to obtain $\hat{\mathbf{b}}$.

Further aspects of the GFDM waveform design as its more general framework concept emulating other prominent waveform candidates, applications in cognitive radio, and nonlinear receiver approach in MIMO systems can be found in [33–37].

### 1.3.4 BFDM

In biorthogonal frequency division multiplexing (BFDM), orthogonality of the set of transmit and receive pulses is replaced with biorthogonality, i.e., they are pairwise (not individually) orthogonal. Another feature is that $TF > 1$ (e.g., $TF = 1.25$), so that there is more flexibility in designing the transmit prototype pulse, e.g., in terms of side-lobe suppression and robustness to time and frequency asynchronisms. Notably, block transmission (say, for low latency) can be easily incorporated by "tail biting" of the prototype pulse. Efficient implementations are available in the literature and use the same FFT size as OFDM with some additional pre/postprocessing. They are discussed in detail in [38–41].

The BFDM approach, together with a suitable waveform, is well suited to sporadic traffic and the one shot transmission approach discussed (refer to Section 1.2) since the PRACH and D-PRACH symbols (recall that D-PRACH is the data carrying physical layer random access channel) are relatively long so that transmission is very robust to (even negative) time offsets. In addition, BFDM is also more robust to frequency offsets in the transmission, which, as is well known, typically sets a limit to the symbol duration in OFDM transmission. Finally, the concatenation of BFDM and OFDM symbols together in a frame requires a good tail behavior of the transmit pulse in order to keep the distortion to the payload carrying physical layer uplink shared channel (PUSCH) small. The excellent and controllable tradeoff between performance degradation due to time and frequency offsets is the main advantage of BFDM with respect to conventional OFDM.

As discussed, conventional OFDM and BFDM can be formulated as a pulse-shaped Gabor multicarrier scheme. For the time-frequency multiplexing, we will adopt a two-dimensional index notation $n = (n_1, n_2) \in \mathbb{Z}^2$. Let j denote the imaginary unit, and $\mu = (\mu_1, \mu_2)$. The baseband transmit signal is then

$$s(t) = \sum_{n \in \mathcal{I}} x_n \gamma_n(t) = \sum_{n \in \mathcal{I}} x_n (S_{\Lambda n} \gamma)(t), \qquad (1.14)$$

where $(S_\mu \gamma)(t) := e^{j 2\pi \mu_2 t} \gamma(t - \mu_1)$ is a time-frequency shifted version of the transmit pulse $\gamma$, i.e., $\gamma_n := S_{\Lambda n} \gamma$ is shifted according to a lattice $\Lambda \mathbb{Z}^2$. The lattice is generated by the $2 \times 2$ real generator matrix $\Lambda$, and the indices $n = (n_1, n_2)$ range over the doubly countable set $\mathcal{I} \subset \mathbb{Z}^2$, referring to the data burst to be transmitted. The coefficients $x_n$ are the random complex data symbols at time instant $n_1$ and subcarrier index $n_2$ with the property $\mathbb{E}\{xx^H\} = \mathbb{I}$ (from now on, $\overline{\cdot}$ always denotes complex conjugate, $\cdot^H$ means conjugate transpose, and $x = (\dots, x_n, \dots)^T$). We will denote the linear time-variant channel by the operator $\mathcal{H}$ and by $n(t)$ an additive distortion (a realization of a noise process).

In practice, $\Lambda$ is usually diagonal, i.e., $\Lambda = \mathrm{diag}(T, F)$, and the time-frequency sampling density is related to the bandwidth efficiency (in complex symbols) of the signaling, i.e., $\epsilon := |\det \Lambda^{-1}| = (TF)^{-1}$. The received signal is

$$r(t) = (\mathcal{H}s)(t) + n(t) = \int_{\mathbb{R}^2} \Sigma(\mu)(S_\mu s)(t) d\mu + n(t) \qquad (1.15)$$

with $\Sigma : \mathbb{R}^2 \to \mathbb{C}$ being a realization of the (causal) channel spreading function of finite support. In the wide-sense stationary uncorrelated scattering (WSSUS) assumption [42], the channel statistics is characterized by the second-order statistics of $\Sigma$, given as the scattering function $C : \mathbb{R}^2 \to \mathbb{R}_+$:

$$\mathbb{E}\{\Sigma(\mu)\overline{\Sigma(\mu')}\} = C(\mu)\delta(\mu - \mu') \qquad (1.16)$$

Moreover, we assume that $\mathbb{E}\{\boldsymbol{\Sigma}(\mu)\} = 0$ and $\|\boldsymbol{C}\|_1 = 1$. To obtain the (unequalized) symbol $\tilde{x}_m$ on time-frequency slot $m \in \mathcal{I}$, the receiver projects on $g_m := \boldsymbol{S}_{\Lambda m} g$, i.e.,

$$\tilde{x}_m := \langle g_m, r \rangle = \int_{\mathbb{R}} e^{-j2\pi(\Lambda m)_2 t} \overline{g(t - (\Lambda m)_1)}\, r(t) dt, \tag{1.17}$$

using the $\mathcal{L}_2$ scalar product $\langle \cdot, \cdot \rangle := \langle \cdot, \cdot \rangle_{\mathcal{L}_2}$. By introducing the elements

$$H_{m,n} := \langle g_m, \mathcal{H}\gamma_n \rangle = \int_{\mathbb{R}^2} \boldsymbol{\Sigma}(\mu) \langle g_m, \boldsymbol{S}_\mu \gamma_n \rangle d\mu \tag{1.18}$$

of the channel matrix $H \in \mathbb{C}^{|\mathcal{I}| \times |\mathcal{I}|}$ the overall transmission can be formulated as a system of linear equations $\tilde{x} = Hx + \tilde{n}$, where $\tilde{n} = (\dots, \langle g_m, n \rangle, \dots)^{\mathrm{T}}$ is the vector of the projected noise. We use the AWGN assumption such that $\tilde{n}$ is a Gaussian random vector with independent components, each having variance $\sigma^2 := \mathbb{E}\{|\langle r_m, n \rangle|^2\}$. The diagonal elements are

$$H_{m,m} = \int_{\mathbb{R}^2} \boldsymbol{\Sigma}(\mu) e^{j2\pi(\mu_1(\lambda m)_2 - \mu_2(\lambda m)_1)} \mathbf{A}_{g\gamma}(\mu) d\mu. \tag{1.19}$$

Here,

$$\mathbf{A}_{g\gamma}(\mu) := \langle g, \boldsymbol{S}_\mu \gamma \rangle = \int_{\mathbb{R}} g(t) \left( \boldsymbol{S}_\mu \gamma \right)(t) dt \tag{1.20}$$

is the well-known cross-ambiguity function of $g$ and $\gamma$.

### 1.3.4.1 *A General Approach to Capture Multiterminal Interference*

Our system model has to capture that many users each occupies a small number of subcarriers and each of them asynchronously (in time, frequency, or both) accesses this resource in an uncoordinated fashion. For a particular time-frequency slot $m = (m_1, m_2)$, we will denote the (random) channel operator as $\mathcal{H}(m)$ and the asynchronism as $\mathcal{D}(m)$. We assume that $\mathcal{H}(m)$ can be estimated using channel estimation procedure whereas $\mathcal{D}(m)$ cannot. Writing the received complex symbol $\tilde{x}_m$ in the absence of additive noise yields

$$\tilde{x}_m = \overline{H}_m x_m + \overbrace{(H_{m,m} - \overline{H}_m)x_m}^{\Delta_m} + \underbrace{\sum_{n \in \mathcal{I}, n \neq m} H_{m,n} x_n}_{\text{ICI}}, \tag{1.21}$$

where we defined $\overline{H}_m := \mathbb{E}_{|\mathcal{H}}\{H_{m,m}\}$, i.e., the mean value conditioned on a fixed channel $\mathcal{H}$.[1] Thus, the transmitted symbol $x_m$ will be multiplied by a constant and disturbed by two zero-mean random variables (RV), $\Delta_m$ and ICI. The first RV $\Delta_m$ represents a distortion, which comes from the randomness of $\mathcal{D}(m)$, and the second term ICI represents both. The mean power of both contributions, conditioned on a fixed channel $\mathcal{H}$, are $D_m := \mathbb{E}_{|\mathcal{H}}\{|\Delta_m|^2\} = P_m - |\overline{H}_m|^2$ where $P_m := \mathbb{E}_{|\mathcal{H}}\{|H_{m,m}|^2\}$ and $I_m := \mathbb{E}\{|\text{ICI}|^2\}$. Each element of the distortion sum $\sum_{n\neq m} H_{m,n}x_n$ is given by

$$H_{m,n} = \langle g_m, \mathcal{D}(n)\,\mathcal{H}(n)\gamma_n\rangle. \tag{1.22}$$

Notably, even if $\mathcal{D}(n)$ is the identity (synchronous access), $\tilde{x}_m$ is affected by all other individual contributions where the operators depend also on the index $n = (n_1, n_2)$. Hence, the performance for individual slots will be quite different, which complicates the situation, and no analytical approach is available in the literature! If all $\mathcal{D}(n)$ and $\mathcal{H}(n)$ are independent of $n$, then standard analysis can be used [43].

To find a tractable way, we consider the following approach. We assume that the $H_{m,m}$ can be estimated, and we consider the distortion sum $\sum_{n\neq m} H_{m,n}x_n$ averaged over all the subcarriers $m$. Obviously, this will average out individual interference for a specific subcarrier, but we can assume that these interference terms do not differ much. Individual performance is then measured by $H_{m,m}$ only! Then, we average over the random operators $\mathcal{D}(n)$ and $\mathcal{H}(n)$ as follows.

The sum can be bounded as [39]

$$\sum_m \sum_{n\neq m} \left|H_{m,n}\right|^2 \leq B_g E_g \sum_n \|\mathcal{D}(n)\,\mathcal{H}(n)\gamma_n\|_2^2 - \sum_m \left|H_{m,m}\right|^2. \tag{1.23}$$

Here, $B_g$ is the Bessel bound of the Gabor family $\mathcal{G}(g, \Lambda)$.[2] In the last step, we see that only the "action" of the operators $\{\mathcal{D}(n), \mathcal{H}(n)\}$ on $\gamma$ is relevant. We have set without loss of generality $\|\gamma\|_2^2 = 1$ and $1 \leq \|g\|_2^2 \leq E_g$ (typically, $E_g \approx 1$). Next, we compute the expectations, and we use $D_m = P_m - |\overline{H}_m|^2$ [39]:

$$\mathbb{E}\sum_m \sum_{n\neq m} \left|H_{m,n}\right|^2 + \mathbb{E}\sum_m D_m \leq B_g E_g \sum_n \mathbb{E}\,\|\mathcal{D}(n)\,\mathcal{H}(n)\gamma_n\|_2^2 - \sum_m \mathbb{E}|\overline{H}_m|^2. \tag{1.25}$$

---

[1] As a matter of fact, the expectations depend only on the marginal distribution of $\mathcal{H}(m)$.

[2] A *Gabor frame* $\mathcal{G}(\gamma, \Lambda^\circ)$ establishes a frame (for $L_2(\mathbb{R})$) if there are frame bounds $0 < A_\gamma \leq B_\gamma < \infty$ such that

$$A_\gamma \|f\|_2^2 \leq \sum_{n\in\mathbb{Z}^2} |\langle S_{\Lambda^\circ n}\gamma, f\rangle|^2 \leq B_\gamma \|f\|_2^2 \tag{1.24}$$

for all $f \in L_2(\mathbb{R})$.

We assume that the asynchronisms cannot increase the received power. For the first term, we estimate

$$\sum_n \mathbb{E} \, \|\mathcal{D}(n)\,\mathcal{H}(n)\gamma_n\|_2^2 \le \sum_n \mathbb{E}\|\mathcal{H}(n)\,\gamma_n\|_2^2 \le \sum_n \|C_n\|_1, \qquad (1.26)$$

according to (1.16). It remains to bound the second term:

$$\sum_m \mathbb{E}|\overline{H}_m|^2 = \sum_m \mathbb{E}\left|\|\mathbb{E}_{|\mathcal{H}}\langle g_m, \mathcal{D}(m)\,\mathcal{H}(m)\gamma_m\rangle\right|^2. \qquad (1.27)$$

For $a, b \in \mathbb{R}^2$, set $[a, b] := a_1 b_2 - a_2 b_1$ (the symplectic form) and define

$$s_m(\mu) := \mathbb{E}_{|\mathcal{H}}\langle g_m, \mathcal{D}(m)\,S_\mu \gamma_m\rangle \qquad (1.28)$$
$$= e^{-j2\pi[\mu, \Lambda m]}\mathbb{E}_{|\mathcal{H}}\langle g, S_{\Lambda m}^* \mathcal{D}(m)\,S_{\Lambda m} S_\mu \gamma\rangle, \qquad (1.29)$$

which essentially contains the distortion of the $\mu$th contribution in terms of the pulses conjugated by $S_{\Lambda m}$, i.e., "shifted" to T-F-slot $m$ in the time-frequency plane. For a fixed channel $\Sigma$, we have $\overline{H}_m = \langle \Sigma, s_m\rangle$, and on average, with respect to $\mathcal{H}(m)$, we have $\mathbb{E}\{|\overline{H}_m|^2\} = \langle C_m, |s_m|^2\rangle$. Hence, fixing the normalization such that $\sum_{m \in \mathcal{I}}\|C_m\|_1/|\mathcal{I}| = 1$, we have proved the following:

**Theorem 1.1.** *Suppose that $\|\gamma_m\|_2^2 = 1$ (without loss of generality), $\|g_m\|_2^2 = E_g$, and $\langle g_m, \gamma_m\rangle = 1$ (perfect reconstruction in noiseless case). The average distortion power per subcarrier is upperbounded by*

$$\frac{1}{|\mathcal{I}|}\mathbb{E}\sum_{m \in \mathcal{I}}(I_m + D_m) \le E_g B_g - \frac{1}{|\mathcal{I}|}\sum_{m \in \mathcal{I}}\langle C_m, |s_m|^2\rangle, \qquad (1.30)$$

*where $s_m$ is defined by $s_m(\mu) = \mathbb{E}_{\mathcal{H}}\langle g_m, \mathcal{D}(m)\,S_\mu \gamma_m\rangle$. Hence, the interference depends on the orthogonality of the Gabor frame represented by $B_g$ and the "average ambiguity" $\frac{1}{|\mathcal{I}|}\sum_{m \in \mathcal{I}}\langle C_m, |s_m|^2\rangle$ depending solely of the prototype pulse and the scattering function.*

As a special case, assume a deterministic time-frequency shift $\mathcal{D}(m) = S_{\nu(m)}$. This distortion is nonrandom and energy-preserving, i.e., $\|\mathcal{D}(m)\,g_m\|_2 = \|g_m\|_2$. Evaluating the function $s_m$ in (1.28) gives, for $\mu := \mu(m)$ and $\nu := \nu(m)$,

$$s_m(\mu) = e^{-j2\pi([\nu+\mu, \Lambda m]+\nu_1 \mu_2)}\mathbf{A}_{g\gamma}(\nu + \mu). \qquad (1.31)$$

Hence, in the AWGN case, we have:

$$\frac{1}{|\mathcal{I}|}\mathbb{E}\sum_{m \in \mathcal{I}}(I_m + D_m) \le E_g B_g - \frac{1}{|\mathcal{I}|}\sum_{m \in \mathcal{I}}|\mathbf{A}_{g\gamma}(\nu(m))|^2. \qquad (1.32)$$

Notably, for analytical purposes, we need to evaluate the cross-ambiguity function $\mathbf{A}_{g\gamma}$ requiring explicit expressions of transmit pulse $\gamma$ and receive $g$. However, often the dual is just the result of some complicated, often iterative, numerical procedure, which prevents simple estimates and (first) conclusions from (1.32).

For this purpose, consider the following bound. Define $\gamma_\mu := S_\mu \gamma$ and $\gamma_\nu := S_\nu \gamma$, and using $\langle g, \gamma \rangle = 1$, $\|g\|_2 = E_g$, and $\|\gamma\|_2 = E_\gamma$ yields

$$|\mathbf{A}_{g\gamma}(\mu)| = |\langle g, \gamma_\mu - \gamma + \gamma \rangle| \tag{1.33}$$
$$= |1 + \langle g, \gamma_\mu - \gamma \rangle| = |1 + \langle g, \gamma_\mu - \gamma_\nu + \gamma_\nu - \gamma \rangle| \tag{1.34}$$
$$\geq 1 - |\langle g, \gamma_\mu - \gamma_\nu \rangle| - |\langle g, \gamma_\nu - \gamma \rangle| \tag{1.35}$$
$$\geq 1 - E_g \|\gamma_\mu - \gamma_\nu\|_2 - E_g \|\gamma_\nu - \gamma\|_2, \tag{1.36}$$

where the RHS constitutes a similarity measure for $\gamma$. The same analysis can be carried out by shifting $g_{-\mu}$, i.e.,

$$|\mathbf{A}_{g\gamma}(\mu)| = |\langle g_{-\mu}, \gamma \rangle| \tag{1.37}$$
$$= |1 + \langle g_{-\mu} - g_{-\nu} + g_{-\nu} - g, \gamma \rangle| \tag{1.38}$$
$$\geq 1 - E_\gamma \|g_{-\mu} - g_{-\nu}\|_2 - E_\gamma \|g_{-\nu} - g\|_2. \tag{1.39}$$

Altogether, we have thus proved the following:

**Theorem 1.2.** *Suppose that* $\|\gamma_m\|_2^2 = E_\gamma = 1$ *(without loss of generality),* $\|g_m\|_2^2 = E_g$, *and that* $\langle g_m, \gamma_m \rangle = 1$ *(perfect reconstruction in noiseless case). Then, we have*

$$|\mathbf{A}_{g\gamma}(\nu)| \geq 1 - \min\left\{ E_g \|\gamma - \gamma(\cdot - \nu_1)\|_2 - E_f \|\hat{\gamma} - \hat{\gamma}(\cdot - \nu_2)\|_2 , \right. \tag{1.40}$$
$$\left. E_\gamma \|g - g(\cdot - \nu_1)\|_2 - g E_f \|\hat{g} - \hat{g}(\cdot - \nu_2)\|_2 \right\}, \tag{1.41}$$

*where* $\hat{\gamma}$ ($\hat{g}$) *denotes the Fourier transform of* $\gamma$ ($g$).

### *1.3.4.2 Numerical Example: OFDM*

Let us consider a numerical example in the context of OFDM. Here, the cross ambiguity function for $\gamma$ (rectangular pulse width $T_u + T_{cp}$) and $g$ (rectangular pulse width $T_u$) can be compactly written as (see [44])

$$\mathbf{A}_{g\gamma}(\nu) = \frac{\sin \pi \nu_2 (T_u - |[\nu_1]_{cp}|)}{\pi \nu_2 T_u} e^{j(\phi_0 - \pi \nu |[\nu_1]_{cp}|)} \chi_{[-T_u, T_u]}([\nu_1]_{cp}), \tag{1.42}$$

where the phase $\phi_0 = \pi \nu T_u$ is related to our choice of time origin, and $[\tau]_{cp} := \tau$ for $\tau \leq 0$, $[\tau]_{cp} := \tau - T_{cp}$ for $\tau \geq T_{cp}$, and zero else. Moreover, $\chi_{[-T_u, T_u]}$ is the characteristic function of the interval $[-T_u, T_u]$. Eventually, we need the Bessel bound, say, for the receive pulse $g$, which is $B_g = 1$, and the energy constant, which is $E_g = \epsilon$, so that altogether

$$\frac{1}{|\mathcal{I}|} \mathbb{E} \sum_m (I_m + D_m) \leq \epsilon - \frac{1}{|\mathcal{I}|} \sum_{m \in \mathcal{I}} \frac{\sin^2(\pi \nu_2^m (T_u - |[\nu_1^m]_{cp}|))}{(\pi \nu_2^m T_u)^2} \chi_{[-T_u, T_u]}([\nu_1^m]_{cp}).$$
$$\tag{1.43}$$

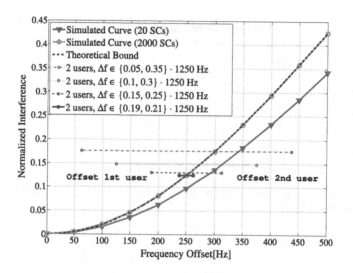

**FIGURE 1.12**

Simulated interference over frequency offset using different numbers of subcarriers, compared to theoretical bound. In addition, the behavior of the two-user case is illustrated, where each user gets half of the available subcarriers and has a different frequency offset. The (aggregate) interference converges to the value of the bound when the differences between the two offsets, which are centered around $0.2 \cdot 1250$ Hz, and their average become smaller.

First, we consider only frequency offsets with no additional delay in time. Fig. 1.12 shows a comparison of this interference bound with simulated curves at different numbers of subcarriers. We can observe that with an increasing number of subcarriers considered, the interference curve gets closer to the theoretical bound. In case of 200, or even more 20,000, subcarriers, the simulations match the bound almost perfectly.

The described curves are based on a single frequency offset only, which is the same for all subcarriers. However, Fig. 1.12 additionally illustrates the behavior in case of multiple different offsets. For illustration, we demonstrate the case of two offsets here, where each offset applies to an equal share of the available subcarriers. It can be observed that with decreasing difference in the offsets, the resulting interference level gets closer to value of the bound at the corresponding average of the offsets.

Let us now consider the case of time delays. Fig. 1.13 compares the interference caused by the asynchronous mode of operation to that predicted by the theoretical bound. Again, we can observe that the numerical results converge to the bound when increasing the number of subcarriers. Note that the CP length is 103 ms; a smaller offset does not produce any interference for OFDM as expected (however, negative delays do so, which is analyzed later).

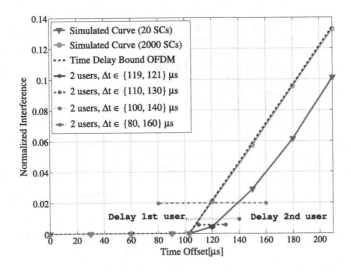

**FIGURE 1.13**

Simulated interference over time offsets using different numbers of subcarriers, compared to theoretical bound.

### 1.3.4.3 *Numerical Example: Spline*

Define the following pulse ("tent or first-order spline function"):[3]

$$g(t) = \left( \frac{\sin(\pi \frac{\alpha t}{T})}{\pi \frac{\alpha t}{T}} \right)^2 \chi_{[c,d]}(t), \tag{1.44}$$

where $2\pi \alpha / T$ is the bandwidth of this pulse related to the critical density, and parameters $c$ and $d$ align the pulse within the transmission frame. A simple estimate following approximation for frequency offsets:

$$A_{g\gamma}((0, \Delta \omega)) \geq 1 - \frac{\sqrt{3}\Delta \omega T}{2\pi \alpha} \sqrt{\left( 1 - \frac{\Delta \omega T}{2\pi \alpha} \right)}. \tag{1.45}$$

Using this approximation, we can calculate the interference part in (1.32). Similarly to frequency offsets, we can derive

$$A_{g\gamma}((\Delta t, 0)) \geq 1 - \frac{2\pi \alpha \Delta t}{\sqrt{20}T}. \tag{1.46}$$

---

[3]"Tent" stems from the triangular shape of its Fourier spectrum.

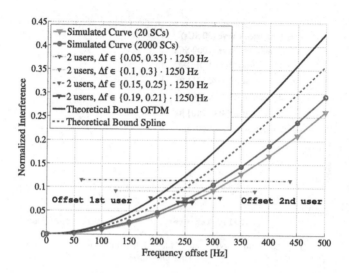

**FIGURE 1.14**

Simulated interference and interference bound vs. frequency offsets for the spline waveform. Numerical results for different numbers of subcarriers are shown. To foster an easy comparison, the interference bound for OFDM is also depicted.

Clearly, the bounds are rather loose, but they analytically prove that, for certain $\alpha$-settings (together with the numerical Bessel bounds), spline outperforms OFDM. This will be numerically evaluated next.

In Fig. 1.14, we show the simulated interference of the spline waveform together with the corresponding bound based on a numerical computation of $A_{g\gamma}$ in (1.32) for frequency offsets. In addition, the figure again shows also the interference bound for OFDM, whereas the results for the spline waveform indicate lower interference than in the OFDM case, and the bound appears to be less tight, even with large numbers of subcarriers.

Let us now consider the case of time offsets. Fig. 1.15 depicts the interference bound, again based on a numerical calculation of $A_{g\gamma}$ in (1.32), and the simulated interference vs. a time offset for the spline waveform. To allow an easy comparison, the bound for OFDM is also shown. It should be noted that although the results do not outperform OFDM for the positive delays considered in Fig. 1.15, the behavior is different for negative delays.

### 1.3.4.4 Reference Simulations

We assume that D-PRACH data resource contains only a very few number of subcarriers (about 5–20 subcarriers) as part of the standard LTE PRACH. In addition, in a 5G system, we can expect that there is a massive number of MTC devices, which will concurrently employ these data resources in an uncoordinated fashion. In the simplest approach, the D-PRACH uses the guard bands between PRACH and PUSCH, which

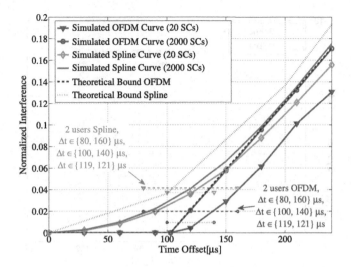

**FIGURE 1.15**

Simulated interference and interference bound vs. time offset for the spline waveform. Numerical results for different numbers of subcarriers are shown. To foster an easy comparison, the interference bound for OFDM is also depicted.

is the focus of this paper. We show that waveform design in such a setting is necessary since the OFDM waveform used in LTE cannot handle the highly asynchronous access of different devices with possible negative delays or delays beyond the CP. Clearly, guards can be introduced between the individual (small) data sections and to the PUSCH, which, though, makes the approach again very inefficient.

Fig. 1.16 illustrates the advantages of the BFDM approach for both frequency offsets and positive and negative time delays using the mentioned *spline or tent pulses*. We plot the D-PRACH symbol error rates over the time offset of a second, asynchronous user. In addition, a constant small frequency offset of 62.5 Hz is applied. The SNR is fixed at 25 dB. Moreover, in Fig. 1.16, we assume perfect channel knowledge for the user of interest. We can be observe that the additional frequency offset has a detrimental effect on both schemes. However, the performance loss of OFDM is significantly higher if the receiver window is shifted symmetrically around the zero.

Eventually, it is worth emphasizing that the guard bands in 4G LTE are relatively large so that the application of ATA is restricted to relatively demanding settings with large time and frequency shifts. However, future 5G systems will have shorter symbol lengths, so that the effects of new waveforms described in this paper are relevant for very typical IoT scenarios.

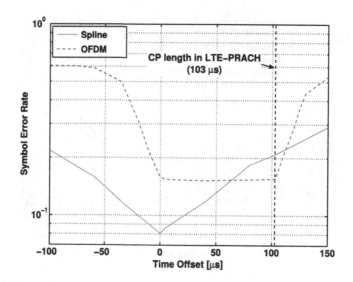

**FIGURE 1.16**

Symbol error rate in PRACH (using 4QAM) with perfect channel knowledge, averaged over 10 out of 20 data subcarriers vs. a varying time offset of a second user with a frequency offset of 62.5 Hz. The SNR is set to 25 dB. The black line shows the CP length in LTE PRACH.

## 1.4 CONCLUDING REMARKS

In this chapter, we presented an overview of the some candidate waveforms for a novel 5G air interface. The aim of the proposed solutions is to overcome the OFDM demerits in terms of poor spectral containment, robustness in highly asynchronous and high mobility scenarios, and inflexibility of parameter choice, which adversely affects the support of diverse services and asynchronous transmission as targeted by 5G. This is obtained by applying different degrees of filtering; in particular, the candidate waveforms have been grouped into subcarrier-wise filtered solutions and subband-wised filtered solutions.

## REFERENCES

[1] F. Schaich, B. Sayrac, S. Elayoubi, I.-P. Belikaidis, M. Caretti, A. Georgakopoulos, X. Gong, E. Kosmatos, H. Lin, P. Demestichas, *et al.*, "FANTASTIC 5G: Flexible air interface for scalable service delivery within wireless communication networks of the 5th generation," *Trans. Emerging Telecommun. Techn.*, vol. 27, no. 9, pp. 1216–1224, 2016.

[2] G. Wunder, *et al.*, "5GNOW: Non-orthogonal, asynchronous waveforms for future mobile applications," *IEEE Commun. Mag.*, vol. 52, no. 2, pp. 97–105, 2014.

[3] G. Wunder, H. Boche, T. Strohmer, and P. Jung, "Sparse signal processing concepts for efficient 5G system design," *IEEE Access*, Dec. 2015. [Online]. Available: http://arxiv.org/abs/1411.0435.

[4] P. P. G. Wunder and C. Stefanovic, "Compressive coded random access for massive MTC traffic in 5G systems," in *49th Annual Asilomar Conference on Signals, Systems, and Computers*, Pacific Grove, USA, Nov. 2015, invited paper.

[5] G. Wunder, P. Jung, and M. Ramadan, "Compressive random access using a common overloaded control channel," in *IEEE Global Communications Conference (Globecom'14) – Workshop on 5G & Beyond*, San Diego, USA, Dec. 2015. [Online]. Available: http://arxiv.org/abs/1504.05318.

[6] G. Wunder, P. Jung, and C. Wang, "Compressive random access for post-LTE systems," in *IEEE International Conf. on Commun. (ICC'14) – Workshop on Massive Uncoordinated Access Protocols*, Sydney, Australia, May 2014.

[7] F. Schaich and T. Wild, "Waveform contenders for 5G – suitability for short packet and low latency transmissions," in *Proc. VTC-Spring*, Seoul, Korea, May 2014.

[8] T. Wild, F. Schaich, and Y. Chen, "5G air interface design based on universal filtered (UF-)OFDM," in *Proc. Intern. Conf. on Digital Signal Process. (DSP)*, Hong Kong, Aug. 2014.

[9] F. Schaich and T. Wild, "Subcarrier spacing – a neglected degree of freedom?," in *Proc. SPAWC*, Stockholm, Sweden, Jun. 2015.

[10] F. Schaich, T. Wild, and R. Ahmed, "Subcarrier spacing – how to make use of this degree of freedom," in *Proc. VTC-Spring*, Nanjing, China, May 2016.

[11] F. Schaich and T. Wild, "Relaxed synchronization support of universal filtered multi-carrier including autonomous timing advance," in *Proc. ISWCS*, Paris, France, Aug. 2014.

[12] S. Saur, A. Weber, and G. Schreiber, "Radio access protocols and preamble design for machine-type communications in 5G," in *49th Asilomar Conference on Signals, Systems and Computers*, Pacific Grove, CA, USA, Nov. 2015.

[13] V. Vakilian, T. Wild, F. Schaich, S. ten Brink, and J.-F. Frigon, "Universal-filtered multi-carrier technique for wireless systems beyond LTE," in *Proc. Globecom Workshops*, Atlanta, GA, USA, Dec. 2013.

[14] Nokia, Alcatel-Lucent Shanghai Bell, "3GPP TSG-RAN1-165014: Subband-wise filtered OFDM for New Radio below 6 GHz." Tech. Rep., May 2016.

[15] Nokia, Alcatel-Lucent Shanghai Bell, "3GPP TSG-RAN1-165013: Initial uplink performance results for the New Radio waveforms below 6 GHz." Tech. Rep., May 2016.

[16] S. Venkatesan and R. A. Valenzuela, "OFDM for 5G: Cyclic prefix vs. zero postfix, and filtering vs. windowing," in *Proc. ICC*, Kuala Lumpur, Malaysia, May 2016.

[17] G. Berardinelli, F. Tavares, T. Sorensen, P. Mogensen, and K. Pajukoski, "OFDM for 5G: Cyclic prefix vs. zero postfix, and filtering vs. windowing," in *Globecom Workshops*, Atlanta, GA, USA, Dec. 2013.

[18] T. Wild and F. Schaich, "A reduced complexity transmitter for UF-OFDM," in *Proc. VTC-Spring*, Glasgow, Scotland, May 2015.

[19] X. Wang, T. Wild, F. Schaich, and S. ten Brink, "Pilot-aided channel estimation for universal filtered multi-carrier," in *Proc. VTC-Fall*, Boston, USA, Sep. 2015.

[20] X. Wang, T. Wild, F. Schaich, and A. Santos, "Pilot-aided channel estimation for universal filtered multi-carrier," in *Proc. European Wireless Conf.*, Barcelona, Spain, May 2014.

[21] X. Wang, T. Wild, and F. Schaich, "Filter optimization for carrier-frequency- and timing-offsets in universal filtered multi-carrier systems," in *Proc. VTC-Spring*, Glasgow, Scotland, May 2015.

[22] Y. Chen, F. Schaich, and T. Wild, "Multiple access and waveforms for 5G: IDMA and universal filtered multi-carrier," in *Proc. VTC-Spring*, Seoul, Korea, May 2014.

[23] R. Ahmed, T. Wild, and F. Schaich, "Co-existence of UF-OFDM and CP-OFDM," in *Proc. VTC-Spring*, Nanjing, China, May 2016.

[24] P. Weitkemper, J. Koppenborg, J. Bazzi, R. Rheinschmitt, K. Kusume, D. Samardzija, R. Fuchs, and A. Benjebbour, "Hardware experiments on multi-carrier waveforms for 5G," in *Proc. WCNC*, Doha, Qatar, Apr. 2016.

[25] M. Bellanger, *et al.*, "FBMC physical layer: A primer," Jun. 2010. [Online]. Available: http://www.ict-phydyas.org.

[26] M. Bellanger, "FS-FBMC: An alternative scheme for filter bank based multicarrier transmission," in *Proc. ISCCSP*, Roma, Italy, May 2012.

[27] J.-B. Doré, V. Berg, N. Cassiau, and D. Kténas, "FBMC receiver for multi-user asynchronous transmission on fragmented spectrum," *EURASIP Journal on Advances in Signal Processing, Special Issue on Advances in Flexible Multicarrier Waveform for Future Wireless Communications*, vol. 2014:41, pp. 1–20, 2014.

[28] J.-B. Doré, V. Berg, and D. Kténas, "Channel estimation techniques for 5G cellular networks: FBMC and multiuser asynchronous fragmented spectrum scenario," *Trans. Emerging Telecommun. Techn.*, vol. 26, no. 1, pp. 15–30, Sep. 2014.

[29] G. Fettweis, M. Krondorf, and S. Bittner, "GFDM – generalized frequency division multiplexing," in *VTC-Spring*, Barcelona, Spain, Apr. 2009.

[30] N. Michailow, M. Matthé, I. Gaspar, A. Navarro Caldevilla, L. L. Mendes, A. Festag, and G. Fettweis, "Generalized frequency division multiplexing for 5th generation cellular networks," *IEEE Trans. Commun.*, 2014.

[31] I. S. Gaspar, M. N. A. Navarro Caldevilla, E. Ohlmer, S. Krone, and G. Fettweis, "Low complexity GFDM receiver based on sparse frequency domain processing," in *VTC-Spring*, Dresden, Germany, Jun. 2013.

[32] E. Bala, J. Li, and R. Yang, "Shaping spectral leakage: A novel low-complexity transceiver architecture for cognitive radio," *IEEE Veh. Technol. Mag.*, vol. 8, no. 3, pp. 38–46, 2013.

[33] I. S. Gaspar, "Waveform advancements and synchronization techniques for generalized frequency division multiplexing," 2016. http://nbn-resolving.de/urn:nbn:de:bsz:14-qucosa-201875.

[34] R. Datta, *Generalized Frequency Division Multiplexing in Cognitive Radio*. Dresden: Jörg Vogt Verlag, 2014.

[35] N. Michailow, *Generalized Frequency Division Multiplexing Transceiver Principles*. Dresden: Jörg Vogt Verlag, 2015.

[36] D. Zhang, L. Mendes, M. Matthe, I. Gaspar, N. Michailow, and G. Fettweis, "Expectation propagation for near-optimum detection of MIMO-GFDM signals," *IEEE Trans. Wireless Commun.*, vol. 15, no. 2, pp. 1045–1062, 2015.

[37] M. Matthé, I. Gaspar, D. Zhang, and G. Fettweis, "Short paper: Near-ML detection for MIMO-GFDM," in *VTC-Fall*, Boston, USA, Sep. 2015.

[38] M. Kasparick, G. Wunder, P. Jung, and D. Maryopi, "Bi-orthogonal waveforms for 5G random access with short message support," in *European Wireless Conference (EW'14)*, Barcelona, Spain, May 2014.

[39] G. Wunder, M. Kasparick, and P. Jung, "Interference analysis for 5G random access with short message support," in *European Wireless Conference (EW'15)*, Budapest, Hungary, May 2015.

[40] G. Wunder, S. A. Gorgani, and S. Ahmed, "Waveform optimization using trapezoidal pulses for 5G random access with short message support," in *SPAWC-2015*, Stockholm, Sweden, Jun. 2015.

[41] G. Wunder, M. Kasparick, P. Jung, T. Wild, F. Schaich, Y. Chen, G. Fettweis, I. Gaspar, N. Michailow, M. Matthé, *et al.*, "New physical-layer waveforms for 5G," in *Towards 5G: Applications, Requirements and Candidate Technologies*, R. Vannithamby and S. Talwar, Eds. Wiley, 2016, pp. 303–341.

[42] P. A. Bello, "Characterization of randomly time-variant linear channels," *IEEE Trans. Commun.*, vol. 11, no. 4, pp. 360–393, 1963.

[43] P. Jung and G. Wunder, "The WSSUS pulse design problem in multicarrier transmission," *IEEE Trans. Commun.*, vol. 55, pp. 1918–1928, Oct. 2007.

[44] P. Jung and G. Wunder, "On time-variant distortions in multicarrier with application to frequency offsets and phase noise," *IEEE Trans. Commun.*, vol. 53, pp. 1561–1570, Sep. 2005. [Online]. Available: http://ieeexplore.ieee.org/xpls/abs_all.jsp?arnumber=1287432.

# TVWS as an Emerging Application of Cognitive Radio

# 2

**Dominique Noguet\*, Ranveer Chandra[†], Markus Mueck[‡]**

*CEA-Leti, Grenoble, France\* Microsoft Corporation, Redmond, WA, United States[†]*
*INTEL Deutschland GmbH, Munich, Germany[‡]*

## CONTENTS

Television White Space (TVWS) communications are based on secondary usage of VHF/UHF TV bands. These bands are characterized by very good electromagnetic propagation. On the other hand, TVWS systems have to meet specific requirements to guarantee nonharmful interference to incumbent radio systems. In this particular context, both orthogonal and nonorthogonal waveforms have been considered for TVWS physical layers. This chapter provides an overview of TVWS scenarios and multicarrier technologies developed in the scope of TVWS communications, highlighting the benefits of each approach. Coexistence management for TVWS systems is also discussed. Finally, a short status of current trials and deployments is made.

## 2.1 REGULATORY CONTEXT OF TVWS

In the recent years, there has been a worldwide concern related to spectrum shortage. As an example, in June 2010, the White House issued a Presidential Memorandum stating that the National Telecommunications and Information Administration (NTIA) will collaborate with FCC to make available a total of 500 MHz of Federal and nonfederal spectrum over the next 10 years, suitable for both mobile and fixed wireless broadband use. One of the means to make new spectrum available is through sharing, and the Digital Switch Over (DSO) in TV bands, which has resulted in making the so-called TV White Space (TVWS) UHF spectrum available, was the first actual example where such a mechanism has been allowed. TVWS availability depends on TV broadcast frequency usage profile, thus changing across time and space. TVWS usage relies on unlicensed secondary Dynamic Spectrum Access (DSA) under the principle on a nonharmful interference with incumbent users [1].

In the USA, the FCC proposed rules for the Unlicensed Operation in the TV Broadcast Bands [2], with the final set of rules in 2009 [3] and an additional notice in 2010 [4], along with an additional memorandum [5]. In Japan, MIC has published rules for secondary operation in TV white space [6]. Similarly, the IDA in Singapore [7] and Industry Canada in Canada [8] published rules for secondary operation in the TVWS. In Europe, the Ofcom UK was the first regulator to establish rules for TVWS usage [9,10]. Subsequently, Ofcom organized a nation-wide trial where several technologies were tested [11]. Following up the approval by the European Parliament and Council of the first Radio Spectrum Policy Programme (RSPP) in March 2012 [12], the European Commission released a Communication [13] in which shared use of TV White Spaces in the 470–790 MHz band is identified as a major opportunity. Then, regulatory actions have taken place in this area. In a first report, European Regulation (CEPT) has defined technical considerations for TVWS operation in Europe [14]. Then, ECC Report 159 established technical and operational requirements for the possible operation of cognitive radio systems in the "white spaces of the [470–790] MHz band" [15]. CEPT thoroughly addressed the way forward in European TV White Spaces, assessing both geolocation database and spectrum sensing as enabling technologies and setting out technical requirements for the use of TVWS. Finally, ETSI established some requirements for TVWS equipment in answer to the R&TTE Directive and delivered the EN 301 598 Harmonized EN [16].

In all countries where TVWS opportunistic access is allowed or considered, noninterference with incumbents relies on an overlay mechanism for which spectrum usage is allowed in channels vacant of any incumbent transmission. Overlay mechanism implies that, before setting an opportunistic transmission, the opportunistic system ensures that the target band is vacant of incumbent operation, with a listen-before-talk mechanism. In fact, the actual constraint is that there should be no victim device (i.e., incumbent receiver) using the frequency targeted by the opportunistic system in the coverage area of this opportunistic system. The likelihood of interference is tied to the choice on parameter values, whether it is sensor sensitivity (when a sensing based approach is used), or positioning accuracy (when the "no talk area" is determined by a geolocation approach), and of course, emission power levels. In any case, incumbent protection has to be traded against the inhibition of opportunistic access. Adjacent channel power is limited as well. For instance, the FCC demands adjacent channel leakage ratio (ACLR) to be at least 55 dB below in band emissions. This stringent rule is 10 dB stricter than for LTE systems and implies specific technology or implementation choices as it will be highlighted hereafter.

## 2.2 SCENARIOS AND APPLICATIONS IN TVWS

The main usage scenarios for TVWS applications are related to Internet broadband access and TVWS indoor WLAN. This section focuses on broadband standards.

### 2.2.1 TVWS BROADBAND ACCESS

A base station is providing Internet access to the end users by utilizing TVWS frequency bands over ranges comparable to today's cellular systems, e.g., in the range of 0 to 10 km. The base station serves customer premises equipment (CPE) to deliver content access (see Fig. 2.1). This is the first scenario associated with TVWS and was the first use case considered in standardization pioneering work, such as the one of IEEE 802.22 wireless regional area network (WRAN). The good propagation properties of sub-GHz UHF frequency and the authorized EIRP for fixed TVWS stations by the FCC (4W) provide TVWS with good assets for this scenario. Wireless broadband access has mainly been considered in low densely populated areas where DSL is not available. For this reason, this scenario is often referred to as Rural Broadband Access. From a technical viewpoint, this scenario also encompasses large campus WLAN access case as illustrated in Fig. 2.1.

In [17], ETSI has analyzed various mobility situations and how this impacts the way devices shall identify vacant channels.

#### 2.2.1.1 Mid-/Long Range, No Mobility

A base station is providing wireless access toward fixed devices, e.g., a nonmobile home base station/access point. The geolocation of both the base station and the fixed device is well known.

#### 2.2.1.2 Mid-/Long Range, Low Mobility

A base station is providing wireless access towards mobile devices where the users have low mobility, e.g., they are staying at their location or are moving at walking speed. In that respect, the mobility of the user does not lead to invalid sensing results

**FIGURE 2.1**

Rural broadband access (left) and campus (right) scenarios.

for primary users retrieved for the current location. The geolocation of the base station is well known. The geolocation of the mobile device must be determined during operation, e.g., via GPS, GLONASS, Galileo, or Beidou system or cellular positioning systems.

### 2.2.1.3 *Mid-/Long Range, High Mobility*

A base station is providing wireless access toward mobile devices and the concerned mobile devices may move at high speed, e.g., because a user is in a car or a train. In that respect, sensing results for primary users retrieved for the current location may get invalid quickly due to the mobility of the user. Thus, this use case leads to challenging constraints for the detection of primary users, and it needs to be evaluated if high mobility can be suitably supported in TVWS at all.

## 2.2.2 TVWS INDOOR WLAN

This scenario considers an indoor access point such as in classical WLAN scenarios for homes of larger buildings. The access point or a base station is providing Internet access to the end users via short-range wireless communication (e.g., in the range of 0 to 100 m) by utilizing TVWS frequency bands. UHF good propagation conditions make it possible to expect indoor to outdoor coverage as illustrated in Fig. 2.2.

In this scenario, the user experience is expected to substantially improve with respect to traditional WLAN systems in 2.4 GHz and 5 GHz bands. The better propagation and wall penetration characteristics of TVWS spectrum will typically ensure that a single TVWS Wi-Fi Access Point is sufficient to provide high-quality access throughout an entire, massively build home. ETSI has identified various situations as far as coexistence management is concerned [17].

### 2.2.2.1 *Networks Without Coexistence Management*

One or more independent networks access TVWS frequency bands. The access points require knowledge on the incumbent users of the spectrum (e.g., via TVWS incum-

**FIGURE 2.2**

Indoor scenario with indoor to outdoor coverage.

bent Geolocation Databases, sensing, etc.). However, in this scenario, the different networks are uncoordinated, and thus they have no knowledge on other secondary networks and other users operating in the TVWS bands.

### 2.2.2.2 Networks With Distributed Coexistence Management

Multiple networks access TVWS frequency bands. The different networks are independent, and different network operators provide the backbone connectivity. Such a scenario can be envisaged, e.g., in an apartment building, where residents independently acquire their own local area access points operating in TVWS frequency bands. Typically, the concerned access points can be operated and maintained by the residents themselves or the Internet Service Providers. To work properly, effective coexistence mechanisms are required for TVWS frequency access.

### 2.2.2.3 Networks With Centralized Coexistence Management

In this case, a TVWS operator operates the TVWS networks in the proximity in coordinated manner. Examples of this kind of usage can be small-scale corporate networks, networks for academic institutions, etc.

### 2.2.2.4 Hybrid of Networks With Distributed and Centralized Coexistence Management

This scenario combines the above two scenarios, i.e., in the same geographic area, there are centralized coexistence management and distributed coexistence management applied. Such a setup typically occurs in combinations of public and private places, like campus areas and shopping malls, where, e.g., the "official" local area networks, operating under centralized coexistence management, are complemented by independent access points set up independently by some individuals.

The overall coexistence management in this scenario is distributed due to the existence of the independent networks.

## 2.3 STANDARD TECHNOLOGIES

A set of standardization bodies developed technology standards to facilitate the implementation of TVWS radios.

### 2.3.1 OFDM-BASED STANDARDS

In December 2008, a new group called the Cognitive Networking Alliance (CogNeA) was initiated. This group is composed of Philips, Samsung, HP, ETRI, GeorgiaTech, and Motorola. The goal of CogNea was to drive the early definition and adoption of industry-wide standards for low-power personal and portable wireless devices to operate in the TVWS. In this regard, a new standard was created in the framework of the European Computer Manufacturers Association (ECMA) in December 2009 [18]. This standard (ECMA 392) covers "PHY and MAC for Operation in TV White Space." The standard target applications are high-speed video streaming and Internet access on personal/portable electronics, home electronics equipment, and computers and peripherals [19]. Then, IEEE 802.22 [20] and, more recently, IEEE 802.11af [21] issued standards for WRAN and WLAN applications, respectively. The main use case of the IEEE 802.22 standard is inherited from the IEEE 802.16e scenarios related to wireless regional area network (WRAN). It corresponds to the typical rural broadband access with a target range of 17 to 30 km between the base station and the customer premises equipment (CPE). The reference architecture of IEEE 802.22 includes a PHY service access point (SAP) and a MAC SAP [22].

On the other hand, ECMA 392 and IEEE 802.11af target WLAN scenarios with extended range, often referred to as Super Wi-Fi or White-Fi. These TVWS broadband standards (ECMA 392, IEEE 802.22, IEEE 802.11af) are based on Cyclic Prefix-Coded Orthogonal Frequency Division Multiplexing physical layer (PHY), thereafter referred to as OFDM.

OFDM has proven to be very effective for mobile wireless communications and has been used in a large number of modern broadband wireless systems. By dividing a frequency selective fading channel into a large number of narrow-band flat fading subchannels, OFDM systems can easily compensate the effects of the channel using a simple one-tap frequency domain equalizer. All these broadband TVWS standards (ECMA 392, IEEE 802.22, IEEE 802.11af) often inherited from previous PHY standard developments. For instance, IEEE 802.22 is derived from IEEE 802.16e, whereas IEEE 802.11af is inspired by IEEE 802.11ac, and adapted to make them suitable for TVWS bands. The PHY layers of these systems mainly differ from a set of parameter values specified to serve the aforementioned scenarios (see Table 2.1).

### 2.3.2 FBMC-BASED STANDARD

Adapting existing standards to TVWS bands was the best way to guarantee fast market readiness. However, OFDM had some difficulties to meet the TVWS specific requirements in terms of interference control [1]. As an example, IEEE 802.11af

**Table 2.1** OFDM-based standard key parameters

| | IEEE 802.22 | IEEE 802.11af | ECMA 392 |
|---|---|---|---|
| Typical scenario | Rural broadband access/WRAN | Super Wi-Fi WLAN | WLAN |
| Modulation | OFDM | OFDM | OFDM |
| FFT size | 2048 | 64, 128, 256 (512, 1024 opt.) | 128 |
| Max. data carriers | 1140 | 96 (or 108 in VHT[a] mode) per BCU[b] | 98 |
| Channelization [MHz] | 6, 7, 8 | 6, 7, 8 | 6, 7, 8 |
| Carrier modulation | QPSK, 16-QAM, 64-QAM | BPSK, QPSK, 16-QAM, 64-QAM, 256-QAM | QPSK, 16-QAM, 64-QAM |
| FEC | Convolutional (Turbo code, LDPC, STBC opt.) | Convolutional | Outer Reed–Solomon Inner Convolutional |
| Max. data rate [Mbps] | 22.69 | 35.6[c] | 23.74 |
| Max. spect. eff. $b \cdot s^{-1} \cdot Hz^{-1}$ | 3.78 | 4.45 | 2.97 |
| Access scheme | TDMA/OFDMA | CSMA-CA | CSMA-CA |
| Mesh support | No | Yes | No |
| Spectrum sensing quiet per. | Yes | No | Yes |
| Interface with TVWS DB | Yes | Yes | Yes |

[a] *Very High Throughput.*
[b] *IEEE 802.11af defines Basic Channel Unit (BCU) as one block of 6, 7, or 8 MHz and allows channel bonding of up to 4 BCUs.*
[c] *In one BCU of 8 MHz.*

only uses $5\frac{1}{3}$ MHz out of a 6-MHz channel in order to allow sharper filtering to achieve 55 dB ACLR [21]. This reduces available bandwidth and implies that additional spectrum shaping filtering is mandatory. In March 2010, IEEE DySPAN Standards Committee (formerly known as SCC41) created an ad hoc group on White Space (WS) Radio "to consider interest in, feasibility of, and necessity of developing standard defining radio interface (medium access control and physical layers) for white space radio system." Subsequently, the 1900.7 working group on "Radio Interface for White Space Dynamic Spectrum Access Radio Systems Supporting Fixed and Mobile Operation" [23] was created.

The IEEE 1900.7-2015 standard is the result of a clean slate technology analysis, where the working group tried to identify the most suitable technology to TVWS requirements. The chosen technology is based on a Filter Bank Multi-Carrier (FBMC) PHY and a contention-based CSMA-CA MAC in order to cover various use cases, among which the ones presented in Section 2.2.

IEEE 1900.7-2015 selected FBMC PHY vs. classical OFDM approaches because of its superior performance on ACLR, without sacrificing spectral efficiency or performance over frequency selective channels. On top of this, FBMC provides a very high level of flexibility, particularly when fragmented spectrum access is concerned by using spectrum polling techniques (i.e., by switching on or off carriers). To our knowledge, the IEEE 1900.7-2015 standard is the first standard to use FBMC as PHY modulation and is therefore discussed in more details hereafter.

FBMC uses a prototype filter that filters each subcarrier. A proper design of the prototype filter enables to trade time and frequency localization and thus to control ACLR. Because the prototype filter's response spreads each subcarrier over several adjacent subcarriers, another dimension is used to "restore" orthogonality. In the framework of IEEE 1900.7, the offset QAM (OQAM) approach was used. OQAM consists in a complex to real conversion where real and imaginary parts of each complex symbol are multiplexed in consecutive time samples into Pulse Amplitude Modulation (PAM) symbols. In order for this preprocessing not to impact data rate, the PAM symbols are up-sampled by a factor of 2. Then, the output real numbers are multiplied by an offset QAM sequence to form a new complex symbol: $h_{k,l} = (-1)^{k.l}(\text{j})^{k+l}e_{k,l}$. These symbols are consequently filtered through a polyphase network $G(z)$.

The IEEE 1900.7-2015 standard specifies two different sizes for the prototype filter: K = 3 or K = 4 governing the level of protection of adjacent channels. Also, for the sake of flexibility, several modes are proposed. They consider different number of carriers and two channelization modes (2 or 8 MHz). Thus, IEEE 1900.7-2015 can be used in any country authorizing TVWS operation and can cover medium to broadband channels.

The block diagram of the IEEE 1900.7-2015 transmitter is shown in Fig. 2.3. The transmitter architecture is composed of two main elements: a forward error correction block and a data mapping and modulation block. Forward error correction (FEC) is implemented using a standard convolutional encoder. The code may be punctured to support variable encoding rates. The convolutional code is segmented by blocks of fixed size. The trellis is closed at the beginning and the end of each FEC block. The output of the encoder is forwarded to a bit interleaver of size multiple of the output length of the encoder.

To allow a high level of flexibility, several profiles have been defined. Channelization is provided with 2-MHz and 8-MHz bands. These channelizations will be considered as elementary blocks that can be aggregated to form various profiles. For instance, 6-MHz US TV channelization can be addressed by three adjacent 2-MHz blocks. The 2-MHz quantum also gives the possibility to address lower data rate use cases like machine-type communication. Finally, several nonadjacent blocks can be used, even with different transmit powers in order to optimize spectrum usage under regulation rules that allow unequal transmit powers across various TVWS channels (e.g., as prescribed by Ofcom in the UK). As an example, Fig. 2.4 shows the spectrum of a DVB-T signal (8-MHz block in the middle) surrounded by two 4-MHz-wide FBMC IEEE 1900.7-2015 blocks.

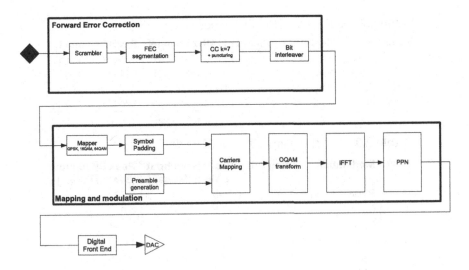

**FIGURE 2.3**

IEEE 1900.7-2015 PHY transmitter block diagram.

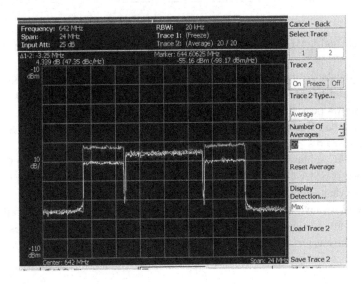

**FIGURE 2.4**

Shared spectrum between DVB-T (middle) and two IEEE 1900.7-2015 blocks.

Also, several carrier spacings are allowed in order to cope with a large variety of channel conditions and thus with a large number of use cases as shown in Table 2.2. The table shows the number of active carriers and the actual signal bandwidth in

**Table 2.2** IEEE 1900.7-2015 PHY modes

| Mode | $N$ | Intercarrier spacing | 2-MHz channel | 8-MHz channel |
|------|------|----------------------|----------------|----------------|
| 4K | 4096 | 3.75 kHz | 504 (1.86 MHz) | 2016 (7.56 MHz) |
| 1K | 1024 | 15.00 kHz | 124 (1.86 MHz) | 504 (7.56 MHz) |
| 0.5K | 512 | 30.00 kHz | 64 (1.92 MHz) | 252 (7.56 MHz) |
| 0.25K | 256 | 60.00 kHz | 32 (1.92 MHz) | 124 (7.44 MHz) |

**Table 2.3** Peak throughput for the 1K mode

| MCS | Modulation | Coding rate | Peak throughput (2 MHz) [Mbps] | Peak throughput (8 MHz) [Mbps] |
|-----|------------|-------------|--------------------------------|--------------------------------|
| 0 | BPSK | 1/2 | 0.93 | 3.78 |
| 1 | QPSK | 1/2 | 1.86 | 7.56 |
| 2 | QPSK | 3/4 | 2.79 | 11.34 |
| 3 | 16QAM | 1/2 | 3.72 | 15.12 |
| 4 | 16QAM | 3/4 | 5.58 | 22.68 |
| 5 | 64QAM | 2/3 | 7.44 | 30.24 |
| 6 | 64QAM | 3/4 | 8.37 | 34.02 |
| 7 | 64QAM | 5/6 | 9.30 | 37.80 |

the 2-MHz and 8-MHz profiles. The exploited band is roughly 95% of the available channel.

The standard suggests several carrier modulations schemes: BPSK, QPSK, 16QAM, and 64QAM, with convolutional coding with rates of 1/2, 3/4, 5/6. Table 2.3 shows peak throughput for each modulation and coding schemes (MCS) in the case of the 1K mode. Maximal spectrum efficiency is obtained for MCS 7 and is equal to $4.725$ b$\cdot$s$^{-1}\cdot$Hz$^{-1}$.

The MAC sublayer is based on a Carrier Sense Multiple Access with Collision Avoidance (CSMA-CA) approach using "Request To Send" and "Clear To Send" (RTS/CTS) handshake mechanism. A basic network operates in a master–slave mode. In the master–slave mode, a device is designated as master (network coordinator), and others are associated with the master as slaves. The master coordinates channel access in the master–slave mode.

## 2.4 TVWS MEDIUM ACCESS CONTROL STANDARDS – A COORDINATED AND AN UNCOORDINATED APPROACH

Protection of incumbent systems is a key regulation requirement. For an efficient usage of the technology, however, it is important to consider the resource sharing among multiple TVWS systems themselves. Otherwise, a chaotic and uncontrolled access to the resource will make the system hard to handle (in particular, in dense environ-

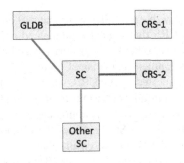

**FIGURE 2.5**

Overview of a TVWS system.

ments). For this purpose, [24] introduces a system architecture and corresponding procedures for coordinated and uncoordinated use of TVWS.

The proposed approach is built on three entities as introduced in [24]: the Cognitive Radio System (CRS), the Spectrum Coordinator (SC), and the Geo-Location Database (GLDB) as shown in Fig. 2.5. The use of SC is specific to coordinated approach.

The cognitive radio system (CRS) consists of a TVWS Device (WSD) or network of WSDs (i.e., a master WSD and some slave WSDs). The CRS uses available TVWS resources, which are identified with the help of Geo-Location Database (GLDB) and/or the CRS uses available TVWS resources by exploiting the GLDB and with additional knowledge of spectrum usage by its neighbor CRSs by the spectrum coordinator (SC).

The operation of a GLDB is mandated or authorized by a regulatory authority and that provides a WSD in a CRS with location specific information on the available frequencies and associated maximum EIRP values. The concerned WSD is permitted to use the service which allows for protection of the incumbent service.

The SC is coordinating spectrum usage of CRSs. For this purpose, information is requested from GLDB and is communicated to CRSs, as well as supplemental spectrum usage data from different CRSs using its service. Different SCs are capable of communicating with each other.

## 2.4.1 COORDINATED USAGE OF TVWS

In this section, we outline how several (distinct) CRSs can be coordinated among themselves to ensure access to required amount of spectrum and protection from harmful interference for secondary systems.

*Spectrum Coordination.* A SC uses spectrum coordination to serve CRSs such that they can operate in available spectrum resources of TVWS without causing harmful interference to each other and having a predictable access to the spectrum resources. The SC coordinates the management of radio resources among a set of CRSs that are potentially interfering with each other (coexistence) and allows for channel assignment requested by a CRS that requests guaranteed access to full capacity of a channel and with priority over other CRSs (priority-based channel assignment). The priority-based channel assignment is managed by the SC based on some minimum protection requirements requested by the CRS, which includes minimum bandwidth, minimum SINR (or maximum allowable interference), and some guaranteed minimum availability time. There are two basic approaches on how to implement priority access: i) In the first approach, the SC translates these requirements into protection criteria, which are used by the GLDB to ensure that the priority-based channel assignment is maintained in the presence of other WSDs not using the SC; ii) In the second approach, the SC stores the spectrum usage of the priority access, and the GLDB checks with the SC before providing available channels to other CRSs not using SC. The algorithms and procedures to enable coexistence and/or priority-based channel assignment are further described in [17]. From the perspective of the CRSs, coexistence and priority-based channel assignment are provided as a set of two available SC services: the information service (for coexistence only) and the management service (for both coexistence and priority-based channel assignment). Each SC provides at least an information service or a management service for CRSs or provides both services.

*Information Service.* CRSs can be subscribed to an information Service. In this case, a SC provides information about selected operational parameters (e.g., the operational parameters of other CRS in the available spectrum resources). In the information service, a SC does not make decision on the operational parameters to be used by those CRSs, but rather, all decisions are made by the CRS itself. However, the SC may process information about the current usage of spectrum to provide it to the CRS in order to support the CRS decision making processes (such as ranking the potential operational parameters according to the resulting expected performance).

*Management Service.* For CRSs that are subscribed to the management service, a SC provides the operational parameters to be used by a CRS based on its requests and, if requested, certain QoS and usage time requirements. A CRS will not make any decision for its operational parameters (e.g., channel and transmit power) that are determined by the SC.

*High-Level Operation Sequence.* An overview of coordinated usage of TVWS is shown in Fig. 2.6. A CRS consists of a master WSD and one or more slave WSDs. The master WSD sends device parameters to a GLDB via the SC. The SC typically acts as relay and can also store the device parameters of the master WSD. The SC, during the process, maintains additional data about spectrum usage of the different CRSs using its service. This additional data contains information on the current state of spectrum usage, including spectrum measurement data from WSDs and usage maps or areas of occupancy of the different CRSs. It also contains information related to the Radio Access Technology of each CRS that facilitates coexistence. A GLDB receives information from the master WSD about the characteristics of that WSD in order to generate operational parameters for that WSD. The GLDB provides

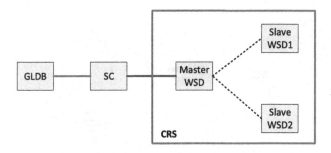

**FIGURE 2.6**

Overview of coordinated usage of TVWS system.

operational parameters to the master WSD via the SC. During this process, the SC determines the operational parameters using the information obtained from the GLDB and the additional data about spectrum usage provided by the different CRSs and sends these operational parameters to the master WSD in response to the request for TVWS access. The operational parameters determined by the SC will not violate the protection criteria of the incumbent and are therefore compliant with the information obtained from the GLDB. The master WSD then sends the selected channel usage parameter to the GLDB via the SC [17]. The SC will also update its additional spectrum usage data based on information sent by the WSD. At any time in the process of assigning channels to the CRSs, the SC can reconfigure the channel usage of the CRSs to ensure an efficient use of spectrum, such as reducing fragmentation in the available spectrum. The GLDB can either access SC or use channel usage parameters sent by the SC to ensure that WSDs can operate in the presence of other WSDs not using the SC.

## 2.4.2 UNCOORDINATED USAGE OF TVWS

Whereas the upper approach introduces coordination among CRSs, the uncoordinated usage of TVWS spectrum ensures a protection of the Incumbents without providing specific protection for interference between the secondary users themselves.

An overview of uncoordinated usage of TVWS is shown in Fig. 2.7. A CRS consists of a master WSD and one or more slave WSDs. The master WSD communicates with a GLDB to obtain its operational parameters in TVWS. A GLDB receives information from a WSD about the characteristics of that WSD in order to generate operational parameter for that WSD. A GLDB maintains a record of the actual usage of the TVWS. This information can be used to enable offending WSDs to be readily identified if interference to incumbent users were to occur, and to allow the GLDB to know the extent to which available TVWS are being used.

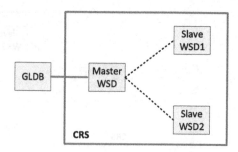

**FIGURE 2.7**

Overview of uncoordinated usage of TVWS system.

## 2.5 AVAILABLE PRODUCTS

There are three categories of products that collectively enable the TVWS ecosystem: the radio vendors, TVWS database providers, and ISP networks that provide Internet access to the end consumer.

Some popular radio vendors are Adaptrum Inc, 6Harmonics, Runcom, Carlson, KTS, etc. Another company, called Neul, that build TVWS radios for IoT was acquired by Huawei. Whereas most of these radios use OFDM, KTS and Neul radios rely on other spread spectrum technologies that trade range with capacity. The above radios all operate in 6 MHz, and most of them can scale to 8 or 10 MHz. Adaptrum, 6Harmonics, and Carlson have plans to enable channel bonding up to four channels and support MIMO based on the IEEE 802.11af WhiteFi standard [25].

Several companies provide TVWS databases. In the US, 10 companies are authorized to provide TVWS database services, and Ericsson, Google, and Spectrum Bridge already provide these services in the US. Ofcom recently certified Microsoft TVWS database for the UK.

The TVWS technology is finally brought forth to the consumer through the ISPs. In the US, these have typically been rural Wireless ISPs (WISPs), whereas outside of the US, these networks have been typically run by startups or government agencies, as described below. In practice, anyone can set up a TVWS network and start providing Internet access, either to their home or office, or to other consumers and businesses.

## 2.6 CURRENT TRIALS WORLDWIDE AND LESSONS LEARNED

TVWS technology has moved beyond research and labs to pilots and real-world deployments. The key benefit of a TVWS network is that it enables long-range networks at a lower cost than any competing technology [26].

Most existing TVWS deployments have focused on the backhaul use case, with a Wi-Fi front end to reach the end consumer devices. These showcase the point-to-multipoint benefits of TVWS technologies.

The first TVWS deployments were carried out simultaneously by Spectrum Bridge and by Microsoft Research in 2009. Spectrum Bridge used KTS radios to set up a TVWS link in rural Virginia. Microsoft used White-Fi and WiMax radios to provide Internet access in its bus shuttles in the Microsoft corporate headquarters [MC2R]. These deployments showed that it was possible to set up broadband TVWS networks without causing interference to the existing users of this spectrum.

Since then, Microsoft, Google, BBC, Arquiva Wireless, and other companies have done trials in the UK. A big trial with over 10 companies was carried out in Cambridge, UK, in 2012. This deployment helped Ofcom drive the rules for operation in the TVWS. A subsequent trial in Glasgow demonstrated the industry readiness for TVWS technology.

There have been several trials in different parts of the world, and these TVWS networks have connected more than half a million people to the Internet. Deployments in Kenya use a combination of point-to-point microwave, coupled with TVWS to provide Internet access to schools, hospitals, and villages. In South Africa, some of these networks extend the coverage from libraries to school. In India, a TVWS network was used in Andhra Pradesh to connect five schools to the Internet. Some of the TVWS links provide up to 1.5 Mbps throughput at about 15-km range, when the TVWS devices are operating at 20 dBm, with a 11-dBi antenna. Given that worldwide regulations allow for more transmission power, we can expect even longer-range networks to enable new scenarios over this emerging technology.

In Europe, the UK is the first country to authorize TVWS operation. Ofcom UK rolled out a large-scale pilot across the country in February 2015. The purpose of the trial was to assess various aspects of the technology, like database access and protection margins. The trial was based upon Ofcom prospective rules, which are consistent with the ETSI harmonized standard for white space devices [16]. Because Primary system transmissions, terrain, and population implantation are very different compared to the US case, the rules established by Ofcom significantly differ from the ones of the FCC. The main difference concerns the allowed maximum power, which is adjusted per channel. In contrast, the FCC considers binary situations where the allowed power is fixed whenever channels are considered as vacant. In other words, the white space database provides the WSD with a map of allowed EIRP per channel rather than occupied/vacant information. The maximum allowed EIRP is calculated based on antenna elevation, primary system powers, terrain information, etc. As a consequence, the UK approach is expected to free a larger number of spectrum opportunities, albeit with restrictions on transmit power. Leveraging on the trial results, Ofcom decided to enable license exempt use of WSD in the 470 to 790 MHz band in December 2015.

# REFERENCES

[1] D. Noguet, M. Gautier, and V. Berg, "Advances in opportunistic radio technologies for TVWS," *EURASIP Journal on Wireless Communications and Networking*, vol. 2011,

no. 1, p. 170, 2011. [Online]. Available: http://jwcn.eurasipjournals.com/content/2011/1/170.

[2] F. proposed rule, "Unlicensed operation in the TV broadcast bands." US Federal Register, Tech. Rep., Jun. 2004, vol. 69, no. 117.

[3] F. final rule, "Unlicensed operation in the TV broadcast bands." US Federal Register, Tech. Rep., Feb. 2009, vol. 74, no. 30.

[4] FCC, "In the matter of unlicensed operation in the TV broadcast bands: Additional spectrum for unlicensed devices below 900 MHz and in the 3 GHz band." Second memorandum opinion and order, Tech. Rep., Sep. 2010.

[5] F. notice, "Unlicensed operation in the TV broadcast bands." US Federal Register, Tech. Rep., Feb. 2011, vol. 76, no. 26.

[6] H. Harada, "Status report on usage of TV white space in Japan." IEEE 802.11-12/677r0, Tech. Rep. 802.11-12/677r0, May 2012.

[7] IDA, "Regulatory framework for TV white space operations in the VHF/UHF bands," Jun. 2014.

[8] I. Canada, "Framework for the use of certain non-broadcasting applications in the television broadcasting bands below 698 MHz," Feb. 2015.

[9] Ofcom, "Digital dividend: Cognitive access, statement on licence exempting cognitive devices using interleaved spectrum." Tech. Rep., OFCOM, Jul. 2009.

[10] OFCOM, "TV white spaces – approach to coexistence." Tech. Rep., OFCOM, Sep. 2013.

[11] OFCOM, "White Spaces Pilot," Dec. 2014. [Online]. Available: http://stakeholders.ofcom.org.uk/spectrum/tv-white-spaces/white-spaces-pilot/.

[12] "Decision No. 243/2012/EU of the European Parliament and of the Council of 14 March 2012 establishing a multiannual radio spectrum policy programme. Text with EEA relevance," 2012. [Online]. Available: http://eur-lex.europa.eu/LexUriServ/LexUriServ.do?uri=OJ:L:2012:081:0007:0017:EN:PDF/20120510_inventory_workshop_invitation.pdf.

[13] "Communication from the Commission to the European Parliament, the Council, the European Economic and Social Committee and the Committee of the Promoting the shared use of radio spectrum resources in the internal market (COM/2012/0478)," 2012. [Online]. Available: http://eur-lex.europa.eu/LexUriServ/LexUriServ.do?uri=OJ:L:2012:081:0007:0017:EN:PDF/20120510_inventory_workshop_invitation.pdf.

[14] Report, "Technical considerations regarding harmonisation options for the Digital Dividend." Tech. Rep. 24, CEPT, 2008. [Online]. Available: http://www.erodocdb.dk/Docs/doc98/official/pdf/CEPTREP024.PDF.

[15] E. Report, "Technical and operational requirements for the possible operation of cognitive radio systems in the "white spaces" of the frequency band 470–790 MHz." Tech. Rep. 159, CEPT, Jan. 2011. [Online]. Available: http://www.erodocdb.dk/Docs/doc98/official/pdf/CEPTREP024.PDF.

[16] ETSI, "White Space Devices (WSD), Wireless Access Systems operating in the 470 MHz to 790 MHz frequency band; Harmonized EN covering the essential requirements of article 3.2 of the R&TTE Directive." Tech. Rep. EN 301 598, ETSI, Apr. 2014.

[17] *Reconfigurable Radio Systems (RRS); Use Cases for Operation in White Space Frequency Bands*, ETSI Std., 2011.

[18] *MAC and PHY for Operation in TV White Space*, ECMA-392 Std., 2009.

[19] J. Wang, M. Song, S. Santhiveeran, K. Lim, G. Ko, K. Kim, S. Hwang, M. Ghosh, V. Gaddam, and K. Challapali, "First cognitive radio networking standard for personal/portable devices," in *DySPAN*, Singapore, May 2010.

[20] *Cognitive Wireless RAN Medium Access Control (MAC) and Physical Layer (PHY) Specifications: Policies and Procedures for Operation in the TV Bands*, IEEE Std., I. standard 802.22, Jul. 2011.

[21] *Wireless LAN Medium Access Control (MAC) and Physical Layer (PHY) Specifications. Amendment 5: Television White Spaces (TVWS) Operation*, IEEE Std., I. standard 802.11af, Dec. 2013.

[22] C. R. Stevenson, "IEEE 802.22: The first cognitive radio wireless regional area network standard," *IEEE Commun. Mag.*, pp. 130–138, Jan. 2009.

[23] I. P. PAR, "Radio interface for white space dynamic spectrum access radio systems supporting fixed and mobile operation." Tech. Rep., IEEE, Jun. 2011.

[24] *Reconfigurable Radio Systems (RRS); System Architecture and High Level Procedures for Coordinated and Uncoordinated Use of TV White Spaces*, ETSI Std., 2015.

[25] "Dynamic spectrum alliance global summit," Apr. 2016, Bogota.

[26] S. Roberts, P. Garnett, and R. Chandra, "Connecting Africa using the TV white spaces: From research to real world deployments," in *International Workshop on Local and Metropolitan Area Networks, LANMAN 2015*, Beijing, China, Apr. 2015.

# Broadband Private Mobile Radio (PMR)/Public Protection and Disaster Relief (PPDR) Services Evolution

3

**Christophe Gruet, Laurent Martinod, Philippe Mège, Xavier Pons-Masbernat**

*AIRBUS Defence and Space, Elancourt, France*

## CONTENTS

Orthogonal Waveforms and Filter Banks for Future Communication Systems. DOI: 10.1016/B978-0-12-810384-5.00003-7

## 3.1 INTRODUCTION

Public Safety and Security (PSS) organizations are using radio communications systems for their day-to-day needs, PP1 (Public Protection in day-to-day mode), for exceptional events, PP2 (Public Protection in exceptional planned events: sports events, cultural events, demonstrations, etc.), and for disaster recovery conditions, DR (Disaster Relief: exceptional unplanned events). These usages are called collectively as PPDR (Public Protection and Disaster Relief).

Today, they are using their radio communications systems (based on dedicated Private Mobile Radio (PMR) standards such as TETRA [1] and TETRAPOL [2] in Europe and in a large part of the world or APCO P25 [3] in North America) for voice communications essentially (individual or group calls) and for low-rate data transmissions (short messaging, user location reporting, basic database access). This is due to the technological limitations of currently deployed PMR/PPDR systems, which have only small frequency bandwidth (up to 25 kHz) and so naturally limited throughput, typically from 4 to 10 kbps. Those PMR standards are very similar in their design to the commercial 2G technologies (like GSM).

Following commercial networks evolution, new services and features are now requested by the public safety community. In recent years, PSS organizations in the UK, France, Netherlands, and others have been conducting trials on PSS high-speed data. There is clearly a strong need amongst these organizations to have in the coming years a nationwide broadband network to allow services to support mission critical applications such as detailed photographic images transfer, ad hoc video camera and surveillance camera real-time information delivery (from the field or to the field), detailed maps and plans exchanges (for use at an incident, e.g., a fire), biometric data off-site monitoring or download, database access, etc.

The ETSI Special Committee EMTEL and project MESA have identified user requirements for future broadband mission critical PP1, PP2, and DR applications. The EMTEL document TS 102 181 [4] on requirements for communication between authorities and organizations during emergencies lists the following situations for effective communication:

- Mobilization of the teams and people;
- Updates on the emergency/situational reports;
- Updates on requirements to other organizations so that they can prepare, e.g., informing hospitals on likely number of casualties and individual patients and their needs;
- Sending of command and control information to the incident area;
- Requesting of information from the incident area, e.g., building plans, chemical information;
- Sending of still and video images from the incident area.

**Table 3.1** Data services attributes from ETSI TS 102 181 [4]

| Service | Throughput | Timeliness | Robustness |
|---|---|---|---|
| E-mail | Medium | Low | Low |
| Imaging | High | Low | Variable |
| Digital mapping/Geographical information services | High | Variable | Variable |
| Location services | Low | High | High |
| Video (real time) | High | High | Low |
| Video (slow scan) | Medium | Low | Low |
| Data base access (remote) | Variable | Variable | High |
| Data base replication | High | Low | High |
| Personnel monitoring | Low | High | High |

The services required to support the above include:

- Voice services (one-to-one and group calling);
- High-level security encryption with multiple keys and Over The Air Rekeying (OTAR);
- Video teleconferencing to assist in coordination between the services and also to provide information from the incident area back to the control rooms;
- Data services (full e-mail, intranet browsing, database downloading, etc.; see Table 3.1 for the attributes of these services), status monitoring and location services, including, for example, measuring exposure to environmental conditions, reporting PSS responders' vital signs and determining their physical proximity, all in real time.

These services obviously require much higher data rates than currently available in deployed PMR PPDR systems.

Fourth-generation (4G) commercial technologies are natural candidates as main building blocks for this PMR systems evolution. A first step has already been made by the US National Public Safety Telecommunications Council (NPSTC) choosing that the next broadband PMR networks generation would be based on the Third-Generation Partnership Project (3GPP) Long-Term Evolution (LTE) technology in the US 700 MHz band now dedicated to Public Safety [5]. This US 700 MHz frequency band has been selected to operate the nationwide FirstNet broadband network considering a 2 × 10 MHz channel bandwidth in Frequency Division Duplexing (FDD) mode. A second step has now been made in Europe too, where CEPT/ECC, PSS organizations, manufacturers, operators, and regulators have all agreed on the choice of 3GPP LTE for future broadband PPDR networks in the 400 MHz and 700 MHz bands.

With this choice, the PMR community will fortunately take great benefits of all the work already done to define, standardize, and implement the 3GPP LTE standard for commercial mobile systems. However, despite the fact that this LTE toolbox ([6,7]) could be envisaged as a very promising solution to address such PMR evolution toward broadband, dedicated and specific adaptations or modifications are still needed, if not mandatory, in order to be compliant with critical key PMR requirements. Those adaptations or evolutions could significantly differ from the features specifically designed and optimized for a commercial usage LTE.

Some of those PMR specificities and their potential impacts on the future broadband choices and solutions are detailed hereafter.

## 3.2  AN IMPERATIVE NEED FOR FREQUENCY RESOURCES

Public Safety and Security is composed of different organizations having similar operational constraints and contributing to the protection and the rescue of the public. These organizations are: Police, Fire Brigades, Ambulances, Military services for Public Safety, Customs, etc.

They have common needs:

- Coverage of the full area under their responsibility: national or regional coverage;
- 100% availability: it will work even if all other communication means are out of service, which has already happened several times when events such as natural disasters, industrial disasters, bombings, riots, etc., are happening;
- Fast access;
- Management of priorities;
- Fully secured communications.

As mentioned in the Introduction, there is a strong need for much higher data rates than currently deployed PMR PPDR systems in order to improve the operational efficiency: real-time video, large file transfer, map transfer, database consultation, telemedicine, etc.

Fig. 3.1 illustrates the data throughput requirements evolution.

The new broadband needs cannot be fulfilled by existing narrowband systems (TETRA, TETRAPOL, APCO P25) using very narrow bandwidth: from 10 to 25 kHz only. This narrow bandwidth, combined with the requirement of large coverage, is strongly reducing the achievable data rates that are in the range of 4 to 10 kbit/s. This is clearly far below the need for broadband services, and only broadband systems (such as LTE) can provide these new high-data-rate services.

This new required capacity can only be achieved in two complementary ways: by obtaining new frequency bands for PMR/PPDR data services and/or by fitting a novel broadband data service within the scarcely available spectrum currently devoted to PMR systems.

This raises the problematic of frequencies: in which frequency band a Broadband PMR system could be deployed? Frequency matters are managed by national regulators and by international bodies. At European level, CEPT/ECC FM 49, dealing with Frequency Management for PPDR, is currently working on the subject of frequencies for Public Safety systems. The PPDR users, regulators and manufacturers, in FM 49

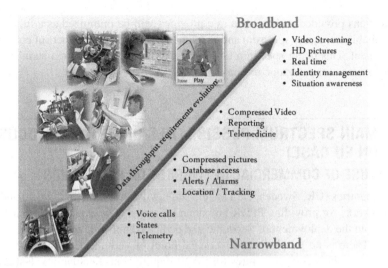

**Broadband**
- Video Streaming
- HD pictures
- Real time
- Identity management
- Situation awareness

- Compressed Video
- Reporting
- Telemedicine

- Compressed pictures
- Database access
- Alerts / Alarms
- Location / Tracking

- Voice calls
- States
- Telemetry

**Narrowband**

**FIGURE 3.1**

Data throughput requirements evolution.

working group, have specified the characteristics of the future frequency bands for broadband PPDR:

- A frequency band has to be dedicated to this usage because this is the only way to ensure a full availability;
- The total amount of frequency needed for broadband PPDR is 2 × 10 MHz;
- Moreover, this frequency band has to be below 1 GHz for economic reasons: low frequencies provide better radio coverage characteristics.

It allows deploying large cells in rural areas and smaller cells in dense urban areas. This also limits the number of sites required, and so the total deployment costs, to ensure a regional or national coverage. FM 49 has issued ECC Report 199 [8] that deals with user requirements and spectrum needs for broadband PPDR.

Without surprise, it is challenging to find frequency bands that meet those requirements, and it is difficult for PSS to get more spectrum, particularly in quite low-frequency bands, due to the competition for accessing to the frequency spectrum, especially with commercial networks operators. In its ECC Report 218 [9], FM 49 has identified two frequency bands for broadband PPDR: the 400 MHz band (which is already a PMR band), and the 700 MHz band (in the frame of the second Digital Dividend).

Clearly, frequency spectrum is a very scarce resource that is hardly accessible to users such as PPDR users. So spectrum efficiency is a must in order to use efficiently the frequency resource and to save it. This implies better spectrum occupancy: the band occupied by one system has to be as much as possible usefully occupied, and so the guard bands have to be minimized. This means also that the number of communications provided and the data rate provided will be optimized as a function of the available frequency spectrum and as a function of the total numbers of cells in the deployment. This better spectrum usage has to be achieved while maintaining good radio coverage for economic reasons.

## 3.3 MAIN SPECTRUM POSSIBILITIES OR OPTIONS (FOCUS ON EU CASE)

### 3.3.1 USE OF COMMERCIAL NETWORKS FOR PPDR NEEDS

Some countries (UK, Sweden, etc.) are envisaging using commercial broadband mobile networks for providing PPDR broadband services. The expected benefit would be to avoid the deployment of a specific and dedicated network and by doing so save money. There is no problem on a technical point of view: LTE is offering or will offer most of the technical features for PPDR requirements (availability, fast access, priority management) and specific needs (push-to-talk group communications, direct mode, etc., in future releases).

The real difficulty is of regulatory nature, as reported in SCF report of "Study on use of commercial mobile networks and equipment for mission-critical high-speed broadband communications in specific sectors" [10] issued for European Commission. The question is: will commercial operators really offer the level of service that PPDR users need at a reasonable cost? For sure, they can offer the service for non-mission critical services. However, serving PPDR mission-critical users requires a stringent priority management, which may even lead to preempt other users' communications: this would be fully contradictory with commercial operators' business model.

### 3.3.2 DEDICATED NETWORKS AND FREQUENCY RESOURCES

The Law Enforcement Working Party (LEWP), which is the European PPDR users group where Ministries of Interior and Ministries of Justice are represented, has clearly expressed PPDR need for a dedicated broadband network. A dedicated network can be operated by a PPDR network operator or by the users themselves. This excludes a solution based on a commercial network. A dedicated one is the only way to insure that PPDR users will be able to use the resource when they need it for mission-critical communications. This also means dedicated frequencies for broadband PPDR.

### 3.3.3 MUTUALIZATION AND COEXISTENCE BETWEEN NARROWBAND PMR SYSTEMS AND BROADBAND PMR SYSTEMS

Different PPDR organizations have their own geographical area under responsibility. Thus, the PPDR network must cover all this area because the radio service has to be available anywhere it can be necessary. This means that, at any given moment, the average traffic in a PPDR network can be very low, but, depending on events, the traffic can grow rapidly, locally, anywhere under the coverage of the network. So there is clearly an advantage to have several PPDR organizations sharing the same network in order to get a better traffic usage of it.

Moreover, today PPDR speech communications are provided by narrowband PMR networks (TETRA, TETRAPOL) in Europe. Since the lifetime of a PMR system is very long and since investments have to be optimized, the narrowband networks will last for long at least until 2030. So NarrowBand (NB) PPDR networks and BroadBand (BB) PPDR networks will have to coexist for a long time.

Even if 3GPP work includes Push-To-Talk speech group communications for PPDR, the migration from narrowband to broadband, in particular for speech communications, will take place but only on long term. Then PPDR speech communications will remain for a long time in the 380–385/390–395 MHz band, where European narrowband PPDR networks are mainly deployed.

### 3.3.4 THE PREFERRED FREQUENCY BAND OPTIONS FOR BROADBAND PPDR

#### 3.3.4.1 *400 MHz Band*

In Europe, nationwide Public Safety Networks (PSN) are located in the 380–400 MHz band. This was made possible by NATO decision (in the 1990s) to allow PSN deployment using part of the NATO spectrum as a harmonized $2 \times 5$ MHz channel bandwidth choice (FDD mode). Similar approach exists in the Middle East and in many APAC and Latin American countries, taking benefits from the existing European ecosystem. Thus, the 400 MHz band is a frequency band that is already dedicated to PMR, and as such, it is the preferred option for nationwide coverage PPDR networks in Europe. Propagation characteristics in the 400 MHz band are particularly interesting and suited for minimizing the number of sites in a national deployment. The 400 MHz band is subdivided in 3 subbands: 380–385/390–395 MHz (dedicated today to narrowband PPDR), 410–420/420–430 MHz, and 450–460/460–470 MHz (both are used for different civil, public safety, and military PMR applications).

FM 49 considers the two last options above as potential candidate bands for broadband PPDR. After CEPT SE 7 studies, the conclusions are that broadband PPDR can be deployed in 450–460/460–470 MHz band and/or in 410–420/420–430 MHz band.

Fig. 3.2 illustrates the schemes proposed by CEPT FM 49 for the deployment of broadband PPDR in the 450–470 MHz band.

#### 3.3.4.2 *700 MHz Band*

The 700 MHz band is also a band of great interest for broadband PPDR. This can be an opportunity, in the frame of the second Digital Dividend in Europe, for getting more frequencies in a band that will be freed by Digital Broadcasting application. The 700 MHz band has the advantage that, after release by broadcasters, the band will be fully available contrary to the 400 MHz band, where incumbent narrowband

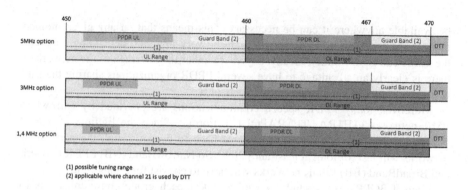

**FIGURE 3.2**

FM 49 options for dedicated PPDR spectrum in 450–470 MHz band.

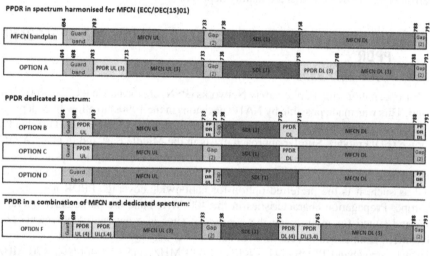

**FIGURE 3.3**

FM 49 options for dedicated PPDR spectrum in 700 MHz band.

PPDR and PMR systems are already present. The 700 MHz band would be more specifically suited for deployment in dense urban areas and for coverage extensions (provided, for instance, by a vehicular repeater). The options taken into account by CEPT/ECC, and included in ECC report 218, about PPDR spectrum in the 700 MHz band are shown in Fig. 3.3.

These options are enabling 2 × 5 MHz (698–703/753–758 MHz) plus 2 × 3 MHz (733–736/788–791 MHz) for dedicated spectrum for broadband PPDR in the 700 MHz band, outside MFCN bands (commercial broadband bands).

### 3.3.5 THE PROBLEMATIC OF THE 400 MHZ BAND: NEED FOR A REFARMING

As mentioned previously, in the 400 MHz band, narrowband PMR incumbent communications are already present. In 410–430 MHz and 450–470 MHz subbands, there are a lot of PMR networks of very different sizes: all parts of the spectrum are used but in a quite sparse manner even in dense urban areas. So the effective spectrum efficiency is low in these bands. The situation is different in 380–385/390–395 MHz subband since it is used for national narrowband Public Safety (PPDR) networks deployments (TETRA, TETRAPOL).

In summary, refarming the 410–430 MHz and 450–470 MHz subbands by reorganizing the frequency channels used by all the systems in those bands will allow freeing a significant part of the spectrum for broadband PPDR introduction. As an example, it is estimated that, in France, at least 2 × 3 MHz could be freed with such a refarming. This would allow deploying a broadband network using LTE technology in the 400 MHz band and thus would constitute the first step for introduction of broadband PPDR in this band. CEPT/ECC SE 7 has carried coexistence studies of LTE in the 410–430 MHz and 450–470 MHz subbands for broadband PPDR, notably with Digital TV and narrowband PMR.

### 3.3.6 ADVANCED NARROWBAND–BROADBAND COEXISTENCE

It is possible to go one step further: better spectrum efficiency can be achieved through a better coexistence management between narrowband and broadband communications. Allocating distinct and disjoint frequency bands to narrowband and broadband PPDR networks is the most straightforward way to ensure coexistence between the two systems. However, it may not be the most spectrum-efficient scheme. Deploying a broadband network in the same band as preexisting narrowband systems, while preserving the service and coverage of the legacy narrowband systems, ensures optimized spectrum efficiency. This is possible by shutting down some subcarriers of the broadband signal corresponding to the channels used by narrowband signals. Of course, in this case, the broadband signal must propose far better spectral containment characteristics than 3GPP LTE modulation scheme based on CP-OFDM.

Here high flexibility and spectral agility, in combination with efficient fragmented spectrum usage, are mandatory requirements for the broadband system as proposed in Fig. 3.4. To reach good spectral efficiency and minimize interferences between the different services, well-contained spectrum, providing improved adjacent channel protection, is essential.

## 3.4 RADIO PLANNING CONSIDERATIONS

To start with PMR constraints, PMR networks and currently deployed technologies for those networks are designed on the basis of a specific capacity-coverage trade-off scheme: the density of users for a typical day-to-day PP1 usage is relatively low,

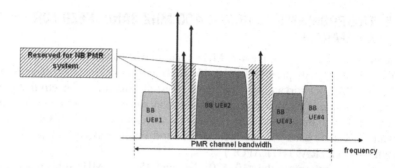

**FIGURE 3.4**

Proposed future BB PMR system in cohabitation with current legacy systems.

and the requirement for coverage is defined as a high percentage of the geographical area (in contrast with commercial networks for which coverage requirements are expressed in terms of percentage of population). Such constraints lead to adopt low-frequency bands (below 1 GHz: typically 400 and 700/800 MHz bands) for their good propagation properties and to boost the user equipment power (handheld or mobile) in order to optimize the global link budget: up to 2 W of output power for the handheld and up to 10 W for the mobile configuration.

## 3.4.1 LTE CHANNEL BANDWIDTH CONFIGURATION

As already mentioned, in Europe, broadband PMR channels allocation in the 400 MHz band exhibit limited bandwidth: up to 5 MHz for the Uplink (UL) and Downlink (DL) bands are generally available (2 × 5 MHz FDD mode). As a consequence, 1.4, 3.0, and 5.0 MHz LTE channel bandwidth (BW) configurations are the only suitable ones. By using only the LTE smallest bandwidth options, overall throughput performances are indeed reduced compared to what is claimed for the LTE commercial case (with a typical 20 MHz BW channel).

The LTE radio planning main assumption relies on a frequency reuse factor of 1 (same Radio Frequency (RF) allocated to all Base Station (BS) sites), and the resulting capacity is then limited by interference level (note that interference coordination and cancellation mechanisms to mitigate inter-cell interferences are available in the LTE standard, e.g., Inter-Cell Interference Coordination or ICIC).

Besides, thanks to the dynamic real-time link adaptation mechanism of LTE using multiple Modulations and Coding Schemes (MCS), the achievable throughput in downlink for LTE 3 MHz varies from 1.2 Mbs/s (mobile at cell edge) up to 6 Mbs/s (mobile close to the base station) assuming a medium network load.

To illustrate typical figures for such PMR scenarios, Fig. 3.5 provides the average performances (Mbs/s) for an LTE network with 1.4, 3, and 5 MHz BW configurations, considering a medium network load. It should be noted that the capacity

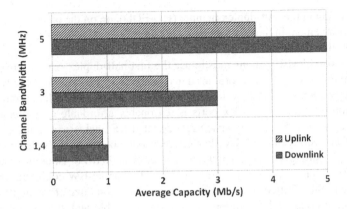

**FIGURE 3.5**

LTE cell average capacity for different bandwidth configurations.

performances presented in Fig. 3.5 are based on the following additional assumptions:

- 2×2 MIMO (Multiple-Input-Multiple-Output) transmission scheme for DL mode,
- 1×2 SIMO (Single-Input-Multiple-Output) transmission scheme for UL mode.

Such MIMO hypotheses are relatively fair because of the sub-GHz frequency constraint driven by the PMR requirements and thus limiting achievable decorrelation in multiple antennas configuration (but assuming vehicular configurations, up to 2 antennas MIMO scheme seems to be realistic). Moreover, taking into account terminal cost considerations, only one power amplifier is considered for smartphones. Once again, recall that large cell coverage is targeted.

### 3.4.2 DUPLEX SEPARATION

One very important issue is also the topic of frequency duplex separation, which must be considered carefully for any future LTE-based PMR system. Considering, for example, the European 400 MHz PMR frequency band case, duplex separation is generally equal to 10 MHz between UL and DL bands. It must be noted as well that in June 2012, Brazilian government, following ITU-R recommendation, has licensed the 450 MHz band according to the following arrangement: 451–458 MHz for UL and 461–468 MHz for DL. 3GPP has then standardized the arrangement for this band, more precisely, for 452.5–457.5 MHz/462.5–467.5 MHz, which does not fully cover the Brazilian government choice.

Such a tight frequency duplex gap may be problematic, especially for a terminal operating in full duplex mode (FDD), because receiver desensitization might occur caused by its own Tx signal leakage received in its Rx band. To avoid Rx desensitization from reduced duplex separation configurations, one solution is to operate Tx/Rx time switching in the half-duplex manner (HD-FDD), or, on the other hand, the other way is to implement in the terminal hardware an RF duplexer isolating locally the Rx band from the Tx band.

Narrowband PMR systems mostly use the first half-duplex option for operation, whereas the second solution is preferred in commercial LTE systems. Mainly because duplex separation is at least 30 MHz, and so, RF duplexer providing around 50 dB of isolation between Tx and Rx ports is reasonably affordable to implement. But, considering a 10 MHz duplex spacing, a detailed RF analysis in terms of LTE Tx unwanted emissions and LTE Rx filtering leads to the fact that RF duplexer specifications must be strongly strengthened: up to 70 or even 90 dB of isolation is needed to prevent Rx desensitization depending on the LTE channel BW configuration. These values are highly challenging in terms of implementation (size of the duplexer), and even if some commercial components are now available and able to provide such huge RF performances, every way to ease or override such requirements are welcome. That is why the HD-FDD mode proposed in the LTE toolbox ([11,12]) is also interesting: here, LTE terminals will declare to the eNodeB on which they are connected their HD-FDD restriction in order not to be scheduled simultaneously in UL and DL. Of course, this implies a significant eNodeB radio resources scheduling complexification.

### 3.4.3 POWER ASPECTS

To propose a smooth and easier transition toward broadband PMR, the idea would be to reuse existing narrowband PMR sites. In that case, the LTE PMR link budget must be carefully analyzed [13]. The frequency band is the same, and the propagation figures will remain identical, but in LTE, cell coverage is UL constrained. Thus, using LTE mobiles limited to +23 dBm output power, a maximum cell range of 1 km in urban case at 400 MHz is achieved, whereas typical narrowband cell range is in the order of 5.4 km (with a 10 W high-power mobile). This represents an important difference. As presented in [13], to improve the situation, three adaptations could be envisaged: boosting the LTE mobile Tx power up to 5 W (37 dBm) for a mobile mounted in a vehicle, using less resource blocks and targeting a lower SNR (but agreeing on expected bitrate reduction), or LTE sites sectorization (leading to increase of the BS antenna gain).

## 3.5 VOICE ASPECTS

### 3.5.1 PMR LOW BITRATE VOCODERS

Some words about voice transmission in the scope of LTE. First releases of LTE clearly did not address the voice transmission specificities and constraints, and these early releases were focused on large data transmission in the most efficient way. Then, different initiatives and work groups were put in place in order to deal with voice over IP transmission topics using LTE networks. It was clearly stated that the usage of these first releases for VoIP transmission over LTE will result in a large waste of radio resources. Specific adaptations and optimizations are a mandatory evolution to allow VoIP transmission over LTE in a very efficient way. This paragraph will not present the outcomes of these items, but it will just highlight the fact that the need to increase efficiency is even more relevant in the PMR context. This is due to the fact that vocoders used for PMR voice transmissions are currently working with a lower bitrate than commercial one. As a result, PMR voice packets issued from these low-bitrate vocoders are even smaller, and the current Transport Block Sizes (TBS) defined in LTE are not adapted at all for such tiny packets.

### 3.5.2 PUSH-TO-TALK (PTT) MECHANISM NETWORK CONSTRAINTS

Another PMR requirement relates to PMR voice services because voice calls management and network resources access are granted based on a Push-to-Talk (PTT) mechanism requiring a specific PMR network infrastructure. In the case of the LTE IP-based network, the presence of 3GPP IMS (Internet Multimedia System) is then mandatory coupled with the use of the Open Mobile Alliance (OMA) Push over Cellular (PoC) functionalities [14]. Unfortunately, OMA PoC standard does not meet the PMR PTT requirements yet (for example, the call setup time is far too long at present), and further adaptations or evolutions are still needed.

Moreover, for group calls involving many PMR users, the number of required radio resources are minimized through the usage of common channels (for both UL and DL) in classical PMR narrowband systems. The LTE basic strategy offering dedicated user channelization is not adapted at all in this purpose, and DL broadcasting would be preferable. In this aim, enhanced Multimedia Broadcast Multicast Service (e-MBMS) LTE feature is a possible way of optimization. In addition, when deployed using a single frequency radio planning scheme, DL e-MBMS channels are transmitted synchronously from several cells and can be received simultaneously by the LTE terminals: this is the MBMS Single Frequency Network (MBSFN) mechanism. Currently, in order to go further in the definition of such features, a 3GPP specific work item devoted to the talk group communication aspects involves user representatives from US (NPSTC) and Europe (TCCA) [15].

## 3.6 DIRECT MODE OF OPERATION (DMO) COMMUNICATION

### 3.6.1 DMO COMMUNICATION IN CURRENT PSN

Another main requirement and very important feature to consider is the ability of PMR users to communicate without going through any network infrastructure: typically, when there is no PMR infrastructure at all, or when this infrastructure is neither available nor reachable. This particular feature is named Direct Mode of Operation (DMO), and specific narrowband channels are devoted to this mode using a dedicated DMO protocol.

Unfortunately, until now, commercial operators have considered this communication mode as a threat. Since the communications are not managed by and routed through their infrastructure, the associated services are out of their control for billing, and their obligation for lawful interception cannot be fulfilled. Indeed, as far as current DMO for PSNs is concerned, the network operator could not have any way to control direct communications between terminals. Establishments of multihop ad hoc networks further increase the threat of loss of control over a large number of communications.

## 3.6.2 DEVICE TO DEVICE (D2D) SOLUTIONS FOR FUTURE LTE-BASED PSN

Historically, direct communications in cellular networks were specifically dedicated to PS services. But, since recently, direct communications are seen by public operators as a potential new communication mode: this open-minded attitude is at least a first step, and despite all the hesitations, vendors and public operators meditate on possible solutions to be implemented in future cellular networks. Of course, from a public networks point of view, the main issue is the control of the licensed frequency bands where direct communications would take place. Nevertheless, direct communication mode introduction in public networks could offer huge benefits: reduced power consumption, data rates increase, dense scenarios offloading schemes, etc. In that sense, public cellular network operators might wish to develop a new business model to protect their market while proposing direct communication features.

In 3GPP LTE, no such feature exists from the origin. That is why 3GPP is currently standardizing ProSe features that have applications for commercial networks and for Public Safety usage. This standardization work is in line with PMR PPDR constraints [16]. With the emerging interest of operators, the adoption of LTE standard for future PSNs and the arrival of the Internet of Things (IoT), direct communications features have been renamed Device-to-Device (D2D). The aim of such a renaming was to include all types of direct communications: from DMO to Machine-to-Machine (M2M) communications.

Whereas dedicated frequencies were specifically reserved for direct communications in existing narrowband PPDR systems, it will not be the case for future broadband systems. Thus, it was decided that the D2D communications will share the frequency allocation of the fixed network: to minimize the hardware impacts on the terminal side and especially on the power amplifier, D2D transmissions will occur in the uplink band (FDD case).

However, when D2D mobiles are within coverage of the infrastructure network, they will access the shared radio resources without creating interferences to other users (D2D users and "regular" network users). For that reason, synchronization of D2D users is derived from the downlink signals transmitted by the base station.

In out-of-coverage scenario, the imperative need of sharing common frequency and time references between users is similar. Hence, at least one D2D user (reference UE) will transmit some new defined signals for that purpose. Then, a fraction or the entire bandwidth of the uplink band can be used by out-of-coverage D2D mobiles to exchange information.

More generally, two modes of allocation for D2D users are defined in the LTE standard:

- Mode 1: Scheduled resource allocation. In this mode, the base station is in charge of dynamically allocating the UL resources requested by a D2D user that plans to establish a direct communication with its group members. This mode is likely to be the only one allowed for commercial users since it solves the issue of billing described above.
- Mode 2: UE autonomous resource selection. In this mode, the D2D mobile does not need to be connected to the base station. It autonomously picks within a resource pool some resources for its next D2D communication. The resource pool is defined as a set of resource blocks that are dedicated for D2D transmissions. The definition of that resource pool is broadcasted by the base station or preconfigured in the mobile.

The actual D2D features described in release 12 will support a first version of functionalities, and the work is continuing in order to address other PPDR requirements and needs, as well as new usages (for instance, driven by M2M demands).

## 3.7 STANDARD WAVEFORMS AND CANDIDATES FOR EVOLUTION

In addition to the previous PMR constraints and in order to be able to deploy a broadband data service in a band already occupied by narrowband PMR systems, a focus on spectrum improvements needs and possible ways to enhance coexistence between future broadband PMR systems and current legacy PMR systems already deployed is also introduced in this chapter. As depicted previously in Fig. 3.4, the most adequate way to optimize spectrum usage would be to achieve high flexibility and spectral agility, in combination with efficient fragmented spectrum allocation as mandatory requirements for the broadband system. To reach good spectral efficiency and minimize interferences between the different services and systems, well-contained spectrum providing improved adjacent channel protection is essential.

Narrowband PMR systems already provide good spectral performances (thanks to the stringent PMR standards specifications), but the broadband signal must propose far better spectral containment characteristics than 3GPP LTE (based on CP-OFDM) modulation scheme.

### 3.7.1 COEXISTENCE SCENARIO

In 3GPP LTE standard, it is theoretically possible to free spectrum "holes" by shutting down the respective Resource Blocks (RB), but the "white space" created by doing so is not usable to insert another system due to the high level of remaining intrinsic noise

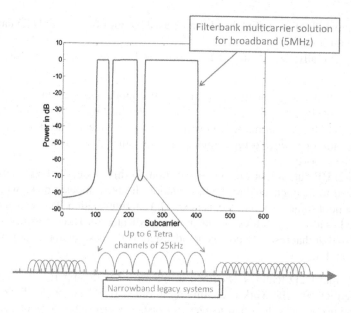

**FIGURE 3.6**

TETRA channels usage in blanked RBs of broadband signal.

of the CP-OFDM (and SC-FDMA) multicarrier modulation scheme. Indeed, the LTE spectrum is decreasing very slowly; this means that a significant frequency guard band is necessary for coexisting with another service, for instance: LTE 1.4 MHz uses only 1.08 MHz, LTE 3 MHz uses only 2.7 MHz, and LTE 5 MHz uses only 4.5 MHz for the useful signal.

By using an "optimized" broadband signal, typically derived from LTE, but implementing more advanced modulation schemes, such as Filtered MultiTone (FMT), Filter Bank MultiCarrier (FBMC), or Generalized Frequency Division Multiplexing (GFDM), the realization of the previous scenario seems to be reasonably achievable. Fig. 3.6 illustrates the shutting-down of some resource blocks or subcarriers in an optimized filterbank LTE signal (from a spectral point of view).

Moreover, an intelligent refarming of the narrowband systems can reduce the number of impacted broadband RBs by grouping the narrowband channels in order to limit the impact on broadband global BW. This has to be done by using frequency planning tools and processes. Only the narrowband cells close to the broadband cell have to be taken into account, so for a regular deployment, only the serving cell and the six surrounding cells have to be considered. In fact, as an example, up to six TETRA channels (25 kHz) can fit into one broadband RB (180 kHz), as shown in Fig. 3.6. This approach implies a flexible reconfigurable approach (shut down of some RBs) and a global frequency planning of all the radio systems deployed in the same frequency band (NB and BB systems).

This is the condition for a spectrum efficient coexistence between narrowband and broadband PMR systems. This approach enables to add broadband capability and new services while maintaining narrowband systems by improving spectrum usage and spectrum efficiency.

### 3.7.2 FUTURE WAVEFORM CANDIDATES

The interest of filterbank approach is that the compactness in the frequency domain is much better than for classical OFDM systems. With filterbank, the spectrum decay is much faster: this enables to reduce guard bands and to improve spectrum occupancy and spectrum efficiency. As close rejection in blanked RBs is much higher, the protection of narrowband communications is enhanced. This allows an optimized coexistence of broadband with TETRA and TETRAPOL communications: as a final result, narrowband services are preserved all over their radio coverage, and new broadband services are introduced. Once again, the utilization of well-designed waveforms provides an efficient and flexible access to fragmented spectrum, leading to a substantial improvement in the overall spectrum efficiency.

The following waveforms are the ideal candidates for such a scenario:

- Generalized Frequency-Division Multiplexing (GFDM). In GFDM, the out-of-band (OOB) radiation of the transmit signal is controlled by an adjustable pulse shaping filter applied to the individual subcarriers: root-raised-cosine (RRC), Xia, or Gaussian prototype filters can be used. Note that there might be intercarrier interference (ICI) among adjacent subcarriers, and intersymbol interference (ISI) might arise too if the transmit and receive filters do not fulfill the Nyquist criteria. Therefore, an appropriate interference canceling is needed at the receiver.
- Filtered Multitone modulation (FMT). Contrary to OFDM scheme, here again, a pulse shaping filter is used during modulation process. FMT scheme exploits a form of poly-phase filterbank modulation where a high level of spectral containment is achieved by subchannel spectral shaping using steep roll-off band-pass filters, resulting in quasi-nonoverlapping subbands (very low residual ICI level).
- Filter Bank Multicarrier (FBMC)/OQAM. With FBMC, modulated data at each subcarrier is shaped by a time-frequency well-localized prototype filter. However, a multicarrier system cannot simultaneously employ well-localized waveforms, keep complex orthogonality, and transmit at the Nyquist rate. In order to breakthrough this bottleneck, a symmetric root-Nyquist filter is used at both transmitter and receiver sides, and the OQAM modulation is implemented by introducing a half-symbol period.

## 3.8 CONCLUDING REMARKS

The PMR and PPDR communications are undergoing significant transformations toward supporting broadband data communications services in their future generations.

Having in mind the widely expected dominance of 3GPP LTE in the next few years, it is natural to consider the LTE system as a reference basis in the PMR/PPDR system development due to its modularity and wide adoption in the civil world. This is then necessary to introduce PMR specific features in LTE, and this is what 3GPP is currently doing. We highlight main particular aspects (coverage, voice services, and direct mode communications) showing that adaptations must be introduced in the LTE standard. However, there is a clear need to go beyond LTE, especially improving the spectral characteristics and striving to maximize the spectral efficiency. That is why we clearly focus on the spectrum optimization topic, as far as waveform design is concerned, because of the scarcity of additional frequency resources dedicated to PMR services for the introduction of the new broadband system. A first overview on innovative technological solutions supporting the development of a flexible and effective broadband system based on adaptable multicarrier waveforms for PPDR radio communications in coexistence with existing PMR networks is provided to facilitate a smooth migration toward future broadband systems. The use of such new waveforms will clearly allow enhanced cohabitation scenarios of broadband and narrowband PMR systems inside the same frequency band allocation.

However, the PMR over LTE evolution must not forget some other additional primary items: infrastructure resiliency reinforcement, advanced end-to-end security mechanisms that will complement the current air interface LTE ciphering/integrity mechanisms, and interoperability with other networks being PMR or commercial.

## ACKNOWLEDGMENTS

This research and study on PMR broadband evolution regarding PHY layer aspects with filter-bank modulation scheme introduction was supported by the European Commission under grant 318362 (FP7 Call 8), the EMPhAtiC project, and by the French National Research Agency under the grant ANR-13-INFR-0007-03, the PROFIL project.

## REFERENCES

[1] TETRA Voice + Data, EN 300 392. Sophia Antipolis: ETSI, 2002.
[2] TETRAPOL Specifications Part: PAS 0001-2, Version 3.0.0. Bois d'Arcy, France: TETRAPOL Forum, Nov. 1999.
[3] *APCO Project 25 Standards for Public Safety Digital Radio*. South Daytona, FL: APCO International, Oct. 1995.
[4] ETSI TS 102 181 Emergency Communications (EMTEL); Requirements for communication between authorities/organizations during emergencies, Feb. 2008.
[5] NPSTC, "700 MHz public safety broadband task force report and recommendations," Sep. 2009.
[6] H. Holma and A. Toskala, *LTE for UMTS*. Wiley, 2009.
[7] S. Sesia, *LTE: The Long Term Evolution of UMTS*. Wiley, 2009.

[8] "User requirements and spectrum needs for future European broadband PPDR systems (Wide Area networks)". ECC Report 199. Approved by CEPT/ECC and published.

[9] "Harmonized conditions and spectrum bands for the implementation of future European broadband PPDR systems". ECC Report 218. Approved by FM 49, to be approved by WG FM for sending to public consultation.

[10] "Is commercial cellular suitable for mission-critical Broadband". Study on use of commercial mobile networks and equipment for mission-critical high-speed broadband communications in specific sectors, SCF Associates LTD, Simon Forge, Robert Howitz and Colin Blackman, 2014.

[11] "R4-080255 Half Duplex FDD Operation in LTE," Feb. 2008. http://www.3gpp.org/ftp/tsg_ran/WG4_Radio/TSGR4_46/Docs/R4-080255.zip.

[12] "R4-080618 Half Duplex Operation of E-UTRA," Apr. 2008. http://www.3gpp.org/ftp/tsg_ran/WG4_Radio/TSGR4_46bis/Docs/R4-080618.zip.

[13] C. Gruet, *et al.*, "The LTE evolution: Private mobile radio networks," *IEEE Veh. Technol. Mag.*, vol. 8, no. 2, pp. 64–70, 2013.

[14] "OMA Push to Talk over Cellular specifications v2.0." http://www.openmobilealliance.org/technical/release_program/poc_v2_0.aspx.

[15] "SP-120421 Description of the WI "Group Communication System Enablers for LTE"," Jun. 2012. http://www.3gpp.org/ftp/tsg_sa/TSG_SA/TSGS_56//Docs/SP-120421.zip.

[16] TS 23.303 Proximity-based services (ProSe); Stage 2, Release 12.

# Application of FBMC in Optical Communications

# 4

**Jessica Fickers\*, Trung-Hien Nguyen\*, Jérôme Louveaux[†], Simon-Pierre Gorza\*, François Horlin\***

*Université libre de Bruxelles, Brussels, Belgium\**
*Université catholique de Louvain, Louvain-la-Neuve, Belgium[†]*

## CONTENTS

## 4.1 EVOLUTION TO OPTICAL COHERENT COMMUNICATIONS

Over the last decades, the reduction of optical fiber losses and the invention of fiber-doped optical amplifiers enabled the use of wavelength-division multiplexing (WDM) technology, which consists in the simultaneous transmission of multiple signals at different wavelengths densely located on the frequency grid. WDM was initially employed for intensity modulation/direct detection (IMDD), which has been quickly limited in capacity. In contrast to the IMDD systems, the systems using advanced modulation formats yield an improvement in capacity and an increase in spectral efficiency (SE). Indeed, advanced modulation format signals exploit not only the amplitude but also the other signal domains (i.e., phase and state of polarization (SOP)) to encode the electrical data onto an optical carrier. Furthermore, the combination of coherent detection and digital signal processing (DSP) allow one not only

Orthogonal Waveforms and Filter Banks for Future Communication Systems. DOI: 10.1016/B978-0-12-810384-5.00004-9

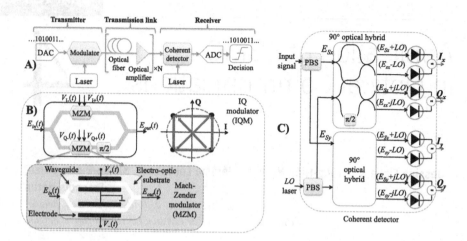

**FIGURE 4.1**

(A) Typical optical communication link. (B) Electro-optic modulator structures. (C) Dual polarization coherent receiver.

to demodulate advanced modulation formats signals but also to mitigate the main fiber impairments. This evolution of optical communications toward modulation formats commonly used in the radio communication domain was made possible thanks to the advents in signal generation (integrated IQ modulator), detection (integrated 90° hybrid), and in DSP implementation (clock speed of silicon chips, enabling the convergence of the silicon chip speed with optical line-rates [1]).

## 4.2 COHERENT COMMUNICATIONS ON OPTICAL FIBERS
### 4.2.1 THE OPTICAL FIBER COMMUNICATION SYSTEM

Fig. 4.1A illustrates a typical optical long-haul communications link. The binary information to be transmitted undergoes DSP before being converted into an analog waveform. Depending on each system configuration, the transmitter DSPs can be composed of channel coding, precompensation of the transmitter impairments, etc. A continuous-wave (CW) laser is then modulated externally with this waveform by an electro-optic modulator (EOM). Fig. 4.1B presents a typical EOM structure, which is the IQ modulator (IQM) composed of two Mach–Zehnder modulators (MZMs) and a 90° phase shifter between the two arms [2]. The IQM is used to imprint the electrical signals onto different degrees of freedom (i.e., amplitude, phase) of optical carriers. As a consequence, different optical modulation format signals can be generated, such as amplitude-shift keying (ASK), phase-shift keying (PSK), quadrature amplitude modulation (QAM), etc. Furthermore, polarization division multiplexing (PDM) is used to double the SE. In dual polarization systems, two IQMs are used and followed by polarizers, which impose a linearly polarized SOP on two orthogonal transverse axes. The two signals are then combined in a polarization combiner to compose the polarization-multiplexed optical signal.

The resulting modulated signal is fed into the optical fiber (1000 to 5000 km for terrestrial transmission links). At the receiver, the optical signal is coherently detected and transformed back into the electrical domain by photodiodes. Fig. 4.1C shows a structure of the dual polarization coherent receiver. The input signal and local oscillator (LO) are passed to a polarization beam splitter (PBS) to separate the $x$- and $y$-polarizations. Considering only $x$-polarization (the principle is similar for $y$-polarization), the data signal $E_{Sx}$ and LO fields $E_{LOx}$ are coupled into a 90° optical hybrid-unit, where they are split and cross-combined, producing four output interference terms. These four outputs are detected by balanced photodiodes. The summation of signals coming from each pair of balanced photodiodes gives the beating of signal and LO fields and allows one to remove the quadratic terms of detection. The two outputs, called the in-phase (I) and quadrature (Q) components, are finally quantized and digitally processed before they are converted back into a binary sequence.

Note that the laser performance is mostly limited by spontaneous emission, resulting in both amplitude and phase noise. The laser frequency noise of telecommunication lasers is commonly assumed a Brownian motion (equivalently, a Lorentzian line shape) with a linewidth in the MHz range [3]. The frequency noise can be computed as a white noise with spectral density equal to $\Delta \nu / \pi$, where $\Delta \nu$ is the full width at half maximum linewidth. When coherent detection is used, the phase noise beating of transmitter and receiver lasers imprints itself on modulated signals. Consequently, the received signal shows a phase drift in accordance with the laser linewidths. This phase drift can be compensated for in the DSP.

### 4.2.2 THE OPTICAL FIBER AS A TRANSMISSION MEDIUM

In optical fibers, the main physical effects leading to the distortion of the wave-packet as it propagates are (1) attenuation, (2) chromatic dispersion (CD), (3) polarization mode dispersion (PMD), and (4) the nonlinear Kerr effect. The evolution of the optical field $\mathbf{A}$ along its propagation axis $z$ is described by a partial differential equation, called the coupled nonlinear Schrödinger equation (NLSE) [4]. By ignoring all right-hand terms in the NLSE except the term of interest we can deduce the analytical solution with respect to each impairment.

1. Keeping only the attenuation effect ($\partial \mathbf{A} / \partial z = -\alpha/2\mathbf{A}$), in which $\alpha$ is the attenuation coefficient, reveals that the power of the electric field will decrease exponentially with transmission distance (i.e., 0.2 dB/km in standard single-mode fiber (SSMF)).

2. We define the inverse Fourier transform of the wave-packet as $\mathbf{A}(z, \tau) = \int_{-\infty}^{\infty} \tilde{\mathbf{A}}(z, \omega) e^{j\omega\tau} d\omega$. When considering only the CD, limited to the third order, the NLSE equation in the frequency domain (FD) can be written as $\tilde{\mathbf{A}}(z, \omega) = \tilde{\mathbf{A}}(0, \omega) \exp\left(-j\left(\frac{\beta_2}{2}\omega^2 + \frac{\beta_3}{6}\omega^3\right)z\right)$, where $\beta_i = \left.\partial^i \beta / \partial \omega^i\right|_{\omega=\omega_0}$, $\omega_0$ is the cen-

tral pulsation, and $\beta(\omega)$ is the propagation constant. It turns out that the CD manifests in the FD as the accumulation of a quadratic phase for the second-order dispersion and a cubic phase for the third-order dispersion. It can thus be inverted in the FD by multiplication with the exponential of the opposite phase. In the fiber-optic community, CD is generally quantified by the dispersion parameter $D$ [ps/nm·km], which is linked to $\beta_2$ and the central wavelength of the signal $\lambda$ as follows: $D = -\left(2\pi c/\lambda^2\right)\beta_2$, where $c$ is the light speed in the vacuum.

3. At each axial position $z$, the fiber is characterized by an eigenmode corresponding to the field polarization with the slowest propagation constant [5]. At the same time, the orthogonal eigenmode is the mode with the highest propagation constant. The distribution of the eigenmodes as a function of distance slowly changes with time on the ms scale [5], which means that PMD parameters must be tracked at the receiver.

4. The optical Kerr effect is the main source of nonlinear distortion in optical communication systems [5], in which the refractive index depends on the power of the electric field propagating through the fiber. In WDM transmissions, the optical Kerr effect can be categorized as the self-phase modulation (SPM), cross-phase modulation (XPM), and four-wave mixing (FWM). More details on these effects can be found in [5].

## 4.2.3 DIGITAL SIGNAL PROCESSING IN COHERENT OPTICAL COMMUNICATIONS SYSTEMS

In a realistic optical long-haul communication setup, the transmitted signal is altered by several impairments that can be compensated for at the receiver. For the widespread single-carrier QAM modulations, receiver DSP can be subdivided into well-defined steps (Fig. 4.2A).

At first, CD is normally compensated. Due to the fact that CD is a polarization-independent phenomenon and its strength can roughly be known at the receiver given the fiber type and distance, we can directly compensate for CD either in the time domain (TD) or in the FD. For a large number of equalizer taps, the complexity of implementation is much lower for the FD equalizer [6]. The FD filter to compensate for CD can be designed as $G\left(\omega\right) = \exp\left(-\mathrm{j} \cdot D\lambda^2 z/4\pi c \cdot \omega^2\right)$. To overcome the cyclic properties of the fast Fourier transform (FFT), the FD equalizer is usually implemented using the overlap and save method (Fig. 4.2B) [6]. The incoming data is transformed with an FFT of length $N$ samples. Then the FFT result is multiplied with the filter impulse response $G$, and finally the data is transformed back to TD. $K/2$ samples are discarded at the beginning and the end of the block, respectively, to yield result vector of size $(N - K)$. In the next step, the FFT is applied to the incoming data vector delayed by $K$ samples. The overlap and save filtering is applied separately on both polarization components.

Polarization demultiplexing is usually carried out afterwards by the adaptive butterfly constant modulus algorithm (CMA), the most universally used algorithm in coherent optical communications [7]. The standard CMA uses the stochastic gradi-

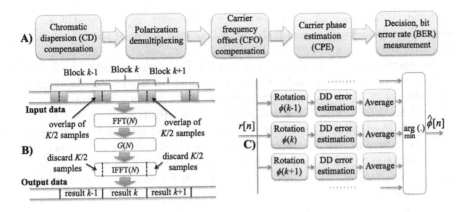

**FIGURE 4.2**

(A) Major DSP blocks for single carrier optical QAM modulation. (B) Principle of overlap and save frequency domain CD compensation. (C) Maximum-likelihood carrier phase estimation.

ent algorithm as its update device. This simple update mechanism has been proven a very robust and versatile algorithm for coherent optical communications. Although it was originally designed for constant modulus modulations such as BPSK or QPSK, the CMA is now used for high-order modulation format such as 64-QAM [8].

The error signal of the CMA is sensitive to the amplitude of the waveforms only. Consequently, phase tracking can be carried out independently in the next step. Carrier frequency offset (CFO) and laser phase noise (PN) result in a rotation of the constellation diagram in the complex plane. In a popular approach [9] for coarse CFO estimation, the incoming signal is processed in a sliding window approach. For each window, the $P$th power, where $P$ is the modulation order, is applied to the signal (i.e., $P$-PSK). By this way the phase modulation of the signal is canceled out. Afterwards, the signal is translated to FD by FFT. The index of the maximum sample is indicative of the frequency offset. For $P$-PSK modulation formats, the Viterbi–Viterbi algorithm [10] is the standard approach for PN estimation. The estimated PN on the $n$th symbol $r[n]$ is calculated by $\hat{\phi}[n] = 1/P \arg\left(\sum_{u=-U/2}^{u=U/2} (r[n+u])^P\right)$. Finally, the phase is unwrapped. For quadrature modulation formats (i.e., 16-QAM), the maximum-likelihood (ML) phase estimator (Fig. 4.2C) is often preferred. The ML criterion refers to the maximization of the probability of manifestation of the received sequence with respect to all possible sent sequences [11]. In practice, test phases are computed from $-\pi/4$ to $\pi/4$ due to the symmetric nature of QAM constellations. The number of test phases determines the phase resolution of the algorithm. The incoming signal is rotated by each test phase. The Euclidean symbol estimation error is computed for each hypothesis, and the results are summed over a certain number of precursor and postcursor symbols to average out the additive noise. Finally, the phase that minimizes the mean square error (MSE) is selected for each symbol. Since the algorithm is blind, the Euclidean estimation error distance is approximated by the distance of the symbol with respect to the nearest constellation point (decision-directed approach, DD).

In optical communication systems, a forward error correction (FEC) code is applied to the bit sequence in order to reduce the bit error rate to very small values (typically, $10^{-14}$) [12]. The most currently used advanced soft decision low-density parity-check (LDPC) FECs work with a bit-error ratio (BER) at approximately $10^{-2}$ with a 20% overhead [13]. In the following discussion, we account for a percentage of FEC overhead to yield the net bitrate. Correspondingly, we use a target pre-FEC quality-factor ($Q^2$-factor) for the considered FEC; $Q^2$-factor (in dB) is defined as $Q^2 = 20\log_{10}\left(\sqrt{2}erfc^{-1}\left(2 \cdot \mathrm{BER}\right)\right)$, where $erfc^{-1}$ is the inverse complementary error function.

## 4.3 TWO APPROACHES TO INCREASE THE SPECTRAL EFFICIENCY

In WDM optical communication systems, increasing the SE is a key target to achieve higher capacities. To this aim, one solution is to reduce the channel spacing defined as the frequency separation between two optical carriers. As the communication channels suffer from intercarrier interference (ICI) when the channel spacing is reduced toward the symbol rate, substantial effort is currently devoted to design orthogonal multichannel systems. Two main approaches have been proposed in the literature to achieve this goal. The first one, referred to as Nyquist wavelength-division multiplexing (N-WDM) [14], consists in using nearly rectangular frequency pulses to limit the channel bandwidth to the symbol rate. The other approach, called coherent orthogonal frequency-division multiplexing (CO-OFDM) [15], results from the application of the OFDM modulation widely used for wireless communications systems to multiple optical channels.

### 4.3.1 NYQUIST WAVELENGTH DIVISION MULTIPLEXING (NYQUIST-WDM)

N-WDM systems operate with shaping pulses having nearly rectangular frequency spectrum of bandwidth equal to the symbol rate [14]. The class of root-raised-cosine (RRC) functions is of particular interest because they satisfy the Nyquist criterion of zero inter-symbol interference (ISI) whatever the roll-off factor when applied at transmitter and receiver. During the last years, RRC filtering with a close to 0 roll-off factor has attracted much research interest in order to maximize SE [16]. However, N-WDM suffers from hardware implementation limitations such as the finite length of the pulse shaping filters, the timing jitter of the data sampling, and the finite resolution of the analog/digital converters (ADC, DAC). These constraints translate into ISI and ICI and therefore affect significantly the performance. Allowing for nonzero roll-off factors relaxes the constraints on the filter length and the tolerable jitter at expense of increasing ICI. Most recent works on N-WDM assume very small roll-off factors [17].

### 4.3.2 MULTICARRIER OFFSET QAM MODULATIONS

The first implementations of OFDM for optical fibers are electrical per channel and aim at low computational complexity equalization of the channel impairments [15]. OFDM divides the wideband channel in a set of narrowband flat subchannels that can be equalized independently with a single-tap equalizer. OFDM is easily combined with multi-input multi-output (MIMO) techniques interesting to support PDM [18]. In the case of CO-OFDM, the number of subchannels remains generally small and the overhead incurred by the guard interval is too large. Ref. [19] proposes to completely or partially remove the guard interval and to replace it with conventional per-channel DSP to cope with the resulting ICI. Minimum, however, not negligible ICI arises when the channel spacing equals the symbol rate on each channel and when the symbols of the modulated channels are time-aligned as demonstrated experimentally in [20]. Filter-bank multicarrier (FBMC, or MC for short) modulations and, more specifically, OFDM-offset quadrature amplitude modulation (OQAM) are seen as an interesting extension to OFDM for future communication systems [21]. Like OFDM, MC-OQAM decomposes the communication channel in a set of lower-bandwidth subchannels that can therefore also be compensated at a low complexity with a single-tap equalizer. The time/frequency resolution of the waveforms is increased resulting in a better utilization of the physical resources and in an improved robustness to channel time variations and frequency offsets. Contrary to OFDM, MC-OQAM does not require the addition of a redundant guard interval, and the created subchannels are only approximately flat and orthogonal [21].

Similarly to OFDM, transmission impairments are bound to break the subcarrier orthogonality of MC-OQAM [21]. The benefit of MC-OQAM therefore strongly depends on the ability to recover subcarrier orthogonality using DSP. Ref. [22] has demonstrated a 224 Gb/s MC-OQAM communication system with seven 16-OQAM subcarriers in a back-to-back (B2B) configuration using a modified blind decision-directed equalization on each subcarrier. Ref. [23] has demonstrated the benefits of MC-OQAM for superchannel crosstalk suppression in a B2B experiment.

## 4.4 KEY PERFORMANCES AND OPEN CHALLENGES

### 4.4.1 PRELIMINARY IMPLEMENTATION OF MULTICARRIER OFFSET QAM ON EXPERIMENTAL SETUP

We have demonstrated the potential of MC-OQAM with coherent long-haul optical communication experiments [24]. The experimental setup is illustrated in Fig. 4.3A. A programmable DAC ($2^{18}$ samples) provides 32.5 GBaud MC-OQPSK or MC-QPSK waveforms at two samples per symbol to modulate the optical field of the CW

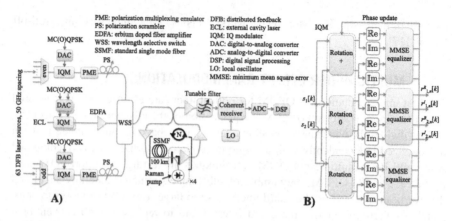

**FIGURE 4.3**

(A) Experimental setup. (B) MC-OQAM phase-corrected MMSE equalizer. (Rotation +) designates a complex-plane rotation of $\psi + \alpha$ degrees, where $\psi$ is the current overall estimated phase shift, and $\alpha$ is the test phase of the tracking algorithm. (Rotation 0) is a rotation of $\psi$, and (Rotation −) is a rotation of $\psi - \alpha$ degrees. $\mathbb{R}e$ and $\mathbb{I}m$ are the real and imaginary part operators, respectively.

lasers. The WDM transmitter involves an array of 64 distributed feedback (DFB) lasers, which fall on the standard 50 GHz ITU frequency grid, being polarization-multiplexed by the polarization division multiplexing emulators (PMEs). The signal is then passed into a low-speed (<10 Hz) polarization scrambler (PS). The central test channel at 1545.72 nm is replaced by an external cavity laser (ECL) and modulated via a polarization-multiplexed IQM. The WDM multiplex is sent into a recirculation loop, composed of four spans of 100 km standard single-mode fiber (SSMF). Fiber loss is compensated for by hybrid Raman–Erbium optical repeaters, which have two back-propagating laser diode pumps at wavelengths 1432 nm and 1457 nm and offer approximately 10 dB on-off gain. At the receiver side, the MC signal is mixed with a LO, sampled, and recorded, by a 80 GSamples/s real-time scope of 33 GHz bandwidth for the next step of DSPs.

Except for the CD compensation (using the well-known FD overlap-and-save method [6]), substantial changes need to be made to the standard algorithms when MC-OQAM is used. Particularly, we build a new DSP architecture to tackle linear impairments such as polarization demultiplexing and phase tracking [24]. We derive the minimum mean square error (MMSE) continuous butterfly per-subchannel equalizers [21] based on the subcarrier estimations. To track channel variations mainly due to polarization effects, we implement an adaptive equalizer based on the stochastic gradient algorithm updated using a decision-directed approach [11]. To ensure the fast-varying phase tracking, we propose a butterfly equalizer using trial phases. The algorithm is detailed in Fig. 4.3B. The method is inspired from ML phase tracking [25] but is less complex to implement. First, trial phases are initialized on the

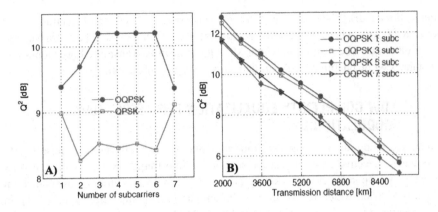

**FIGURE 4.4**

(A) Assessment of the subcarrier performance after 4000 km of DU SSMF for 7-subcarrier MC-QPSK and MC-OQPSK using a 0.3 roll-off factor. (B) Performance of MC-OQPSK as a function of distance over dispersion-unmanaged SSMF. [reproduced from [24]]

three arms of the equalizer as $[-\alpha\ 0\ \alpha]$, where $\alpha$ is a free parameter representing the phase resolution of the algorithm. After each block of $N$ symbols, the block-wise average square error is calculated with respect to the nearest constellation points. The phase value minimizing the average square error is selected. The test phases are then updated so that the phase minimizing the error migrates to the central arm and the corresponding equalizer is copied on the three arms. In this updated version, the butterfly equalizer has three convergence parameters: the conventional $\beta$ of the stochastic gradient, the phase resolution $\alpha$, and the block size $N$ of the fast phase tracking. The choice of block size $N$ represents a trade-off between phase noise dynamics and noise averaging. More details on DSPs should be referred to [24]. Due to the space constraint, we report here the measured MC-OQPSK waveform sets for different subcarrier counts 1, 3, 5, or 7 in the long transmission distance. To evaluate the performance, we have measured the BER on each subcarrier, calculated the average BER (or, equivalently, $Q^2$-factor) over subcarriers for the MC signal.

We study the performance of each subcarrier (0.3 roll-off factor) separately over 4000 km of dispersion unmanaged SSMF. Fig. 4.4A illustrates the $Q^2$-factor of seven-subcarrier MC-QPSK/MC-OQPSK. In the MC-QPSK case, the central subcarriers suffer from twice as much crosstalk as the edge subcarriers because of no subcarrier orthogonality. In the MC-OQPSK case, the overall performance averaged over subcarriers is better. We observe that there is a $Q^2$-factor deterioration of 0.8 dB on the edge subcarriers, indicating that the transmitter DAC and/or receiver ADC impairments are more stringent on higher frequencies. The performance distribution is inverted with respect to MC-QPSK. We conclude that in the case of MC-OQPSK, the receiver DSP has been able to recover the subcarrier orthogonality. In Fig. 4.4B, we finally study the performance of MC-OQPSK as a function of transmission distance. For a 6 dB $Q^2$-factor, one- and three-subcarrier OQPSK have been transmitted over 8400 km, whereas five- and seven-subcarrier OQPSK have been transmitted over 6800 km. The results show that MC-OQAM can be transmitted over long-haul distances while enabling DSP parallelization over the subcarriers.

## 4.4.2 LOW COMPLEXITY EQUALIZATION FOR OFFSET QAM MODULATIONS

In wireless communications, MC modulations are frequently used because highly frequency selective channels can be subdivided into frequency slots with quasi-flat frequency responses. As a consequence, one-tap equalizers can be implemented on each subcarrier, which results in overall computational complexity savings [21]. In this section, we apply this principle to MC-OQAM modulations in coherent optical communications. In contrast to using small number of subcarriers as in the previous section, the number of subcarriers is increased up to 8192 subcarriers inside the 30 GHz range of one optical channel. By this way the CD can be equalized by multiplying each output of the analysis filter bank (AFB) with a single complex coefficient. In this case, instead of rough CD compensation before the AFB as in the single carrier (SC) modulation [6], the CD compensators can be put after the AFB in the MC modulation.

However, per-subcarrier treatment makes the transmission scheme more prone to variable phase offsets, a fact that is well known in the wireless OFDM communications [18]. Thus, MC-OQAM is likely to suffer increasingly from phase related impairments when the number of subcarriers is increased at a constant aggregate symbol rate. To verify this argument, we use numerical simulations to assess the performance of MC-OQAM optical fiber transmissions using FD equalizers under the phase noise impacts. At the receiver, the equalizers are put after the AFB in the poly-phase network (PPN) implementation [26]. For all simulations, to simulate the inline impairments such as CD, we suppose an aggregate symbol rate of 30 GHz, the roll-off factor of 1, no CFO, and no sample timing jitter. Moreover, the error vector magnitude (EVM) is used as a metric for comparison among different fiber length transmissions and laser phase noise strengths.

We study the combined effect of CD and phase noise in Fig. 4.5. It is clear that the system requirements on the number of subcarriers $M$ for CD and phase noise compensation conflict. More specifically, for a small number of subcarriers, the CD limitation dominates because the FFT filters are not narrow enough to resolve the channel. For a large number of subcarriers, the sensitivity of OQAM modulation to phase offsets increases with the number of subcarriers because the dynamics of the phase variations catch up with the symbol duration of the individual subcarriers. As a consequence, even for a moderate laser linewidth of 500 kHz and a propagation distance of 1000 km, the EVM cannot be reduced to an acceptable level. The limited laser phase noise therefore represents a real obstacle to the use of large subcarrier numbers in MC-OQAM modulations.

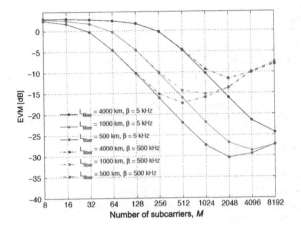

**FIGURE 4.5**

EVM versus the number of subcarriers for different fiber lengths/phase noise linewidths.

### 4.4.3 OPTICAL IMPLEMENTATION OF MULTICARRIER OFFSET QAM

The number of subcarriers per optical channel should be kept small enough to reduce the phase noise influence, whereas the band gap between two optical channels can be reduced by keeping adjacent channels overlapped with OQAM signals. The implementation of optical multichannel MC-OQAM systems is challenging since it requires the generation of a set of frequency-locked and synchronously modulated optical carriers [21]. Alternative hardware architectures are considered to this end: the cascaded MZMs for small numbers of carriers [27] and the recirculating frequency shifter when the number of carriers is larger [28]. The frequency-locked carriers are separated by a wavelength demultiplexer, before being individually modulated by an IQ modulator, and combined by an optical coupler. In such systems, the coherence between two adjacent subcarriers of two optical carriers should be guaranteed to ensure the orthogonality.

We can alleviate the coherent requirement between two adjacent subcarriers of two optical channels by increasing the number of subcarriers in each optical carrier and then removing one subcarrier at the boundary between two optical channels. However, the subcarrier suppression can reduce the SE. To maximize the SE in such systems, the number of subcarriers should be maximized as much as possible. As aforementioned, the phase noise is the major limitation in the use of large number of subcarriers. When designing the optical MC-OQAM systems, we should carefully take into account this trade-off between the CD-induced maximum reachable distance and the strength of the laser phase noise.

# REFERENCES

[1] A. E. Willner, S. Khaleghi, M. R. Chitgarha, and O. F. Yilmaz, "All-optical signal processing," *J. Lightw. Technol.*, vol. 32, no. 4, pp. 660–680, Feb. 2014.

[2] M. Nakazawa, K. Kikuchi, and T. Miyazaki, *High Spectral Density Optical Communication Technologies*. Heidelberg: Springer, 2010.

[3] S. Bartalini, *et al.*, "Measuring frequency noise and intrinsic linewidth of a room-temperature DFB quantum cascade laser," *Opt. Express*, vol. 19, no. 19, pp. 17996–18003, Sep. 2011.

[4] G. P. Agrawal, *Fiber-Optic Communication Systems*. New York: John Wiley and Sons, 2002.

[5] R. W. Boyd, *Nonlinear Optics*. San Diego, CA: Academic Press, 2006.

[6] R. Kudo, *et al.*, "Coherent optical single carrier transmission using overlap frequency domain equalization for long-haul optical systems," *J. Lightw. Technol.*, vol. 27, no. 16, pp. 3721–3728, Aug. 2009.

[7] S. J. Savory, "Digital coherent optical receivers: Algorithms and subsystems," *IEEE J. Sel. Topics Quantum Electron.*, vol. 16, no. 5, pp. 1164–1179, Oct. 2010.

[8] S. Randel, S. Corteselli, and P. J. Winzer, "Generation of a digitally shaped 55-GBd 64-QAM single-carrier signal using novel high-speed DACs," in *Proc. OFC 2014*, San Francisco, CA, USA, Mar. 2014.

[9] L. Li, *et al.*, "Wide-range, accurate and simple digital frequency offset compensator for optical coherent receivers," in *Proc. OFC 2008*, San Diego, CA, USA, Feb. 2008.

[10] A. Viterbi and A. Viterbi, "Nonlinear estimation of PSK-modulated carrier phase with application to burst digital transmission," *IEEE Trans. Inf. Theory*, vol. 19, pp. 543–551, Jul. 1983.

[11] J. Proakis, *Digital Communications*. New York: McGraw-Hill, 2001.

[12] I. Djordevic, *Coding for Optical Channels*. New York: Springer, 2010.

[13] X. Zhou, *et al.*, "High spectral efficiency 400 Gb/s transmission using PDM time-domain hybrid 32–64 QAM and training-assisted carrier recovery," *J. Lightw. Technol.*, vol. 31, no. 7, pp. 999–1005, Apr. 2013.

[14] R. Schmogrow, *et al.*, "Real-time Nyquist pulse generation beyond 100 Gbit/s and its relation to OFDM," *Opt. Express*, vol. 20, no. 1, pp. 317–337, Dec. 2011.

[15] W. Shieh, *et al.*, "Coherent optical OFDM: Has its time come?," *J. Opt. Netw.*, vol. 7, no. 3, pp. 234–255, Mar. 2008.

[16] X. Zhou, *et al.*, "8 × 450-Gb/s, 50-GHz spaced, PDM-32QAM transmission over 400 km and one 50 GHz-grid ROADM," in *Proc. OFC 2011*, Los Angeles, CA, USA, Mar. 2011.

[17] J. Fickers, *et al.*, "Design rules for pulse shaping in PDM-QPSK and PDM-16QAM Nyquist-WDM coherent optical transmission systems," in *Proc. ECOC 2012*, Amsterdam, Netherlands, Sep. 2012.

[18] F. Horlin, *et al.*, "Polarization division multiplexing for SC-FDE communications over dispersive optical fibers," in *Proc. ICC 2010*, Cape Town, South Africa, May 2010.

[19] A. Sano, *et al.*, "No-guard-interval coherent optical OFDM for 100-Gb/s long-haul WDM transmission," *J. Lightw. Technol.*, vol. 27, no. 16, pp. 3705–3713, Aug. 2009.

[20] S. Chandrasekhar and X. Liu, "Experimental investigation on the performance of closely spaced multi-carrier PDM-QPSK with digital coherent detection," *Opt. Express*, vol. 17, no. 24, pp. 21350–21361, Nov. 2009.

[21] F. Horlin, *et al.*, "Dual-polarization OFDM-OQAM for communications over optical fibers with coherent detection," *Opt. Express*, vol. 21, no. 5, pp. 6409–6421, Mar. 2013.

[22] S. Randel, *et al.*, "Generation of 224-Gb/s multicarrier offset-QAM using a real-time transmitter," in *Proc. OFC 2012*, Los Angeles, CA, USA, Mar. 2012.

[23] Z. Li, *et al.*, "Experimental demonstration of 110-Gb/s unsynchronized bandmultiplexed superchannel coherent optical OFDM/OQAM system," *Opt. Express*, vol. 21, no. 19, pp. 21924–21931, Sep. 2013.

[24] J. Fickers, *et al.*, "Multicarrier offset-QAM for long-haul coherent optical communications," *J. Lightw. Technol.*, vol. 32, no. 24, pp. 4671–4678, Dec. 2014.

[25] P. Y. Kam, "Maximum likelihood carrier phase recovery for linear suppressed-carrier digital data modulations," *IEEE Trans. Commun.*, vol. 34, pp. 522–527, Jun. 1986.

[26] P. Siohan, C. Siclet, and N. Lacaille, "Analysis and design of OFDM/OQAM systems based on filterbank theory," *IEEE Trans. Signal Process.*, vol. 50, pp. 1170–1183, May 2002.

[27] T. Healyet, *et al.*, "Multi-wavelength source using low drive-voltage amplitude modulators for optical communications," *Opt. Express*, vol. 15, no. 6, pp. 2981–2986, Mar. 2007.

[28] Y. Ma, *et al.*, "1-Tb/s per channel coherent optical OFDM transmission with subwavelength bandwidth access," in *Proc. OFC 2009*, San Diego, CA, USA, Mar. 2009.

# Filterbank Systems for Communications: Theory and Design

# Multirate Signal Processing and Filterbanks

5

**Markku Renfors, Juha Yli-Kaakinen**
*Tampere University of Technology, Tampere, Finland*

## CONTENTS

## 5.1 INTRODUCTION

This chapter introduces certain important theories and signal processing tools as background for later developments in this book. After a brief introduction to real and complex linear systems, spectral models for discrete-time systems are formulated in a generalized way and conditions for alias-free sampling are formulated for real and complex baseband and passband systems. Next, multirate filtering concepts are introduced, again with emphasis on complex and bandpass signal models. Also polyphase filters and filterbanks are introduced in this context. Then the Nyquist pulse shaping principle is explained as a central element of both classical communication theory and filterbank based waveforms. Finally, after a brief introduction to the Discrete Fourier Transform (DFT), the basic form of effective uniform filterbanks, the DFT FilterBank (DFT-FB), is introduced, and general filterbank concepts and classifications are summarized.

*Orthogonal Waveforms and Filter Banks for Future Communication Systems.* DOI: 10.1016/B978-0-12-810384-5.00005-0

## 5.2 REAL AND COMPLEX LINEAR SYSTEMS

We start by recalling that Linear Time-Invariant (LTI) systems or filters[1] are defined in time domain by their impulse responses, denoted as $h(t)$ and $h[n]$ in continuous-time and discrete-time cases, respectively. The corresponding transfer functions are obtained through the Laplace and $z$-transforms as $\overline{H}(s) = \mathcal{L}\{h(t)\}$ and $H(z) = \mathcal{Z}\{h[n]\}$, respectively. In the continuous-time case, the frequency response is obtained from the $s$-transfer function on the imaginary axis or through the Fourier transform of the impulse response, $H(f) = \overline{H}(j2\pi f) = \mathcal{F}\{h(t)\}$. The discrete-time frequency response is obtained from the $z$-transfer function at the unit circle, $H(e^{j\omega})$.

In many signal processing applications, it is sufficient to consider only real filters, with real impulse response and frequency response satisfying Hermitian (complex conjugate) symmetry, i.e., symmetric magnitude response and antisymmetric phase response. However, in communications signal processing, there are good reasons for considering complex systems/filters having complex impulse responses and no specific symmetries in the frequency response. The main reason is that complex signal models make it possible to apply equivalent baseband system models in analytical and simulation studies for arbitrary passband systems. Notably, the frequency responses of practical communication channels cannot be expected to obey any symmetry relations with respect to the channel center frequency. It is also worth mentioning that noncausal filter/system models are commonly applied in analytic studies, especially when the system impulse response becomes a symmetric real function, in which case also the frequency response becomes real and nonnegative.

One important class of complex bandpass filters is obtained by frequency shifting the frequency response of a real lowpass filter to an arbitrary angular center frequency $\omega_c$ [11]. In the discrete-time case, this can be expressed in terms of $z$-transform, frequency response, and impulse response as follows:

$$H_{BP}(z) = H_{LP}(ze^{-j\omega_c}), \tag{5.1a}$$

$$H_{BP}(e^{j\omega}) = H_{LP}(e^{j(\omega-\omega_c)}), \tag{5.1b}$$

$$h_{BP}[n] = e^{jn\omega_c}h_{LP}[n]. \tag{5.1c}$$

In any digital (Finite Impulse Response (FIR) or Infinite Impulse Response (IIR)) filter block diagram, such frequency translation can be realized by associating the complex coefficient $e^{j\omega_c}$ with every unit delay. For an arbitrary center frequency, such a bandpass filter becomes rather complicated to implement, but with certain convenient values of $\omega_c$, the implementation becomes straightforward. One particular case is the lowpass to highpass transformation with $\omega_c = \pi$ and $e^{j\omega_c} = -1$, and another one is $\omega_c = \pi/2$, i.e., the center frequency is quarter of the sampling rate, and $e^{j\omega_c} = j$. In multirate signal processing, convenient cases are found when the

---

[1] For detailed treatise see, e.g., [16].

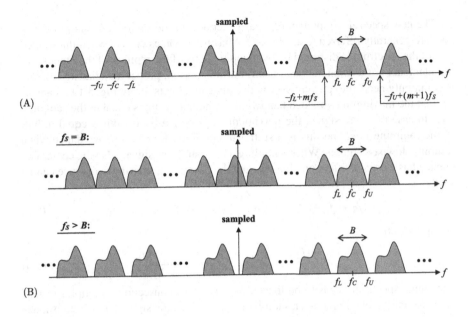

**FIGURE 5.1**

Spectrum models for discrete-time signals. (A) Real signal. (B) Complex signal with two different sampling rates.

bandpass center frequency is a multiple of the low sampling rate or half of it. Furthermore, the frequency translation-based bandpass filter model is a core element of the modulation-based uniform filterbanks to be discussed in the sequel.

## 5.3 SAMPLING, RESAMPLING, AND MULTIRATE FILTERING
### 5.3.1 REAL, COMPLEX, BASEBAND, AND BANDPASS SAMPLING

When dealing with discrete-time baseband and bandpass signals, the aliasing and imaging effects and the constraints for sampling related parameters become apparent when we consider the periodicity of the spectrum models [4,8,14,21]; see Fig. 5.1.

- In case of a complex In-phase/Quadrature (I/Q) signal, the spectrum is periodic with a period equal to the sample rate. There are no additional symmetry constraints for the spectrum. In the initial sampling process, the continuous-time signal spectrum is replicated with shifts corresponding to all integer multiples of the sample rate.
- In case of real signal, the initial spectrum with Hermitian symmetry is replicated with shifts corresponding to all integer multiples of the sampling rate. Naturally, also the resulting spectrum exhibits Hermitian symmetry.

The new spectral component replicas generated by the sampling/resampling process are generally referred to as aliases.[2] Aliasing becomes critical when the wanted part of the spectrum is distorted by overlapping parts of the replicas of the other spectral elements or the wanted signal itself. For this reason, proper antialiasing filtering is commonly needed to prevent destructive aliasing effects. From Fig. 5.1 it is apparent that the maximum two-sided bandwidth without aliasing is equal to the sampling rate. In case of a real signal, the maximum single-sided bandwidth is equal to half of the sampling rate, and this is a sufficient condition for aliasing avoidance when sampling lowpass signals. When sampling a real bandpass signal, also the center frequency has a critical effect on aliasing, and the criteria for aliasing avoidance can be written as

$$m f_S - f_L \le f_L \quad \text{and} \quad (m+1) f_S - f_U \ge f_U \tag{5.2}$$

or, equivalently,

$$f_L \ge \frac{m f_S}{2} \quad \text{and} \quad f_U \le \frac{(m+1) f_S}{2}. \tag{5.3}$$

The signal spectrum needs to be located between two consecutive multiples of half of the sampling rate. For a given sampling rate, the feasible signal bandwidth is maximized when the signal is centered at an odd multiple of quarter of the sampling rate, $f_C = (2m+1) f_S/4$. The same conditions apply also when taking the real part of an I/Q signal, which can be seen as a sampling rate reduction operation by two, and generally for real (re)sampling of I/Q signals.

In I/Q sampling of a complex bandlimited bandpass signal, it suffices that the sampling rate is equal to the signal bandwidth, independently of the center frequency.

### 5.3.2 MULTIRATE FILTERING

The sample rate can be altered digitally by two fundamental multirate operations, namely, decimation and interpolation (see Fig. 5.2). From the communication receivers and/or transmitters point of view, they also offer an interesting alternative to mixing in performing frequency translations. Detailed introductions to these topics can be found in [9,10,12,20].

The basic block-diagram of a decimator, reducing the sample rate of an I/Q signal by an integer factor $M$, is presented in Fig. 5.3, together with an illustration of its operation in frequency domain. First, a digital lowpass or bandpass antialiasing filter is applied to select the wanted part of the spectrum and suppress the unwanted parts. The filter bandwidth and sampling rate should satisfy the criteria for alias-free resampling. The second stage, down-sampler by $M$, picks up every $M$th sample to form the output sequence, and the new sample rate becomes $f_S/M$. After this process, the periodic images of the target frequency band appear in the spectrum, and the

---

[2]In multirate signal processing context, a discrete-time signal may exhibit spectral periodicity within the Nyquist zone. Such spectral replicas are often called images.

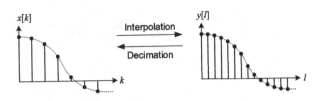

**FIGURE 5.2**

The general idea of decimation and interpolation in multirate signal processing.

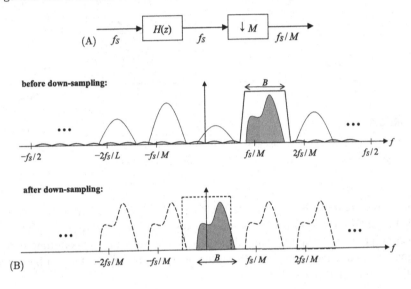

**FIGURE 5.3**

(A) Decimator realized as a cascade of digital antialiasing filter and down-sampler.
(B) Spectrum model for decimation operation with bandpass I/Q signal.

unwanted spectral components are suppressed to a level that depends on the stopband attenuation of the decimation filter.

In Fig. 5.3, decimation is applied to an I/Q signal, but the same model applies also to real signal cases. Now the additional constraints of (5.3) have to be respected so that the symmetric elements of the spectrum of the real signal find their places without causing harmful aliasing.

The basic block-diagram of an interpolator increasing the sample rate of an I/Q signal by an integer factor $M$ is presented in Fig. 5.4, together with an illustration of its operation in frequency domain. The first stage, up-sampler[3] by $M$, inserts $M - 1$ zeros between consecutive input samples, increasing the sample rate to $Mf_S$. The pe-

---

[3] Also called expander [9,20].

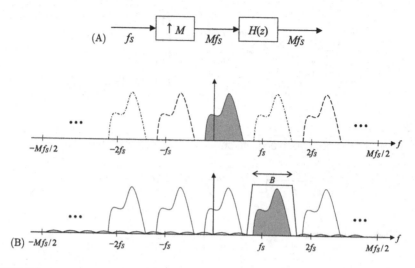

**FIGURE 5.4**

(A) Interpolator realized as a cascade of up-sampler and digital anti-imaging filter.
(B) Spectrum model for interpolation operation with bandpass I/Q signal.

riodic structure remains the same, but the Nyquist zone is expanded by the factor $M$. Then a digital lowpass or bandpass anti-imaging filter is applied to select the wanted spectral replica and suppress the other ones. This results in a smooth time-domain signal. The same model applies also to the interpolation of a real signal by using a real lowpass or bandpass filter.

In the context of multirate I/Q signal processing, conversions between real and complex formats can also be seen as multirate operations:

- Taking the real part of a complex signal reduces the rate of real-valued signal samples by the factor of two. Before applying this operation, it must be secured, by filtering and by the choice of essential parameters, that the generated mirror images of spectral components will not cause harmful aliasing. The process is similar to decimation by two.
- Complex single-sided (analytic) filtering can be used for removing the mirror images from the spectrum of a real signal. The resulting complex signal has double rate of real-valued samples compared to the initial real signal, and this process is similar to interpolation by two.

In addition to sampling rate conversion, bandpass decimation and interpolation can also be used for translating the wanted signal spectrum in frequency, basically by integer multiples of the lowest sampling rate applied in the processing chain. The most straightforward approach for down-conversion and channelization filtering in a communications receiver consists of I/Q down-conversion to baseband and low-

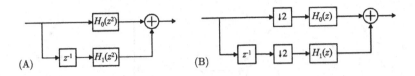

**FIGURE 5.5**

(A) Two-path polyphase structure. (B) Polyphase structure for decimation by two.

pass decimation filtering. An alternative scheme may consist of bandpass sampling, bandpass decimation, and fine-tuning I/Q mixing to move the signal to baseband.

### 5.3.3 POLYPHASE STRUCTURES FOR MULTIRATE FILTERS

One of the corner-stones of multirate filtering is that spectacular savings in computational complexity can be achieved by proper design. Good examples of this can be found in Chapter 6. Generally, it is beneficial to minimize the sampling rate in any computationally intensive digital signal processing module.

FIR filters are commonly used in multirate filtering. The most basic low-complexity realizations can be achieved by using direct-form 1 and direct-form 2 FIR filter structures for decimators and interpolators, respectively, and avoiding unnecessary computations (e.g., in case of decimator, not calculating samples that are thrown away by the down-sampler) [9]. Another important tool, in case of composite sampling rate conversion factors, is multistage realization of decimators and interpolators, which gives significant savings in both computation rates and total number of filter coefficients, with the cost of somewhat increased processing delay [9,10,17].

A systematic approach for low-complexity decimators/interpolators (also for multistage realizations) is to use the polyphase FIR filter structures [2,3,9,11]. The $M$-path (type I) polyphase decomposition of a FIR transfer function $H(z)$ can be expressed as

$$H(z) = \sum_{l=0}^{L-1} h[l]z^{-l} = \sum_{\ell=0}^{M-1} z^{-\ell} H_\ell(z^M), \qquad (5.4a)$$

where

$$H_\ell(z) = \sum_{k=0}^{L_\ell-1} h[\ell + kM]z^{-k}. \qquad (5.4b)$$

Here $H_\ell(z)$ contains every $M$th term of the transfer function, starting from index $\ell$. $L$ is the filter length, and $L_\ell \in \{\lceil L/M \rceil, \lfloor L/M \rfloor\}$ are the lengths of the polyphase path filters. Fig. 5.5A shows the basic two-path polyphase structure. The down-sampling and up-sampling operations by the factor $M$ can be effectively combined with the polyphase filter structure. Fig. 5.5B shows an efficient structure for decimation by two. The structure makes use of the fact that filtering by $H(z^2)$ fol-

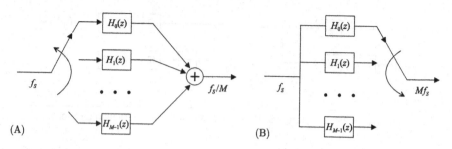

**FIGURE 5.6**

Commutator structures for sampling rate conversion by the factor $M$. (A) Decimator. (B) Interpolator.

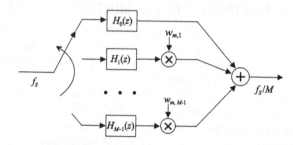

**FIGURE 5.7**

Commutator structure for bandpass decimation by $M$ with center frequency of $2\pi m/M$. Here, $w_{m,\ell} = e^{2j\pi m\ell/M}$.

lowed by down sampling by two is equivalent to first down sampling by two and then filtering by $H(z)$. The operations in front of the path filters basically correspond to passing the odd input samples to the lower path and even samples to the upper path. These ideas can be generalized in the form of the so-called commutator models for decimation and interpolation, shown in Fig. 5.6. In these structures, all path filters are operating at the lower sampling rate.

Furthermore, frequency translation of a lowpass frequency response to the center frequencies of $\omega_m = m \cdot 2\pi/M$ can be easily combined with the polyphase structures because these frequency translations do not affect the polyphase path filters. The resulting commutative decimator structure is shown in Fig. 5.7.

## 5.4 NYQUIST PULSE-SHAPING PRINCIPLE

This section gives a compact introduction to the elements of a basic Single-Carrier (SC) data communication scheme based on the famous Nyquist pulse shaping principle. For more detailed presentation of this topic, see, e.g., [1,10].

The basic signal model (represented as an equivalent complex baseband model) for a data transmission link using linear digital SC modulation schemes can be expressed as

$$y(t) = \sum_{k=-\infty}^{\infty} d_k p(t - kT) + f(t) \star v(t), \qquad (5.5)$$

where $d_k$ is the transmitted symbol sequence, usually taken from a complex Quadrature Amplitude Modulation (QAM) or real Pulse Amplitude Modulation (PAM) alphabet, and

$$p(t) = g(t) \star h(t) \star f(t) \qquad (5.6)$$

with $g(t)$, $h(t)$, and $f(t)$ representing the impulse responses of the transmit filter, channel, and receive filter, respectively, $T$ is the symbol interval, $v(t)$ is a white complex Gaussian channel noise, and $y(t)$ is the signal at the receiver filter output. For simplicity, we assume in most of this chapter that the channel is ideal (frequency flat and the path loss is normalized to unity, $h(t) = \delta(t)$, the Dirac delta function).

In this basic SC transmission context, orthogonality corresponds to the zero Inter-Symbol Interference (ISI) condition, meaning that at proper symbol timing instances, and in the absence of channel noise, the transmitted symbol values can be precisely measured in the receiver. A trivial way to reach this is to use short nonoverlapping symbol pulse shapes $p(t)$. However, such a choice for the pulse shape wastes the spectrum, i.e., the signal bandwidth becomes significantly wider than necessary. Using a properly designed overlapping symbol pulse shape, the bandwidth can be reduced/minimized. The Nyquist pulse shaping principle [6] leads to clear constraints on the pulse shape design while the ISI is completely avoided. In time domain, the zero-ISI condition can be expressed (in the noncausal zero-phase model) as

$$p(kT) = \begin{cases} 1, & k = 0, \\ 0 & \text{otherwise.} \end{cases} \qquad (5.7)$$

The equivalent frequency-domain criterion follows directly from the Nyquist sampling theorem:

$$\sum_{m=-\infty}^{\infty} P(f - mF) = T, \qquad (5.8)$$

where $F = 1/T$ is the symbol rate, and $P(f)$ is the Fourier transform of $p(t)$. The Nyquist pulse-shaping criteria are illustrated in Fig. 5.8. The frequency response $P(f) = G(f) \cdot F(f)$ (assuming that $H(f) = 1$) has typically flat passband and a symmetric transition band around $F/2$. In the baseband model, the stopband edge can be expressed as $(1+\alpha) \cdot F/2$, and the total bandwidth used for carrier-modulated transmission (including transition bands) is $(1 + \alpha) \cdot F$. Here $\alpha$ is the roll-off factor with $0 \le \alpha \le 1$. It follows that the minimum bandwidth in ISI-free carrier-modulated

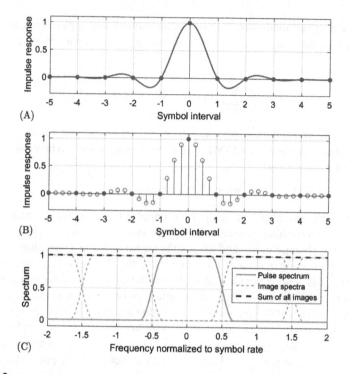

**FIGURE 5.8**

Nyquist pulse-shaping principle for zero-ISI using RRC-filter with 33% roll-off.
(A) Continuous-time impulse response. (B) Discrete-time impulse response with
4×-oversampling. (C) Normalized baseband spectrum and its images, illustrating the
frequency-domain Nyquist criterion.

transmission is equal to the symbol rate $F$. However, the practical choice for the roll-off depends on various metrics regarding the link performance and implementation related aspects, typically leading to roll-off values of $0.1 \ldots 0.4$ in traditional SC systems.

The Nyquist pulse-shaping criteria involve all the linear filtering elements in the transmission chain, but the zero-ISI condition is essential only at the symbol detection stage in the receiver. The division of the overall frequency response affects the channelization on the Transmitter (TX) and Receiver (RX) sides, i.e., the transmitted Power Spectral Density (PSD) and the attenuation of the out-of-band spectral components in the receiver. Furthermore, it determines how the channel noise is treated in the receiver. The so-called matched filter principle says that Signal-to-Noise Ratio (SNR) at the receiver filter output is maximized by choosing (in the general case, with frequency selective channel) $F(f) = G^*(f)H^*(f)$. With this choice, the continuous-time signal at the receive filter output can be sampled at symbol rate without loosing any useful information.

One commonly used pulse shaping filter design uses Raised Cosine (RC) frequency response for $P(f)$. Splitting this frequency response to two equal parts (i.e., taking the square root of the RC frequency response) leads to the use of the same square Root Raised Cosine (RRC) filter for RX and TX. In the case of frequency-flat channel, this solution satisfies both the zero-ISI condition (the first Nyquist criterion) and the matched filter condition. The normalized time and frequency expressions of the RRC pulse are [6]

$$g_{RRC}(t) = \sqrt{F} \frac{4\alpha F t \cos\left(\pi(1+\alpha)Ft\right) + \sin\left(\pi(1-\alpha)Ft\right)}{\left(1 - (4\alpha Ft)^2\right)\pi Ft}, \tag{5.9}$$

$$G_{RRC}(f) = \begin{cases} \frac{1}{\sqrt{F}}, & |f| \le (1-\alpha)\frac{F}{2}, \\ \frac{1}{\sqrt{F}} \cos\left(\frac{\pi}{2\alpha}\left(\frac{|f|}{F} - \frac{1-\alpha}{2}\right)\right), & (1-\alpha)\frac{F}{2} < |f| \le (1+\alpha)\frac{F}{2}, \\ 0, & (1+\alpha)\frac{F}{2} < |f|. \end{cases} \tag{5.10}$$

In practical FIR filter-based implementation, the impulse response can be obtained by sampling $g_{RRC}(t)$ at a rate $f_s \ge (1+\alpha)F$; typically, oversampling by a small integer factor is used. Additionally, the sampled sequence has to be truncated to a finite length and time-delayed to obtain a causal impulse response. Alternatively, RRC-type "half-Nyquist" filters can be designed by digital FIR filter optimization techniques with additional constraints to guarantee zero-ISI [6,10,13,18].

From the multirate signal processing perspective, discrete-time Nyquist filters with $f_S = MF$ (i.e., $M$ times oversampling with respect to the symbol rate) are often referred to as $M$th-band filters. Their 6-dB bandwidth is $f_S/M$, and their impulse responses satisfy the discrete-time version of condition (5.7), which is illustrated in Fig. 5.8B, i.e., every $M$th sample of the impulse response is 0, except for the center one. This results in computational saving, which becomes significant especially in the half-band case with $M = 2$. In the polyphase structure of a half-band filter, one of the paths contains just a pure delay without any filter coefficients. A cascade of multirate half-band filters is a common low-complexity solution for decimation and interpolation by a power-of-two factor.

## 5.5 ABOUT THE DISCRETE FOURIER TRANSFORM

Another corner-stone of the later developments in this book is the DFT and its effective implementation using the Fast Fourier Transform (FFT). These important signal processing tools are briefly introduced in this section. See, e.g., [16] for a detailed introduction to DFT.

For finite-length discrete-time sequences, the connections between time- and frequency-domain representations are established by the transform pair of DFT and

Inverse Discrete Fourier Transform (IDFT), expressed as follows:

$$X[m] = \sum_{l=0}^{M-1} x[l] e^{-j2\pi ml/M} \tag{5.11}$$

and

$$x[l] = \frac{1}{M} \sum_{m=0}^{M-1} X[m] e^{j2\pi ml/M}, \tag{5.12}$$

respectively, where $M$ is the transform length.

It is straightforward to show that the complex exponential basis functions of DFT are orthogonal, even in the presence of arbitrary phase-shifts:

$$\left\langle e^{-j(2\pi m_1 l/M + \phi_1)}, \; e^{-j(2\pi m_2 l/M + \phi_2)} \right\rangle = \sum_{l=0}^{M-1} e^{-j(2\pi m_1 l/M + \phi_1)} \cdot e^{j(2\pi m_2 l/M + \phi_2)}$$

$$= e^{j(\phi_2 - \phi_1)} \cdot \sum_{l=0}^{M-1} e^{-j2\pi(m_1 - m_2)l/M} = \begin{cases} e^{j(\phi_2 - \phi_1)}, & m_1 = m_2, \\ 0 & \text{otherwise.} \end{cases} \tag{5.13}$$

The DFT basis functions are used as subcarriers in Orthogonal Frequency-Division Multiplexing (OFDM) systems, and their orthogonality is the key to the robustness of the Cyclic Prefix Orthogonal Frequency-Division Multiplexing (CP-OFDM) scheme. Considering a multipath transmission channel with all multipath delays smaller than the Cyclic Prefix (CP) length, the channel effect can be seen as the cyclic convolution of the basis functions with the channel impulse response. In other words, a received subcarrier signal is the weighted sum of phase-shifted variants of the corresponding basis function, which affects the amplitude and phase of the subcarrier but does not harm the orthogonality of subcarriers.

FFT and Inverse Fast Fourier Transform (IFFT) are computationally efficient implementations of DFT and Inverse Discrete Fourier Transform (IDFT), respectively, when the transform length is a power of two [7]. In recent OFDM system developments, like Long-Term Evolution (LTE), also other transform lengths have been introduced. Computationally effective DFT algorithms exist for various non-power-of-two transform lengths, especially for cases where the length is a product of small integers [5,19].

## 5.6 MULTIRATE FILTERBANKS

### 5.6.1 DFT FILTERBANK

Considering the commutator structures for bandpass decimation (Fig. 5.7), we can notice that the sets of complex frequency-shifts in the path coefficients correspond

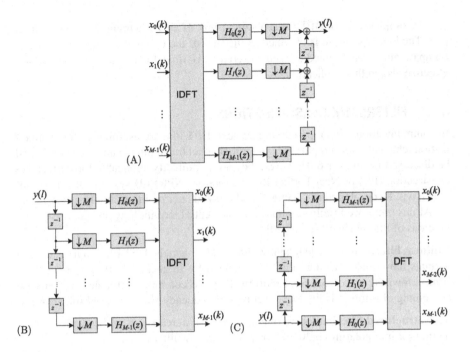

**FIGURE 5.9**

DFT filterbanks. (A) SFB. (B) IDFT-based AFB. (C) DFT-based AFB.

to an IDFT. Therefore, connecting the polyphase path outputs to an $M$-point IDFT, a bank of bandpass filters with center frequencies at $\omega_m = m \cdot 2\pi/M$, $m = 0, 1, \ldots, M - 1$, can be implemented effectively (see Fig. 5.9B). This is an Analysis FilterBank (AFB) splitting a wideband signals with high sample rate ($f_S$) to narrow subbands with sample rate $f_S/M$.

Likewise, we can apply frequency translation to the same bandpass center frequencies in the interpolator structure of Fig. 5.6B with frequency-shifting coefficients on the input side of the polyphase paths. These sets of coefficients correspond to an IDFT as well. Then an interpolating Synthesis FilterBank (SFB) can be constructed by performing IDFTs for sets of $M$ samples, one from each subband at a time, and then connecting the IDFT outputs to the polyphase path filters, as shown in Fig. 5.9A. A SFB combines $M$ low-rate subband signals into a single wideband high-rate signal. Both of these FilterBanks (FBs) are called DFT-FBs [2,9,11]. Whereas IDFT can be considered as a natural choice for the transform of a synthesis bank, it is often preferred to use DFT as the transform on the AFB side. This can be achieved through a modified (type II) polyphase decomposition of the prototype filter, $H(z) = \sum_{\ell=0}^{M-1} z^{-\ell} H_{M-1-\ell}(z^M)$ [9,11]. The resulting structure is shown in Fig. 5.9C.

The simplest case of DFT-FB is obtained with $M = 2$ as a highpass–lowpass filter pair. The lowpass output is obtained as the sum of the two path signals, whereas the highpass output is obtained as their difference. Halfband FIR filters are particularly effective also in this application.

## 5.6.2 FILTERBANK CLASSIFICATIONS

In communications signal processing context, DFT-FBs are useful, e.g., for channelization (channel filtering) purposes in multichannel transmitters and receivers, as will be discussed in Chapter 6. However, they are not directly applicable for Perfect Reconstruction (PR) or Near Perfect Reconstruction (NPR) FB systems, the theory of which will be developed in Chapter 7 and utilized later on in this book.

At this point, we briefly summarize some central terminology and general classifications of filterbanks [9–11,15,20].

**Uniform filterbank** has equal bandwidth for all channels. DFT-FB and its enhanced variants have subband filters that are obtained by frequency shifting (modulation) from a lowpass prototype. Real uniform FBs are commonly obtained by pairwise combining of symmetrically located complex frequency-shifting based subchannels.

**Tree structured filterbanks** are obtained through a tree-structured cascade of decimating (or interpolating) halfband lowpass–highpass filter pairs.

**Real and complex filterbanks:** DFT-FB and its variants have complex signal models both for the subband signals and for the combined high-rate signal. Such a model is preferred for waveform processing in carrier-modulated communication systems. Real high-rate signal model is applicable in baseband transmission systems, e.g., in wireline communications, and this can be reached by applying subcarrier signal level conjugate symmetry, i.e., $x_{M-1-m}[k] = x_m^*[k]$. In various other applications, like subband based source coding, real signal format is needed for both subband and combined signals. The commonly used Cosine-Modulated FilterBanks (CMFBs) have also a close connection to DFT-FBs.

**Critically sampled filterbanks**, like the DFT-FB, have the same total number of samples in the subband signals as in the combined high-rate signal, i.e., the subband sampling rate is $f_S/M$. In various applications, oversampling of subband signals by a small integer factor, typically 2, is required or beneficial. Especially, the Offset Quadrature Amplitude Modulation (OQAM) signal model used for the FilterBank MultiCarrier with Offset-QAM subcarrier modulation (FBMC/OQAM) waveform necessitates the use of two times oversampled model.

**Even/odd stacked subchannels:** The DFT-FB has even-stacked subband arrangement, which means that one of the subbands is zero-frequency (DC)-centered. In the odd-stacked case, two subchannels are symmetrically located around DC, as illustrated in Fig. 5.10.

**Analysis-synthesis configuration:** A FB system where an AFB splits a high-rate signal into subbands, and after application specific subband processing, a high-rate

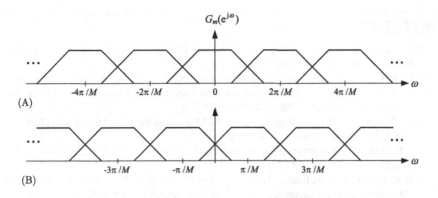

**FIGURE 5.10**

(A) Even and (B) odd subchannel stacking in uniform complex FBs.

wideband signal is regenerated using a SFB. Applications include subband-based source coding and frequency-domain (channel) equalization.

**Transmultiplexer (TMUX) configuration** is applied in FilterBank MultiCarrier (FBMC) systems. On the transmitter side, a SFB is used to generate the multicarrier signal from (typically) QAM-modulated subcarriers. On the receiver side, an AFB is used for separating the subcarriers.

**Perfect reconstruction (PR)** TMUX (or analysis-synthesis) system regenerates the original subband signals (or high-rate signal) precisely in the back-to-back connection of the SFB and AFB. In the TMUX context, this is also referred to as an orthogonal system. In a NPR FB system, the reconstruction is achieved as a good approximation.

**Paraunitary FB system** is a PR system using the same prototype filter for SFB and AFB but in complex conjugate form on the other side.

**Biorthogonal FB system** is a PR system using different prototype filters for SFB and AFB. Such designs may be useful if the filter selectivity requirements are different on the two sides, and it may also help to reduce the latency of FB systems.

## 5.7 CONCLUDING REMARKS

This chapter introduced certain important theories and signal processing tools as a basis for later developments in this book. One core element was to extend the discrete-time signal and spectrum models for complex signals and passband systems. The polyphase structured DFT-FB can be seen as the first step in developing adequate filter bank structures and designs for FBMC communication systems.

# REFERENCES

[1] J. R. Barry, E. A. Lee, and D. G. Messerschmitt, *Digital Communication*, 3rd ed. Boston: Kluwer Academic Publishers, 2004.

[2] M. Bellanger, G. Bonnerot, and M. Coudreuse, "Digital filtering by polyphase network: Application to sample-rate alteration and filter banks," *IEEE Trans. Acoust., Speech, Signal Process.*, vol. 24, pp. 109–114, Apr. 1976.

[3] M. Bellanger and J. L. Daguet, "TDM-FDM transmultiplexer: Digital polyphase and FFT," *IEEE Trans. Commun.*, vol. 22, Sep. 1974.

[4] J. L. Brown, "On quadrature sampling of bandpass signals," *IEEE Trans. Aerosp. Electron. Syst.*, vol. 15, pp. 366–371, May 1979.

[5] B. C. S. Burrus, "Fast Fourier Transforms. OpenStax CNX," Nov. 2012. [Online]. Available: http://cnx.org/contents/82e6ba6f-b828-42ef-9db1-8de4b448b869@22.1.

[6] P. R. Chevillat and G. Ungerboeck, "Optimum FIR transmitter and receiver filters for data transmission over band-limited channels," *IEEE Trans. Commun.*, vol. 30, pp. 1909–1915, Aug. 1982.

[7] J. W. Cooley and J. W. Tukey, "An algorithm for the machine calculation of complex Fourier series," *Math. Comput.*, vol. 19, no. 90, pp. 297–301, Apr. 1965.

[8] A. J. Coulson, R. G. Vaughan, and M. A. Poletti, "Frequency-shifting using bandpass sampling," *IEEE Trans. Signal Process.*, vol. 42, pp. 1556–1559, Jun. 1994.

[9] R. Crochiere and L. Rabiner, *Multirate Digital Signal Processing*. Prentice-Hall, 1983.

[10] B. Farhang-Boroujeny, *Signal Processing Techniques for Software Radios*, 2nd ed. Lulu, 2011.

[11] N. J. Fliege, *Multirate Digital Signal Processing*. Wiley, 1994.

[12] F. Harris, *Multirate Signal Processing for Communication Systems*. Prentice Hall, 2004.

[13] F. Harris and C. Dick, "An alternate design technique for square root Nyquist shaping filters," in *Proc. SDR WinnCom-2015 Conference*, San Diego, CA, Mar. 2015, pp. 15–17.

[14] A. J. Jerri, "The Shannon sampling theorem – Its various extensions and applications: A tutorial review," *Proc. IEEE*, vol. 65, no. 11, pp. 1565–1596, Nov. 1977.

[15] H. S. Malvar, *Signal Processing with Lapped Transforms*. Boston, MA: Artech House, 1992.

[16] A. Oppenheim and R. Schafer, *Discrete-Time Signal Processing*, 3rd ed. Pearson, 2014.

[17] T. Saramäki, "A class of linear-phase FIR filters for decimation, interpolation, and narrowband filtering," *IEEE Trans. Acoust., Speech, Signal Process.*, vol. 32, pp. 1023–1036, Oct. 1984.

[18] P. Siohan and F. Moreau de Saint-Martin, "New designs of linear-phase transmitter and receiver filters for digital transmission systems," *IEEE Trans. Circuits Syst. II*, vol. 46, pp. 428–433, Apr. 1999.

[19] H. V. Sorensen and C. S. Burrus, "Fast DFT and convolution algorithms," in *Handbook for Digital Signal Processing*, S. K. Mitra and J. F. Kaiser, Eds. New York: John Wiley and Sons, Inc., 1993, pp. 491–610, ch. 8.

[20] P. P. Vaidyanathan, *Multirate Systems and Filter Banks*. Englewood Cliffs, NJ: Prentice Hall, 1993.

[21] R. G. Vaughan, N. L. Scott, and D. R. White, "The theory of bandpass sampling," *IEEE Trans. Signal Process.*, vol. 39, pp. 1973–1984, Sep. 1991.

# Filter Banks for Software-Defined Radio

# 6

**Fred Harris\*, Xiaofei Chen\*, Elettra Venosa†**

*San Diego State University, San Diego, CA, United States\**
*Space Micro, San Diego, CA, United States†*

## CONTENTS

## 6.1 INTRODUCTION

In this chapter, we examine filter banks as channelizers for DSP-intensive radio transmitters and receivers. On the transmitter side, a synthesis channelizer forms a composite broadband output signal from a set of narrowband baseband input signals. The analysis channelizer reverses the process in the receiver forming a set of narrowband baseband output signals from composite broadband input signals. The two types of filter banks are each other's duals. The filter banks are remarkable in their capability, flexibility, and efficiency in the tasks they perform. Central to their use is their ability to change sample rate while changing bandwidth and to move signals between different spectral regions using aliasing and to separate aliases with phase-coherent sums.

In the following sections, we examine an *M*-path partition of a low-pass filter and learn how bandwidth reduction with embedded sample rate reduction can achieve significant reductions in computational workload. By cascading downsampling and

Orthogonal Waveforms and Filter Banks for Future Communication Systems. DOI: 10.1016/B978-0-12-810384-5.00006-2

upsampling $M$-path filters we can achieve comparable workload reduction while preserving a fixed sample rate. We then extend the downsampling and upsampling process to band-centered filters, which by judicious choice of center frequency are aliased to and from baseband by the resampling. In the $M$-path filter, every multiple of the output sample rate aliases to baseband and any of the aliased bands can be separated from other aliased bands by a phase coherent sum aligned with the unique phase profile of the selected alias. With minor restriction, we extend the process to multiple band centered filters and use the IFFT to perform the required phase coherent sums. The $M$-path filter with IFFT aligned phase profiles is called a polyphase filter bank [1]. We then apply a cascade of polyphase downsampling, analysis filter bank and a polyphase upsampling synthesis filter bank to synthesize multiple wideband polyphase filters from the narrow band analysis set of filters. Details of the design process are to be found in other chapters.

## 6.2 $M$-PATH FILTERS

Here we start to develop understanding of how $M$-path filters morph from single-channel filters through polyphase decomposition to multiple-fixed-bandwidth filters and then to flexible multiple-variable-bandwidth channelizers. To do so, it is useful to first examine and learn how an $M$-path filter uses resampling to implement an efficient single-bandwidth filter. The entry point to this process is the design and implementation of a filter when there is a large ratio of sample rate to bandwidth. Fig. 6.1 presents an example of such a filter.

When we have a large ratio of sample rate to bandwidth, the filter has a large number of coefficients, and a large number of arithmetic operations are required to implement it. We now examine a number of options that implement these filters with reduced workload. Since the problem is caused by the high sample rate relative to the filter bandwidth, the obvious solution is to reduce the sample rate. We can do this with an $M$-path polyphase filter that reduces the sample rate as part of the filtering process. System considerations may require both bandwidth reduction and equal input and output sample rates. We satisfy the two requirements with two filters; the first reduces the sample rate while reducing the bandwidth, and the second increases the sample rate while preserving the bandwidth. We have been asked the question "Why would two filters be better than one filter?" The answer is because there are two problems here and we should treat them as such. The block diagram of the polyphase downsampler and the polyphase upsampler is shown in Fig. 6.2. In this example, the prototype filter is partitioned into a 20-path polyphase filter with 20 coefficients per path. The input and output sample rates of the filter are 1000 kHz and 50 kHz, respectively, and the two-sided bandwidth of the filter, at its –80 dB stopband level, is 40 kHz. A second 20-path filter with different weights is designed to use the 10 kHz excess sample rate as its transition bandwidth when upsampling the 50 kHz sample rate back to the 1000 kHz sample rate. Note the cascade filters require 20 operations per input sample and 20 operations per output sample for a total of 40 operations

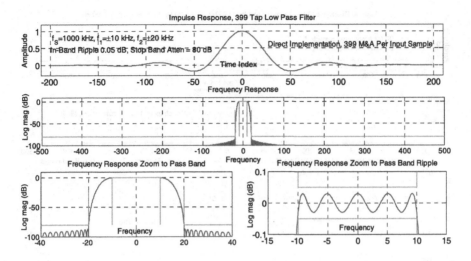

**FIGURE 6.1**

Time and frequency response of 399-tap FIR filter with large ratio of sample rate to bandwidth.

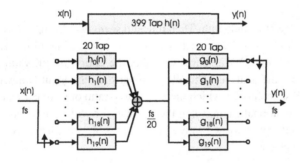

**FIGURE 6.2**

Implementing a narrow bandwidth filter as a cascade of *M*-path downsampling and upsampling filters.

per input–output sample pair. This is a 10 to 1 reduction in workload to implement this filter. Fig. 6.3 shows the time and frequency response of the cascade filter. The significant aspects of the spectral responses are essentially identical to that seen in the direct implementation. The obvious difference in the two implementations is the time delay of the impulse response. The delay is seen to be approximately twice the original interval, 380 samples rather than 199 samples. This extra delay is the consequence of passing the signal through two filters.

A requirement to access this significant workload reduction is a sample rate reduction as part of the bandwidth reduction, a condition assured when there is large

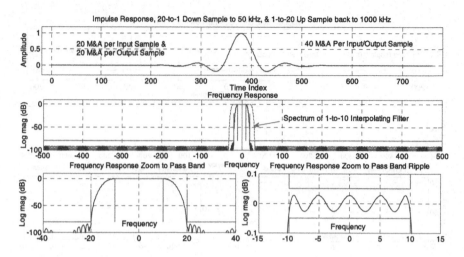

**FIGURE 6.3**

Time and frequency response of cascade 20-to-1 down-sampling and 1-to-20 upsampling $M$-path filters.

ratio of sample rate to bandwidth. Then it would seem that this option is not available when this condition is not met, such as when the sample rate to bandwidth ratio is small, such as 1.5 or 2.2. The surprise is that the option is still valid for this later case. We can use the analysis channelizer to partition the input bandwidth into narrow bandwidth segments for which there is a large ratio of sample rate to bandwidth. Thus, the computational savings can then be had for wide bandwidth signals partitioned temporarily into narrow bandwidth signals, which are then reassembled by the synthesis channelizer.

## 6.3 FILTER BANKS

Fig. 6.4 illustrates the tasks performed by the two types of channelizers. We start with a set of narrow bandwidth baseband signal sequences, each sampled at a common low sample rate slightly higher than the signal's two-sided bandwidth. For instance, the signals may have a two-sided bandwidth of 16 MHz at a 20-MHz sample rate. Having the sample rate exceeding the two-sided bandwidth by 20 to 25 percent is an important consideration in the design of channel filters in the sampled data signal domain. The multiple baseband sequences, say $M$ of them, are presented to the synthesis channelizer. The channelizer interpolates each sequence to raise the sample rate by a factor of $M$, the rate necessary to satisfy the Nyquist criterion for the wider bandwidth of its composite output signal. With access to the higher sample rate, the

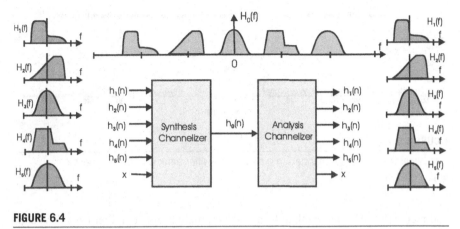

**FIGURE 6.4**

Input and output spectra for synthesis and analysis channelizers.

synthesizer can up convert each signal to its assigned center frequency and sum their up-converted components to form the composite output signal.

As a specific example, suppose we form a composite output signal containing five baseband signals sampled at 20 MHz and arrange for the channel centers to be multiples of 20 MHz. When the channel spacing equals the channel sample rate, the channelizer is known as a maximally decimated filter bank. The bandwidth of the composite signal, spanning five 20 MHz bands, is approximately 100 MHz. Keeping in mind the need to have a sample rate slightly higher than the signal bandwidth, we design the synthesizer for an output sample rate of 120 MHz, the sample rate equivalent to having six 20 MHz channels. We design the synthesis channelizer to accept six 20 MHz input signals with one channel being a null channel. We use the null channel to raise the sample rate to 120 MHz and reserve the extra bandwidth of the null channel for the transition bandwidth of the analog filters following the conversion from the sampled data representation to the continuous domain. In a similar fashion, we collect and present the composite signal at the input to the analysis channelizer at a sample rate that supports null channels for the benefit of the analog antialiasing filter.

An example of spectral response for maximally decimated filter bank is shown in Fig. 6.5. Here the output sample rate is 20 MHz, same as the channel spacing, the filter passband bandwidth is 16 MHz with transition bandwidths of 4 MHz. We observe that each channel response overlaps their two adjacent channels. When a channel is filtered and downsampled to 20 MHz, the transition bandwidth extending beyond the 10 MHz folding frequency folds or aliases into its own transition band. Whereas there are many scenarios where this folding is acceptable, we are interested in a variation of the channelizer in which the transition bands that extend into the adjacent spectral interval do not fold due to resampling. We avoid the band edge folding by raising the sample rate beyond the two-sided bandwidth that includes the transition bandwidth out to their stopband edges. Raising the output sample rate so that it no longer performs *M-to-1* or *1-to-M* resampling changes the filter architecture slightly

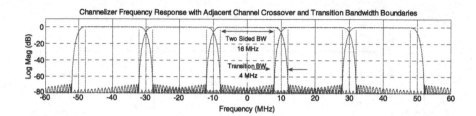

**FIGURE 6.5**

Spectra of five occupied channels in a six-channel filter bank with overlap between adjacent channels.

and reclassifies the filter bank to be a nonmaximally decimated filter bank. The supplemental material to this chapter contains MATLAB script for both the maximally decimated and nonmaximally decimated filter banks.

## 6.4 CASCADE POLYPHASE ANALYSIS AND SYNTHESIS CHANNELIZERS

Channelizers can be stand-alone processes and in fact often are. For instance, a synthesis channelizer can form a broadband signal from multiple baseband signals as a modulator embedded in a transmitter of a communication system. Its signal is delivered through a channel to a receiver that contains an analysis channelizer that extracts the multiple narrow bandwidth components preceding the detection process. In the example just cited, while the analysis and synthesis channelizers are in cascade, they reside at the two ends of a communication link. The channelizers we now study operate as a cascade-coupled pair with the pair residing at both ends of the communication link.

### 6.4.1 NONMAXIMALLY DECIMATED $M$-PATH POLYPHASE ANALYSIS AND SYNTHESIS CHANNELIZERS

The maximally decimated $M$-path analysis filter bank accepts $M$ input samples prior to computing its output vector from its $M$ output ports. Fig. 6.6 shows the essential components of an $M$-path analysis and an $M$-path synthesis channelizer. Here we can clearly see their dual structures. The analysis filter bank is formed by an input $M$-port commutator that delivers $M$ input samples to the length $M$ input data buffer, an $M$-path polyphase filter, and an $M$-point IFFT that outputs successive samples from the $M$-output channels. The synthesis filter is formed by an $M$-point IFFT that accepts input time samples from $M$ parallel input sources, an $M$-path polyphase filter, and a length $M$ output data buffer accessed by the $M$-port output commutator that delivers $M$ output time samples. The analysis filter bank performs an $M$-to-1 down

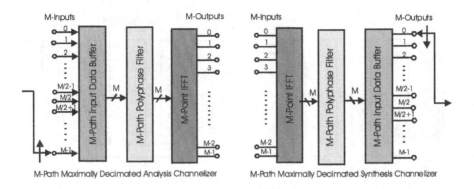

**FIGURE 6.6**

Essential components of *M*-path maximally decimated analysis and synthesis filter banks.

sampling, forming an *M*-point output vector at the rate $f_S/M$, which matches the channel frequency spacing of $f_S/M$.

We can raise the analysis channelizer output sample rate by altering the filtering process to accept fewer than *M* samples prior to computing the *M*-point output vector. For instance, in the six-path channelizer we introduced earlier as our ongoing example, we can form a six-point vector output for every five input samples, which would result in an output sample rate of 24 MHz (120/5) as opposed to the original 20 MHz (120/6) sample rate. Another option is to obtain a six-point vector output for every four input samples, or every three input samples, which would result in output rates of 30 MHz (120/4) or 40 MHz (120/3), respectively. The downsampling embedded in the polyphase filter is responsible for the spectral aliasing. The aliasing causes all multiples of the output sample rate to alias to baseband. We process the aliased signal components to separate the aliases to different output channels. If the sample rate is less than the channel's two-sided bandwidth, then the filter's transition bandwidth aliases within the band. These aliasing terms can also be canceled, but we elect to modify the channelizer to avoid the transition bandwidth folding by increasing the sample rate of the separate channels.

When the output sample rate matches the channel spacing, all channels are aliased to baseband, which is a wonderful attribute of the maximally decimated filter bank. With an increased output sample rate, the channel centers no longer alias to baseband, but they alias to some known offset frequency. Successive samples from each channel port are spinning due to the frequency offset, and we can complete the conversion to baseband by simply despinning the IFFT successive output samples with the conjugate complex phasor. For instance, had we elected to do 4-to-1 downsampling in our six-channel analysis channelizer to obtain a 30 MHz output sample rate, we would know that the signal centered at 20 MHz has aliased to $-10$ MHz with a 30 MHz sample rate and is thus spinning $-2\pi/3$ radians per sample, whereas the signal centered at 40 MHz has aliased to $+10$ MHz and is thus spinning at $+2\pi/3$ radians/sample. A simple state machine can track the successive angle corrections to

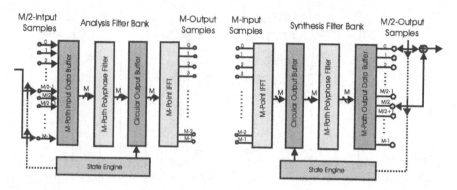

**FIGURE 6.7**

Essential components of $M$-path nonmaximally decimated analysis and synthesis filter banks for $2\times$ oversampled subbands.

be applied to each offset alias output channel to despin its signal back to baseband. An alternate approach to the despinning task is to recall that time delay at the input to the transform causes phase shift in the output of the transform. State machine controlled circular shifts at the input to the transform, an operation performed with a pointer, performs the desired frequency-dependent phase correct without multiplies. This option is described in the supplemental materials. The amount of downsampling of particular interest to us in this section is $M/2$ input samples per output sample. For ease of implementation, we select $M$ to be an even integer. For the $M/2$ down sample, the output sample rate becomes $2f_S/M$. Thus, the output sample rate is twice the channel spacing in the $M$-path channelizer. For the example we have been using in this section, we will perform 3-to-1 downsampling in our six-path channelizer, i.e., the output sample rate is 40 MHz with channel spacing of 20 MHz, for which the adjacent channel filter responses cross at $\pm 10$ MHz. The even multiple of 20 MHz, which are $\pm 40$ MHz, alias to baseband, whereas the odd multiples, $\pm 20$ and 60 MHz, alias to the half sample rate. The supplemental materials to this chapter present the MATLAB scripts for a 6-to-1 and for a 3-to-1 down sampled six-path synthesis and analysis channelizer. It is instructive to see the difference in the script files. The modification to the $M$-path channelizers when they are nonmaximally decimated is shown in Fig. 6.7. Compare this to the versions shown in Fig. 6.6. The primary difference is seen in the sum at the output of the synthesis channelizer that merges the first and second halves of the output data buffer. To emphasize and preserve the duality of the two channelizers, this summation is echoed at the input to the analysis filter. The derivation of the dual flow diagram can be found in [2].

## 6.4.2 FILTER DESIGN FOR PERFECT RECONSTRUCTION

We will shortly examine an extremely versatile structure where transmitter and receiver channelizers are constructed by a cascade of nonmaximally decimated analysis and synthesis filter banks. To make use of this configuration, the channel filters must be designed to realize a Nyquist frequency response. Nyquist filter frequency responses for channelizers require the adjacent channel cross over point to be 0.5 (or −6 dB) and have an odd symmetric transition bandwidth about the crossover frequency. Our first response to meet this requirement is to design the channelizer filter to be a square-root (SQRT) Nyquist filter. This seems reasonable since we pass through the same filter twice, once in the analysis filter and then again in the synthesis filter as discussed in Section 5.4. The problem here is that the cosine tapered SQRT Nyquist filter has a terrible frequency response characteristic. Hard to believe, but it is true! The cosine taper is not sufficiently smooth to obtain the spectral side lobe levels or in-band ripple levels we require for our channelizers. We know how to design SQRT Nyquist filters with other tapers that will support the desired side lobe and in band ripple levels [3,4]. We have used the SQRT Nyquist filters designed with these alternate techniques to implement the cascade filter banks, and they significantly improve the channel frequency response in side lobe levels and in-band ripple level. The problem with the SQRT Nyquist filters is that when we merge adjacent channels to obtain perfect reconstruction, the composite spectrum exhibits significant ripple levels at the channel crossover frequencies.

In our current application, it is not necessary to form a Nyquist filter as a cascade of two SQRT Nyquist filters. The reason we use the two filters in a communication application is that the matched filter minimizes the channel noise introduced between transmitter and receiver. In our cascade channelizers, the two filters are not separated by a noisy channel. Thus, we are free to make one filter of the cascade filter pair a Nyquist filter and design the other filter to have a wider passband that does not distort the passband or transition band of the Nyquist filter. We normally place the Nyquist filter in the analysis filter bank because there are times we want to observe the channelized signals between the analysis and synthesis banks.

Fig. 6.8 shows the frequency response limits for the Nyquist filter design that will reside in the analysis filter bank and for the wider reconstruction filter that will reside in the synthesis filter bank. The top subplot shows the Nyquist spectrum at baseband and at the 20 MHz offset center frequencies either side of baseband. The adjacent channels crossover at 10 MHz offsets at amplitude 0.5 or −6 dB. The transition bandwidth of the Nyquist filter should not extend more than half way to the folding frequency of its band because we have to leave a reasonable span of transition bandwidth for the following synthesis filter. Reducing the transition bandwidth of the Nyquist filter lengthens the filter, which increases the workload and delay through the filter. A reduced transition bandwidth may offer benefit when examining the signal content at the analysis channelizer output. The center subplot shows the spectral response of the baseband analysis filter with a nominal 6 dB bandwidth of 20 MHz with 40 MHz sample rate. The bottom subplot shows the spectral replicas of the analysis filter output at baseband and at 40 MHz offsets. The synthesis channelizer is designed with a passband that spans the two-sided passband and transition bandwidth of the Nyquist spectrum and rejects the edges of the spectral replicas centered at 40 MHz offsets. In practice, the stopband edge of the synthesis filter is permitted to have a

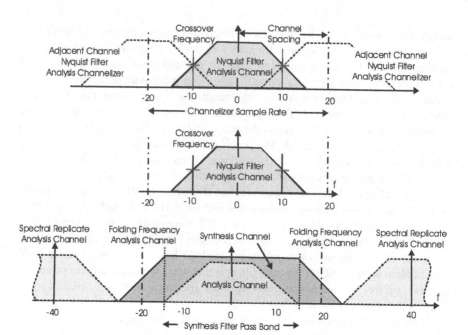

**FIGURE 6.8**

Frequency response for Nyquist filter in analysis channelizer, analysis channelizer output, and wider synthesizer channelizer reconstruction filter.

slight overlap with the edge of the analysis filter spectral replica. The design of the Nyquist filter is performed as a windowed sinc sequence, and the reconstruction filter is designed with a modified Remez algorithm. Details of this procedure are included in the supplemental materials.

## 6.5 CASCADE CHANNELIZERS

We are now prepared to join the analysis and the synthesis filter banks. Fig. 6.9 shows how to connect the analysis and synthesis filter banks for two demonstrations. To review the process, the analysis filter bank partitions the input spectrum into multiple contiguous overlapped narrow band channels that are down sampled and down converted to baseband. The synthesis channelizer accepts multiple input narrow-band baseband channels that are up sampled and up converted to contiguous overlapped channels. We can form a selectable bandwidth filter by presenting a subset of the baseband channelized time series from the analysis filter bank to the corresponding input terminals of the synthesis filter bank. The output of the synthesis channelizer is

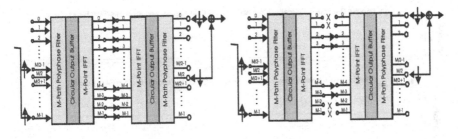

**FIGURE 6.9**

Cascade nonmaximally decimated analysis and synthesis filter banks. Pair on left fully connected for verification test; pair on right partially connected to merge channels for wider super channel.

a super channel formed by seamlessly merging the multiple narrowband input channels. This is the option illustrated on the right side of Fig. 6.9.

The cascade channelizer is designed to be a perfect reconstruction filter process, a process enabled by the use of the Nyquist filter-based analysis filter bank. The Nyquist filter exhibits imperfect response related to nonzero levels of in-band ripple and finite stopband attenuation levels. An interesting question is how good is the perfect reconstruction process when implemented with imperfect Nyquist filters? The simplest test to measure the deviation from perfect reconstruction is the system's impulse response. We apply a single impulse to the cascade filter bank with a fully connected analysis and synthesis filter as shown on the left side of Fig. 6.9. This is a valid test in spite of the fact that a multirate system does not have a transfer function and in fact has multiple impulse responses. We show and discuss the results of the impulse response probe here and present details of this process in the supplemental materials.

Fig. 6.10 presents the result of the impulse response probe of the fully connected cascade of nonmaximally decimated analysis and synthesis filter banks. The top subplot shows the expected response of the cascaded 71-tap six-path filter banks. Note that we do not see the impulse response of the filters embedded in the channelizer, but rather we see the impulse response of perfect reconstruction filter system, which reproduces the input at its output. The pulse has been reconstructed after a 71 sample delay. The center subplot shows the time domain artifacts by zooming in, with high magnification, to the low-level components. The largest artifact at position 35 of amplitude $1.7 \cdot 10^{-5}$, is an artifact 5 orders of magnitude below the desired signal. The bottom subplot is the spectrum of the unit impulse, which if there were no artifacts, would be a constant 0 dB over the spectral interval. We see instead a periodic ripple pattern with peak amplitude $1.8 \cdot 10^{-4}$ dB, which is approximately 20.7 parts per million. This ripple is traceable to both the in-band ripple levels and the out-of-band ripple levels of the two prototype filters in the analysis and synthesis filter banks.

These artifacts can be made arbitrarily small by reducing the in-band and out-of-band ripple levels in both filters. Testing all six responses, we found that the

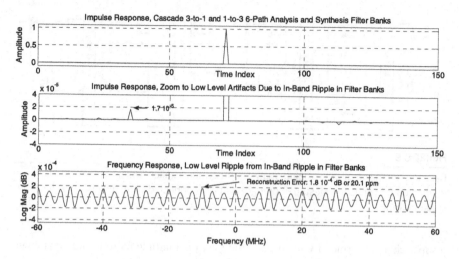

**FIGURE 6.10**

Impulse response of cascaded 71-tap six-path analysis and synthesis filter banks (top), zoom to low-level time domain artifacts (center), zoom to spectrum ripple levels (bottom).

worst-case periodic ripple pattern had a peak amplitude of $2.6 \cdot 10^{-4}$ dB, or 30.3 ppm maximum reconstruction error. We can expect any super channel formed by a subset of analysis filters bands reassembled by the synthesis filter, as shown in the right side of Fig. 6.9, to exhibit comparable levels of reconstruction error. Verification is shown in the supplemental materials section.

## 6.5.1 CASCADE CHANNELIZERS FOR VARIABLE BANDWIDTH FILTERS

In the previous section, we examined a cascade of six-path analysis and synthesis channelizers operating as a 3-to-1 and 1-to-3 resampling filter bank. We had selected a small number of stages to illustrate the frequency and time domain properties of the prototype filters and how they interact in the cascade. We now examine a 30-to-1 and 1-to-30 resampling sixty channel channelizer to better illustrate the flexibility and versatility offered by having access to more degrees of freedom. In particular, we designed a 60-channel channelizer. As done for the six-path filter, a 719-tap analysis filter was designed for 80 dB stopband attenuation. Since the Nyquist filter has equal passband and stopband ripple, its passband is absurdly small, $2.8 \cdot 10^{-4}$ dB. Using the Remez algorithm penalty weights, we were able to design a 699-tap synthesis filter with the same 80 dB attenuation and with a reasonable 0.005 dB passband ripple. Designing the synthesis filter to relaxed passband ripple specifications resulted in the reduced length filter, which in turn reduced the computational workload of the filter and the group delay through the filter.

We will not show the result of the impulse response test for the cascade of the 60-path analysis and synthesis filter banks, but we did conduct the test to validate

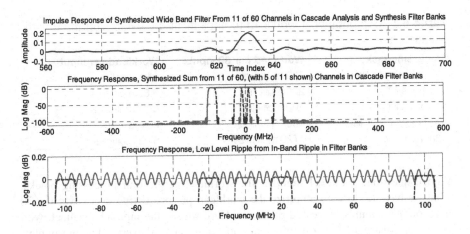

**FIGURE 6.11**

Impulse response of 11-subchannel synthesized super channel (top), frequency response of synthesized channel and analysis channel subchannels (center), and zoom detail of passband ripple (bottom).

proper operation of our script. What we will show is the result of the reduced bandwidth design presented in the right side of Fig. 6.9. In this design, we coupled 11 of the output ports from the analysis filter bank to their corresponding ports at the input to the synthesis filter bank to form a filter with two-sided bandwidth 11/60 of the input sample rate. Fig. 6.11 shows the impulse response and the frequency response of the synthesized super channel formed from the 11 selected filter bank channels. Also, shown is the in-band ripple of the supper channel, which has inherited the 0.005 dB design ripple level of the synthesis channelizer bank. Note in particular that the equal level ripple of the synthesized super channel has two distinct components. The ripple aligned with the channel pass band has a different period than the ripple spanning the interval between the channel pass bands. We have an option to select the 11 channels from any offset spectral interval of the analysis filter band and deliver them from the offset frequency span to the baseband span of the synthesis filter. Had we elected this option, we would have had a free spectral translation in the filter bank. We have a few more clever options available to us, but we first want to call attention here to the computational economy of implementing the filter as a super filter formed by narrow-band subchannels.

Suppose we are tasked to implement a single FIR filter with the same spectral response as the superfilter we just synthesized by merging 11-subchannels in the 60-path cascade-analysis synthesis engine. The filter meeting the same specifications of transition bandwidth, passband ripple, and stopband attenuation level of the synthesis filter requires 601 taps.

Now we might ask why would we implement the 601-tap filter with a 719-tap analysis and 599-tap synthesis filter on the input and output of two 60-point IFFTs?

**Table 6.1** Computational workload to implement a 601-tap filter in 60-path channelizer

| Processing task | Workload per 30 inputs | Workload per input |
|---|---|---|
| 720-tap analysis filter | 720 multiplies | 24 multiplies |
| 60-point IFFT | 200 multiplies | 6.67 multiplies |
| 60-point IFFT | 200 multiplies | 6.67 multiplies |
| 600-tap synthesis filter | 600 multiplies | 20 multiplies |
| Total workload | 1720 multiplies | 57.3 multiplies |

A good question! Table 6.1 itemizes the workload of the four processing blocks. What we have to do is to amortize the workload per processing cycle over the 30 input–output samples processed per cycle. Here we see the input channelizer, with its 720 taps distributed over its 60 input ports and exercised every 30 input samples, which requires 24 multiplies per input, and similarly the output channelizer, with its 600 taps distributed over its 60 output ports and exercised every 30 output samples, which requires 20 multiplies per output. We are up to 44 multiplies per input; what remains is the workload for two 60-point IFFTs. The 60-point IFFT is implemented as a Good-Thomas, or Prime Factor, algorithm with factors 3, 4, and 5. When the short factor IFFTs are performed by a set of unnested Winograd transforms, the workload for the 60-point IFFT is 200 real multiples for complex input samples. The nested version of the same algorithm would require 188 real multiplies. The workload for the pair of IFFTs is 13.3 multiplies per input. Thus, the total workload for the cascade polyphaser implementation of the 601-tap filter is 57.3 multiplies per input. This workload is less than 10% of the workload for the direct implementation. So this is the reason we might consider the cascade polyphaser filters to perform a filtering task. The cascade filters are *Green*, and they offer an order of magnitude reduction in workload to perform the filtering task! Fig. 6.12 presents the frequency responses of the two implementations. The reduced workload of the channelizer implementation does have a processing cost. The cost is to be seen in the causality delay of the two filter realizations. The delay to the center tap of the channelizer impulse response is 630 samples, and the delay of the 601-tap filter is 300 samples. The cascade channelizer has the signal propagating through both the input and output channelizer filters, so we would expect the additional delay in the cascade implementation.

An additional clever thing we can do with the cascade channelizer is to change the sample rate while forming reduced bandwidth superchannels from the narrow bandwidth subchannels. We may have cause to do this when the bandwidth of the synthesized channel is significantly narrower than the sample rate of the channelizer. This is true for the channel we just synthesized from 11 subchannels of the available 60 channels. There the bandwidth of the synthesized channel was $\pm 106$ MHz, and the sample rate was maintained at 1200 MHz. We have the option to reduce the sample rate, say to 400 MHz by reducing the size of the output IFFT and $M$-path synthesis channelizer from 60 paths to 20 paths. The block diagram of the synthesized and

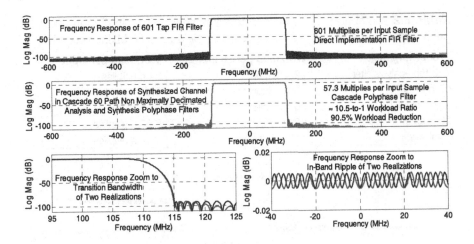

**FIGURE 6.12**

Frequency response of 601-tap direct implementation of low-pass filter and the corresponding cascade channelizer with detailed comparisons of transition bandwidths and stop-band and pass-band ripple levels.

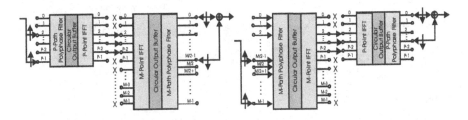

**FIGURE 6.13**

Cascade of nonmaximally decimated $P$-path analysis, $M$-path synthesis upsampling filter bank and $M$-path analysis, $P$-path downsampling filter bank.

down sampled cascade is shown on right side of Fig. 6.13. The left side of the same figure presents the synthesized and up sampled cascade, the dual of the version we now examine. We will discuss the use of the dual resampling channelizers in the next section.

Table 6.2 itemizes the workload of the two processing blocks in the now different size analysis and synthesis channelizers. We still amortize the workload per processing cycle over the 30 input samples presented per work cycle. The input channelizer, with its 720 taps, the output channelizer with its 200 taps, and the two IFFTs, the 60- and 20-point IFFTs requiring 200 and 40 multiplies, respectively, resulting in a total of 1160 multiplies per 30 input samples. Thus, the total workload for the cascade polyphase implementation and 3-to-1 down sampled version

**Table 6.2** Computational workload to implement a 601-tap filter in 60-path channelizer with embedded 3-to-1 down-sample from 1200 MHz to 400 MHz

| Processing task | Workload per 30 inputs | Workload per input |
|---|---|---|
| 720-tap analysis filter | 720 multiplies | 24 multiplies |
| 60-point IFFT | 200 multiplies | 6.67 multiplies |
| 20-point IFFT | 40 multiplies | 1.33 multiplies |
| 200-tap synthesis filter | 200 multiplies | 6.67 multiplies |
| Total workload | 1160 multiplies | 38.7 multiplies |

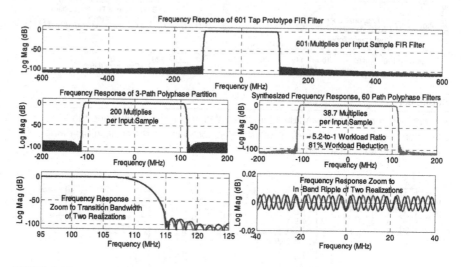

**FIGURE 6.14**

Frequency responses of 601-tap direct implementation of a low-pass filter and the corresponding cascade with detailed comparisons of transition bandwidths and stopband and passband ripple levels.

of the 601-tap filter is 1160/30 or 38.7 multiplies per input. We now compare this workload to the 3-to-1 down sampled 3-path polyphase partition of the 601-tap direct implementation, which has a workload of 200 multiplies per input sample. The resampled channelizer workload of 38.7 multiplies is 19.35% (about 1/5) of the workload for the resampled direct implementation, which is still a significant workload reduction. Fig. 6.14 presents the frequency responses of the two down sampled implementations. Interestingly, whereas both versions of the filters meet the stopband specifications, due to stopband aliasing, they are seen to have different rates of stopband roll off. The impulse responses of the two filters have delays of 211 and 100 samples, which of course is the same delay interval, but with 1/3 of the clock samples at clocks with 3-times longer periods.

### 6.5.2 CASCADE CHANNELIZERS FOR MULTIPLE SIMULTANEOUS VARIABLE BANDWIDTH FILTERS

In this section, we apply the resampling $M$-path channelizers to assemble and disassemble composite waveforms containing multiple arbitrary bandwidth signal components. We implement the assembly process and show a number of useful variants for different input bandwidths [5–7]. Due to limited space, we do not illustrate the dual disassembly process but trust the reader to see that it is performed in a similar manner. We use our 60-channel synthesis channelizer as the framework for the demonstrations. Our 60-path synthesis channelizer forms a composite output signal containing the components from 60 subchannels separated by 20 MHz center frequencies. Input signal samples presented to the $k$th port of the synthesis channelizer are upsampled by a factor of 30 and translated to the $k$th center frequency of the composite output signal. The sample rate of each input sequence is 40 MHz, which is twice the spacing between adjacent channels. We want to be clear now what we mean by input signal bandwidth, and it would be useful to examine the spectra presented in Fig. 6.10. The signal bandwidth to the channelizer includes its passband width plus the width of both transition bandwidths to their stopband edge. Since the input sample rate to each port is 40 MHz, the unaliased input signal bandwidths must be below 40 MHz, and to accommodate the synthesizer filter's transition bandwidth, we restrict the input bandwidth to be 30 MHz.

Fig. 6.15 presents the spectra of four different QPSK modulated waveforms we want to present to the synthesis channelizer. In the upper left, the spectrum of our first signal input signal is seen to be a shaped baseband modulation signal with a 10 MHz symbol rate exhibiting a 15 MHz bandwidth, sampled at four samples per symbol to obtain the desired 40 MHz sample rate. We simply present these samples to a channelizer input port, which we will do shortly. In the upper right, the spectrum of our second input signal is seen to be a shaped baseband modulation signal with a 20 MHz symbol rate exhibiting a 30 MHz bandwidth, sampled at two samples per symbol to also obtain the desired 40 MHz sample rate. Here too, we simply present these samples to a channelizer input port.

In the bottom row of Fig. 6.15, the spectrum of our fourth input signal is seen to be a shaped baseband modulated signal with an 80 MHz symbol rate exhibiting a 120 MHz bandwidth sampled at 160 MHz. This signal requires processing in an 8-path, 4-to-1 down sampling analysis channelizer. This too will form 20 MHz bandwidth output channels at a 40 MHz output sample rate with channels separated by 20 MHz centers. The spectra of the 8-path analysis filter bank channel bandwidths and the spectra of the time series from each channel are shown in Fig. 6.16. Note that seven of the eight channels contain spectral components from the analyzed input signal. The time series from the seven channels of the analysis filter bank will be presented to seven ports of the synthesis filter bank, in which they will be reassembled in the selected seven frequency offset channels of the synthesizer.

Fig. 6.17 shows the spectra formed by the 60-path synthesis channelizer at its output sample rate of 1200 MHz along with the spectral segments delivered to it at their

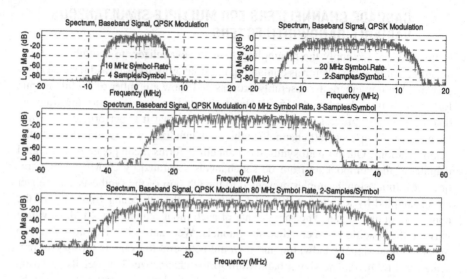

**FIGURE 6.15**

Spectra of four baseband modulated signals, with symbol rates of 10, 20, 40, and 80 MHz sampled at 40, 40, 120, and 160 MHz, respectively. Signals to be processed and presented to 60-path synthesis channelizer.

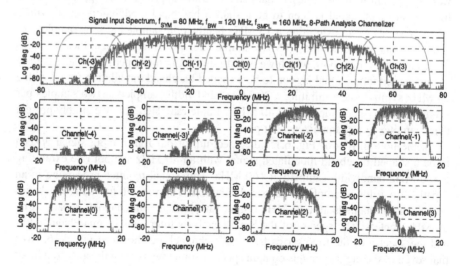

**FIGURE 6.16**

Spectra of input signal and channel frequency responses of 8-path analysis channelizer and filtered, base-banded and resampled spectra from eight channel filters illustrating the spectral decomposition performed by 8-path analysis filter bank.

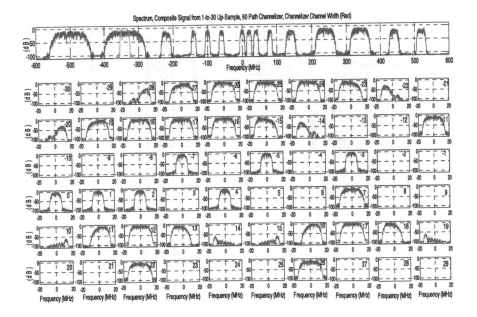

**FIGURE 6.17**

Spectra of input signal and spectral components input to 60-path synthesis channelizer illustrating upsampling alias translation and perfect reconstruction performed by 60-path synthesis filter bank.

40 MHz sample rates. The narrow bandwidth signals were presented to the channelizer without preprocessing, whereas wide bandwidth signals were preprocessed to form channelized segments by the six-path and eight-path analysis filters. The spectrum of the channelizer channel is overlaid in red (in the web version) to show the relative sizes of the synthesized supper channels and the channel bandwidth which assembled them.

## 6.5.3 CASCADE CHANNELIZER ENHANCEMENTS FOR INCREASED FLEXIBILITY

The cascade $M$-path analysis and synthesis filter banks, while impressive by themselves, can offer increased capabilities through the aid of minor signal processing tasks residing at the input or output of the bank and between the analysis and synthesis channelizers [8–10]. We will discuss some of these enhancements. Fig. 6.18 shows a synthesis filter bank being accessed through a range of signal conditioning processing blocks connecting the external signal source to the channelizer input ports. The first processing block we encounter starting at the top of the input chain is the analysis filter block that channelizes wide band input signals into narrow chan-

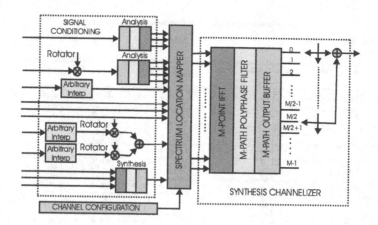

**FIGURE 6.18**

Possible signal conditioning options at input to cascade $M$-path channelizer.

nels that match the synthesizer's channel bandwidth and spacing. We have already covered this option!

The next option is interesting because it removes one of the synthesis channelizer's constraints. The channelizer aliases baseband signal spectra to center frequencies that are multiples of $f_S/M$. This usually corresponds to multiples of the input sample rate, but with filters performing $M/2$-to-1 resampling, the center frequencies are multiples of half the input sample rate. We can heterodyne the input signal off of baseband with a complex heterodyne operating at the input sample rate. The amount of frequency shift we apply here is always less than half the channel spacing. If we want a larger shift, then we would simply use the next input port as the reference frequency. The amount we shift the input signal off of DC is the amount the spectrum is offset relative to the frequency bin center selected as the input port of the $M$-point IFFT. We call this the worm-hole effect. We perform a small frequency offset at baseband at a low input sample rate, and the spectrum nominally at the $k$th center frequency operating at the high output sample rate experiences the same frequency offset. The dual graph benefits from the same coupled offset property. If a signal center frequency is offset from one of the Nyquist zone center frequencies that aliases to baseband when we alias the center frequency to baseband in an analysis channelizer, then the same offset frequency becomes the offset from baseband's DC. A final heterodyne down conversion at the low output sample rate will complete the base-banding down conversion.

The third signal conditioning option is an arbitrary interpolator, which converts the sample rate of the input signal to the required input sample rate of the synthesis channelizer. An example of the use of this option follows. Suppose we are presented with an input signal with a 10.24 MHz symbol rate formed at four samples per symbol, which gives us a 40.96 MHz sample rate. The channelizer wants a 40.0 MHz sample rate, so we use the arbitrary interpolator to change 40.96 to 40.0. A matching interpolator at the dual analysis channelizer would perform the reverse interpolation, converting the 40.0 MHz output sample rate back to the 40.96 MHz sample rate.

The fourth option of signal conditioning applied to the synthesizer input stream is to do nothing. The signal arrives with properly confined bandwidth and with the correct sample rate and can be passed directly to the synthesizer input ports.

The fifth useful option for signal conditioning is first use arbitrary interpolators to bring multiple input sample rates to the correct common output rate to be input to the synthesis channelizer and then frequency offset and add their components. This is in fact a mini channel synthesizer, which combines multiple narrow bandwidth signals into a composite wider bandwidth signal more closely matched to the channelizer bandwidth.

The last of the useful options we identify extends the process of the previous option to use a single low-rate synthesis filter bank to perform the interpolation and frequency offsets in a single process more efficiently than the direct implementation. After all, isn't what this chapter is about? The use of cascade synthesis channelizers is reminiscent of a dual conversion block converter.

## 6.6 CONCLUDING REMARKS

Filter banks are amazing processing engines. The task performed by a filter bank are combinations of the common operations of spectral translation, bandwidth reduction, and sample rate changes. What we do in filter bank designs is coupling the processes, embedding one within another, and rearranging the order in which these operations are performed. The net result of these modifications is that we obtain systems that require fewer resources with reduced processing costs to implement many of our desired communication system tasks. This chapter has presented a short journey along a learning path with mile posts that identified important ideas and perspectives to help the reader understand the workings of filter banks. We have avoided abstract math descriptions and have emphasized the straightforward implementation processes along the way [11].

One of the key points in filter bank design is the concept of embedding sample rate changes in the filtering process. We have two choices, to reduce the sample rate or to increase it. We also have two routes to consider. We can change the sample rate of the time series when reducing the signal bandwidth, or we can change the sample rate while preserving the signal bandwidth. We do both! When we change the signal's sample rate, we have the opportunity to interact with the periodicity of sampled data spectra. We can do clever things with spectra residing in different Nyquist zones such as enter a process with a spectral span centered in one Nyquist zone and leave the process with a version of the same spectral span centered in a different Nyquist zone. We intentionally alias to obtain frequency translation between Nyquist zones. Aliasing can be your friend. When we upsample, we move spectra from baseband to higher Nyquist zones, and when we downsample, we move spectra from higher Nyquist zones to baseband. How neat!

We started our journey by demonstrating that when we want to extract a narrow-band signal from a broadband signal which has a high sample rate to accommodate the broadband aggregate spectrum, we have an option to reduce the sample rate as part of the bandwidth reduction. By embedding an $M$-to-1 resample operation in an $N$-tap FIR filter we formed an $M$-path filter that collectively performed the bandwidth reduction at the reduced workload of $N/M$ operations per input sample to obtain output samples at 1-$M$th of the input sample rate. We then upsampled the narrow-band signal with the dual of the $M$-path downsampling filter, the $M$-path up-sampling filter. This second filter also required $N/M$ operations per output sample. The cascade of downsample and upsample in a pair of $M$-path filters reduces the workload from $N$ operations per input–output pair to $2N/M$ operations per input–output pair. Great trade!

At first, we were a bit sad when we realized that the workload reduction benefit required a large ratio of sample rate to bandwidth. We recovered our good spirits when we realized that the aliasing of spectral spans to baseband from band centers at multiples of the reduced output sample rate could be separated by their unique phase profiles associated with successive delays in the $M$-path filter. The IFFT unwraps the multiple aliases and gives us our reduced sample rate base-banded narrowband signals. The $M$-path prototype filter, which formed the input $M$-path analysis filter bank, was designed to be a Nyquist capable of remerging the multiple base-banded spectral spans. We modified the input $M$-path filter to down sample by $M/2$ rather than by $M$ to avoid aliasing the transition bandwidth of the channel filters. We note that it is not necessary to avoid transition bandwidth aliasing; there are versions of filter banks that operate with aliased transition bands and cancel that alias as part of the reconstruction process, specifically the PR Cosine Modulated Filter banks. We chose to avoid the transition band aliasing to access wider range of capabilities and options. Our next modification was to form the dual of the down sampling filter with a spectral response that preserved the Nyquist spectra of the analysis filter bank. The altered output filter bank is the synthesis bank. The cascade of the two banks performed perfect reconstruction of user selected narrowband segments. Other sample rate options wrapped around the filter bank pair make this engine a formidable tool in our communication system signal processing toolbox. Welcome to it and enjoy the pleasure of work well done at reasonable costs.

# REFERENCES

[1] f. harris, *Multirate Signal; Processing for Communication Systems*. Prentice Hall, 2004.

[2] f. harris, C. Dick, X. Chen, and E. Venosa, "Wideband 160 channel polyphase filter bank cable TV channelizer," *IET Signal Process., Special Issue on Multirate Signal Processing*, vol. 5, no. 3, Jun. 2011, pp. 325–332.

[3] f. harris, C. Dick, and S. Seshigiri, "An improved square-root Nyquist shaping filter," in *SDR Conference-2005*, Anaheim, CA, Nov. 2005.

[4] f. harris and C. Dick, "An alternate design technique for square root Nyquist shaping filters," in *SDR WinnCom-2015 Conference*, San Diego, CA, Mar. 2015.

[5] f. harris, E. Venosa, and X. Chen, "Polyphase analysis filter bank down-converts unequal channel bandwidths with arbitrary center frequencies, Design I," in *Software Defined Radio Conference (SDR'10)*, Washington, DC, Nov.–Dec. 2010.

[6] f. harris, E. Venosa, and X. Chen, "Polyphase synthesis filter bank up-converts unequal channel bandwidths with arbitrary center frequencies, Design I," in *Software Defined Radio Conference (SDR'10)*, Washington, DC, Nov.–Dec. 2010.

[7] M. Renfors and f. harris, "Highly adjustable multirate digital filters based on fast convolution," in *European Conference on Circuit Theory and Design, ECCTD2011*, Linkoping, Sweden, Aug. 2011.

[8] f. harris, C. Dick, X. Chen, and E. Venosa, "Comparing LTE channelizers implemented with linear phase recursive filters and FIR filters." SDR-13, Washington, DC, 8–10 Jan. 2013.

[9] f. harris, C. Dick, X. Chen, and E. Venosa, "Interleaving different bandwidth narrowband channels in perfect reconstruction cascade polyphase filter banks for efficient flexible variable bandwidth filters in wideband digital transceivers," in *DSP-2015 Conference*, Singapore, Jul. 2015.

[10] R. D. Koilpillai and P. P. Vaidyanathan, "Cosine-modulated FIR filter banks satisfying perfect reconstruction," *IEEE Trans. Signal Proc.*, vol. 40, pp. 770–783, Apr. 1992.

[11] P. P. Vaidyanathan, *Multirate Systems and Filter Banks*. Englewood Cliffs: Prentice-Hall, 1993.

CHAPTER

# Orthogonal Communication Waveforms

7

**Pierre Siohan\*, Markku Renfors†**

*Independent Researcher, Rennes, France\**
*Tampere University of Technology, Tampere, Finland†*

## CONTENTS

## 7.1 INTRODUCTION

Multicarrier schemes, most often using the Orthogonal Frequency-Division Multiplexing (OFDM) modulation, are now widely adopted in a large number of communication standards. Indeed, to face the constant need of higher and higher data rates, new communication systems always operate on broader bandwidths. Then, multicar-

Orthogonal Waveforms and Filter Banks for Future Communication Systems. DOI: 10.1016/B978-0-12-810384-5.00007-4

rier schemes present clear advantages when transmitting through channels that are heavily frequency selective. However, since the basic communication principles have been founded on the Nyquist criterion, the road was long to build our modern communication systems and to propose viable alternatives to OFDM. In this chapter, we review this evolution from Nyquist pulse shaping to the most advanced orthogonal communication waveforms. In particular, we illustrate some of their important links with Gabor and filterbank theories.

Orthogonality is the simplest way to insure a perfect discrimination between different signals. The orthogonality principle is therefore commonly used in the communication domain and for many signal processing purposes as well. In this chapter, we focus on waveform orthogonality showing how this fundamental concept is now evolving rapidly from the traditional communication knowledge, as founded by the Nyquist criterion, to the most recent and efficient communication systems taking advantage of the most advanced signal processing knowledge.

To start with a basic fact, let us recall that a fundamental advance for spectrum usage results from a smart combination between communication theory and signal processing. Indeed, originally, the application of Nyquist criterion only permitted one to build optimal orthogonal Single-Carrier (SC) systems. Keeping this optimality, in terms of Shannon's capacity, for a frequency multiplex of SC signals, was not possible before the advent of a multicarrier system based on the Discrete Fourier Transform (DFT), allowing one to preserve orthogonality in spite of the frequency overlapping between different subcarrier signals. This multicarrier system, known either as OFDM or Discrete MultiTone (DMT), is nowadays still the modulation scheme of reference because of its efficiency and implementation simplicity for basic single-user systems and also for multiple-user systems based on the principle of Orthogonal Frequency-Division Multiple Access (OFDMA). However, OFDM is not the ultimate step in terms of multicarrier modulation. In this chapter, we review the multicarrier waveform theory and the resulting waveform alternatives.

## 7.2 MULTICARRIER WAVEFORM THEORY

### 7.2.1 EVOLUTION OF ORTHOGONAL MULTICARRIER WAVEFORMS

#### 7.2.1.1 Nyquist-Pulse Shaping Applications

As mentioned in our introduction, Frequency-Division-Multiplexing (FDM) transmission without orthogonality means wasting a large amount of spectrum in order to avoid Inter-Carrier Interference (ICI); see, e.g., [90] for an overview of some of the pioneering works on FDM. The discovery of the DFT, and especially its efficient Fast Fourier Transform (FFT) implementation in 1965 [22], was a key feature for the advent of consumer-oriented applications using transmission schemes known as OFDM or DMT. The use of block transforms, like DFT, implicitly corresponds to rectangular windows in waveform processing, leading to widely spread frequency spectra, e.g., with the $\sin x/x$ spectrum shape. It took a certain time for the communication community to commute to a digital signal processing approach being able to fully exploit these new tools. Firstly, at that time, the continuous-time formalism was the usual way to describe transmission systems. Secondly, cost-efficient digital circuitry was still lacking. Therefore, the idea to play with the pulse shape, making it non-rectangular in discrete-time, took time to develop. Along with the invention of FFT, also the first multicarrier proposals based on nonrectangular waveforms date back to the mid-1960s. They were elaborated using a continuous-time formalism, i.e., a framework more appropriate for analog realizations. Nevertheless, some of these first multicarrier studies set up time-frequency orthogonality principles still actively used nowadays. Firstly, to avoid the drawback of previously proposed orthogonal multi-plex [62] based on time-limited orthogonal signals leading to wide spectra, Chang [15] and Saltzberg [74] preferred to use a class of band-limited orthogonal signals. Secondly, both authors operate at the maximum transmission rate using a set of par-allel transmission filters satisfying time-frequency orthogonality requirements and being all derived from one unique filter, i.e., corresponding to what is now called a prototype function. In reference [15], the proposed multicarrier system is described for transmission of real data symbols, e.g., Pulse Amplitude Modulation (PAM) sym-bols. Saltzberg [74] extends Chang's principle to permit the transmission of complex symbols, e.g., Quadrature Amplitude Modulation (QAM) symbols. The idea is to build in time and frequency a smart multiplex consisting of pairs of real symbols. Such data symbols, corresponding either to the real or imaginary part of QAM sym-bols, are now called Offset Quadrature Amplitude Modulation (OQAM) symbols.

Figs. 7.1 and 7.2 give, in the absence of any channel or system impairments, a sim-ple illustration of time-frequency orthogonality with time and band-limited functions, respectively. In the case of a time-limited signal, as there is no overlapping between successive symbols, it is clear that the time orthogonality is satisfied, i.e., there is no Inter-Symbol Interference (ISI). In the frequency domain, the rectangular win-dows give rise to $\sin x / x$ functions. If the multiplex operation is carried out with a frequency spacing ($F$) being the exact inverse of the symbol duration ($T$), then we get a maximum transmission rate. In spite of the subcarrier overlapping, we also get orthogonality in frequency since we exactly satisfy the Nyquist criterion in the fre-quency domain, i.e., there is no ICI.

Unfortunately, the OFDM/DMT-type orthogonality of Fig. 7.1 only works with a rectangular window, and then, when transmitting over a band-limited channel, strong interferences naturally arise. This is why Chang, with a PAM-based system [15], soon followed by Saltzberg [74] with staggered, or offset QAM, i.e., OQAM, preferred to build multiplexes of band-limited signals. Fig. 7.2 illustrates how the Saltzberg time-frequency staggering rule applies to a full roll-off function. Due to the QAM splitting, orthogonality is ensured in time and frequency between the real and imag-inary parts of the complex QAM components. Then, still assuming the case of a modulator–demodulator back-to-back scheme, we only need to take care of the sub-set of time symbols and frequency subcarriers corresponding either to the purely real or purely imaginary components. So, in Fig. 7.2, we can see that the sequence of overlapping time symbols satisfies the Nyquist criterion and therefore provides time-

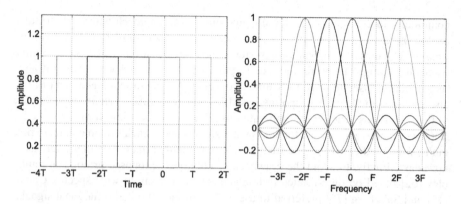

**FIGURE 7.1**

Orthogonality in time (at the left) and frequency (at the right) with a multiplex of time-limited signals.

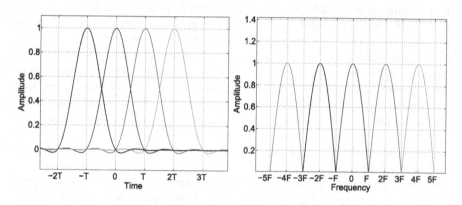

**FIGURE 7.2**

Orthogonality in time (at the left) and frequency (at the right) with a multiplex of band-limited signals (only the real-valued components are represented).

orthogonality. In frequency, with the band-limited cosine function, there is no overlap (since here we only represent the even-indexed components). So, at the end, we get a perfect time-frequency orthogonality in real domain.

In a recent paper [31], Farhang-Boroujeny and Yuen have revisited the pioneering works of Chang [15] and Saltzberg [74]. In their tutorial review, they named these schemes Cosine-modulated MultiTone (CMT) and Staggered-modulated Multi-Tone (SMT) for the transmission of PAM and QAM symbols, respectively. Reference [31] rederives the orthogonality conditions for CMT and SMT assuming that the continuous-time transmit filters are zero-phase and real-valued in time. Then, to get orthogonality, applying the matched filter rule, the receive filters are identical to the transmit ones.

### 7.2.1.2 *The OFDM/DMT Big Wave*

It took a long time for OFDM to impose itself as a viable technology before becoming the big wave we know since the 1990s. OFDM/DMT has been selected for a large bunch of communication standards and deployed in prominent mass market applications: Asymmetric Digital Subscriber Line (ADSL), TV broadcast, WiFi, Long-Term Evolution-Advanced (LTE-A), etc. The first OFDM predecessor appeared in the mid-1950s with a system known as Kineplex [29,62], and it was dedicated to high-frequency military systems. Ten years later another high-frequency radio transceiver, named KATHRYN, was built using the DFT in an analog hardware implementation [96]. As recalled by Weinstein himself [90], their idea with Ebert in 1971 [91] was to take a full advantage of the FFT to build a multicarrier system able to successfully compete with the SC approaches. Also aware of the interference problem created by this block transform approach, they added a "guard space" between adjacent transmission blocks, together with a smoothing of these transition zones. This naturally led to a reduction of the spectral efficiency, together with a loss of orthogonality due to the windowing when considering the back-to-back connection of modulator/demodulator; these drawbacks still exist. But the main issue was the implementation cost since special purpose hardware was then required.

So, one had to wait nearly another ten years before getting a first presentation of the DMT/OFDM as we know it today. Indeed, even if the DMT and OFDM acronyms were not yet introduced,[1] in their 1980 paper, Peled and Ruiz [64] provide all the main ingredients still characterizing these MultiCarrier Modulation (MCM) schemes, namely:

- use of Inverse Discrete Fourier Transform (IDFT) and DFT for implementing the modulation and demodulation, respectively;
- a cyclic extension in the beginning of each OFDM symbol consisting of the copy of a number of IDFT output samples from the end of the symbol, the number of copied samples being at least equal to the channel memory;
- a removal of these corrupted samples at the receiver side after the synchronization stage;
- a simple division by each subchannel coefficient, i.e., a Zero Forcing (ZF) equalizer, directly exploiting the circular convolution property in the frequency domain, permitting one to exactly recover the transmitted symbols (in the absence of noise).

---

[1] According to Siclet's thesis [78], the OFDM acronym seems to be mentioned at first in Keasler's thesis [47] and in a journal paper authored by Cimini [18], whereas, as mentioned in Chapter 4 of a web book on Cioffi's web page [19], the DMT acronym emanated from Standford university, but also relatively late.

Furthermore, as the MCM system of [64] was dedicated to baseband transmission over telephone lines, they also introduced two features more specific to DMT transmission:

- in their DMT-like construction rule, the QAM symbols of the IDFT inputs satisfy an Hermitian symmetry property guaranteeing the generation of a real-valued baseband signal;
- the wired transmission channel, being nearly static, only needs to be sounded from time-to-time to get knowledge of the Signal-to-Noise Ratio (SNR) in different subchannels. Then, using a feedback loop, an optimal subcarrier power allocation strategy can be put in place.

For time-varying channels, e.g., those related to portable or mobile wireless applications, the overhead of a feedback loop may be a crippling drawback. So, in the absence of SNR knowledge per subcarrier on the Transmitter (TX) side, any MCM system becomes more sensitive to the losses due to deep fading or narrow-band interference. Therefore, appropriate channel coding strategies have to be adopted in conjunction with MCM. In this line of thought, starting from the mid-1980s, CCETT (Joint Research Centre for Broadcasting and Telecommunications, part of the R&D branch of France Télécom and Télédiffusion de France) proposed a MCM system named Coded Orthogonal Frequency-Division Multiplexing (COFDM), a coded system in which the Forward Error Correction (FEC) and interleaving were designed to prevent from the negative impact of deep fading and narrow-band interferences. The basic principles were stated in [5] and, soon after, in the context of the Eureka project EU 147 Digital Audio Broadcasting (DAB), COFDM was proposed as the transmission means for Digital Audio Broadcasting (DAB) [30]. References [48,49,97] recall the different birth stages of DAB and Digital Video Broadcasting-Terrestrial (DVB-T), also based on COFDM. As in practice, MCM transmission systems now systematically use FEC, and also interleaving, the C of COFDM is implicitly assumed and tends to disappear from this acronym. So, from now on, we also leave out it, but without forgetting it.

All these positive features have made over the years OFDM/DMT multicarrier candidates superior over different SC proposals and be adopted for a large number of communication standards. Indeed beyond ADSL, DAB, and DVB-T, OFDM/DMT also constitutes the radio interface for the latest standards in broadcasting (DVB-T2, DVB-NGH), in wired access (ADSL2+, VDSL2), fixed radio access (IEEE 802.16a for WiMAX, IEEE 802.22), mobile access (IEEE 802.16e, 3GPP LTE-A), indoor connectivity with the WiFi (IEEE 802.11a/g/n, IEEE 802.11vht) and the Power Line Communications (PLC) (IEEE P1901).

In these different cases, the main arguments in favor of OFDM/DMT versus SC were: (i) simple and robust channel equalization; (ii) effective resource allocation; and (iii) Multiple-Input Multiple-Output (MIMO) capability. Detailed comparisons in terms of transmission capacity and computational complexity can be found, for instance, in [20,21]. Note also that OFDM is not only a key transmission technology but is also attractive, with the OFDMA, from the multiple access point of view [75].

Nevertheless, the single vs. multiple carrier debate is not yet closed, and it will probably reappear with each emerging new application. In particular, the fact that MCM naturally leads to high Peak-to-Average Power Ratio (PAPR) values (see Chapter 18 of this book for more details) still constitutes an open problem. This has been recently illustrated with the choice of a SC-like technique, the Single-Carrier Frequency-Division Multiple Access (SC-FDMA), for the Physical layer (PHY) of UpLink (UL) of LTE-A.

### 7.2.1.3 *The Comeback of Nonrectangular Waveforms*

Though the OFDM/DMT had been adopted by many academics and big players from the industry, in the mid-1990s, some researchers agreed on the fact that this transmission technique nevertheless could still be improved while keeping the intrinsic advantages of an FFT-based MCM scheme. Indeed, as the spectrum had to be shared by a growing number of applications, the spectral efficiency loss in the time domain due to the Cyclic Prefix (CP) became more and more annoying. Furthermore, the insufficient attenuation of the rectangular window made also OFDM/DMT inefficient in the frequency domain, leading to ICI problems and to another loss in spectral efficiency. Indeed, for coexistence reasons with other communication applications, a large portion of subcarriers had to be switched off, either to satisfy spectrum masks by insertion of "guard bands" or by inserting frequency notches to avoid interferences with narrow band applications sharing the same frequency band.

So, nearly at the same period of time when OFDM/DMT was taking off, the ideas of Chang and Saltzberg were given a second life. It is also worth noting that in the meantime Hirosaki had presented OFDM with OQAM, a scheme he named Orthogonally Multiplexed QAM (O-QAM) [43], which could be efficiently implemented by DFT. Fliege [35] went a step further with a system he named Orthogonal Multiple Carrier (OMC) using a polyphase structure [11] for the filtering part. But instead of keeping the Nyquist cosine waveform in the frequency domain, as proposed by [15,43,74], authors then took different directions. Li and Stette [52] proposed a dual form with respect to Chang, i.e., the square Root Raised Cosine (RRC) function is applied in the time domain. So they obtain a Time-Limited Orthogonal (TLO) MCM scheme with time-frequency orthogonality features that are dual to the ones depicted in Fig. 7.2, i.e., time and frequency representations are exchanged.

However, the most remarkable proposal at that time appeared in the Proceedings of the IEEE with a paper authored by Le Floch, Alard, and Berrou [48], which presents a simplified version of an invention by Alard [4]. Similarly to [52], [48] introduces a TLO scheme named OFDM/Minimum-Shift Keying (MSK), but the major contribution is the introduction of the Isotropic Orthogonal Transform Algorithm (IOTA) prototype function [4]. IOTA is a continuous-time function resulting from a double orthogonalization of the Gaussian function (more details are provided in Chapter 8). Among its other specific features, we can notice that the IOTA function is identical to its Fourier transform and, as conjectured by Alard, could be the optimal

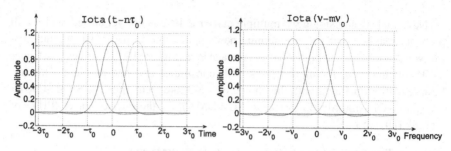

**FIGURE 7.3**

Orthogonality in time (at the left) and frequency (at the right) with a multiplex of overlapping IOTA signals.

prototype function in terms of time-frequency localization,[2] among the set of all the perfectly Orthogonal Frequency-Division Multiplexing with Offset-QAM subcarrier modulation (OFDM/OQAM)[3] schemes. Fig. 7.3, in which $\tau_0 = T/2$ and $\nu_0 = F$ denote the OQAM subsymbol rate and frequency spacing, respectively, illustrates the fact that the time-frequency orthogonality can be obtained in spite of overlapping in time and frequency.

Independently from the IOTA study, Haas and Belfiore [41,42] have also proposed OFDM/OQAM systems that could be well localized in time and frequency. Instead of starting from the Gaussian function, their construction mode is based on the selection of an appropriate subset of Hermite functions, all being identical to their respective Fourier transform, guaranteeing that, in the end, their linear combination will provide a prototype function with this same property. Then, it is sufficient to have the orthogonality satisfied in one single domain to get a full time-frequency orthogonality.

If, differently from these IOTA and Hermite-based systems, we want to get a complex orthogonality, as with OFDM, but together with a good localization in time and frequency, we have to relax the maximum spectral efficiency requirement.[4] To the best of our knowledge, the first proposal of this type, named "fraction-spaced multicarrier," has to be attributed to Vallet and Haj Taieb [88], but it went almost unnoticed. The idea, popularized later on, with the Filtered MultiTone (FMT) system (see Section 7.3.1), was to fractionally oversample the input QAM symbols in order to increase the spacing between subcarriers. Then, it became possible to get a realizable nearly Frequency-Limited Orthogonal (FLO) MCM system, essentially producing no ICI.

At this step, we can state that in the mid-1990s all the ingredients of the modern MCM systems were already existing. Concerning the modulation and demodulation

---

[2]A simple counterexample is provided in [50], but not presenting the nice features of IOTA, e.g., no Fourier transform invariance.

[3]Joint acronym also coined in this pioneering paper [48].

[4]The corresponding formal expression is given in Section 7.2.2.

functions, the tracks of progress to come have then been found in the link between MCM and the theory of Gabor, or equivalently Weyl–Heisenberg, families, from one side and the filterbank theory on the other side.

## 7.2.2 TIME-FREQUENCY ANALYSIS AND FILTERBANK DUALITY

### 7.2.2.1 Gabor Theory

A particularly useful tool to analyze the different MCM schemes we have just recalled is given by the Gabor, also named Weyl–Heisenberg, theory. Let us just here summarize some salient features of this theory, initiated by Gabor in 1946 [37]. For interested readers, deeper descriptions can be found, for instance, in [33].

A Gabor family may be characterized by a triplet including a function $g(t) \in \mathcal{L}^2(\mathbb{R})$ and two strictly positive numbers $T$ and $F$. Then, the Gabor family of functions $\{g_{m,k}(t)\}$ with $(m, k) \in \mathbb{Z}^2$ may be expressed as

$$g_{m,k}(t) = g(t - kT)e^{j2\pi m F t}e^{j\psi_{m,k}}, \tag{7.1}$$

including a phase term $\psi_{m,k}$ generally set to zero as it has no impact on the orthogonality in the complex domain. The function $g$ is called the prototype function, and $d = \frac{1}{TF}$ is the density of the Gabor family. Then, the transmitted baseband signal writes as a linear combination of the $\{g_{m,k}(t)\}$ family, i.e.,

$$s(t) = \sum_{m,k} c_{m,k} g_{m,k}(t), \tag{7.2}$$

where the $c_{m,k}$ are the symbols to be transmitted.

As proved in [28], to be an orthogonal basis (a Riesz basis) of its spanned space, the Gabor family has to be such that $d \leq 1$, and the basis functions have to satisfy the condition

$$\int_{-\infty}^{\infty} g_{m,k}(t)g^*_{m',k'}(t)dt = \delta_{m,m'}\delta_{k,k'} \tag{7.3}$$

with $\delta_{i,j}$ the Kronecker delta.

This orthogonality condition can be rewritten with respect to the ambiguity function of $g$, i.e.,

$$A_g(\tau, \nu) = \int_{-\infty}^{\infty} g\left(t - \frac{\tau}{2}\right)g^*\left(t + \frac{\tau}{2}\right)e^{-j2\pi \nu t}dt. \tag{7.4}$$

Then, the orthonormality condition, also called the generalized Nyquist criterion [9], becomes

$$A_g(kT, mF) = \begin{cases} 1, & k = m = 0, \\ 0 & \text{otherwise.} \end{cases} \tag{7.5}$$

In addition to the orthogonality property, the time and frequency localization of the prototype function is another important characteristic. The usual localization measure for continuous-time function directly corresponds to its second-order moments in time and frequency domains. Denoting by $G(f)$ the Fourier transform of $g(t)$ and by $m^{(i)}(g)$, $\mathfrak{M}^{(i)}(g)$ the $i$th-order moments in time and frequency, respectively, we can write

$$m^{(1)}(g) = \frac{1}{\|g\|^2} \int_{-\infty}^{\infty} t|g(t)|^2 dt, \qquad (7.6)$$

$$\mathfrak{M}^{(1)}(g) = \frac{1}{\|g\|^2} \int_{-\infty}^{\infty} f|G(f)|^2 df, \qquad (7.7)$$

$$m^{(2)}(g) = \frac{1}{\|g\|^2} \int_{-\infty}^{\infty} (t - m^{(1)}(g))^2 |g(t)|^2 dt, \qquad (7.8)$$

$$\mathfrak{M}^{(2)}(g) = \frac{1}{\|g\|^2} \int_{-\infty}^{\infty} (f - \mathfrak{M}^{(1)}(g))^2 |G(f)|^2 df. \qquad (7.9)$$

Then, to get the time-frequency localization measure of the prototype function $g$, we can either measure it by the product of the second-order moments in time and frequency, i.e., measure if by the quantity $m^{(2)}(g)\mathfrak{M}^{(2)}(g)$ or, as proposed in [48], use the following time-frequency localization measure:

$$\xi(g) = \frac{1}{4\pi \sqrt{m^{(2)}(g)\mathfrak{M}^{(2)}(g)}}. \qquad (7.10)$$

This normalization guarantees that the measure will be always such that $0 \leq \xi(g) \leq 1$, the optimum value of 1 being reached by the Gaussian function, i.e., setting $g(t) = Ce^{-\lambda t^2}$ with $C \in \mathcal{C}$ and $\lambda > 0$.

**Theorem 1.** *Balian–Low Theorem (BLT) [10,57]. For the critical density, i.e., $d = 1$, if $g_{m,k}(t)$ constitutes an orthogonal basis of functions, then $m^{(2)}(g)\mathfrak{M}^{(2)}(g) = +\infty$.*

A direct consequence of BLT is the impossibility for an MCM to simultaneously get the three following properties:

1. Orthogonality in the complex domain.
2. Maximum spectral efficiency, i.e., $d = 1$ or, equivalently, $TF = 1$.
3. A prototype function being well localized in time and frequency.

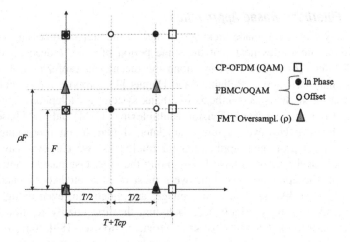

**FIGURE 7.4**

Time-frequency lattices for the three main Gabor-like MCM schemes.

Reanalyzing the evolution of orthogonal MCM waveforms in relation with Gabor's theory, we can see how the three main categories of Gabor-type proposals are connected with the BLT.

- OFDM/DMT, without CP, satisfies conditions 1) and 2) but not 3) because of the unbounded second-order moment in frequency of the $\sin x/x$ function. When adding a CP, condition 2) is no longer satisfied.
- FMT satisfies 1) and 3) but not 2) because of the spacing increase in frequency.
- OFDM/OQAM or, equivalently, FilterBank MultiCarrier (FBMC)/OQAM[5] satisfies conditions 2) and 3), but the orthogonality condition is only satisfied in the real domain, meaning that the orthogonality condition (7.3) has to be rewritten, by a Gabor-like family $\{g_{m,n}\}$ defined on a time-frequency lattice of density 2 (see Fig. 7.4), taking the real part of the following scalar product:

$$\mathbb{Re}\{\langle g_{m,n}, g_{m',n'}\rangle\} = \delta_{m,m'}\delta_{n,n'} \qquad (7.11)$$

Otherwise said, $\langle g_{m,n}, g_{m',n'}\rangle$ has to be a pure imaginary term or zero if $(m, n) \neq (m', n')$.

Fig. 7.4 depicts the three different situations, denoting by $\rho$ the oversampling factor of the FMT system. Note also that $\rho = 1/d$.

---

[5]These acronyms are sometimes written using - as separator, e.g. FBMC-OQAM.

### 7.2.2.2 *Filterbank-Based Approaches*

Filterbank systems have made their appearance in the signal processing and communication communities nearly at the same period of time. Indeed, in the early 1970s, Schafer [76] includes them among the recent and useful tools for speech spectrum analysis, whereas Bellanger and Daguet [11] propose a transmultiplexer, i.e., the dual of an analysis–synthesis filterbank system, for telephony applications (for conversions between Time-Division-Multiplexing (TDM) and FDM based transmission schemes). However, a few years later, digital signal processing clearly became the most prominent application for filterbanks. We can even say that filterbank and wavelet theories have been among the most fashionable topics during the nineties. The applications behind were most often related to subband coding for source signals like speech, audio, image, and video. So, even though the filterbank theory was fully established, including the first duality relation between Perfect Reconstruction (PR) filterbanks and transmultiplexers [89], transmission applications took a longer time than subband coding to benefit from their links with filterbanks and Gabor theories. Indeed, it was only in 1994 that Tzannes et al. [87] proposed to build a uniform critically decimated transmission system, i.e., decimation and expansion factors are all equal and also equal to the number of subbands, taking advantage of a duality property of analysis–synthesis filterbank and transmultiplexer configurations. This transmultiplexer system, named Discrete Wavelet MultiTone (DWMT), corresponds to a fully digital implementation of Chang's modem, the one used for PAM transmission and named CMT in [31]. The term wavelet is a little bit misleading here because wavelets are normally used for recursive decomposition and reconstruction of signals,[6] whereas DWMT corresponds to a nonrecursive implementation of critically decimated Cosine-modulated FilterBanks (CMFBs).

In [44], the authors extend the duality theory of PR systems to uniform oversampled filterbanks, meaning that in this case the decimation and expansion factors are higher than the number of subbands. This procedure also corresponds to what will be done later on with the FMT scheme.

For OFDM/OQAM, the link with filterbanks took a particular path. Indeed, it is firstly related to a class of filterbanks named Modified Discrete Fourier Transform (MDFT), introduced by Fliege [36]. MDFT involves a two-step decimation processing in each subband and a decomposition between the real and imaginary parts of the polyphase components. The MDFT filterbank is a critically sampled complex modulated filter bank for which the PR conditions are given in [46]. References [81,82] show that Cosine-modulated FilterBank (CMFB) and MDFT filterbanks share the same PR conditions. A proof of the duality between MDFT filterbank and OFDM/OQAM transmultiplexer is provided in [79] together with the link to the Gabor–Weyl–Heisenberg theory. Another analysis is also reported in [14], which es-

---

[6]Wavelets and wavelet packets are also used for transmission; see, for instance, [73] for an overview, but they are out of the scope of this book.

**Table 7.1** Some milestones in the multicarrier modulation history

| decade → lattice density ↓ | 50 | 60 | 70 | 80 | 90 | 00 | 10 |
|---|---|---|---|---|---|---|---|
| 1 (PAM) | | CMT [15] | | | DWMT [87] | DTWE [86] | |
| 1 (QAM) | | SMT [74] | | O-QAM [43] | OMC [35] OFDM/ OQAM [13,48,81] | FBMC/ OQAM [8,82] | COQAM [1,38,56] |
| ≥ 1 | [29,62] | | [91] | CP-OFDM [64] COFDM [5] | DMT [19] | GFDM [34] | |
| > 1 | | | | | Over. OFDM [23,44,88] FMT [17] | CB-FMT [84] | |

tablishes the link with Gabor and Wilson expansions for OFDM/OQAM, whereas in [23] the author illustrates the connection of oversampled OFDM with respect to Gabor frames.

OFDM/OQAM, and more precisely the acronym FBMC/OQAM, has also connections with another class of filterbanks. Indeed, soon after the duality between MDFT and OFDM/OQAM was stated, another family of filter banks was proposed to process complex signals using CMFB and Sine-modulated FilterBank (SMFB). This PR system, named Exponentially Modulated Filterbanks (EMFB) [7], has strong similarities with MDFT filterbanks. Its dual form gave rise to the FBMC transmultiplexer [8]. However, now the acronym FBMC is generally dedicated to designate an OFDM/OQAM, i.e., an even-stacked filterbank, rather than an EMFB transmultiplexer, i.e., an odd-stacked-one.

### 7.2.2.3 *Recent Trends*

Three Gabor-type MCM options have been extended in the recent years. Indeed, instead of generating the modulated signal by a linear convolution, as expressed by Eq. (7.2), transmission blocks are created permitting to replace the linear convolution processing by a circular one. In this way, the number of Gabor-like schemes is increased by a factor of two. These recent MCM schemes, named Generalized Frequency-Division Multiplexing (GFDM), Cyclic Block-Filtered MultiTone (CB-FMT), and Circular Offset Quadrature Amplitude Modulation (COQAM), are presented in Section 7.3 in addition to the classical linear convolution-based Gabor schemes.

Table 7.1 reports the evolution of the different categories of orthogonal[7] communication waveforms according to the density of their associated time-frequency lattices. The circular convolution-based schemes appear in italic characters. In the next section, we limit our presentation to the MCM schemes related to Gabor families, so we will not discuss the CMT scheme devoted to the transmission of PAM symbols nor its circular convolution-based implementation, which is connected to Discrete Time Wilson Expansion (DTWE) [86].

## 7.3 WAVEFORM ALTERNATIVES

OFDM and its alternatives all correspond to MCM schemes generated by a Gabor family of functions, i.e., they can be defined using (7.1)–(7.2). For OFDM, the unique possibility to get orthogonality, i.e., to satisfy condition (7.5), is to set $g(t) = \Pi_T(t)$ with $\Pi_T(t)$ the rectangular window of length $T$ and $TF = 1$. Then, at sampling rate $f_s$, setting $g[l] = \Pi_M[l]$ and $T = MT_s = \frac{M}{f_s}$, the discrete-time version of (7.2) leads for the OFDM symbol of index $k$, with $0 \leq l \leq M-1$, to

$$s_k[l] = \sum_{m=0}^{M-1} c_{m,k} e^{j\frac{2\pi ml}{M}}, \tag{7.12}$$

where the $c_{m,k}$ coefficients correspond to the QAM symbols to transmit. Note that, differently from the continuous-time case, all discrete windows such that $\Pi_M[l] = \pm 1$ can provide orthogonality. But, in any case, OFDM cannot escape from the third consequence of BLT.[8] On the contrary, the alternative waveforms to be presented now have all the advantage to be, at least, correctly localized in time and frequency.

To make the comparisons between different Gabor triplets $(T, F, g)$ easier, we use for OFDM the triplet $(T_0, F_0, \Pi)$ with

$$T = T_0 = MT_s = \frac{M}{f_s} = \frac{1}{F_0} = \frac{1}{F}. \tag{7.13}$$

We will also suppose that the prototype filters are of the Finite Impulse Response (FIR) type with length $L_g$ and real-valued coefficients

$$G(z) = \sum_{l=0}^{L_g-1} g[l]z^{-l}. \tag{7.14}$$

Since the considered MCM systems are orthogonal, the receiver is directly deduced by the matched filter principle. Thus, we will only focus on the transmitter side.

---

[7]With the exception of GFDM, which is nonorthogonal as discussed later on.
[8]Strictly speaking, the BLT only concerns the continuous-time Gabor functions. However, its practical implications also apply to the discrete-time MCM versions.

### 7.3.1 **FMT**

As previously said, what has been called FMT in [17] and popularized in relation with the very-high-bit-rate DSL (VDSL) application, was known before under different names: fraction-spaced MCM [88] and oversampled OFDM [23,44]. In the continuation, the FMT acronym applies to all these equivalent systems. All these MCM variants are such that $TF = \frac{N}{M}$ with $N > M$. For simplicity, we will omit here the impact of the prototype filter length,[9] so that we can write the discrete-time noncausal baseband signal as follows:

$$s[l] = \sum_k \sum_{m=0}^{M-1} c_{m,k} g[l - kN] e^{j\frac{2\pi ml}{M}}. \tag{7.15}$$

This expression also corresponds to the output of a uniform $M$-subband Synthesis FilterBank (SFB) with expansion factor $N$ [80].

Originally, the aim with FMT is to relax the maximum spectral efficiency constraint ($TF = 1$) by increasing the subcarrier spacing compared to OFDM and setting $F = \frac{N}{M} F_0$. Consequently, the bandwidth is increased, but the resulting ICI decreases for two reasons: (i) the spacing between subcarriers is increased, and (ii) there is some degree of freedom to design a frequency selective prototype filter. In this case, we can have the same symbol duration as for OFDM, but the sampling period $T_s^{(1)}$ needs to be different:

$$T = NT_s^{(1)} = MT_s = T_0 \Rightarrow T_s^{(1)} < T_s. \tag{7.16}$$

Using the RRC filter (see Section 5.4) with roll-off $\alpha$, we constitute an appropriate and usual solution, setting $f_s = (1 + \alpha)F_0$, to get an FMT system with oversampling $\rho = \alpha + 1$.

On the other hand, using the same formalism, we can target another objective, keeping the same frequency spacing ($F = F_0$) and the same sampling period ($T_s$) by setting $T(= NT_s) > T_0(= MT_s)$. This could be advantageous when considering the compatibility of FMT with Cyclic Prefix Orthogonal Frequency-Division Multiplexing (CP-OFDM) systems. This approach has led to systems known as Weighted Cyclic Prefix OFDM (WCP-OFDM) [72] and Pulse-shaped OFDM (P-OFDM) [95].

To check the orthogonality of an FMT system, we can either use the discrete-time version of the ambiguity function in (7.4) or take advantage of some results provided in the filterbank literature. For example, as reported in [24], the PR condition for the prototype filter may be written as

$$\sum_v g[l + vM] \, g[l + vM + sN] = \delta_s, \quad 0 \le l \le M - 1, \ s \ge 0, \tag{7.17}$$

---

[9]A causal description for the orthogonal and biorthogonal cases is reported in [80].

where $\delta_s = 1$ if $s = 0$ and 0 if $s \neq 0$. Pinchon and Siohan [65] provide a simple analytical expression, which satisfies this orthogonality condition for the prototype filter length of $L_g = N$:

$$g[l] = \begin{cases} \sin\frac{(2l+1)\pi}{4\Delta}, & 0 \leq l \leq \Delta - 1, \\ 1, & \Delta \leq l \leq M - 1, \\ \sin\frac{(2(\Delta-l)-1)\pi}{4\Delta}, & M \leq l \leq N - 1, \end{cases} \tag{7.18}$$

with $\Delta = N - M$. For this short prototype filter, we naturally have a significant overlapping at the transition bands. This is not the case when using an appropriate RRC prototype filter. However then, due to the FIR length constraint, we can only get a Near Perfect Reconstruction (NPR) system.

### 7.3.2 FBMC/OQAM

Differently from OFDM and FMT, the FBMC/OQAM scheme does not directly transmit the $c_{m,k}$ QAM symbols over $M$ subcarriers. In a first step, these complex symbols are decomposed into their real and imaginary parts using a transformation now often called C2R [83]. Starting from $T$-spaced $c_{m,k}$ samples, where $m$ and $k$ denote the frequency and time index, respectively, the real and imaginary parts are interleaved with a relative time offset of $T/2$ (hence offset QAM) producing real-valued symbols:

$$d_{m,2k} = \begin{cases} \mathbb{Re}\{c_{m,k}\}, & m \text{ even}, \\ \mathbb{Im}\{c_{m,k}\}, & m \text{ odd}, \end{cases} \tag{7.19}$$

$$d_{m,2k+1} = \begin{cases} \mathbb{Im}\{c_{m,k}\}, & m \text{ even}, \\ \mathbb{Re}\{c_{m,k}\}, & m \text{ odd}. \end{cases} \tag{7.20}$$

According to different authors, OFDM/OQAM has been presented either using a mathematical formulation in which real and imaginary components always appear explicitly, as in [14], or, as proposed in [48], is seen as an MCM scheme transmitting real data symbols associated with a phase term $\varphi_{m,n}$. Then, the FBMC/OQAM signal is expressed as

$$s(t) = \sum_n \sum_{m=0}^{M-1} d_{m,n} \underbrace{g\left(t - n\frac{T}{2}\right) e^{j2\pi mFt} e^{j\varphi_{m,n}}}_{g_{m,n}(t)}. \tag{7.21}$$

The phase term cannot be chosen arbitrarily. Indeed, if we introduce a phase rotation $\theta_{m,n}$ of the basis functions, then the orthogonality conditions become

$$\mathbb{Re}\left\{e^{j(\theta_{m',n'}-\theta_{m,n})}\langle g_{m,n}, g_{m',n'}\rangle\right\} = \delta_{m,m'}\delta_{n,n'}. \tag{7.22}$$

Actually, a quadrature rule has to be imposed in time and frequency such that [78]

$$\varphi_{m,n} = \begin{cases} 0 \text{ or } \pi & \text{if } m \text{ and } n \text{ have the same parity,} \\ \pm\frac{\pi}{2} & \text{if } m \text{ and } n \text{ have different parity.} \end{cases} \quad (7.23)$$

This condition is satisfied, as in [48], by setting $\varphi_{m,n} = \frac{\pi}{2}(m+n)$.

When going to a discrete-time realization, to keep the advantage of maximum spectral efficiency, together with the real orthogonality property, the sampling period has to be such that $T_s = \frac{T}{M}$. A precise and complete derivation of the resulting FBMC/OQAM system, based on the filterbank theory, is reported in [82]:

$$s[l] = \sum_{n}\sum_{m=0}^{M-1} d_{m,n} g[l - nM/2] e^{j\frac{2\pi m}{M}(l-\frac{L_g-1}{2})} e^{j\varphi_{m,n}}, \quad (7.24)$$

where, in order to keep a close link with MDFT filterbanks, the phase term is such that $\varphi_{m,n} = (m+n)\frac{\pi}{2} - mn\pi$.

Again the orthogonality condition can be checked using the ambiguity function [4,6]. We can also use the link with filterbanks, as it has been shown in [82] that the PR conditions are the same for the CMFB and MDFT filterbanks and for the FBMC/OQAM transmultiplexer. Therefore, the PR condition [58,67] for a linear-phase CMFB prototype filter with length $L_g = KM$ reads for FBMC/OQAM as follows:

$$\sum_{v} g\left[l + v\frac{M}{2}\right] g\left[l + v\frac{M}{2} + sM\right] = \delta_s, \quad 0 \le l \le M - 1. \quad (7.25)$$

For short prototype filters, simple analytical solutions have been found by Malvar for what is called a Modulated Lapped Transform (MLT). For FBMC/OQAM, the corresponding solution for $L_g = M$ writes as [58]

$$g[l] = \sin\left(\frac{\pi}{M}\left(l + \frac{1}{2}\right)\right), \quad (7.26)$$

whereas for $L_g = 2M$, the orthogonality is insured with the following impulse response [59]:

$$g[l] = \frac{1}{2\sqrt{2}} - \frac{1}{2}\cos\left(\frac{\pi}{M}\left(l + \frac{1}{2}\right)\right). \quad (7.27)$$

As shown in [82], using the link with the PR CMFB presented in [63], the orthogonality condition (7.25) can be extended to cover the case of arbitrary length linear-phase real-valued prototype filters.[10]

---

[10]In Siclet's thesis [78], the orthogonality and biorthogonality conditions are derived for FBMC/OQAM systems having prototype filters of arbitrary length, symmetrical or not, with complex-valued coefficients.

Using the $M$-path polyphase decomposition of $G(z)$, we get

$$G(z) = \sum_{i=0}^{M-1} z^{-i} E_i(z^M).\qquad(7.28)$$

Then, the PR conditions with respect to the polyphase components $E_i(z)$ for $0 \leq i \leq \frac{M}{2}$ are expressed as

$$E_i(z)E_i(z^{-1}) + E_{i+\frac{M}{2}}(z)E_{i+\frac{M}{2}}(z^{-1}) = \frac{1}{M}.\qquad(7.29)$$

## 7.3.3 CIRCULAR (OR CYCLIC) CONVOLUTION (CC)-BASED ALTERNATIVES

Differently from OFDM, which is a block processing MCM, the Gabor-based alternatives we have examined correspond to overlapped transforms.[11] This means that the successive MCM symbols are no longer independent, which makes more complicated the introduction of a CP and the application of Space Time Block Code (STBC) in the case of multiantenna systems. A simple manner to come back to block processing with Gabor-based MCM systems is to replace the linear convolution by a circular one. Even though the principle of CC is inherent to the basics of digital signal processing (see, e.g., [69]) and is also applied to get CP-OFDM, its application to nonrectangular waveforms is relatively recent.

The difference with CP-OFDM is that, for CC-based systems, the CP can be shared by $P$ consecutive OFDM subsymbols, which limits the loss in spectral efficiency. When applied to OFDM, FMT and FBMC/OQAM, CC produces new schemes named GFDM, CB-FMT and FBMC/COQAM, respectively. All being block transforms, the CP insertion works similarly as for CP-OFDM, but considering a $PM$-length macro-block. Therefore, we will not present the CP appending operation in here. But, before entering into individual short presentations of these CC-based MCM schemes, let us present in Table 7.2 the difference of nature between the signals we have to deal with in the continuous- and discrete-time cases, when using linear convolution, and in the periodic, in time and frequency, discrete case when using circular convolution.

Finally, we notice that for CC-based schemes, the filtering can be operated in the frequency domain, which gives new possibilities both for the implementation and for the prototype filter design (see Chapter 8).

---

[11] With the exception of FMT when equipped with a short-length prototype filter ($L_g = N$).

**Table 7.2** Typical expressions of signals according to the formalism used

| | Continuous time | Discrete time | Discrete time (L-periodicity) |
|---|---|---|---|
| Time | $x(t)$ | $x[n]$ | $x_L[n]$ |
| Frequency | $X(f)$ | $X(e^{j\omega})$ | $X_L[m]$ |

### 7.3.3.1 *GFDM Scheme*

The GFDM scheme has been introduced in [34]. Its baseband expression, for a block of length $PM$, clearly shows that it is a member of the Gabor family of MCM:

$$s[l] = \sum_{k=0}^{P-1}\sum_{m=0}^{M-1} c_{m,k}\tilde{g}[l - kM]e^{j\frac{2\pi ml}{M}} \qquad (7.30)$$

for $0 \leq l \leq PM - 1$. The symbol $\tilde{\ }$ means that a periodic repetition of the prototype filter is applied, i.e.,

$$\tilde{g}[l] = g[\mathrm{mod}(l, PM)]. \qquad (7.31)$$

So for GFDM, at the TX side, we operate a CC, the cyclic processing referred to as tail-biting in [34]. It is shown in [61] that GFDM can be implemented by $M$ subcarrier-wise FFTs of size $P$, followed by a filtering operation in the frequency domain, and finally a length-$PM$ Inverse Fast Fourier Transform (IFFT) providing the baseband signal.

Being a direct extension of OFDM, GFDM can only be orthogonal if $g[l] = \Pi_M[l]$. This is clearly not the aim, and nonrectangular waveforms are used leading to better spectrum but also to nonorthogonality. The resulting self-interference needs to be compensated, and this can be done, with the cost of some additional computational complexity, using a Successive Interference Cancellation (SIC) method [27].

### 7.3.3.2 *CB-FMT Scheme*

CB-FMT has been invented [84] nearly at the same period of time as GFDM. However, the publications only came out a few years later [40,85]. FMT, due to its oversampling, can provide orthogonality with well-localized waveforms, so the gain expected from a CC-based implementation is here different. It is related to the fact that, as for GFDM, a block processing allows insertion of a CP. This CP can make CB-FMT resilient with longer delay spread channels. The baseband transmitted signal for a CB-FMT transmission block writes as

$$s[l] = \sum_{k=0}^{P-1}\sum_{m=0}^{M-1} c_{m,k}\tilde{g}[l - nN]e^{j\frac{2\pi ml}{M}} \qquad (7.32)$$

with

$$\tilde{g}[l] = g[\mathrm{mod}(l, PN)]. \qquad (7.33)$$

Setting $N = M$, we recover the GFDM scheme, and the orthogonality property is lost. Note also that, as shown in [40] and in a more detailed manner in [66], when $N > M$, the PR conditions are different from the FMT ones. Nevertheless, if one only targets nearly orthogonal systems, as shown in [56], then this objective can be reached using a long RRC prototype filter.

### 7.3.3.3 *FBMC/COQAM Scheme*

FBMC/OQAM provides orthogonality in the real domain together with the possibility of selecting prototype filters with good localization in time and frequency. However, as recalled before, the absence of a CP makes it, like FMT, sensitive to long delay spread channels, requiring somewhat complex equalizers (see Chapter 12) and specific STBC procedures [70]. Furthermore, being an overlapped MCM system, it may require specific processing to avoid interference at the frame borders in the case of packet transmission [12,26]. So, for the FBMC/OQAM scheme, CC could also help to solve these problems, in particular, with a simple introduction of a CP. Indeed, it had been shown earlier that a CP could be inserted in a classical FBMC/OQAM scheme solving, at the price of a loss in spectral efficiency, the robustness problem with respect to multipath fading [55], and the problems with Alamouti coding [54]. However, the CC processing introduced in [38] was a different way to do this.

The baseband FBMC/COQAM signal in a block interval can be expressed in a noncausal form as follows [56]:

$$s[l] = \sum_{n=0}^{P'-1} \sum_{m=0}^{M-1} d_{m,n} \tilde{g}\left[l - n\frac{M}{2}\right] e^{j\frac{2\pi ml}{M}} e^{j\varphi_{m,n}} \qquad (7.34)$$

in which $\tilde{g}[l] = g[\mod(l, PM)]$. As the real OQAM symbols $d_{m,n}$ have a half duration compared to the QAM symbols $c_{m,n}$, we set $P' = 2P$ to keep the same block duration as for GFDM.

Note that for the FBMC/COQAM schemes proposed in [38] and [56], the authors add afterwards a CP. This is not the case for the FBMC/COQAM scheme proposed in [1], the goal of which is only solving the border effect.

A weakness of the FBMC/COQAM schemes proposed in [38] and [56], which was also present in [54,55], is the discontinuity between consecutive blocks, which has a negative impact on the Power Spectral Density (PSD). In [56], this problem is solved by the introduction of a windowing at the border of the blocks, leading to a scheme named Windowed Cyclic Prefix (WCP)-OQAM, whereas in [16] different FBMC/COQAM variants are compared that do not require windowing.

The PSD issue, which becomes critical for small values of $P$, also concerns GFDM and CB-FMT. So, Fettweis' team has recently proposed a GFDM variant also including different types of windows [60].

### 7.3.4 FILTERED OFDM

FBMC/OQAM and FMT schemes provide good spectrum localization for each subcarrier, but this is often not required since resource allocation and adaptive coding and modulation schemes are commonly applied with a group of subcarriers, or Physical Resource Block (PRB), as the basic unit. Recently, various schemes utilizing filtering at resource block level have been proposed for future wireless communications. These include Universal-Filtered OFDM (UF-OFDM) (also known as Universal Filtered MultiCarrier (UFMC)) [77], Resource block Filtered-OFDM (RB-F-OFDM) [51], and f-OFDM [2]. In these studies, window-based time-domain filtering has been used for UF-OFDM and f-OFDM, whereas uniform polyphase filter bank structures are considered for RB-F-OFDM. Recently, an effective Fast-Convolution (FC) based implementation scheme for Filtered OFDM (F-OFDM) has also been proposed [71].

This class of MultiCarrier (MC) systems can generally be called F-OFDM. They maintain high level of commonality with legacy OFDM systems, including simple and robust channel equalization, and allow direct application of multiantenna techniques developed for OFDM. However, sharp filtering to suppress the sidelobes of an OFDM resource block destroys the orthogonality, especially for subcarriers that are close to the edge of the PRB [32]. This introduces inband interference to the OFDM signal. Consequently, the design of an F-OFDM waveform involves a trade-off between guard bandwidth between filtered PRBs, inband interference level, and implementation complexity. It was demonstrated in [71] that using a guardband of 3–5 subcarrier spacings enables to reach about $-30$ dB Mean-Squared Error (MSE) (or Error Vector Magnitude (EVM)) level in the edge subcarriers.

### 7.3.5 FILTERBANK-BASED SINGLE-CARRIER WAVEFORMS

Apart from the poor frequency localization, the main limitation of OFDM is the strong envelope variation of the modulated waveform, i.e., high PAPR. For this reason, a DFT-precoded OFDM scheme, called SC-FDMA, has been adopted for the Long-Term Evolution (LTE) UL [25]. The precoding greatly reduces the envelope variations leading to PAPR characteristics similar to SC-waveforms with very small roll-off. Although SC-FDMA has high level of commonality with OFDM and supports quasi-synchronous multiple access with zero excess bandwidth (without guardbands), it suffers from limited frequency localization in the same way as OFDM.

For improved frequency localization, a flexible FilterBank Single-Carrier (FB-SC) transmission scheme, with similar structure as SC-FDMA, was proposed in [45]. In [39], GFDM schemes with reduced PAPR and SC-like characteristics were investigated. The paper [92] presented an effective and flexible FC-based implementation of traditional SC waveforms with adjustable roll-off starting from very small values.

### 7.3.6 SOME OTHER RECENT WAVEFORM DEVELOPMENTS

All previously presented modulation schemes have attracted a great attention for future radio links, mostly in the perspective of the 5th Generation (5G) multiservice communication system. Although an exhaustive presentation of all the modulation candidates is beyond the scope of this book, we would like to briefly comment on some other recent waveform proposals, which are also of great interest for 5G.

A limitation of FBMC/OQAM is the fact that orthogonality only holds in the real field, which leads to interferences when transmitting through realistic complex channels. These interferences may be difficult to compensate and necessitate signal processing not compatible with existing LTE technology, for instance, for MIMO and channel estimation. Nevertheless, the FBMC/OQAM scheme can be amended while keeping some of its strong advantages, good frequency localization, and maximum spectral efficiency.

To avoid the difficulties due to the OQAM signal model, Quadrature Amplitude Modulation-FilterBank MultiCarrier (QAM-FBMC) proposals have appeared recently [93,94]. The latter reference makes use of multiple FilterBank (FB) bases and maintains the maximum spectral efficiency while providing a good PSD thanks to the higher degree of freedom offered in the design of the multiple (at least 2) prototype filters. However, these nice features come at the price of a loss of the orthogonality property.

Another MCM scheme has been proposed in [53] to bypass the lack of compatibility between the signal processing blocks of the FBMC/OQAM and LTE-OFDM. In this scheme, named Flexible Configured OFDM (FC-OFDM), the transceiver can commute between two main configurations. One corresponds to conventional OFDM, whereas the other one corresponds to a specific implementation of a particular instance of FBMC/OQAM, in which filtering is carried out in the frequency domain. In this approach, the transceiver selects the most appropriate configuration according to the type of service to deliver, e.g., OFDM configuration for Mobile BroadBand (MBB) and FBMC/OQAM for Machine Type Communication (MTC).

We finally note that a solution involving only very simple modifications with respect to the LTE-OFDM scheme has also been proposed under the name of Weighted OverLap and Add (WOLA) [68]. WOLA consists of the application of time windowing at the TX and/or Receiver (RX) sides. The principle is therefore similar to what has been standardized within IEEE P1901 for PLC. Using CP-OFDM plus WOLA for LTE in DownLink (DL) and SC-FDM plus WOLA for LTE in UL can indeed simply solve the upgrade problem from 4th Generation (4G) to 5G. However, as shown, for instance, in [3], the resulting spectral efficiency may abruptly decrease for all services using fragmented spectrum. Generally, effective spectrum control with WOLA increases the guard-period length between OFDM symbols, thus reducing the spectrum efficiency.

## 7.4 CONCLUDING REMARKS

In this chapter, starting from the basics of Nyquist transmission principles, we have presented an up-to-date overview of the most prominent SC and MC modulation schemes. We have seen that, in addition to OFDM and SC-FDMA, there are numerous alternatives waveforms, each with its attractive features. However, as illustrated in some other chapters of our book, even if the intrinsic differences at the modulation level (CP or no CP, real or complex orthogonality, linear or circular convolution) have significant impacts, the alternative modulation schemes need to be evaluated at a higher level taking into account the precise features of the application at hand.

# REFERENCES

[1] J. Abdoli, M. Jia, and J. Ma, "Weighted circularly convolved filtering in OFDM/OQAM," in *Proc. IEEE International Symposium on Personal, Indoor, and Mobile Radio Communications (PIMRC)*, London, UK, Sep. 2013, pp. 657–661.

[2] J. Abdoli, M. Jia, and J. Ma, "Filtered OFDM: A new waveform for future wireless systems," in *Proc. IEEE International Workshop on Signal Processing Advances in Wireless Communications (SPAWC)*, Stockholm, Sweden, Jun. 2015, pp. 66–70.

[3] P. Achaichia, M. Le Bot, and P. Siohan, "Windowed OFDM versus OFDM/OQAM: A transmission capacity comparison in the HomePlug AV context," in *Proc. IEEE International Symposium on Power Line Communications and Its Applications (ISPLC)*, Apr. 2011, pp. 405–410.

[4] M. Alard, "Construction of a multicarrier signal." Patent WO 96/35278, 1996.

[5] M. Alard and R. Lassalle, "Principles of modulation and channel coding for digital broadcasting for mobile receivers," *EBU Rev.-Tech.*, vol. 224, pp. 168–190, Aug. 1987.

[6] M. Alard, C. Roche, and P. Siohan, "A new family of functions with a nearly optimal time-frequency localization." Technical report of the RNRT Modyr project, 1999.

[7] J. Alhava, A. Viholainen, and M. Renfors, "Efficient implementation of complex exponentially-modulated filter banks," in *Proc. IEEE International Symposium on Circuits and Systems (ISCAS)*, Bangkok, Thailand, 2003, pp. 157–160. http://dx.doi.org/10.1109/ISCAS.2003.1205797 [Online].

[8] J. Alhava and M. Renfors, "Exponentially-modulated filter bank-based transmultiplexer," in *Proc. IEEE International Symposium on Circuits and Systems (ISCAS)*, Bangkok, Thailand, 2003, pp. 233–236. http://dx.doi.org/10.1109/ISCAS.2003.1205816 [Online].

[9] P. Amini and B. Farhang-Boroujeny, "Design and performance evaluation of filtered multitone (FMT) in doubly dispersive channels," in *Proc. IEEE International Conference on Communications (ICC)*, Kyoto, Japan, Jun. 2011, pp. 1–5.

[10] R. Balian, "Un principe d'incertitude fort en théorie du signal ou en mécanique quantique," in *Comptes-Rendus de l'Académie des Sciences*, Paris, France, Jun. 1981, pp. 1357–1362.

[11] M. Bellanger and J. L. Daguet, "TDM-FDM transmultiplexer: Digital polyphase and FFT," *IEEE Trans. Commun.*, vol. 22, Sep. 1974.

[12] M. Bellanger, M. Renfors, T. Ihalainen, and C. A. F. da Rocha, "OFDM and FBMC transmission techniques: A compatible high performance proposal for broadband power line communications," in *Proc. 2010 IEEE International Symposium on Power Line Communications and Its Applications (ISPLC)*, Rio de Janeiro, Brazil, Mar. 2010, pp. 154–159.

[13] H. Boelcskei, "Efficient design of pulse shaping filters for OFDM systems," in *Proc. SPIE Conference*, Denver, USA, Jul. 1999, pp. 625–636.

[14] H. Boelcskei, "Orthogonal frequency division multiplexing based on offset QAM," in *Advances in Gabor Theory*. Birkhäuser, 2003.

[15] R. W. Chang, "Synthesis of band-limited orthogonal signals for multi-channel data transmission," *Bell Syst. Tech. J.*, vol. 45, pp. 1775–1796, Dec. 1966.

[16] D. Chen, X. G. Xia, T. Jiang, and X. Gao, "Properties and power spectral densities of CP based OQAM-OFDM systems," *IEEE Trans. Signal Process.*, vol. 63, pp. 3561–3575, Jul. 2015.

[17] G. Cherubini, E. Eleftheriou, and S. Ölçer, "Filtered multitone modulation for VDSL," in *Proc. IEEE Global Telecommunication Conference (GLOBECOM)*, Rio de Janeiro, Brazil, 1999, pp. 1139–1144.

[18] L. J. Cimini, "Analysis and simulations of a digital mobile channel using orthogonal frequency division multiplexing," *IEEE Trans. Commun.*, vol. 33, pp. 665–675, Jul. 1985.

[19] J. M. Cioffi, *Electronic book from J. Cioffi's web page*. [Online]. Available: https://web.stanford.edu/group/cioffi/doc/book/chap4.pdf.

[20] J. M. Cioffi, G. P. Dudevoir, M. V. Eyuboglu, and G. D. Forney, "MMSE decision-feedback equalizers and coding. I. Equalization results," *IEEE Trans. Commun.*, vol. 43, pp. 2582–2594, Oct. 1995.

[21] J. M. Cioffi, G. P. Dudevoir, M. V. Eyuboglu, and G. D. Forney, "MMSE decision-feedback equalizers and coding. II. Coding result," *IEEE Trans. Commun.*, vol. 43, pp. 2595–2604, Oct. 1995.

[22] J. W. Cooley and J. W. Tukey, "An algorithm for the machine calculation of complex Fourier series," *Math. Comput.*, vol. 19, no. 90, pp. 297–301, Apr. 1965.

[23] Z. Cvetković, "Oversampled modulated filter banks and tight Gabor frames in $\mathcal{L}^2(\mathbb{Z})$," in *Proc. IEEE International Conference on Acoustics, Speech, and Signal Processing (ICASSP)*, vol. 2, Detroit, USA, May 1995, pp. 1456–1459.

[24] Z. Cvetković, "Modulating waveforms for OFDM," in *Proc. IEEE International Conference on Acoustics, Speech, and Signal Processing (ICASSP)*, vol. II, Phoenix, USA, 1999, pp. 2463–2466.

[25] E. Dahlman, S. Parkvall, and J. Sköld, *4G LTE/LTE-Advanced for Mobile Broadband*. Academic Press, 2011.

[26] Y. Dandach and P. Siohan, "Packet transmission for overlapped offset QAM," in *Proc. International Conference on Wireless Communications and Signal Processing (IC-WCSP)*, Suzhou, China, May 2010.

[27] R. Datta, N. Michailow, M. Lentmaier, and G. Fettweis, "GFDM interference cancellation for flexible cognitive radio PHY design," in *Proc. IEEE Vehicular Technology Conference, VTC-Fall*, Quebec city, Canada, Sep. 2012.

[28] I. Daubechies, "The wavelet transform, time-frequency localization and signal analysis," *IEEE Trans. Inf. Theory*, vol. 36, no. 5, pp. 961–1005, 1990.

[29] M. L. Doelz, E. T. Heald, and D. L. Martin, "Binary data transmission techniques for linear systems," in *Proceedings of IRE*, vol. 45, May 1957, pp. 656–661.

[30] C. Dosch, P. Ratliff, and D. Pommier, "First public demonstrations of COFDM/MAS-CAM: A milestone for the future of radio broadcasting," *EBU Rev.-Tech.*, vol. 232, pp. 275–283, 1988.

[31] B. Farhang-Boroujeny and C. Yuen, "Cosine modulated and offset QAM filter bank multicarrier techniques: A continuous-time prospect," *EURASIP J. Adv. Signal Process.*, vol. 2010, no. 8, pp. 1–16, 2008.

[32] M. Faulkner, "The effect of filtering on the performance of OFDM systems," *IEEE Trans. Veh. Technol.*, vol. 49, pp. 1877–1884, Sep. 2000.

[33] H. Feichtinger, *et al. Gabor Analysis and Algorithm – Theory and Applications*. Boston–Basel–Berlin: Birkhäuser, 1998.

[34] G. Fettweis, M. Krondorf, and S. Bittner, "GFDM – Generalized frequency division multiplexing," in *Proc. IEEE Vehicular Technology Conference, VTC-Spring*, Barcelona, Spain, Apr. 2009, pp. 26–29.

[35] N. J. Fliege, "Orthogonal multiple carrier data transmission," *Eur. Trans. Telecommun.*, vol. 3, no. 3, pp. 255–264, May–Jun. 1992.

[36] N. J. Fliege, "Computational efficiency of modified DFT polyphase filter banks," in *Proc. Annual Asilomar Conference on Signals, Systems and Computers (ASILOMAR)*, vol. 2, Nov. 1993, pp. 1296–1300.

[37] D. Gabor, "Theory of communications," *J. IEE*, vol. 93, pp. 429–457, 1946.

[38] X. Gao, W. Wang, X. G. Xia, E. K. S. Au, and X. You, "Cyclic prefixed OQAM-OFDM and its application to single-carrier FDMA," *IEEE Trans. Commun.*, vol. 59, pp. 1467–1480, May 2011.

[39] I. Gaspar, N. Michailow, A. Navarro, E. Ohlmer, S. Krone, and G. Fettweis, "Low complexity GFDM receiver based on sparse frequency domain processing," in *Proc. IEEE Vehicular Technology Conference, VTC-Spring*, Dresden, Germany, Jun. 2013.

[40] M. Girotto and A. M. Tonello, "Orthogonal design of cyclic block filtered multitone modulation," in *Proc. European Wireless Conference (EW)*, Barcelona, Spain, May 2014, pp. 1–6.

[41] R. Haas and J.-C. Belfiore, "Multiple carrier transmission with time-frequency well-localized impulses," in *Proc. 2nd IEEE Symposium on Communications and Vehicular Technology in the Benelux*, Louvain-la-Neuve, Belgium, 1994, pp. 187–193.

[42] R. Haas and J.-C. Belfiore, "A time-frequency well-localized pulse for multiple carrier transmission," *Wirel. Pers. Commun.*, vol. 5, pp. 1–18, 1997.

[43] B. Hirosaki, "An orthogonally multiplexed QAM system using the discrete Fourier transform," *IEEE Trans. Commun.*, vol. 29, pp. 982–989, Jul. 1981.

[44] R. Hleiss, P. Duhamel, and M. Charbit, "Oversampled OFDM systems," in *Proc. International Conference on Digital Signal Processing*, Santorini, Greece, Jul. 1997.

[45] T. Ihalainen, A. Viholainen, T. H. Stitz, M. Renfors, and M. Bellanger, "Filter bank based multi-mode multiple access scheme for wireless uplink," in *Proc. EURASIP European Signal Processing Conference (EUSIPCO)*, Glasgow, Scotland, Aug. 2009, pp. 1354–1358.

[46] T. Karp and N. J. Fliege, "MDFT filter banks with perfect reconstruction," in *Proc. IEEE International Symposium on Circuits and Systems (ISCAS)*, Seattle, USA, May 1995, pp. 744–747.

[47] W. E. J. Keasler, "Reliable data communications over the voice bandwidth telephone channel using orthogonal frequency division multiplexing." Ph.D. dissertation. Urbana-Champaign: University of Illinois, 1982.

[48] B. Le Floch, M. Alard, and C. Berrou, "Coded orthogonal frequency division multiplex," *Proc. IEEE*, vol. 83, no. 6, pp. 982–996, Jun. 1995.

[49] B. Le Floch, J.-C. Rault, P. Siohan, R. Legouable, and C. Gallard, "The birth of digital terrestrial broadcasting in Europe: A brief history of the creation and the standardization phases of digital audio broadcasting (DAB) and digital terrestrial TV broadcasting (DVB-T)," in *Proc. IEEE Region 8 Conference on the History of Telecommunications Conference (HISTELCON)*, Madrid, Spain, 2010, pp. 1–6.

[50] S. Lee, W. Kang, and J. Seo, "Performance enhancement of OFDM-$SQ^2AM$ in distorted channel environments," *IEICE Electron. Express*, vol. 7, no. 14, pp. 1020–1026, 2010.

[51] J. Li, E. Bala, and R. Yang, "Resource block filtered-OFDM for future spectrally agile and power efficient systems," *Phys. Commun.*, vol. 14, pp. 36–55, Jun. 2014.

[52] R. Li and G. Stette, "Time-limited orthogonal multicarrier modulation schemes," *IEEE Trans. Commun.*, vol. 43, pp. 1269–1272, Feb./Mar./Apr. 1995.

[53] H. Lin, "Flexible configured OFDM for 5G air interface," *IEEE Access*, vol. 3, pp. 1861–1870, Oct. 2015.

[54] H. Lin, C. Lélé, and P. Siohan, "A pseudo Alamouti transceiver design for OFDM/OQAM modulation with cyclic prefix," in *Proc. IEEE International Workshop on Signal Processing Advances in Wireless Communications (SPAWC)*, Perugia, Italy, Jun. 2009.

[55] H. Lin and P. Siohan, "A new transceiver system for the OFDM/OQAM modulation with cyclic prefix," in *Proc. IEEE International Symposium on Personal, Indoor, and Mobile Radio Communications (PIMRC)*, Cannes, France, Sep. 2008, pp. 1–5.

[56] H. Lin and P. Siohan, "Multi-carrier modulation analysis and WCP-COQAM proposal," *EURASIP J. Adv. Signal Process.*, vol. 2014, no. 1, pp. 1–19, 2014. http://dx.doi.org/10.1186/1687-6180-2014-79 [Online].

[57] F. Low, "Complete sets of wave packets," in *A Passion for Physics – Essay in Honor of Geoffrey Chew*, C. DeTar, Ed. Singapore: World Scientific, Jun. 1985, pp. 17–22.

[58] H. S. Malvar, "Modulated QMF filter banks with perfect reconstruction," *Electron. Lett.*, vol. 26, no. 13, pp. 906–907, 1990.

[59] H. S. Malvar, "Fast algorithm for modulated lapped transforms," *Electron. Lett.*, vol. 27, no. 9, pp. 775–776, 1991.

[60] N. Michailow, M. Matthé, I. S. Gaspar, A. N. Caldevilla, L. L. Mendes, A. Festag, and G. Fettweis, "Generalized frequency division multiplexing for 5th generation cellular networks," *IEEE Trans. Commun.*, vol. 62, pp. 3045–3061, Sep. 2014.

[61] N. Michailow, I. Gaspar, S. Krone, M. Lentmaier, and G. Fettweis, "Generalized frequency division multiplexing: Analysis of an alternative multi-carrier technique for next generation cellular systems," in *Proc. IEEE International Symposium on Wireless Communications Systems (ISWCS)*, Oct. 2012, pp. 171–175.

[62] R. R. Mosier, "A data transmission system using pulse phase modulation," in *IRE Convention Record of First National Convention on Military Electronics*, Washington, USA, Jun. 1957, pp. 233–238.

[63] T. Q. Nguyen and R. D. Koilpillai, "The theory and design of arbitrary-length cosine-modulated filter banks and wavelets, satisfying perfect reconstruction," *IEEE Trans. Signal Process.*, vol. 44, pp. 473–483, Mar. 1996.

[64] A. Peled and A. Ruiz, "Frequency domain data transmission using reduced computational complexity algorithms," in *Proc. IEEE International Conference on Acoustics, Speech, and Signal Processing (ICASSP)*, Denver, USA, vol. 3, Apr. 1980, pp. 964–967.

[65] D. Pinchon and P. Siohan, "Closed form expression of optimal short PR FMT prototype filters," in *Proc. IEEE Global Telecommunication Conference (GLOBECOM)*, Houston, USA, Dec. 2011.

[66] D. Pinchon and P. Siohan, "A general analysis of cyclic block transmultiplexers with cyclic convolution," in *Proc. IEEE International Workshop on Signal Processing Advances in Wireless Communications (SPAWC)*, Stockholm, Sweden, Jun. 2015, pp. 106–110.

[67] J. Princen and A. B. Bradley, "Analysis/synthesis filter bank based on time domain aliasing cancellation," *IEEE Trans. Acoust., Speech, Signal Process.*, vol. 34, Oct. 1986.

[68] Qualcomm Technologies, Inc., *5G Waveform and Multiple Access Techniques*, [Online]. Available: https://www.qualcomm.com/documents/5g-research-waveform-and-multiple-access-techniques.

[69] L. R. Rabiner and B. Gold, *Theory and Application of Digital Signal Processing.* Englewood Cliffs, NJ: Prentice-Hall, 1975.

[70] M. Renfors, T. Ihalainen, and T. Stitz, "A block-Alamouti scheme for filter bank based multicarrier transmission," in *Proc. European Wireless Conference (EW)*, Lucca, Italy, Apr. 2010.

[71] M. Renfors, J. Yli-Kaakinen, T. Levanen, M. Valkama, T. Ihalainen, and J. Vihriälä, "Efficient fast-convolution implementation of filtered CP-OFDM waveform processing for 5G," in *Proc. IEEE Global Telecommunication Conference (GLOBECOM)*, San Diego, CA, USA, Dec. 2015.

[72] D. Roque, C. Siclet, J. M. Brossier, and P. Siohan, "Weighted cyclic prefix OFDM: PAPR analysis and performances comparison with DFT-precoding," in *Proc. Annual Asilomar Conference on Signals, Systems and Computers (ASILOMAR)*, Nov. 2012, pp. 1065–1068.

[73] M. Sablatash, "Designs and architectures of filter bank trees for spectrally efficient multiuser communications: Review, modifications and extensions of wavelet packet filter bank trees," *Signal Image Video Process.*, vol. 2, no. 1, pp. 9–37, 2008. http://dx.doi.org/10.1007/s11760-007-0033-4 [Online].

[74] B. R. Saltzberg, "Performance of an efficient parallel data transmission system," *IEEE Trans. Commun. Technol.*, vol. 15, pp. 805–811, Dec. 1967.

[75] H. Sari, Y. Levy, and G. Karam, "Orthogonal frequency-division multiple access for the return channel on CATV networks," in *Proc. IEEE International Conference on Telecommunications (ICT)*, Istanbul, Turkey, Apr. 1996, pp. 52–57.

[76] R. Schafer, "A survey of digital speech processing techniques," *IEEE Trans. Audio Electroacoust.*, vol. 20, pp. 28–35, Mar. 1972.

[77] F. Schaich, T. Wild, and Y. Chen, "Waveform contenders for 5G – Suitability for short packet and low latency transmissions," in *Proc. IEEE Vehicular Technology Conference, VTC-Spring*, May 2014, pp. 1–5.

[78] C. Siclet, "Application de la théorie des bancs de filtres à l'analyse et à la conception de modulations multiporteuses orthogonales et biorthogonales." Ph.D. dissertation. France: Université de Rennes 1, 2002.

[79] C. Siclet and P. Siohan, "Weyl–Heisenberg signal expansions over $\mathbb{R}$ in $\mathcal{L}^2(\mathbb{Z})$ and duality relations involving MDFT filter banks," in *Proc. IEEE International Conference on Acoustics, Speech, and Signal Processing (ICASSP)*, Hong Kong, China, Apr. 2003.

[80] C. Siclet, P. Siohan, and D. Pinchon, "Analysis and design of OFDM/QAM and BFDM/QAM oversampled orthogonal and biorthogonal multicarrier modulations," in *Proc. IEEE International Conference on Acoustics, Speech, and Signal Processing (ICASSP)*, Orlando, USA, May 2002.

[81] P. Siohan and N. Lacaille, "Analysis of OFDM/OQAM systems based on the filterbank theory," in *Proc. IEEE Global Telecommunication Conference (GLOBECOM)*, Rio de Janeiro, Brazil, Dec. 1999, pp. 2279–2284.

[82] P. Siohan, C. Siclet, and N. Lacaille, "Analysis and design of OFDM/OQAM systems based on filterbank theory," *IEEE Trans. Signal Process.*, vol. 50, pp. 1170–1183, May 2002.

[83] T. Stitz, T. Ihalainen, A. Viholainen, and M. Renfors, "Pilot-based synchronization and equalization in filter bank multicarrier communications," *EURASIP J. Adv. Signal Process.*, vol. 2010, no. 1, pp. 1–18, 2010. http://dx.doi.org/10.1155/2010/741429 [Online].

[84] A. Tonello, "Method and apparatus for filtered multitone modulation using circular convolution." Patent WO 2009/135886 A1, 2009.

[85] A. M. Tonello, "A novel multi-carrier scheme: Cyclic block filtered multitone modulation," in *Proc. IEEE International Conference on Communications (ICC)*, Budapest, Hungary, Jun. 2013, pp. 5263–5267.

[86] P. Turcza, "New transmultiplexer for xDSL based on Wilson expansion," in *Proc. IEEE International Conference on Communications (ICC)*, Paris, France, vol. 1, Jun. 2004, pp. 26–30.

[87] M. A. Tzannes, M. C. Tzannes, J. Proakis, and P. N. Heller, "DMT systems, DWMT systems and digital filter banks," in *Proc. IEEE International Conference on Communications (ICC)*, New Orleans, USA, vol. 1, May 1994, pp. 311–315.

[88] R. Vallet and K. H. Taieb, "Fraction spaced multi-carrier modulation transmission," *Wirel. Pers. Commun.*, vol. 2, pp. 97–103, 1995.

[89] M. Vetterli, "Perfect transmultiplexers," in *Proc. IEEE International Conference on Acoustics, Speech, and Signal Processing (ICASSP)*, Tokyo, Japan, Apr. 1986, pp. 2567–2570.

[90] S. B. Weinstein, "The history of orthogonal frequency-division multiplexing," *IEEE Commun. Mag.*, vol. 47, no. 11, pp. 26–35, Nov. 2009.

[91] S. B. Weinstein and P. M. Ebert, "Data transmission by frequency-division multiplexing using the discrete Fourier transform," *IEEE Trans. Commun. Technol.*, vol. COM-19, pp. 628–634, Oct. 1971.

[92] J. Yli-Kaakinen and M. Renfors, "Flexible fast-convolution implementation of single-carrier waveform processing," in *Proc. IEEE International Conference on Communications, ICC2015 Workshops*, London, UK, Jun. 2015.

[93] Y. H. Yun, C. Kim, K. Kim, Z. Ho, B. Lee, and J.-Y. Seol, "A new waveform enabling enhanced QAM-FBMC systems," in *Proc. IEEE International Workshop on Signal Processing Advances in Wireless Communications (SPAWC)*, Stockholm, Sweden, Jun.–Jul. 2015.

[94] R. Zakaria and D. Le Ruyet, "On maximum likelihood MIMO detection in QAM-FBMC systems," in *Proc. IEEE International Symposium on Personal, Indoor, and Mobile Radio Communications (PIMRC)*, Istanbul, Turkey, Sep. 2010.

[95] Z. Zhao, M. Schellmann, Q. Wang, X. Gong, R. Boehnke, and W. Xu, "Pulse shaped OFDM for asynchronous uplink access," in *Proc. Annual Asilomar Conference on Signals, Systems and Computers (ASILOMAR)*, Nov. 2015, pp. 3–7.

[96] M. S. Zimmerman and A. L. Kirsch, "The AN/GSC-10 (kathryn) variable rate data modem for HF radio," *IEEE Trans. Commun. Technol.*, vol. 15, pp. 197–205, Apr. 1967.

[97] W. Y. Zou and Y. Wu, "COFDM: An overview," *IEEE Trans. Broadcast.*, vol. 41, pp. 1–8, Mar. 1995.

CHAPTER

# FBMC Design and Implementation

**Juha Yli-Kaakinen\*, Pierre Siohan†, Markku Renfors\***

*Tampere University of Technology, Tampere, Finland\**
*Independent Researcher, Rennes, France†*

## CONTENTS

## 8.1 INTRODUCTION

Whereas Chapter 7 introduced the general multicarrier waveform theory and various alternative orthogonal waveforms, the focus in this chapter is on effective practical implementation schemes for the needed filterbanks, with main focus on the Filter-Bank MultiCarrier with Offset-QAM subcarrier modulation (FBMC/OQAM) case. In the first part, effective modulation-based uniform filterbank structures are used, and the designs are defined by the lowpass prototype filter. Two alternative paths are considered. The first approach starts from continuous-time formulation of the prototype filter responses, which are then discretized. Alternatively, the design problem can be formulated as a lowpass digital filter optimization problem with additional constraints to guarantee the orthogonality of subcarriers. While introducing the effective filterbank structures, the main emphasis is on the polyphase FilterBank (FB) model, but the lattice structure and lapped transform-based structure are briefly introduced as well.

One important trend in current wireless communication system development is to exploit the radio spectrum more effectively. This leads to dynamic spectrum use scenarios where, for instance, relatively narrow fragments of the radio spectrum are aggregated for high-data-rate communications, depending on what is left unused by the prioritized users/services. In such cases, high flexibility (agility) of the used waveforms and signal processing techniques become critical. To avoid cross-interferences between the dynamic spectrum users and primary users, sharp transition bands and high stopband attenuation of the channelization filters are also crucial requirements. Similar requirements are also imposed by the 5th Generation (5G) system development targeting at an integrated multiservice network with quite different requirements for the physical layer system parameterization between different services.

Although uniform FB-based waveform processing schemes support well many of these targets, there is a clear interest for additional flexibility. It is worth mentioning that the modulation-based approach has been extended also for the design of nonuniform FBs with Near Perfect Reconstruction (NPR) characteristics [20,54]. Such nonuniform designs have found applications, for example, in audio applications [20,72]. The techniques of partial spectrum reconstruction [48] and frequency band allocation/reallocation [3,28,31] are important additional features of many variable filterbank applications.

In the later part of this chapter, an alternative scheme for implementing uniform and nonuniform filterbanks in effective and flexible manner is introduced. This scheme is based on fast convolution, an old idea from the 1960s to implement high-order digital Finite Impulse Response (FIR) filters effectively in the Discrete Fourier Transform (DFT) domain.

## 8.2 PROTOTYPE FILTER DESIGN AND FILTERBANK STRUCTURES

### 8.2.1 PROTOTYPE FILTER DESIGN

Various existing techniques used for designing prototype filters have more or less followed the evolution described in Chapter 7 concerning orthogonal communication waveforms. That is to say, at the beginning, Nyquist's criterion applied in continuous-time was the unique guideline leading to use of the raised cosine function or its square root as the filter frequency response. Afterwards, the link with filterbank and Gabor theories have brought new waveform models, suggesting new ideas to build more elaborated prototype filters. Let us review some parts of this evolution with a particular focus on the design of FBMC/OQAM prototype filters.

### 8.2.1.1 *The Continuous-Time Track*

Undoubtedly, one of the main reference in terms of prototype functions is the Raised Cosine (RC) or its root, the square Root Raised Cosine (RRC) function. Indeed, the RC function, denoted by $g_{RC}$ and $G_{RC}$ in time and frequency domains, respectively, satisfies the first Nyquist criterion. Hence, for a symbol rate $F = \frac{1}{T}$, its impulse response is such that $g_{RC}(lT) = 0$ for $l \neq 0$, whereas its linear-phase frequency response exhibits odd symmetry with respect to $f = \frac{1}{2T}$. Taking the square root of $G_{RC}$ and setting $\|g_{RRC}\| = 1$, we obtain the normalized time and frequency expressions of the RRC prototype function, which were introduced in Section 5.4.

The way the RRC prototype function also satisfies the Nyquist conditions for the FBMC/OQAM setting was illustrated in Chapter 7. The OQAM time and frequency staggering rule makes that the orthogonality condition needs only to be satisfied by the Gabor triplet $(T, 2F, g_{RRC})$. The Inter-Symbol Interference (ISI)-free condition is satisfied at rate $\frac{1}{T}$ by the combination of the transmit and receive filters, i.e., $g_{RRC}(t) \star g_{RRC}^*(-t)$, as illustrated in Fig. 7.2 of Chapter 7. On the other hand, as the RRC band-limited function does not extend beyond the interval $[-F, F]$, there will be no Inter-Carrier Interference (ICI) with any value of the roll-off factor. This means that, assuming ideal transmission conditions, a sharp transition band is not required for the FBMC/OQAM prototype filter. But, in realistic transmission scenarios, a perfect separation between adjacent subchannels is no longer guaranteed, and then we have to take care of the compaction in time and/or frequency of the prototype filter.

As explained in Chapter 7, the optimum time-frequency compaction is obtained by the Gauss function, i.e., the product of its second-order moments in time and frequency constitutes a lower bound. However, the lack of orthogonality of the resulting Gaussian base functions needs to be compensated at the receiver side, e.g., by using a Viterbi decoder [37]. Using another approach, which attracted more attention, Alard [5,39] has proposed an orthogonalization method for the Gaussian function. It results in the so-called Isotropic Orthogonal Transform Algorithm (IOTA) prototype function. The IOTA method is a two-step orthogonalization procedure of the Gaussian function $g_\lambda(t) = (2\lambda)^{1/4} e^{-\pi \lambda t^2}$. Denoting by $\boldsymbol{F}$ the Fourier transform operator, for a $(\frac{T}{2}, F)$ time-frequency lattice, the IOTA prototype function is expressed as follows:

$$z_{\lambda, F, \frac{T}{2}}(t) = \boldsymbol{O}_{\frac{T}{2}} \boldsymbol{F}^{-1} \boldsymbol{O}_F \boldsymbol{F} g_\lambda(t), \tag{8.1}$$

where $\boldsymbol{O}_a$ with $a > 0$ is the orthogonalization operator defined as

$$(\boldsymbol{O}_a f)(u) = \frac{f(u)}{\sqrt{a \sum_k |f(u - ka)|^2}}. \tag{8.2}$$

The IOTA prototype function corresponds to the isotropic configuration, i.e., $\lambda = 1$, $F = \frac{T}{2} = \frac{1}{\sqrt{2}}$, and is denoted by the symbol $\mathcal{J}$. Otherwise said, $\mathcal{J}$ reads as

$$\mathcal{J}(t) = z_{1,\frac{1}{\sqrt{2}},\frac{1}{\sqrt{2}}}(t) = \boldsymbol{O}_{\frac{1}{\sqrt{2}}} \boldsymbol{F}^{-1} \boldsymbol{O}_{\frac{1}{\sqrt{2}}} \boldsymbol{F} g_1(t). \tag{8.3}$$

In a practical setting, the values of $T$ and $F$ should be selected properly to keep the same time-frequency scale as the delay and Doppler spread of the transmission channel itself [39].

A proof of IOTA orthogonality together with its other main features is reported in [5,6]. In particular, it is shown that $\boldsymbol{F}\mathcal{J} = \mathcal{J}$. However, this property does not guarantee a perfect time-frequency isotropy, as it is the case with the Gaussian function; $\mathcal{J}$ is simply nearly isotropic. Furthermore, it has an excellent concentration in time and frequency since its Time Frequency Localization (TFL) parameter (see (7.10) in Chapter 7) is such that

$$\xi(\mathcal{J}) = \frac{1}{4\pi \mathrm{m}^{(2)}(\mathcal{J})} = 0.977. \tag{8.4}$$

This localization measure is therefore close to the maximum one, equal to 1 for the Gaussian function, and also 13% higher than that of the "best" RRC function with $\alpha = 1$.

To avoid the computational burden related to the $\boldsymbol{O}$ and $\boldsymbol{F}$ operators, Eq. (8.1) can be rewritten differently. Indeed, after some mathematical manipulations, $z_{\lambda,F,\frac{T}{2}}(t)$ can be expressed as an Extended Gaussian Function (EGF) corresponding to the following series expansion [62,68]:

$$z_{\lambda,F,\frac{T}{2}}(t) = \frac{1}{2} \sum_{k=0}^{\infty} d_{k,\lambda,F} \left[ g_\lambda \left( t + \frac{k}{F} \right) + g_\lambda \left( t - \frac{k}{F} \right) \right]$$

$$\times \sum_{l=0}^{\infty} d_{l,1/\lambda,\frac{T}{2}} \cos \left( 4\pi l \frac{t}{T} \right), \tag{8.5}$$

where the $d_{k,\lambda,F}$ are real-valued coefficients reported in the above-mentioned references. It has been shown that EGFs are orthogonal in the range $\alpha_m \leq \alpha \leq 1/\alpha_m$ with $\alpha_m \approx 0.528 F^2$. To recover $\mathcal{J}$ with an accuracy around $10^{-5}$, we set $\lambda = 1$ and $F = \frac{T}{2} = \frac{1}{\sqrt{2}}$, and then only the first six $d_k$ are needed.

Fig. 8.1 clearly illustrates the fact that the EGFs decay linearly on the logarithmic scale. Based on the fact that, for given $T$ and $F$ values, $\boldsymbol{F} z_\lambda = z_{\frac{1}{\lambda}}$, it is clear that $\lambda$ can be used as an adaptivity parameter to appropriately deal with the time and frequency dispersion of mobile transmission channels [27].

Orthogonal Frequency-Division Multiplexing (OFDM)/OQAM can be interpreted in terms of Gabor (or Weyl–Heisenberg) frames (see, e.g., [14,36]), and then it

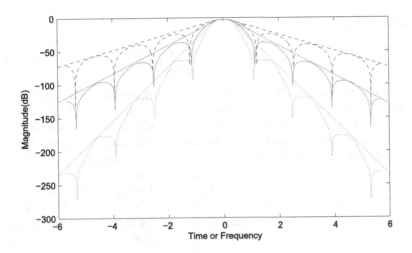

**FIGURE 8.1**

Time-frequency representations of some orthogonal EGFs. Dashed line: $\lambda = \frac{1}{2}$ (in time).
Solid line: $\lambda = 1$. Dotted line: $\lambda = 2$ (in time).

can be shown that the design of an OFDM/OQAM linear-phase pulse shaping function is equivalent to the design of a Tight Gabor Frame (TGF) with oversampling
by two. Then, using the Zak Transform (ZT) [35], it has been shown by Janssen and
Boelcskei [36] that, starting from the Gaussian function, the ZT and IOTA approaches
are two equivalent methods to produce TGFs with oversampling of two. Furthermore,
in [36], it is indicated that the ZT also works for some other initial functions, but no
example is exhibited showing better time-frequency localization value than that obtained with IOTA. In [69], it is shown that, compared to the direct approach of [62],
the ZT constitutes a more efficient manner to rederive the EGF expression. On another hand, note also that the frame-based orthogonalization method is reused in [75],
where the authors extend its application to the case of hexagonal time-frequency lattices. By the way, they show that the TFL can then go up to 0.986.

In parallel, and independently of the IOTA and EGF studies, another approach
proposed by Haas and Belfiore also attempted to get prototype functions with good
TFL starting from a subset of Hermite functions. Again, the idea is to get a prototype
function identical to its Fourier transform [32,33]. Then, denoting by $g_h(t)$ this function, we must have $\boldsymbol{F} g_h = g_h$. Implicitly, the authors use the fact, elaborated more
in details later on in [14], that the orthogonalization of the Gabor family defined on
a time-frequency lattice of density 1/2, i.e., corresponding to an OFDM/Quadrature
Amplitude Modulation (QAM) system oversampled by 2, will also constitute the solution for the OFDM/OQAM pulse shape. The Hermite polynomials of index $n$ can

be expressed as

$$D_n(t) = (-1)^n \frac{e^{\pi t^2}}{(2\pi)^{\frac{n}{2}}} \frac{d^n}{dt^n} e^{-2\pi t^2}. \tag{8.6}$$

For $n = 4k$ with integer $k$, $D_n(t)$ is equal to its Fourier transform. Consequently, any linear combination $h$ of $D_{4k}$ functions satisfies the condition $Fh = h$. Then, instead of orthogonalizing successively in frequency and time as in [39], the authors use the fact that there is only one function to orthogonalize. Then, as explained in Chapter 7, the orthogonality constraints can be expressed using the ambiguity function of the function $h$, meaning that $A_h(\tau, \nu)$ has to satisfy the generalized Nyquist condition (7.5). Using the fact that $A_h(\tau, \nu) = A_h(\nu, \tau)$ and assuming that $A_h(\tau, \nu)$ takes negligible values outside a ring of forced zeros, the solution in [32,33] is obtained for $h$ defined by a linear combination of $N_h$ terms as the solution of a system of $N_h$ quadratic equations, setting $N_h = 5$. The resulting prototype after normalization writes as $g_h(t) = 2^{\frac{1}{4}} h(\sqrt{2}t)$. It leads to a TFL measure $\xi(g_h) = 0.98$, but, differently from IOTA, as some constraints are ignored, it is only nearly orthogonal [6]. On the other hand, the comparison in [29] of the IOTA and Hermite pulses shows that a Hermite pulse has a slightly faster decay than IOTA. Note, finally, that the Hermite pulses have been also studied in the case of hexagonal time-frequency lattices [65] leading to a TFL measure nearly identical (0.986) to the one found in [75].

Due to their faster decay, one advantage of pulses with good TFL over the RRC prototype function is the possibility of pulse truncation without impacting too much the resulting performance. This is naturally interesting from the implementation point of view. In this respect, a simple solution is provided with the dual of the RRC prototype function. Indeed, setting $g_{\text{RRC}}^{(d)}(t) = G_{\text{RRC}}(t)$, we get a Time-Limited Orthogonal (TLO) prototype function, which for $\alpha = 1$ writes as

$$g_{\text{RRC}}^{(d)}(t) = \sqrt{\frac{2}{T}} \cos\left(2\pi \frac{t}{T}\right), \quad \text{if } |t| \le \frac{T}{2}. \tag{8.7}$$

For this TLO prototype function [40], leading to the MultiCarrier Modulation (MCM) system, named OFDM-Minimum-Shift Keying (MSK) in [39], as the filtering is limited at the Transmitter (TX) and Receiver (RX) sides, to an OQAM symbol duration, the implementation cost is minimal.

However, if short pulses are interesting with regard to implementation cost and delay aspects, the shortest ones cannot provide a sufficient frequency localization. On the other hand, pulses of infinite duration, such as the RRC function, have to be truncated to be realizable. Then, to get good approximations, the resulting pulse length may be too long, leading to complexity and delay issues. This problem is avoided if, as proposed in [79,80], the prototype function optimization is carried out with pulses of finite, and relatively low, duration. Adding the orthogonality constraints, Orthogonal Finite Duration Pulse (OFDP) are then obtained by a maximization of the prototype function energy in its passband region. Then, the resulting prototype function can be expressed as

$$g_{OFDP}(t) = \sum_{i=0}^{\infty} a_{2i} \Psi_{2i}(t), \qquad (8.8)$$

where $\Psi_i$ is the $i$th time limited Prolate Spheroidal Wave Function (PSWF) [71], and $a_{2i}$ are real-valued coefficients resulting from the optimization procedure. The coefficients corresponding to the durations $2T$, $3T$, and $4T$ are reported in [80].

### 8.2.1.2 *The Discrete-Time Track*

Generally, the prototype filters are implemented as FIR filters, meaning that when starting from a continuous-time prototype function, such as RRC, IOTA, EGF, or Hermite, not only truncation is necessary, but also discretization. Only this last operation is necessary for OFDP. To keep the maximum spectral efficiency for FBMC/OQAM, the discretization must be operated at the critical sampling rate, i.e., $T_s = \frac{T}{M}$. Then, the discrete-time prototype filter, for $l \in [0, L_g - 1]$, is obtained by setting

$$g[l] = \sqrt{T_s} g(lT_s). \qquad (8.9)$$

Naturally, if we start with a noncausal symmetrical prototype function, a shift of $-\frac{L_g-1}{2}$ is necessary to maintain causality.

Note here that, in the absence of truncation, the orthogonality properties (7.25) can be preserved in discrete-time (see, e.g., the EGF case [68]); however, this is not a general rule. In any case, usually, the design problem is directly handled in discrete-time, and several design methods are available; see, for instance, [63] for a recent overview. Here, our aim is simply to review some of the most prominent ones. For that, let us start with methods not using explicit objective functions for their design. As recalled in Chapter 7, these design methods are closely related to the filterbank and Gabor frame theories. Then, the MCM orthogonality and near orthogonality correspond either to the Perfect Reconstruction (PR), meaning that the output symbols are delayed versions of the input ones, or to the NPR property, meaning that these output samples are only approximations of the input, respectively. Naturally, there exist also simple windowing methods that can be applied with Cyclic Prefix (CP)-OFDM at the TX and/or RX sides. Our analysis will not cover these approaches, for which an overview can be found in [41].

The PHYDYAS prototype filter has been popularized by the European PHYDYAS project [2]. It can be obtained by design methods presented in [13,43,46]. The common feature between these three descriptions is the attempt to get a modulated transmultiplexer system with a minimum energy overlap between adjacent subcarriers. Then, denoting by $K$ the overlapping factor, the impulse response of the prototype filter writes, for $l \in [0, KM - 1]$, as

$$g_\phi[l] = A[0] + 2 \sum_{k=0}^{K-1} (-1)^k A[k] \cos\left(\frac{2\pi k}{KM}(l+1)\right). \qquad (8.10)$$

Especially, for $K = 4$,

$$A[0] = 1; \quad A[1] = 0.97195983; \quad A[2] = \frac{\sqrt{2}}{2};$$

$$A[3] = \sqrt{1 - A[1]} = 0.23514695. \tag{8.11}$$

In [13], these coefficients are computed using the frequency sampling design method [8,56]. The advantage of this closed-form solution is that the coefficients $A[k]$ depend only on $K$ but not on the number of subcarriers, thus making this approach scalable. However, the PHYDYAS prototype filter is only NPR.

The PHYDYAS example shows that usual filter design methods can also be used and adapted for designing MCM systems. Another example is presented in [38] for a MCM scheme named conjugate OFDM/OQAM. Then, the idea is to factorize a Nyquist filter obtained with the classical Remez algorithm [8,56]. Thus, we get a nonlinear-phase half-Nyquist filter at TX and RX sides having, for a given length, higher stopband attenuation than that achievable with a linear-phase prototype filter. However, as for PHYDYAS prototype filters, we only get a NPR system.

When the aim is to design FBMC/OQAM and Filtered MultiTone (FMT) systems with PR characteristics, the design problem for the prototype filter is equivalent to those already encountered when designing some PR modulated filterbank systems in Chapter 7. However, a question that often arises is why designing PR MCM systems if in realistic transmission cases orthogonality does not hold anymore? In addition, PR constraints reduce the degree of freedom for the design. This is true, but it has also been shown, for example, in Power Line Communications (PLC) that, in favorable transmission conditions, PR designs may outperform NPR ones [4]. Furthermore, as we will see in the next subsection, PR systems can lead to some specific and valuable implementation schemes.

Eqs. (7.26) and (7.27) give simplest closed-form expressions for $L_g = M$ and $2M$, respectively. They can be directly employed in conventional FBMC/OQAM schemes or constitute the kernel of what is called Lapped-OFDM in [11].

We already mentioned that the ZT could be used for the orthogonalization of continuous-time functions. Its discrete version, the Discrete Zak Transform (DZT), can also be used to derive orthogonal prototype filters [14,15]. Starting from an arbitrary nonorthogonal lowpass filter, say $g_0$, the DZT can indeed provide an orthogonal solution for a relatively low implementation cost since the DZT is essentially computed using Fast Fourier Transforms (FFTs). By the way, PR FBMC/OQAM systems up to 256 subcarriers and with filter length up to 2048 could be designed using the DZT [15]. However, even though the conditions to be checked by $g_0$ to get good TFL have been theoretically studied [76], to the best of our knowledge, no numerical measures for localization have been yet provided for the DZT method. Actually, to insure an accurate control of the final result, direct optimization methods are necessary in which proper design criteria are imposed.

A first classical criterion consists of the minimization of the out-of-band energy. Denoting by $f_c$ the cut-off frequency and $E(f_c)$ the out-of-band energy in the normalized ($T_s = 1$) frequency range $[f_c, \frac{1}{2}]$, the optimization problem to find the set of coefficients $\boldsymbol{g} = (g[l])_{l=0,\dots,L_g-1}$ may be stated as

$$\min_{\boldsymbol{g}} J_g(\boldsymbol{g}) = \min_{\boldsymbol{g}} \frac{E(f_c)}{E(0)} \quad \text{with} \quad E(f_c) = \int_{f_c}^{\frac{1}{2}} \left| G(e^{j2\pi \nu}) \right|^2 d\nu, \qquad (8.12)$$

under PR constraints (7.25).

This type of criterion and its variants [83] provide frequency selective prototype filters for overlapping factor $K = 3$ and beyond. This is particularly appropriate for relatively static channels, e.g., Asymmetric Digital Subscriber Line (ADSL) and PLC, when channel delay spread is the main limiting factor.

In any case, due to the nonlinearity of the objective functions for the usual design criteria, a general problem with direct design methods is the high number of variables to optimize. Indeed, since the number of subcarriers may be very large, for instance, 2048 for Long-Term Evolution-Advanced (LTE-A), the number of variables may be huge. Even when reducing it by a factor of two by using the prototype filter symmetry, this number may go beyond the possibilities of most of the optimization procedures. For instance, in [83], using the method initially proposed for modulated filterbanks [18], the maximum number of subcarriers considered is 256 and together with overlapping factors up to 5. Though optimization problems with 640 variables were solved, this may be insufficient for certain other situations.

Let us now assume that the length of the linear-phase prototype filter is $L_g = KM$ with even $M$ and $K$ (the case where $M$ is a multiple of 4 is elaborated later on). Then, using the $M$th-order polyphase decomposition of $G(z)$ (see (7.28)), the PR property can be expressed using a set of only $K\frac{M}{4}$ angular parameters $\phi_{k,i}$. This reduces again the number of variables to optimize and further gives rise to lattice structure realizations; see Section 8.2.2. The minimization of the out-of-band energy then writes as

$$\min_{\boldsymbol{\phi}} J_g(\boldsymbol{\phi}) \qquad (8.13)$$

with $\boldsymbol{\phi} = (\phi_{k,i})_{k=0,\dots,K-1, i=0,\dots,\frac{M}{4}-1}$.

It has been observed [53] that the best solutions of this unconstrained nonlinear problem were characterized by always very smooth functions $i \mapsto \phi_{k,i}$. Otherwise said, these angles can accurately be represented by different sorts of polynomial expressions containing a limited number of coefficients, setting, for example,

$$\phi_{k,i} = \sum_{l=0}^{L-1} x_{k,l} \left( \frac{2i+1}{M} \right)^l. \qquad (8.14)$$

Then $x = (x_{k,l})_{k=0,\ldots,K-1,l=0,\ldots,L-1}$ is called the compact representation, and the unconstrained optimization problem reads as

$$\min_{x} J_g(x). \tag{8.15}$$

Since, in general, a value of $L$ around 5 or 6 is enough to get accurate optimization results, we have $KL \ll KM$, and the number of variables then becomes considerably less that what is required when optimizing with respect to $g$ or $\theta$. Then, the design of prototype filters for FBMC/OQAM systems up to (and even beyond) 2048 subcarriers and with length going up to 32768 taps becomes possible [53].

The compact representation also works well when optimizing with respect to the TFL criterion. This criterion is generally selected for time-frequency dispersive channel. However, then the second-order moments in time and frequency have to be defined taking into account the discrete nature of the prototype filter. This is made possible using, as in [53], the second-order moment definitions proposed in [26] for discrete-time sequences. By the way, we can again get a localization measure, denoted $\xi_{\text{mod}}(g)$, such that $0 \leq \xi_{\text{mod}}(g) \leq 1$, 1 being the optimum. A great advantage of this direct optimization with the TFL criterion appears for short prototype filters. Indeed, using it for $L = M$, we are able to get a modified localization measure $\xi_{\text{mod}}(g) = 0.906$ for a 2048-subcarrier system, whereas then the truncation of the IOTA prototype function only provides a measure $\xi_{\text{mod}}(g) = 0.1442$ [70]. Furthermore, it has been shown that, for these short lengths, very accurate closed-form expressions can be derived from the compact representation using low-order ($L = 2$) approximations [52].

### 8.2.1.3 *Design Methods for the FMT System*

The design of FMT MCM systems has more or less followed the evolution we have described for the design of FBMC/OQAM systems. Let us recall that, for the complex orthogonality, the rectangular window constitutes a unique solution if $TF = 1$. For FMT, the oversampling factor $\rho$ (see Chapter 7) is greater than 1, i.e., $TF = \rho > 1$. A consequence for the continuous-time RRC design is that the roll-off has to be chosen accordingly, i.e., setting $\alpha = \rho - 1$. When dealing with the discrete-time case, denoting by $N$ the expansion/decimation factor of the corresponding transmultiplexer, we must set $TF = \rho = \frac{N}{M}$, and to get PR FMT systems, PR conditions as (7.17), or equivalent ones have to be satisfied. Again, the methods presented before can be adapted, which is the case, for instance, of the DZT approach [15], and of some other constrained or unconstrained (e.g., [51,57]) direct optimization methods.

## 8.2.2 REALIZATION STRUCTURES

The most popular implementation for FBMC/OQAM systems, as defined by Eq. (7.21), use the type I $M$th-order polyphase decomposition of the prototype filter, $G(z) = \sum_{i=0}^{M-1} z^{-i} E_i(z^M)$, in addition to the Inverse Fast Fourier Transform (IFFT) or FFT algorithm. As detailed in [70] and reused in the PHYDYAS [2] and

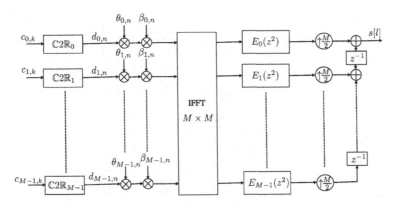

**FIGURE 8.2**

FBMC/OQAM modulator using a type I polyphase decomposition of the prototype filter.

EMPHATIC [1] projects, the corresponding Synthesis FilterBank (SFB) at the TX side can be represented as depicted in Fig. 8.2. In this scheme, the OQAM symbol mapping corresponds, for each subcarrier of index $m$, to the $\mathbb{C}2\mathbb{R}_m$ block delivering the real data symbols $d_{m,n}$ at rate $\frac{2}{T}$, as described in Section 7.3.2. In this causal realization scheme, the phase term and the prototype filter length are taken into account with the two columns of multiplicative factors $\theta_{m,n}$ and $\beta_{m,n}$ at the IFFT inputs

$$\theta_{m,n} = e^{j\frac{\pi}{2}(m+n)}, \tag{8.16}$$

$$\beta_{m,n} = (-1)^{mn} e^{-j\frac{2\pi m}{M}\frac{L_g-1}{2}}. \tag{8.17}$$

Differently from OFDM, in which the IFFT operates at rate $\frac{1}{T}$, in FBMC/OQAM, the IFFT rate is doubled. Different architectures have been proposed to reduce this complexity at the TX side [19,82]. In [24], it is shown that for causal FBMC/OQAM systems with prototype filter of arbitrary length, using the Hermitian symmetry at the IFFT output, it is possible to get a computational complexity nearly equivalent to that of OFDM. The reduction comes from the possibility to replace the full IFFT by a pruned one.

At the RX side, the dual polyphase Analysis FilterBank (AFB) structure can be used, consisting of the processing blocks of Fig. 8.2 but located in the reverse order. In this structure, to create closer commonality with OFDM, the transform is FFT. The path filters are now based on the type II polyphase decomposition of the prototype filter, i.e., $F_m(z) = E_{M-1-m}(z)$, and the $\beta$ and $\theta$ coefficients are conjugated.

However, as explained in Section 5.6.1, also an IFFT-based polyphase AFB structure is available. The connections of IFFT- and FFT-based AFB structures are explained in details in [66,83]. One important issue is that the overall Transmultiplexer (TMUX) delay should be a multiple of $M/2$. There are various alternative ways to

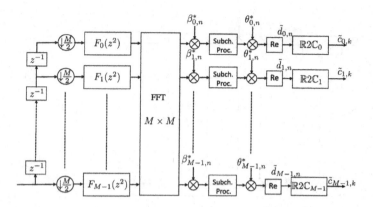

**FIGURE 8.3**

FBMC/OQAM demodulator using a type II polyphase decomposition of the prototype filter.

include additional delays in the TMUX chain to reach this target [67,70,83]. It is also useful to know that with proper delay adjustments, the $\beta$-coefficients can be made trivial ($\pm1$), e.g., for $L_g = KM$ or $L_g = KM - 1$, as with the PHYDYAS prototype of (8.10).

As shown in Fig. 8.3, we take into account the fact that, due to the impairments caused by the channel and the analog Radio Frequency (RF) system (not represented in here), each branch may include a subchannel processing module running at $2\times$ symbol rate. This is used, e.g., for synchronization, channel estimation, and channel equalization purposes, as will explained in Chapters 10, 11, and 12, respectively. In the ideal case, i.e., where all these impairments are negligible, the output of the real part extraction block (Re) will be such that $\tilde{d}_{m,n} = d_{m,n}$, and then the $\mathbb{R}2\mathbb{C}_m$ block will deliver QAM symbols such that $\tilde{c}_{m,n} = c_{m,n}$.

The implementation of the polyphase components can be based either on the classical FIR transversal structure [8,56] or, in the case of a PR system, take advantage of orthogonal lattice structures; see, e.g., [81]. Indeed, we have seen in Chapter 7 that the PR conditions could be written using the polyphase components (7.29). A first step to use this possibility is to pair the polyphase components of index $i$ and $i + \frac{M}{2}$ as shown in Fig. 8.4, setting $\gamma_{m,n} = \theta_{m,n}\beta_{m,n}$.

Then, each pair of polyphase components $(E_i(z), E_{i+\frac{M}{2}}(z))$ can be rewritten as a product of delay and rotation matrices. This is illustrated in Fig. 8.5 considering a prototype filter of length $L_g = KM$ and denoting the rotation matrices as follows:

$$R_{k,i}(\phi) = \begin{pmatrix} \cos(\phi_{k,i}) & \sin(\phi_{k,i}) \\ \sin(\phi_{k,i}) & -\cos(\phi_{k,i}) \end{pmatrix}. \tag{8.18}$$

One advantage of the lattice realization is to structurally guarantee the PR property even after quantization of the angular parameters.

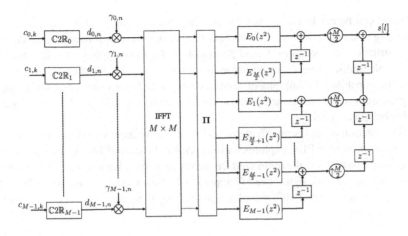

**FIGURE 8.4**

FBMC/OQAM modulator using a specific pairing of the polyphase components.

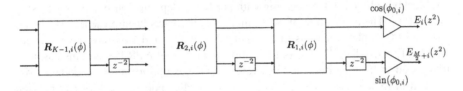

**FIGURE 8.5**

Lattice implementation of the pairs of polyphase components.

At the RX side, with the exception of the complexity reduction (with a pruned IFFT), which is no longer applicable, we recover a similar implementation scheme including lattice-type polyphase network filtering and FFT [70].

Some other types or realizations, inspired by the lapped transforms [42] and introduced first for subband audio coding to avoid the blocking effect of the discrete cosine transform, can also be adapted to the OFDM/OQAM system. This is the direct consequence of the duality between filterbanks and transmultiplexers, and this possibility has been illustrated recently with a system named lapped-OFDM [11].

## 8.3 NOVEL REALIZATION STRUCTURES

The Fast-Convolution FilterBank (FC-FB) is an efficient implementation for realizing highly tunable multirate filterbank configurations [59]. FC-FBs are based on FFT-IFFT pair with overlapped block processing, and they offer a straightforward way to adjust the filters' frequency-domain characteristics, even in real time. Each subband can be easily configured for different bandwidths, center frequencies, and sampling rate conversion factors, including also partial or full-band nearly perfect-reconstruction NPR systems. Such FB systems find various applications, for example, as flexible multichannel channelization filters for software defined radios. They have also the capability to implement simultaneous waveform processing for multiple single-carrier and/or multicarrier transmission channels with nonuniform bandwidths and subchannel spacings.

This section develops an effective model for the frequency response analysis and optimization of such FBs, explores various practical issues of Fast-Convolution (FC) processing, and reports the optimized frequency response characteristics versus direct raised cosine based designs. The FC approach is shown to be a competitive basis for FB-based MultiCarrier (MC) waveform processing in terms of spectral containment and complexity.

## 8.3.1 FAST-CONVOLUTION FILTERBANK

The basic idea of FC-based filters and FBs is to approximate linear convolution using blockwise FFT-IFFT processing with partly overlapping consecutive processing blocks [49]. The application of FC to multirate filters has been presented in [16,47, 54], and FC implementations of channelization filters and FBs have been considered in [17,55,78,88]. The idea of FC implementation of NPR FB systems has been introduced in [58], and detailed analysis and FC-FB optimization methods are developed in [59,85]. The frequency spreading FBMC approach of [10] can be seen as a very special case of FC-FB where the IFFT size is $KM$, and only one sample is utilized from each IFFT block. In [30], a Generalized Frequency-Division Multiplexing (GFDM) technique is proposed for implementing flexible MC transmission systems. Especially, the variant of GFDM presented in [45] exploits similar ideas as the FC-FB scheme under discussion.

In the basic form, FC implements linear convolution between two finite-length sequences as follows:

$$y[n] = h[n] \circledast x[n] = \text{IFFT}(\text{FFT}(h[n]) \cdot \text{FFT}(x[n])). \qquad (8.19)$$

This transform processing implements circular convolution by nature; however, in case of finite block length, linear convolution is obtained if the transform length is high enough. Assuming that the lengths of the sequences are $N_h$ and $N_x$, the length of the linear convolution is $N = N_h + N_x - 1$. The lengths of the FFT and IFFT operations in (8.19) have to be at least $N$; otherwise, distortion will appear in the result [50]. To process long sequences, blockwise FC can be used in conjunction with overlap-add or overlap-save processing [34,49,74]. We use the latter one because of its better properties in finite wordlength implementation [23]. This is due to transients at the beginning and end of each overlap-add processing block. These transients often exceed the numeric range of the final output signal, whereas the transients of consecutive overlapping blocks are canceling each other.

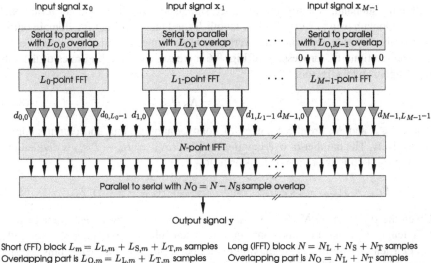

Short (FFT) block $L_m = L_{L,m} + L_{S,m} + L_{T,m}$ samples    Long (IFFT) block $N = N_L + N_S + N_T$ samples
Overlapping part is $L_{O,m} = L_{L,m} + L_{T,m}$ samples    Overlapping part is $N_O = N_L + N_T$ samples

| Overlap $L_{L,m}$ samples | Non-overlapping part $L_{S,m}$ samples | Overlap $L_{T,m}$ samples |
|---|---|---|

| Overlap $N_L$ samples | Non-overlapping part $N_S$ samples | Overlap $N_T$ samples |
|---|---|---|

**FIGURE 8.6**

Fast-convolution-based flexible SFB using overlap-save processing. In this structure, the interpolation factor of subband $m$ is equal to the ratio of the IFFT length and FFT length. Also, the notations used for the number of samples in different parts of the overlap-save blocks are denoted.

### 8.3.1.1 *Multirate Filtering and Filterbanks*

The structure of the proposed adjustable SFB is illustrated in Fig. 8.6. The general idea of the structure is a multirate version of fast convolution [16,17]. We consider the case where multiple low-rate narrowband signals $x_m[n]$ for $m = 0, 1, \ldots, M - 1$ are to be filtered with adjustable frequency responses and possibly adjustable sampling rates and then combined into single wideband signal $y[l]$, following the frequency-division multiplexing principle. We are interested in the cases where the input signals are critically sampled or oversampled by a small factor. We also note that different subbands may be overlapping. The dual AFB structure of Fig. 8.6 can be used for splitting the incoming high-rate wideband signal into several low-rate narrowband signals.

In this structure, each of the $M$ incoming signals is first segmented into overlapping blocks of length $L_m$ for $m = 0, 1, \ldots, M - 1$. Then, each input block is transformed to frequency-domain using DFT of length $L_m$. Typically, the DFTs are implemented using the FFT algorithms. The frequency-domain bin values of the converted signal are multiplied by the weight values corresponding to the DFT of the finite-length linear filter impulse response $d_{m,\ell} = \text{DFT}(h_m[n])$ for $m = 0, 1, \ldots, M - 1$ and $\ell = 0, 1, \ldots, L - 1$. Here, $m$ is the subband index, and $\ell$ is the DFT bin index within the subband.[1] Finally, the weighted signals are combined and converted back to time-domain using inverse DFT of length $N$, and the resulting time-domain output blocks are concatenated using the overlap-save principle [21,56].

The multirate FC-processing of Fig. 8.6 increases the sampling rates of the subband signals by the factors

$$R_m = N/L_m = N_S/L_{S,m}, \tag{8.20}$$

where $L_{S,m}$ and $N_S$ are the numbers of nonoverlapping input and output samples, respectively. The number of overlapping samples $L_{O,m} = L_m - L_{S,m}$ is divided into leading and tailing overlapping parts as follows:

$$L_{L,m} = \lceil (L_m - L_{S,m})/2 \rceil \quad \text{and} \quad L_{T,m} = \lfloor (L_m - L_{S,m})/2 \rfloor. \tag{8.21}$$

Given the IFFT length $N$, the sampling rate conversion factor is determined by the FFT length $L_m$, and it can be configured for each subband individually.[2] Naturally, the FFT length determines the maximum number of nonzero frequency bins, i.e., the bandwidth of the subband. It is also possible to reduce the sampling rate conversion factor by increasing the FFT length by adding zero-valued bins outside the wanted subband frequency range. In communication applications, a subband would contain a communications waveform with specific symbol rate, and the input sampling rate is normally chosen as its (small) integer multiple.

Regardless of the specific processing applied for each subband, the interpolation factor of the structure of Fig. 8.6 is given by (8.20) [16]. This means that the length of the FC output block is increased by the same factor, and so is also the length of the overlapping part in the overlap-save processing. The input and output block lengths have to exactly match, taking into account the sampling rate conversion factor. Generally, we can write $N = aG$ and $N_S = bG$, where $a$ and $b$ are two relatively prime integers, $G = \gcf(N, N_S)$, and $\gcf(\cdot)$ stands for the greatest common factor. Then, for narrowest possible subband case satisfying the integer-length criterion, $L_m = a$ and $L_{S,m} = b$. Generally, $L_m$ has to be a multiple of $N/G$. For example, if $N_S = 3N/4$, $L_m$ has to be a multiple of 4 or if $N_S = 4N/5$, then $L_m$ has to be a multiple of 5. If $f_s$ is the input sampling rate, then the possible output sampling rates are multiples of $4f_s/N$ and $5f_s/N$, respectively. So the configurability of the output sampling rate depends greatly on the choice of $N$ and $N_S$.

### 8.3.1.2 *Linear Periodically Time-Varying (LPTV) Model of the FC-FB*

In the FC SFB case, the block processing of $m$th subcarrier signal $\mathbf{x}_m$ of length $T_m$ can be represented as [7]

$$\mathbf{w}_m = \mathbf{F}_m \mathbf{x}_m, \tag{8.22a}$$

---

[1] For convenience of notation, we use the "FFT-shifted" indexing scheme in this context, i.e., index 0 corresponds to the lower edge of the subband.
[2] Generally, there is no need to restrict the sampling rate conversion factor to take integer values.

where $\mathbf{F}_m$ is the block diagonal transform matrix of the form

$$\mathbf{F}_m = \text{diag}(\mathbf{F}_{m,0}, \mathbf{F}_{m,1}, \ldots, \mathbf{F}_{m,R-1}) = \begin{bmatrix} \mathbf{F}_{m,0} & & & \\ & \mathbf{F}_{m,1} & & \mathbf{0} \\ & & \cdot & \\ & \mathbf{0} & & \cdot \\ & & & & \mathbf{F}_{m,R-1} \end{bmatrix}.$$

$$(8.22b)$$

Here, the dimensions and locations of the $\mathbf{F}_{m,r}$s are determined by the overlapping factor of the overlap-save processing defined as

$$\lambda = 1 - L_{S,m}/L_m = 1 - N_S/N, \qquad (8.23)$$

where $L_{S,m}$ and $N_S$ are the numbers of nonoverlapping input and output samples, respectively.

The multirate version of the FC SFB can be represented using block processing by decomposing the $\mathbf{F}_{m,r}$s as the $N_S \times L_m$ matrix

$$\mathbf{F}_{m,r} = \mathbf{S}_N \mathbf{W}_N^{-1} \mathbf{M}_{m,r} \mathbf{D}_m \mathbf{P}_{L_m}^{(L_m/2)} \mathbf{W}_{L_m}. \qquad (8.24)$$

Here, $\mathbf{W}_{L_m}$ and $\mathbf{W}_N^{-1}$ are the $L_m \times L_m$ DFT matrix (with $[\mathbf{W}_{L_m}]_{p,q} = e^{-j2\pi(p-1)(q-1)/L_m}$ for $p = 1, 2, \ldots, L_m$ and $q = 1, 2, \ldots, L_m$) and the $N \times N$ inverse DFT matrix, respectively. The DFT shift matrix $\mathbf{P}_{L_m}^{(L_m/2)}$ is circulant permutation matrix obtained by cyclically left shifting the $L_m \times L_m$ identity matrix by $L_m/2$ positions. $\mathbf{D}_m$ is the $L_m \times L_m$ diagonal matrix with the frequency-domain weights of the subband $m$ on its diagonal, whereas $\mathbf{M}_{m,r}$ and $\mathbf{S}_N$ are the frequency-domain mapping and time-domain selection matrices, respectively. The $N \times L_m$ frequency-domain mapping matrix maps $L_m$ frequency-domain bins of the input signal to frequency-domain bins $(c_m - \lceil L_m/2 \rceil + \ell)_N$ for $\ell = 0, 1, \ldots, L_m - 1$ of the output signal. Here $c_m$ is the center bin of the subband $m$, and $(\cdot)_N$ denotes the modulo-$N$ operation. In addition, this matrix rotates the phases of the block by

$$\theta_m(r) = \exp(j2\pi r \theta_m) \quad \text{with} \quad \theta_m = c_m L_{S,m}/L_m \qquad (8.25)$$

to maintain the phase continuity between the consecutive overlapping processing blocks. In the basic FC-FB model, each subcarrier phase starts from zero in the beginning of each nonoverlapping FC processing block. However, the nonoverlapping block length corresponds to integer number of subcarrier cycles only in special cases, and this effect has to be compensated by a constant phase rotation for each FC processing block. The needed phase rotation depends on the subcarrier center frequency and the nonoverlapping block length in subcarrier samples. In (8.25), the $N_S \times N$ time-domain selection matrix $\mathbf{S}_N$ selects the desired $N_S$ output samples from the inverse transformed signal corresponding to overlap-save processing.

In the above processing, the bandwidth of each subband can be easily tuned by adjusting the $L_m$ frequency-domain weights in $\mathbf{D}_m$. Here, the maximum available bandwidth is determined by $L_m$. On the other hand, the center frequency of the subband can be adjusted by only cyclically shifting the columns of the frequency-domain mapping matrix $\mathbf{M}_{m,r}$ to their desired locations, whereas the interpolation factor is determined simply by the ratio of $N$ and $L_m$. Therefore, this FC SFB provides an easily configurable basis for flexible multichannel channelization FB for flexible radio access schemes.

In the AFB case, the corresponding analysis subblock matrix of size $L_{S,m} \times N$ can be decomposed as

$$\mathbf{G}_{m,r}^T = \mathbf{S}_{L_m} \mathbf{W}_{L_m}^{-1} \mathbf{P}_N^{(N/2)} \mathbf{D}_m \mathbf{M}_{m,r}^T \mathbf{W}_N, \tag{8.26}$$

where $\mathbf{P}_N^{(N/2)}$ is the inverse Fourier-shift matrix, and the $L_{S,m} \times L_m$ selection matrix $\mathbf{S}_{L_m}$ selects the desired $L_{S,m}$ output samples from the inverse transformed output signal.

In general, the above FC-based synthesis and analysis filterbanks are Linear Periodically Shift Variant (LPSV) systems with period of $L_{S,m}$, that is, the systems have $L_{S,m}$ different impulse responses. In the SFB case, the shift-variant impulse responses are given by the $L_{S,m}$ shift-invariant columns of the $\mathbf{F}_m$ as illustrated in Fig. 8.7. However, it should be pointed out that not all the $\mathbf{F}_{m,r}$s are necessarily the same. Especially, the beginning block $\mathbf{F}_{m,0}$ and the end block $\mathbf{F}_{m,R-1}$ may be different from the remaining $\mathbf{F}_{m,r}$s to properly process the beginning and end transients.

To carry out the processing such that all the incoming samples are facing the same set of impulse responses, the input signal $\mathbf{x}_m$ has to be padded at the beginning by $S_F = L_m - L_{S,m}$ zero-valued samples. Therefore, the number of blocks to be processed is given as

$$R_m = \lceil (T_m - L_m)/L_{S,m} \rceil + 1. \tag{8.27}$$

Further, $\mathbf{x}_m$ has to be appended at the end by $R_m L_{S,m} - T_m$ zero-valued samples to match the sizes of $\mathbf{F}_m$ and $\mathbf{x}_m$, that is, to correctly process all the incoming samples.

### 8.3.1.3 *Example FC SFB Design*

This example illustrates the formulation of the block diagonal transform matrix in (8.24) for $M = 1$, $N = 8$, $L_0 = 4$, and $L_{S,0} = 1$. In this case, the DFT matrix and DFT-shift permutation matrix are expressed as

$$\mathbf{W}_4 = \begin{bmatrix} 1 & 1 & 1 & 1 \\ 1 & -j & -1 & j \\ 1 & -1 & 1 & -1 \\ 1 & j & -1 & -j \end{bmatrix} \quad \text{and} \quad \mathbf{P}_4^{(2)} = \begin{bmatrix} 0 & 0 & 1 & 0 \\ 0 & 0 & 0 & 1 \\ 1 & 0 & 0 & 0 \\ 0 & 1 & 0 & 0 \end{bmatrix}, \tag{8.28}$$

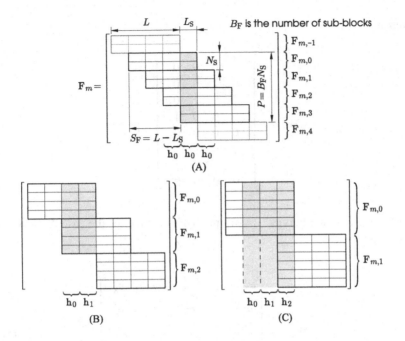

**FIGURE 8.7**

Structure of block-diagonal synthesis matrix $\mathbf{F}_m$ for $L_m = 4$, $N = 8$, and $L_{S,m} = 1, 2, 3$. The gray columns illustrate the elements of the synthesis matrix that form the shift-invariant impulse responses of the synthesis filterbank. (A) For $L_{S,m} = 1$, the system has only one impulse response of eight samples. (B) For $L_{S,m} = 2$, the system has two eight-samples long responses. (C) For $L_{S,m} = 3$, two among the three responses are six samples long and the remaining one is 12 samples long.

respectively. For the $L = 4$ case, the weights for the ideal RRC design are given as $\mathbf{d} = \begin{bmatrix} 0 & 1/\sqrt{2} & 1 & 1/\sqrt{2} \end{bmatrix}^T$, and, correspondingly, the diagonal frequency-domain masking matrix is given as

$$\mathbf{D}_0 = \begin{bmatrix} 0 & 0 & 0 & 0 \\ 0 & 1/\sqrt{2} & 0 & 0 \\ 0 & 0 & 1 & 0 \\ 0 & 0 & 0 & 1/\sqrt{2} \end{bmatrix}. \qquad (8.29)$$

In this example, we desire to shift the incoming signal two frequency bins to the right $[c_0 = 2$ in (8.25)], and, therefore, the $8 \times 4$ frequency-domain mapping matrix is given as

$$
\mathbf{M}_{0,r} = \begin{bmatrix} 1 & 0 & 0 & 0 \\ 0 & 1 & 0 & 0 \\ 0 & 0 & 1 & 0 \\ 0 & 0 & 0 & 1 \\ 0 & 0 & 0 & 0 \\ \vdots & \vdots & \vdots & \vdots \\ 0 & 0 & 0 & 0 \end{bmatrix} \exp(j\pi r). \tag{8.30}
$$

The Inverse Discrete Fourier Transform (IDFT) matrix for the case $N = 8$ is expressed as

$$
\mathbf{W}_8^{-1} = \begin{bmatrix} \omega_8^0 & \omega_8^0 & \omega_8^0 & \cdots & \omega_8^0 \\ \omega_8^0 & \omega_8^1 & \omega_8^2 & \cdots & \omega_8^7 \\ \omega_8^0 & \omega_8^2 & \omega_8^4 & \cdots & \omega_8^{2\times7} \\ \vdots & \vdots & \vdots & \ddots & \vdots \\ \omega_8^0 & \omega_8^7 & \omega_8^{2\times7} & \cdots & \omega_8^{7^2} \end{bmatrix} \tag{8.31}
$$

with $\omega_8 = 1/8 e^{j2\pi/8}$, and the time-domain selection matrix with $N_S = L_S N / L_0 = 2$ is given by

$$
\mathbf{S}_8 = \begin{bmatrix} 0 & 0 & 0 & 1 & 0 & 0 & 0 & 0 \\ 0 & 0 & 0 & 0 & 1 & 0 & 0 & 0 \end{bmatrix}. \tag{8.32}
$$

Finally, the subblock of the block diagonal transform matrix can be expressed as

$$
\mathbf{F}_{0,r} = \mathbf{S}_8 \mathbf{W}_8^{-1} \mathbf{M}_{0,r} \mathbf{D}_0 \mathbf{P}_4^{(2)} \mathbf{W}_4 \tag{8.33a}
$$

$$
= \begin{bmatrix} h_1\omega_8^3 + h_2\omega_8^6 + h_3\omega_8^9 & h_1\omega_8^4 + h_2\omega_8^8 + h_3\omega_8^{12} \\ jh_1\omega_8^3 + h_2\omega_8^6 - jh_3\omega_8^9 & jh_1\omega_8^4 + h_2\omega_8^8 - jh_3\omega_8^{12} \\ -h_1\omega_8^3 + h_2\omega_8^6 - h_3\omega_8^9 & -h_1\omega_8^4 + h_2\omega_8^8 - h_3\omega_8^{12} \\ -jh_1\omega_8^3 + h_2\omega_8^6 + jh_3\omega_8^9 & -jh_1\omega_8^4 + h_2\omega_8^8 + jh_3\omega_8^{12} \end{bmatrix}^{\mathrm{T}} \exp(j\pi r),
$$

$$
\tag{8.33b}
$$

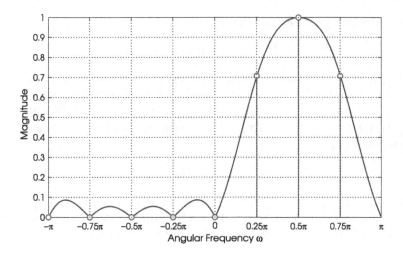

**FIGURE 8.8**

Magnitude response for the FC SFB with $M = 1$, $N = 8$, $L_0 = 4$, and $L_S = 1$.

and the corresponding impulse response (see Fig. 8.7A) is expressed as

$$
\mathbf{h} =
\begin{bmatrix}
-jh_1\omega_8^3 + h_2\omega_8^6 + jh_3\omega_8^9 \\
-jh_1\omega_8^4 + h_2\omega_8^8 + jh_3\omega_8^{12} \\
(-h_1\omega_8^3 + h_2\omega_8^6 - h_3\omega_8^9)\exp(j\pi) \\
(-h_1\omega_8^4 + h_2\omega_8^8 - h_3\omega_8^{12})\exp(j\pi) \\
(jh_1\omega_8^3 + h_2\omega_8^6 - jh_3\omega_8^9)\exp(j2\pi) \\
(jh_1\omega_8^4 + h_2\omega_8^8 - jh_3\omega_8^{12})\exp(j2\pi) \\
(h_1\omega_8^3 + h_2\omega_8^6 + h_3\omega_8^9)\exp(j3\pi) \\
(h_1\omega_8^4 + h_2\omega_8^8 + h_3\omega_8^{12})\exp(j3\pi)
\end{bmatrix}
= \frac{1}{8}
\begin{bmatrix}
0 \\
1 \\
2j \\
-(1 + 2/\sqrt{2}) \\
-2j \\
1 \\
0 \\
-(1 - 2/\sqrt{2})
\end{bmatrix}.
\tag{8.34}
$$

The magnitude response for this one-channel SFB is shown in Fig. 8.8. As can be seen from this figure, the resulting magnitude response perfectly resamples the desired one on the frequency points $\omega_n = 2\pi n/N$ for $n = 0, 1, \ldots, N - 1$.

### 8.3.1.4 *FC-FB Optimization*

In the case where all the $\mathbf{F}_{m,r}$s are the same (neglecting the phase difference), the $P \times L_{S,m}$ submatrix $\mathbf{F}_m^{(sub)}$ with $P = N_S B_F$ for $B_F = \lceil L_m/L_{S,m} \rceil$ containing all the $L_{S,m}$ possible impulse responses of the SFB can be obtained by selecting the desired rows and columns from $\mathbf{F}_m$ as follows:

$$\left[\mathbf{F}_m^{(\text{sub})}\right]_{p,q} = [\mathbf{F}_m]_{p,q} + S_{\text{F},m} \tag{8.35}$$

with $S_{\text{F},m} = L - L_{\text{S},m}$ for $p = 1, 2, \ldots, P$ and $q = 1, 2, \ldots, L_{\text{S},m}$. These submatrices are denoted in Fig. 8.7 by colored areas (gray in the print version).

For the AFB case, the corresponding $Q_m \times P$ submatrix $\mathbf{G}_m^{(\text{sub})}$ with $Q_m = L_{\text{S},m} B_{\text{G}}$ for $B_{\text{G}} = \lfloor (P + N)/N_{\text{S}} - (B_{\text{F}})2 \rfloor$ containing all the possible impulse responses AFB is given by

$$\left[\mathbf{G}_m^{(\text{sub})}\right]_{q,p} = [\mathbf{G}_m]_{q,p+S_{\text{G}}} \tag{8.36}$$

with $S_{\text{G}} = \lfloor (N + [B_{\text{G}} - 1]N_{\text{S}} - P)/2 \rfloor$ for $q = 1, 2, \ldots, Q_m$ and $p = 1, 2, \ldots, P$.

For the transmultiplexer, that is, the synthesis-analysis filterbank configuration, the transfer function from subchannel $m$ to subchannel $n$ is expressed as

$$\mathbf{T}_{m,n} = \mathbf{G}_n^{(\text{sub})} \mathbf{F}_m^{(\text{sub})}. \tag{8.37}$$

In the communications signal processing, the reconstruction error, i.e., the difference between the input signal and the (possibly delayed) output signal, on subchannel $m$ can be considered as ISI. In this case, the mean-squared ISI can be expressed as

$$\varepsilon_{\text{ISI}} = \left\| \mathbf{I} - \mathbf{T}_{m,m} \right\|_{\text{F}}^2, \tag{8.38}$$

where $\mathbf{I} = \begin{bmatrix} \mathbf{0}_{U_m \times L_{\text{S},m}} & \mathbf{E}_{L_{\text{S},m}} & \mathbf{0}_{V_m \times L_{\text{S},m}} \end{bmatrix}^{\text{T}}$ with $U_m = \lfloor S_{\text{G}} L_m / N \rfloor$ and $V_m = P - U_m - L_{\text{S},m}$. Here, $\mathbf{0}_{U_m \times L_{\text{S},m}}$ and $\mathbf{0}_{V_m \times L_{\text{S},m}}$ are the $U_m \times L_{\text{S},m}$ and $V_m \times L_{\text{S},m}$ zero matrices, respectively, $\mathbf{E}_{L_{\text{S},m}}$ is the $L_{\text{S},m} \times L_{\text{S},m}$ identity matrix, and the squared Frobenius norm of an $m \times n$ matrix $\mathbf{A}$ is defined as

$$\|\mathbf{A}\|_{\text{F}}^2 = \sum_{i=1}^{m} \sum_{j=1}^{n} |a_{i,j}|^2 = \text{tr}(\mathbf{A}\mathbf{A}^H). \tag{8.39}$$

Correspondingly, the mean-squared ICI can be considered as an interference from input subchannel $m$ to all the output subchannels $n$ for $n \neq m$ expressed as

$$\varepsilon_{\text{ICI}} = \sum_{n=0, n \neq m}^{M-1} \left\| \mathbf{T}_{n,m} \right\|_{\text{F}}^2. \tag{8.40}$$

The FC-FB transmultiplexer design can be stated as a problem of finding the optimal values of the frequency-domain window (diagonal of $\mathbf{D}_m$ in (8.24) and (8.26)) to minimize

$$\varepsilon = \varepsilon_{\text{ISI}} + \gamma \varepsilon_{\text{ICI}}, \tag{8.41}$$

where $\gamma$ is the desired weighting factor for trading between the ISI and ICI, whereas $\varepsilon_{\text{ISI}}$ and $\varepsilon_{\text{ICI}}$ are given by (8.38) and (8.39), respectively. In the following, we use

$\gamma = 1$. This problem can be straightforwardly solved using nonlinear optimization algorithm since the number of parameters for the optimization is typically around ten when taking into account the symmetrical transition bands.

As an example, we focus here on a 5 MHz Long-Term Evolution (LTE) like MC system utilizing the FBMC/OQAM waveform and using FC-FB-based transmultiplexers. For the 5 MHz LTE case, the number of subcarriers is $M = 512$, out of which $M_{used} = 300$ active subcarriers are used [22]. The 300 active subcarriers are scheduled in Physical Resource Blocks (PRBs) of 12 subcarriers, i.e., there are 25 PRBs. In time domain, the PRB have the length of 0.5 ms, i.e., 3840 samples at the used sampling rate of 7.68 MHz. The PRBs are scheduled to different users in 1 ms subframes of two PRBs. Especially in the uplink, the minimum transmission burst length is 1 ms, consisting of 7680 samples. In the basic LTE transmission scheme, 14 OFDM symbols are transmitted during one subframe. For simplicity, we consider the case where all the short transform lengths $L_m$ for $m = 0, 1, \ldots, M - 1$ are the same, and, therefore, the subband index $m$ is excluded from the block-length expressions.

Here the approach for selecting the main parameters of the FC-FB scheme is to use the same sampling rate, the same subcarrier spacing, and the same parameters for the PRBs as in the 5 MHz LTE system. Consequently, the FFT length can be determined as $N = ML/2$, where $L$ is the IFFT length, and the subcarrier spacing is $L/2$ FFT bins. (For clarity of the discussion, we focus here on the AFB of the receiver.) The choice of the FFT length, together with the overlap factor, defines the time interval of each processing block and has a significant effect on the performance of the scheme with fast-fading channels. However, this depends greatly on the used channel equalization structure, and the analysis of the fading sensitivity is left as a topic for future studies. On the other hand, increasing the FFT length gives better chances to shape the spectrum to improve spectral containment. In any case, we assume that the FFT length should be shorter than the PRB length, and we look for a good trade-off between the spectrum control and fading sensitivity. A third element in the tradeoff is implementation complexity. Increasing the overlap factor improves the spectral containment for a given IFFT length, but it also increases the computational complexity as the ratio of the useful part of FFT to the FFT length is reduced.

For the two-times oversampled case, the suboptimal design can be obtained by simply using the RRC-type filter with roll-off of one. Fig. 8.9 compares the performance of the optimized FC-FB designs with the direct RRC FC-FB and polyphase FBMC designs using the PHYDYAS prototype [83] in terms of simulated MSE between the transmitted and received signals. As can be seen from this figure, the length of the overlapping part $L_O = L - L_S$ determines the minimum achievable MSE, and the performance is similar for the different values of $L$. In addition, the RRC design is nearly optimal for the values of the overlapping part smaller than or equal to four, and the overlap has to be at least six samples for FC-FB design to perform better than PHYDYAS polyphase FBMC with $K = 3$ or eight samples for $K = 4$.

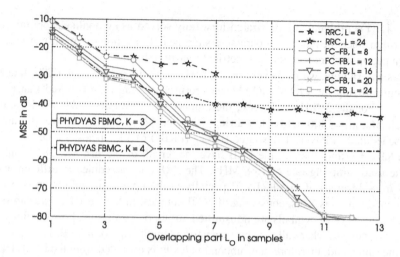

**FIGURE 8.9**

Simulated MSE as a function of the overlapping part $L_O = L - L_S$ of the input samples for FC-FB with $L = 8, 12, 16, 20, 24$. For comparison, the MSE of RRC design for $L = 8$ and $L = 24$ and PHYDYAS polyphase FBMC [83] with $K = 3$ and $K = 4$ are also shown.

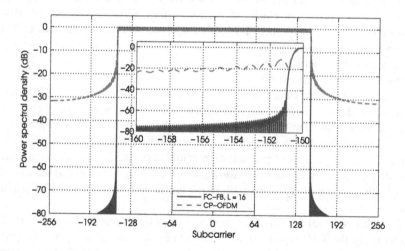

**FIGURE 8.10**

PSD of the FC-FB waveform for $L = 16$ and $L_S = L_O = 8$. For comparison, the PSD of CP-OFDM design with $L_{OFDM} = 512$ and $L_{CP} = 36$ is also shown.

Fig. 8.10 shows the Power Spectral Density (PSD) plot of the optimized FC-FB waveform for $L = 16$ and $L_S = L_O = 8$. The PSD of Cyclic Prefix Orthogonal Frequency-Division Multiplexing (CP-OFDM) design with IFFT length of $L_{OFDM} =$

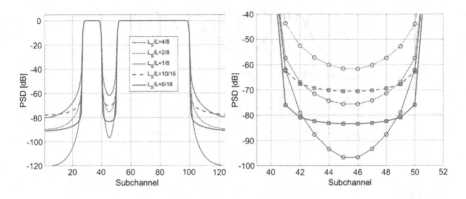

**FIGURE 8.11**

Alternative FC-based FBMC/OQAM designs for a noncontiguous LTE-like scenario with 72 subcarriers, out of which one resource block of 12 subcarriers is deactivated. The designs include cases with $L = \{8, 16\}$ FFT bins per subcarrier and different overlap factors $\lambda = 1 - L_S/L$.

512 and CP length of $L_{CP} = 36$ is also shown for comparison. This figure illustrates the reduced out-of-band emissions of the FC-FB waveform when compared with the basic CP-OFDM.

## 8.3.2 FREQUENCY-SPREADING FBMC

In the special case of maximum overlap in FC processing, i.e., where $L_S = 1$ and $\lambda = \frac{L-1}{L}$, FC processing becomes time-invariant, and the additional interference effects due to FC processing disappear. In this case, the FC implementation corresponds closely (ignoring finite wordlength effects) to a polyphase implementation with specific prototype filter design.[3] This special case of FC-FB has been presented earlier as the so-called Frequency Spreading FBMC (FS-FBMC) scheme [10,25,44], initially in [2] before the introduction of FC-FB-based FBMC processing. This approach gives uncompromising FC processing quality, but with the cost of highly increased computational complexity, even though the high overlap allows us to use relatively short transform lengths.

Fig. 8.11 shows PSDs[4] for LTE-like design cases with an active frequency band of 72 subcarriers. A noncontiguous spectrum use case is assumed, consisting of one PRB of 12 active subcarriers, and another set of 48 active subcarriers, intervened by

---

[3]Actually, the presented FC-FB optimization scheme could be used for optimizing prototype filters for polyphase implementation.
[4]Here PSDs are obtained by an FC-FB with the same subchannel bandwidth and RRC-type frequency responses as the target system, but with high number of FFT bins and high overlap, so that the PSD of the target signal is dominating.

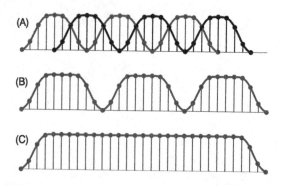

**FIGURE 8.12**

Principle of generating weight masks for different waveforms using a fixed transition band weight mask. (A) FBMC/OQAM. (B) FMT. (C) Traditional single-carrier waveform.

one PRB with deactivated subcarriers. Designs with 4 or 8 FFT bins per subcarrier spacing (corresponding to $L = 8$ or $L = 16$, respectively) are included with different overlap factors. The case with $L_S/L = 1/8$ corresponds to frequency-spread FBMC case with polyphase overlap factor $K = 4$, exhibiting excellent spectrum localization. With increasing FC overlap, the PSD is rapidly degraded. Also the inband Signal-to-Interference Ratio (SIR) is reduced from 50 dB with $L_S = 1$ and 45 dB with $L_S = 2$ to 28 dB with $L_S = 4$. With $L = 16$, the corresponding values are 58 dB and 40 dB with $L_S = 8$ and $L_S = 10$, respectively.

### 8.3.3 FC-FB AS A GENERIC WAVEFORM PROCESSING ENGINE

In addition to FBMC/OQAM, FC-FB can be used for generating and detection of FMT-type multicarrier waveforms [64] and traditional Nyquist pulse shaping-based Single-Carrier (SC) waveforms [86,89], referred to as FilterBank Single-Carrier (FB-SC). FC-FB can also be used for resource block level filtering for filtered OFDM waveforms [60,84] and for channelization filtering with arbitrary waveforms. Basically, the same processing structure, but without overlap processing, can be also used for implementing circular multicarrier waveforms introduced in Section 7.3.3, like GFDM, Cyclic Block-Filtered MultiTone (CB-FMT), and FilterBank MultiCarrier with Circular Offset Quadrature Amplitude Modulation (FBMC/COQAM).

Fig. 8.12 illustrates how different waveforms can be constructed using a fixed transition band weight mask. The (sub)channel frequency responses are constructed by adding one-valued weights between symmetric transition bands. In this way, the bandwidth and roll-off can be adjusted in a flexible manner. Whereas the weights can be optimized separately for each case, the performance reduction is usually rather small when using a well-optimized fixed weight mask. In addition, zero-valued weights are added around the passband to match the short transform size, usually to achieve two times oversampling of the (sub)band signals with respect to the symbol

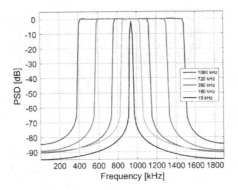

**FIGURE 8.13**

PSDs of SC waveforms with different bandwidths and roll-off factors obtained using a fixed transition band weight mask, which is optimized for the roll-off of 1 with $L = 16$ and $L_S = 10$. The FFT bin spacing is 15/8 kHz.

rate. Fig. 8.13 shows PSDs for FB-SC waveforms with different bandwidths obtained using a transition band weight mask from optimized FBMC/OQAM design with $L = 16$.

As explained in Section 12.5, channel equalization for FBMC and FB-SC waveforms can be realized in a unified manner by combining the channel equalization weights with the (sub)channel weight masks in FFT domain. Also synchronization-related functions can be effectively combined with the FC-FB processing structure, as discussed in Section 10.6.

In the case of filtered OFDM waveform, the FC-FB is applied for Physical Resource Block (PRB) level filtering while utilizing normal CP-OFDM waveform for the PRBs [60,61,84]. One application is in cellular uplink scenarios, in which the different User Equipments (UEs) utilize different PRBs for their transmissions. UEs may utilize single-channel FC-filtering or the corresponding time-domain filtering to shape their spectra. For good isolation of different users' PRBs, a few subcarriers are needed as guardbands between the PRBs. Good Out-of-Band Emission (OBE) characteristics allow us to relax the tight synchronization requirements of traditional uplink multiuser Orthogonal Frequency-Division Multiple Access (OFDMA). On the base-station side, an FC-based analysis filterbank is used for separating the PRBs of different UEs. Furthermore, it becomes possible to parameterize the waveforms differently for different UEs, e.g., using different subcarrier spacings, CP-lengths, and/or frame structures [61,84]. Concerning cellular downlink scenarios, synchronization is not an issue, but the filtered OFDM idea would still make it possible to parameterize individually different users' signals in different groups of PRBs.

With FC-FB, it is easy to adjust the filtering bandwidth for the subbands individually. This is very useful in PRB-filtered OFDM because there is no need to realize filter transition bands and guardbands between equally parameterized synchronous

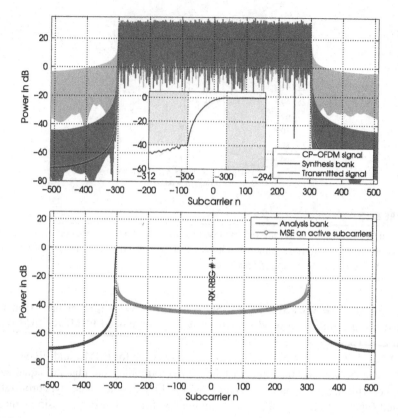

**FIGURE 8.14**

Upper figure: PSD of the generated FC-filtered OFDM signal in the case of 50 active PRBs of 12 subcarriers. The overlap factor for FC processing is $\lambda = 1/4$, and the required stopband attenuation is 40 dB. Lower figure: Simulated in-band subcarrier MSE in the case of 50 active PRBs.

PRBs. In the extreme case, as illustrated in Fig. 8.14, the group of filtered PRBs can cover the full carrier bandwidth, and FC filtering would implement tight channelization filtering for the whole carrier.

## 8.4 PRACTICAL IMPLEMENTATION ASPECTS

### 8.4.1 OQAM PREPROCESSING IN FREQUENCY-DOMAIN

In case of FBMC/OQAM, the OQAM pre- and postprocessing stages need to be combined with the basic FC-FB structure. The time-domain preprocessing block, which utilizes the transformation between QAM and OQAM symbols, is shown in Fig. 8.15.

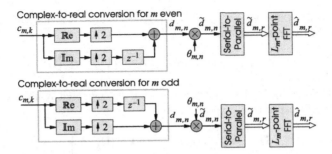

**FIGURE 8.15**

Time-domain OQAM preprocessing.

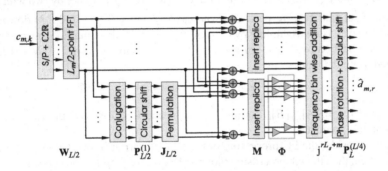

**FIGURE 8.16**

Frequency-domain OQAM preprocessing.

As seen from this figure, the first operation is a simple complex-to-real conversion, where the real and imaginary parts of a complex-valued symbol $c_{m,k}$ are separated to form two new symbols $d_{m,2k}$ and $d_{m,2k+1}$ (this operation is also called staggering). The order of these new symbols depends on the subchannel number, i.e., the conversion is different for even- and odd-numbered subchannels. The complex-to-real conversion increases the sample rate by a factor of two. The second operation is the multiplication by $\theta_{m,n}$ as defined in (8.16).

Alternatively, the OQAM preprocessing can be carried out in frequency-domain by utilizing the symmetries of the discrete-time Fourier transform (Fig. 8.16). For the ease of notation, let $c_{m,k}$ for $m = 0, 1, \ldots, M-1$ and $k = 0, 1, \ldots, S-1$ be the complex-valued symbols given by

$$c_{m,k} = [\mathbb{Re}(c_{m,k}) + j\,\mathbb{Im}(c_{m,k})], \tag{8.42}$$

where $M$ and $S$ are the numbers of subcarriers and symbols, respectively. Furthermore, let $N = 2S$ be the number of OQAM symbols. In the sequel, we denote the $M \times S$ matrices containing the real and imaginary parts of the symbols by

$\mathbf{C}_{\text{re}} = [\mathbb{Re}(c_{m,k})]$ and $\mathbf{C}_{\text{im}} = [\mathbb{Im}(c_{m,k})]$, respectively. Now, the OQAM preprocessing can be expressed as

$$\mathbf{D}_{\text{OQAM}} = \boldsymbol{\Theta} \circ \left[ (\mathbf{V}_M^{(0)})^{\mathsf{T}} \mathbf{D}_e + (\mathbf{V}_M^{(1)})^{\mathsf{T}} \mathbf{D}_o \right] \tag{8.43a}$$

with $\mathbf{D}_e$ and $\mathbf{D}_o$ being $M/2 \times 2S$ matrices giving the staggered symbols on even and odd subcarriers, respectively, expressed as

$$\mathbf{D}_e = \mathbf{V}_M^{(0)} \left[ \mathbf{C}_{\text{re}} (\mathbf{U}_S^{(0)})^{\mathsf{T}} + \mathbf{C}_{\text{im}} (\mathbf{U}_S^{(1)})^{\mathsf{T}} \right], \tag{8.43b}$$

$$\mathbf{D}_o = \mathbf{V}_M^{(1)} \left[ \mathbf{C}_{\text{im}} (\mathbf{U}_S^{(0)})^{\mathsf{T}} + \mathbf{C}_{\text{re}} (\mathbf{U}_S^{(1)})^{\mathsf{T}} \right]. \tag{8.43c}$$

Here, $\mathbf{U}_S^{(0)}$ and $\mathbf{U}_S^{(1)}$ are $2S \times S$ upsampling by two and upsampling by two with one-sample offset matrices,[5] respectively, whereas $\mathbf{V}_M^{(0)}$ and $\mathbf{V}_M^{(1)}$ are $M/2 \times M$ downsampling by two and downsampling by two with one-sample offset matrices, respectively. The multiplication by $\theta_{m,n}$ for $m = 0, 1, \ldots, M-1$ and $n = 0, 1, \ldots, N-1$ is expressed using the matrix

$$[\boldsymbol{\Theta}]_{m+1,n+1} = e^{j\frac{\pi}{2}(m+n)}, \tag{8.43d}$$

and $[A \circ B]_{i,j} = [A]_{i,j} [B]_{ij}$ for all $1 \le i \le n$ and $1 \le j \le m$ is the Hadamard or elementwise product.

Now, consider the FC-processing of the $m$th subcarrier in overlapping segments. If $m$ is even, then the $r$th processing block of $\mathbf{D}_{\text{OQAM}}$ of length $L$ on subcarrier $m$ can be expressed as

$$[\mathbf{d}_{m,r}]_{p+1} = [\mathbf{D}_{\text{OQAM}}]_{m+1,p+1+B_r} \tag{8.44a}$$

$$= [\boldsymbol{\theta}_{m,r}]_{p+1} \circ \left[\mathbf{d}_{m/2,r}^{(e)}\right]_{p+1} \tag{8.44b}$$

for $p = 0, 1, \ldots, L-1$ with $B_r = r(L - L_O)$. Here, $\boldsymbol{\theta}_{m,r}$ and $\mathbf{d}_{m/2,r}^{(e)}$ are the corresponding segments of $\boldsymbol{\Theta}$ and $\mathbf{D}_e$, respectively, given by

$$[\boldsymbol{\theta}_{m,r}]_{p+1} = [\boldsymbol{\Theta}]_{m+1,p+1+B_r}, \tag{8.44c}$$

$$\left[\mathbf{d}_{m/2,r}^{(e)}\right]_{p+1} = [\mathbf{D}_e]_{m/2+1,p+1+B_r}, \tag{8.44d}$$

for $p = 0, 1, \ldots, L-1$. In FC SFB processing, these blocks are transformed to frequency domain using DFT. This process can be expressed as

$$\widehat{\mathbf{d}}_{m,r} = \mathbf{W}_L \mathbf{d}_{m,r} \tag{8.45a}$$

$$= j^{rL_S+m} \mathbf{P}_L^{(L/4)} \mathbf{W}_L \mathbf{d}_{m/2,r}^{(e)}, \tag{8.45b}$$

---

[5] The upsampling by two matrix is identically the $2S \times 2S$ identity matrix with even-numbered columns removed whereas the upsampling by two with one-sample offset matrix is the corresponding identity matrix with odd-numbered columns removed.

where $\mathbf{P}_L^{(q)}$ is a circular permutation matrix obtained by cyclically left-shifting the $L \times L$ identity matrix by $q$ positions. The DFT of a real sequence $\mathbf{d}_{m/2,r}^{(e)}$ can be written using the symmetries of DFT as [49]

$$\mathbf{W}_L \mathbf{d}_{m/2,r}^{(e)} = \frac{1}{2}\left[\mathbf{M}\big(\widehat{\mathbf{c}}_{m/2,r} + \mathbf{J}_{L/2}\mathbf{P}_{L/2}^{(1)}\widehat{\mathbf{c}}_{m/2,r}^*\big) - \boldsymbol{\Phi}\mathbf{M}\big(\widehat{\mathbf{c}}_{m/2,r} - \mathbf{J}_{L/2}\mathbf{P}_{L/2}^{(1)}\widehat{\mathbf{c}}_{m/2,r}^*\big)\right],$$

(8.46a)

where $\widehat{\mathbf{c}}_{m/2,r}$ is the length $L/2$ DFT of $\mathbf{d}_{m/2,r}^{(e)}$ with even samples substituted on the real part and odd samples on the imaginary part expressed as

$$\widehat{\mathbf{c}}_{m/2,r} = \mathbf{W}_{L/2}\left[\mathbf{V}_L^{(0)}\mathbf{d}_{m/2,r}^{(e)} + j\mathbf{V}_L^{(1)}\mathbf{d}_{m/2,r}^{(e)}\right],$$

(8.46b)

whereas $\mathbf{J}_{L/2}$ is the $L/2 \times L/2$ reverse identity matrix, $\mathbf{M}_L = \begin{bmatrix}\mathbf{I}_{L/2} & \mathbf{I}_{L/2}\end{bmatrix}^\mathsf{T}$, and

$$[\boldsymbol{\Phi}]_{p+1,p+1} = \exp\left(\frac{j\pi(1/2 - p)}{L/2}\right)$$

(8.46c)

for $p = 0, 1, \ldots, L - 1$.

Combining (8.45) and (8.46) and performing the same derivations for the odd subcarriers, the block-wise FC processing of each $m$ OQAM subcarrier signal can be carried out by taking the length $L/2$ DFT of the incoming processing blocks and then shifting and rotating the transformed block by the trivial sequence resulting in a simplified processing for the incoming OQAM signals.

## 8.4.2 COMPUTATIONAL COMPLEXITY

In this section, we evaluate the computational complexity of the FC-FB scheme and compare it with the complexity of the polyphase filterbank structure. The focus here is on the FBMC/OQAM-type AFB, and we use the number of real multiplications and real additions per received complex symbol as the metric for the complexity.

Since FFT and IFFT are the core modules in both types of filterbanks, we start with their complexity. For a given transform length, FFT and IFFT have the same complexity, so that we talk only about the FFT complexity. When the transform length is a power of two, the split-radix algorithm is commonly considered to be the most efficient one [73], and the number of real multiplications and additions needed for the transforms are given in Table 8.1. Here we also consider transform lengths of the form $3\hat{N}$ and $5\hat{N}$, where $\hat{N}$ is a power of two. Generally, the availability of transform lengths other than powers of two increases greatly the flexibility of waveform parameterization.

Here, we assume that $M_{\text{used}}$ out of the $M = 2N/L_m$ subchannels of the FBMC/OQAM-type FC-FB structure are in use. Further, we assume that generic complex FFT-domain weights are used. Real weights and certain trivial weight values could be utilized to reduce the complexity, but this limits the possibilities in FC-FB

**Table 8.1** The number of real multiplications and additions for the transform lengths of $N = \hat{N}$, $N = 3\hat{N}$, and $N = 5\hat{N}$, where $\hat{N}$ is a power of two

| $N$ | Number of real multiplications | Number of real additions |
|---|---|---|
| $\hat{N}$ | $C_M = \hat{N}(\log_2(\hat{N}) - 3) + 4$ | $C_A = \hat{N}(3\log_2(\hat{N}) - 3) + 4$ |
| $3\hat{N}$ | $C_M = \hat{N}(3\log_2(\hat{N}) - 7) + 12$ | $C_A = \hat{N}(9\log_2(\hat{N}) + 1) + 12$ |
| $5\hat{N}$ | $C_M = \hat{N}(5\log_2(\hat{N}) - 11) + 20$ | $C_A = \hat{N}(25\log_2(\hat{N}) + 9) + 20$ |

optimization. Additionally, the complex FFT-weights can also be used for subchannel equalization purposes. With these assumptions, an FC AFB implementation includes a) FFT of length $N$, b) $M_{used}(L - 1)$ complex weight coefficients, each assumed to take four real multiplications and two additions, and c) $M_{used}$ IFFTs of length $L$. These calculations for one FFT block produce $M_{used}Ls/2$ complex symbols because the FC-processing produces two times oversampled output sequences.

Let us next consider the arithmetic complexity of a polyphase AFB with the same number of subbands $M = 2N/L$ as in the FC-FB. This implementation includes a) FFT of length $M$, b) $KM$ real coefficients in the polyphase filter structure; the multiplication rate is double due to complex input, c) $2(K - 1)M$ real adders, and d) complex weights as single-tap subcarrier equalizers, each assumed to take four real multiplications and two additions. The equalizer taps are implemented at the subcarrier symbol rate; otherwise, the mentioned processing functions need to be implemented two times for each symbol interval to produce $M_{used}$ complex symbols. Here single-tap subcarrier equalizer is included for both approaches. The complex FFT-domain weights of FC-FB can be also used for implementing multitap subcarrier equalizers with no additional complexity, whereas multitap subcarrier equalization would increase the complexity in the polyphase structure.

Fig. 8.17 gives a comparison of the two filterbank approaches in terms of arithmetic complexity in an example case with 5 MHz LTE like parameters [77]. Also the complexity of basic OFDM receiver processing is included for comparison. It is seen that the FC-FB has significantly lower complexity than the polyphase structure for the same MSE values. On the other hand, FC-FB has the complexity of about two times the complexity of OFDM. Naturally, OFDM has inferior spectral leakage performance compared to the other schemes.

### 8.4.3 BURST TRUNCATION EFFECTS

One important consideration in FBMC systems is the increase in the transmission burst length due to the "filtering tails." This is critical especially in the case of Time-Division Duplex (TDD) operation and tight asynchronous time-domain multiplexing. Generally, when transmitting a burst of FBMC/OQAM symbols through a polyphase SFB system with overlapping factor of $K$, the burst length is increased by $(K - \frac{1}{2})T$.

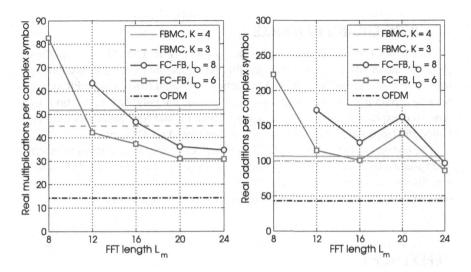

**FIGURE 8.17**

Multiplication and addition rates for FBMC/OQAM-type FC-FBs, polyphase FBs, and basic OFDM processing for $M = 512$ subchannels, out of which $M_{used} = 300$ are used. For FC-FB, the cases with $L_O = 6$ and $L_O = 8$ are shown.

For example, with $K = 4$, filtering tails of 1.75 OQAM symbol intervals are generated on both sides of the transmission burst. However, significant truncation of the tails is possible with reasonable level of inband interference and out-of-band emissions. This issue has been studied in [9,12,87]. Tail truncation to half of the OQAM symbol interval (i.e., $T/2$ on both sides can be considered as a convenient solution, providing inband MSE below $-40$ dB and out-of-band emission level below $-50$ dB. Depending on the target Signal-to-Noise Ratio (SNR) and spectral containment, even $T/4$ might be considered as the tail length after truncation [9].

An efficient way of combining OFDM-type training preamble with the truncated FBMC/OQAM transmission burst using the so-called memory preloading idea with polyphase FB has been presented in [9].

Considering the receiver signal processing side, the filtering tail effect introduces interference also between consecutive closely spaced asynchronous transmission bursts, e.g., in a base-station receiver. In such cases, burst truncation should be done also on the receiver side, instead of running the AFB in a continuous manner over the transition period between consecutive bursts. Zero padding is usually needed for the first and last input data block of each burst. In case of polyphase FB, the filtering of a new burst can be initialized effectively by preloading data to the polyphase FB memory. With FC-FB-based realization, consecutive asynchronous bursts need to be processed in different FC processing blocks.

## 8.5 CONCLUDING REMARKS

In this chapter, different prototype filter designs for FBMC/OQAM were introduced, including alternative time-frequency-localized designs, Frequency-Limited Orthogonal (FLO) designs, and basic TLO approaches. The main focus was on PR systems, but also certain important NPR designs were introduced. Polyphase realization structures were explained, and they can be used for both PR and NPR cases. Also, effective lattice and lapped transform structures were introduced for PR banks.

The last part of the chapter focused on the effective and flexible realization structure based on fast convolution processing. It is applicable for different FBMC and SC waveforms and for effective filtered-OFDM realizations, facilitating also simultaneous processing of different or differently parameterized waveforms.

## REFERENCES

[1] Enhanced multicarrier techniques for professional ad-hoc and cell-based communications (FP7-ICT Project 318362 EMPhAtiC). [Online]. Available: http://www.ict-emphatic.eu/.

[2] Physical layer for dynamic spectrum access and cognitive radio (FP7-ICT Project 211887 PHYDYAS). [Online]. Available: http://www.ict-phydyas.org/.

[3] M. N. Abdulazim, T. Kurbiel, and H. G. Göckler, "Modified DFT SBC-FDFMUX filter bank systems for flexible frequency reallocation," in *Proc. EURASIP European Signal Processing Conference (EUSIPCO)*, Poznań, Poland, 2007, pp. 60–64.

[4] P. Achaichia, M. Le Bot, and P. Siohan, "Windowed OFDM versus OFDM/OQAM: A transmission capacity comparison in the HomePlug AV context," in *Proc. IEEE International Symposium on Power Line Communications and Its Applications (ISPLC)*, Apr. 2011, pp. 405–410.

[5] M. Alard, "Construction of a multicarrier signal." Patent WO 96/35278, 1996.

[6] M. Alard, C. Roche, and P. Siohan, "A new family of functions with a nearly optimal time-frequency localization." Technical report of the RNRT Modyr project, 1999.

[7] D. M. Baylon and J. S. Lim, "Transform/subband analysis and synthesis of signals." Tech. Rep. 559. Research Laboratory of Electronics, Massachusetts Institute of Technology, Jun. 1990.

[8] M. Bellanger, Ed., *Digital Signal Processing: Theory and Practice*, 3rd ed. Wiley, 2000.

[9] M. Bellanger, "Efficiency of filter bank multicarrier techniques in burst radio transmission," in *Proc. IEEE Global Telecommunication Conference (GLOBECOM)*, Dec. 2010, pp. 1–4.

[10] M. Bellanger, "FS-FBMC: A flexible robust scheme for efficient multicarrier broadband wireless access," in *Proc. IEEE Globecom Workshops (GC Workshops)*, Anaheim, CA, USA, Dec. 2012, pp. 192–196.

[11] M. Bellanger, D. Mattera, and M. Tanda, "Lapped-OFDM as an alternative to CP-OFDM for 5G asynchronous access and cognitive radio," in *Proc. IEEE Vehicular Technology Conference, VTC-Spring*, May 2015, pp. 1–5.

[12] M. Bellanger, M. Renfors, T. Ihalainen, and C. A. F. da Rocha, "OFDM and FBMC transmission techniques: A compatible high performance proposal for broadband power line

communications," in *Proc. 2010 IEEE International Symposium on Power Line Communications and Its Applications (ISPLC)*, Rio de Janeiro, Brazil, Mar. 2010, pp. 154–159.

[13] M. Bellanger, "Specification and design of a prototype filter for filter bank based multicarrier transmission," in *Proc. IEEE International Conference on Acoustics, Speech, and Signal Processing (ICASSP)*, vol. 4, Salt Lake City, UT, USA, May 2001, pp. 2417–2420.

[14] H. Boelcskei, "Efficient design of pulse shaping filters for OFDM systems," in *Proc. SPIE Conference*, Denver, USA, Jul. 1999, pp. 625–636.

[15] H. Boelcskei, "Orthogonal frequency division multiplexing based on offset QAM," in *Advances in Gabor Theory*. Birkhäuser, 2003.

[16] M. Borgerding, "Turning overlap-save into a multiband mixing, downsampling filter bank," *IEEE Signal Process. Mag.*, vol. 23, no. 2, pp. 158–162, Mar. 2006.

[17] M.-L. Boucheret, I. Mortensen, and H. Favaro, "Fast convolution filter banks for satellite payloads with on-board processing," *IEEE J. Sel. Areas Commun.*, vol. 17, no. 2, pp. 238–248, Feb. 1999.

[18] R. Bregović and T. Saramäki, "A systematic technique for designing prototype filters for perfect-reconstruction cosine modulated and modified DFT filter banks," in *Proc. IEEE International Symposium on Circuits and Systems (ISCAS)*, vol. II, May 2001, pp. 33–36.

[19] G. Cariolaro and F. Vagliani, "An OFDM scheme with a half complexity," *IEEE J. Sel. Areas Commun.*, vol. 13, no. 9, pp. 1586–1599, Dec. 1995.

[20] Z. Cvetković and J. D. Johnston, "Nonuniform oversampled filter banks for audio signal processing," *IEEE Trans. Acoust., Speech, Signal Process.*, vol. 11, no. 5, pp. 393–399, Sep. 2003.

[21] A. Daher, E.-H. Baghious, G. Burel, and E. Radoi, "Overlap-save and overlap-add filters: Optimal design and comparison," *IEEE Trans. Signal Process.*, vol. 58, no. 6, pp. 3066–3075, Jun. 2010.

[22] E. Dahlman, S. Parkvall, and J. Sköld, *4G LTE/LTE-Advanced for Mobile Broadband*. Academic Press, 2011.

[23] M. Daloglu and E. Serpedin, "Quantization errors in overlapped block digital filtering methods," in *Proc. Advanced International Conference on Telecommunications (AICT)*, Rome, Italy, Jun. 2013.

[24] Y. Dandach and P. Siohan, "FBMC/OQAM modulators with half complexity," in *Proc. IEEE Global Telecommunication Conference (GLOBECOM)*, Houston, USA, Dec. 2011.

[25] J. B. Doré, V. Berg, N. Cassiau, and D. Kténas, "FBMC receiver for multiuser asynchronous transmission on fragmented spectrum," *EURASIP J. Adv. Signal Process., Special Issue on Advances in Flexible Multicarrier Waveform for Future Wireless Communications*, vol. 2014, no. 41, pp. 1–21, Mar. 2014.

[26] M. I. Doroslovački, "Product of second moments in time and frequency for discrete-time signals and the uncertainty limit," *Signal Process.*, vol. 67, no. 1, May 1998.

[27] J. Du and S. Signell, "Pulse shape adaptivity in OFDM/OQAM systems," in *Proc. 2008 Int. Conf. on Advanced Infocomm Technology (ICAIT)*, Shenzhen, China, 2008.

[28] A. Eghbali, H. Johansson, P. Löwenborg, and H. G. Göckler, "Dynamic frequency-band reallocation and allocation: From satellite-based communication systems to cognitive radios," *J. Signal Process. Syst.*, vol. 62, no. 2, pp. 187–203, Feb. 2009.

[29] B. Farhang-Boroujeny and C. Yuen, "Cosine modulated and offset QAM filter bank multicarrier techniques: A continuous-time prospect," *EURASIP J. Adv. Signal Process.*, vol. 2010, no. 8, pp. 1–16, 2010.

[30] G. Fettweis, M. Krondorf, and S. Bittner, "GFDM – Generalized frequency division multiplexing," in *Proc. IEEE Vehicular Technology Conference, VTC-Spring*, Barcelona, Spain, Apr. 2009, pp. 26–29.

[31] A. Groth, F. Budke, H. G. Göckler, and G. Evangelista, "Filter design for multirate systems based on fast convolution," in *Proc. EURASIP European Signal Processing Conference (EUSIPCO)*, vol. IV, Tampere, Finland, Sep. 2000, pp. 1901–1905.

[32] R. Haas and J.-C. Belfiore, "Multiple carrier transmission with time-frequency well-localized impulses," in *Proc. 2nd IEEE Symposium on Communications and Vehicular Technology in the Benelux*, Louvain-la-Neuve, Belgium, 1994, pp. 187–193.

[33] R. Haas and J.-C. Belfiore, "A time-frequency well-localized pulse for multiple carrier transmission," *Wirel. Pers. Commun.*, vol. 5, pp. 1–18, 1997.

[34] H. D. Helms, "Fast Fourier transform method of computing difference equations and simulating filters," *IEEE Trans. Audio Electroacoust.*, vol. AU-15, no. 2, pp. 85–90, Jun. 1967.

[35] A. J. E. M. Janssen, "The Zak transform: A signal transform for sampled time-continuous signals," *Philips J. Res.*, vol. 43, no. 1, pp. 23–69, 1988.

[36] A. J. E. M. Janssen and H. Boelcskei, "Equivalence of two methods for constructing tight Gabor frames," *IEEE Signal Process. Lett.*, vol. 7, no. 4, pp. 79–82, Apr. 2000.

[37] K. D. Kammeyer, U. Tuisel, U. Schulse, and H. Bochmann, "Digital multicarrier-transmission of audio signals over mobile radio channels," *Eur. Trans. Telecommun.*, vol. 3, no. 3, pp. 243–253, May/Jun. 1992.

[38] N. Laurenti and L. Vangelista, "Filter design for the conjugate OFDM-OQAM system," in *Proc. 1st International Workshop on Image and Signal Processing and Analysis (IWISPA)*, 2000, pp. 267–272.

[39] B. Le Floch, M. Alard, and C. Berrou, "Coded orthogonal frequency division multiplex," *Proc. IEEE*, vol. 83, no. 6, pp. 982–996, Jun. 1995.

[40] R. Li and G. Stette, "Time-limited orthogonal multicarrier modulation schemes," *IEEE Trans. Commun.*, vol. 43, no. 2/3/4, pp. 1269–1272, Feb./Mar./Apr. 1995.

[41] Y. P. Lin, C. C. Li, and S. M. Phoong, "A filterbank approach to window designs for multicarrier systems," *IEEE Circuits Syst. Mag.*, vol. 7, no. 1, pp. 19–30, Mar. 2007.

[42] H. S. Malvar, *Signal Processing with Lapped Transforms*. Boston, MA, USA: Artech House, 1992.

[43] K. Martin, "Small side-lobe filter design for multitone data-communication applications," *IEEE Trans. Circuits Syst. II*, vol. 45, no. 8, pp. 1155–1161, Aug. 1998.

[44] D. Mattera, M. Tanda, and M. Bellanger, "Frequency domain CFO compensation for FBMC systems," *Signal Process.*, vol. 114, pp. 183–197, Sep. 2015.

[45] N. Michailow, I. Gaspar, S. Krone, M. Lentmaier, and G. Fettweis, "Generalized frequency division multiplexing: Analysis of an alternative multi-carrier technique for next generation cellular systems," in *Proc. IEEE International Symposium on Wireless Communications Systems (ISWCS)*, Oct. 2012, pp. 171–175.

[46] S. Mirabbasi and K. Martin, "Overlapped complex-modulated transmultiplexer filters with simplified design and superior stopbands," *IEEE Trans. Circuits Syst. II*, vol. 50, no. 8, pp. 456–469, Aug. 2003.

[47] S. Muramatsu and H. Kiya, "Extended overlap-add and -save methods for multirate signal processing," *IEEE Trans. Signal Process.*, vol. 45, no. 9, pp. 2376–2380, Sep. 1997.

[48] T. Q. Nguyen, "Partial spectrum reconstruction using digital filter banks," *IEEE Trans. Signal Process.*, vol. 41, no. 9, pp. 2778–2795, Sep. 1993.

[49] A. Oppenheim and R. Schafer, *Discrete-Time Signal Processing*. Prentice-Hall, 1989.

[50] L. Pelkowitz, "Frequency domain analysis of wraparound error in fast convolution algorithms," *IEEE Trans. Acoust., Speech, Signal Process.*, vol. ASSP-29, no. 3, pp. 413–422, Jun. 1981.

[51] D. Pinchon and P. Siohan, "Oversampled paraunitary DFT filter banks: A general construction algorithm and some specific solutions," *IEEE Trans. Signal Process.*, vol. 59, no. 7, pp. 3058–3070, Jul. 2011.

[52] D. Pinchon and P. Siohan, "Derivation of analytical expressions for flexible PR low complexity FBMC systems," in *Proc. EURASIP European Signal Processing Conference (EUSIPCO)*, Marrakech, Morocco, Sep. 2013.

[53] D. Pinchon, P. Siohan, and C. Siclet, "Design techniques for orthogonal modulated filterbanks based on a compact representation," *IEEE Trans. Signal Process.*, vol. 52, no. 6, pp. 1682–1692, Jun. 2004.

[54] J. Princen, "The design of nonuniform modulated filter banks," *IEEE Trans. Signal Process.*, vol. 43, no. 11, pp. 2550–2560, Nov. 1995.

[55] L. Pucker, "Channelization techniques for software defined radio," in *Proc. Software Defined Radio Technical Conference (SDR)*, Orlando, FL, USA, Nov. 2003.

[56] L. R. Rabiner and B. Gold, *Theory and Application of Digital Signal Processing*. Englewood Cliffs, NJ: Prentice-Hall, 1975.

[57] S. Rahimi and B. Champagne, "Oversampled perfect reconstruction DFT modulated filter banks for multi-carrier transceiver systems," *Signal Process.*, vol. 93, no. 11, pp. 2942–2955, 2013.

[58] M. Renfors and f. harris, "Highly adjustable multirate digital filters based on fast convolution," in *Proc. European Conference on Circuit Theory Design (ECCTD)*, Linköping, Sweden, Aug. 2011, pp. 9–12.

[59] M. Renfors, J. Yli-Kaakinen, and f. harris, "Analysis and design of efficient and flexible fast-convolution based multirate filter banks," *IEEE Trans. Signal Process.*, vol. 62, no. 15, pp. 3768–3783, Aug. 2014.

[60] M. Renfors, J. Yli-Kaakinen, T. Levanen, M. Valkama, T. Ihalainen, and J. Vihriälä, "Efficient fast-convolution implementation of filtered CP-OFDM waveform processing for 5G," in *Proc. IEEE Global Telecommunication Conference (GLOBECOM)*, San Diego, CA, USA, Dec. 2015.

[61] M. Renfors, J. Yli-Kaakinen, T. Levanen, and M. Valkama, "Fast-convolution filtered OFDM waveforms with adjustable CP length," in *Proc. IEEE Global Conference on Signal and Information Processing (GlobalSIP)*, Greater Washington, DC, USA, Dec. 2016.

[62] C. Roche and P. Siohan, "A family of Extended Gaussian Functions with a nearly optimal localization property," in *Proc. First Int. Workshop on Multi-Carrier Spread-Spectrum*, Oberpfaffenhofen, Germany, Apr. 1997, pp. 179–186.

[63] A. Sahin, I. Guvenc, and H. Arslan, "A survey on multicarrier communications: Prototype filters, lattice structures, and implementation aspects," *IEEE Commun. Surv. Tutor.*, vol. 16, no. 3, pp. 1312–1338, 2014.

[64] K. Shao, J. Alhava, J. Yli-Kaakinen, and M. Renfors, "Fast-convolution implementation of filter bank multicarrier waveform processing," in *Proc. IEEE International Symposium on Circuits and Systems (ISCAS)*, Lisbon, Portugal, May 2015.

[65] M. Siala and A. Yongacoglu, "Prototype waveform optimization for an OFDM/OQAM system with hexagonal time-frequency lattice structure," in *Proc. IEEE International Symposium on Signal Processing and Its Applications (ISSPA)*, Feb. 2007, pp. 1–4.

[66] C. Siclet, "Application de la théorie des bancs de filtres à l'analyse et à la conception de modulations multiporteuses orthogonales et biorthogonales." Ph.D. dissertation. France: Université de Rennes 1, 2002.

[67] P. Siohan and N. Lacaille, "Analysis of OFDM/OQAM systems based on the filterbank theory," in *Proc. IEEE Global Telecommunication Conference (GLOBECOM)*, Rio de Janeiro, Brazil, Dec. 1999, pp. 2279–2284.

[68] P. Siohan and C. Roche, "Cosine modulated filterbanks based on Extended Gaussian Functions," *IEEE Trans. Signal Process.*, vol. 48, no. 11, pp. 3052–3061, Nov. 2000.

[69] P. Siohan and C. Roche, "Derivation of extended Gaussian functions based on the Zak transform," *IEEE Signal Process. Lett.*, vol. 11, no. 3, pp. 401–403, 2004.

[70] P. Siohan, C. Siclet, and N. Lacaille, "Analysis and design of OFDM/OQAM systems based on filterbank theory," *IEEE Trans. Signal Process.*, vol. 50, no. 5, pp. 1170–1183, May 2002.

[71] D. Slepian and H. O. Pollak, "Prolate spheroidal wave functions, Fourier analysis and uncertainty – I," *Bell Syst. Tech. J.*, vol. 40, no. 1, pp. 43–63, Jan. 1961.

[72] J. O. Smith, "Audio FFT filter banks," in *Proc. International Conference on Digital Audio Effects (DAFx-09)*, Como, Italy, Sep. 2009, pp. 1–4.

[73] H. V. Sorensen and C. S. Burrus, "Fast DFT and convolution algorithms," in *Handbook for Digital Signal Processing*, S. K. Mitra and J. F. Kaiser, Eds. New York: John Wiley and Sons, Inc., 1993, pp. 491–610, ch. 8.

[74] T. G. Stockham Jr., "High-speed convolution and correlation," in *Proc. AFIPS Spring Joint Computer Conference*, vol. 28, New York, NY, USA, 1966, pp. 229–233. [Online]. Available: http://doi.acm.org/10.1145/1464182.1464209.

[75] T. Strohmer and S. Beaver, "Optimal OFDM design for time-frequency dispersive channels," *IEEE Trans. Commun.*, vol. 51, no. 7, pp. 1111–1122, Jul. 2003.

[76] T. Strohmer, "Approximation of dual Gabor frames, window decay, and wireless communications," *Appl. Comput. Harmon. Anal.*, vol. 11, no. 2, pp. 243–262, 2001. [Online]. Available: http://www.sciencedirect.com/science/article/pii/S1063520301903574.

[77] A. Toskala and H. Holma, Eds., *LTE for UMTS – OFDMA and SC-FDMA Based Radio Access.* Wiley, 2009.

[78] M. Umehira and M. Tanabe, "Performance analysis of overlap FFT filter-bank for dynamic spectrum access applications," in *Proc. Asia-Pacific Conference on Communications (APCC)*, Auckland, New Zealand, Oct.–Nov. 2010, pp. 424–428.

[79] A. Vahlin and N. Holte, "Optimal finite duration pulses for OFDM," *IEEE Trans. Commun.*, vol. 44, no. 1, pp. 10–14, Jan. 1996.

[80] A. Vahlin and N. Holte, "Optimal finite duration pulses for OFDM," in *Proc. IEEE Global Telecommunication Conference (GLOBECOM)*, San Francisco, USA, Nov. 1994, pp. 258–262.

[81] P. P. Vaidyanathan, *Multirate Systems and Filter Banks.* Englewood Cliffs, NJ: Prentice Hall, 1993.

[82] L. Vangelista and N. Laurenti, "Efficient implementations and alternative architectures for OFDM-OQAM systems," *IEEE Trans. Commun.*, vol. 49, no. 4, pp. 664–675, Apr. 2001.

[83] A. Viholainen, M. Bellanger, and M. Huchard, "Prototype filter and structure optimization." ICT-211887 Project PHYDYAS (Physical Layer for Dynamic Access and Cognitive Radio) Technical Report D5.1, Jan. 2009. [Online]. Available: http://www.ict-phydyas.org/.

[84] J. Yli-Kaakinen, T. Levanen, S. Valkonen, K. Pajukoski, J. Pirskanen, M. Renfors, and M. Valkama, "Efficient fast-convolution based waveform processing for 5G physical layer," *IEEE J. Select. Areas Commun.*, vol. 35, no. 8, Aug. 2017.

[85] J. Yli-Kaakinen and M. Renfors, "Optimized reconfigurable fast convolution based trans-multiplexers for flexible radio access," *IEEE Trans. Circuits Syst. II*, vol. 64, 2017, to be published.

[86] J. Yli-Kaakinen and M. Renfors, "Flexible fast-convolution implementation of single-carrier waveform processing," in *Proc. IEEE International Conference on Communications, ICC2015 Workshops*, London, UK, Jun. 2015.

[87] J. Yli-Kaakinen and M. Renfors, "Optimized burst truncation in fast-convolution filter bank based waveform generation," in *Proc. IEEE International Workshop on Signal Processing Advances in Wireless Communications (SPAWC)*, Jun. 2015, pp. 71–75.

[88] C. Zhang and Z. Wang, "A fast frequency domain filter bank realization algorithm," in *Proc. International Conference on Signal Processing*, vol. 1, Beijing, China, Aug. 2000, pp. 130–132.

[89] J. Zhao, W. Wang, and X. Gao, "Transceiver design for fast-convolution multicarrier systems in multipath fading channels," in *2015 Int. Conference on Wireless Communications Signal Processing (WCSP)*, Oct. 2015, pp. 1–5.

# FBMC Signal Processing

# FBMC Over Frequency Selective Channels

# 9

**David Gregoratti, Xavier Mestre**

*Centre Tecnològic de Telecomunicacions de Catalunya (CTTC/CERCA), Barcelona, Spain*

## CONTENTS

## 9.1 INTRODUCTION

The previous chapters of this book have established that FilterBank MultiCarrier with Offset-QAM subcarrier modulation (FBMC/OQAM) schemes can achieve full spectral efficiency of one complex symbol per subcarrier per channel access thanks to a careful alternation of purely real and purely imaginary symbols, in both time and frequency domains. This performance, however, is achieved only with well-shaped prototype pulses (which allow perfect reconstruction) when the communication channel is characterized by a flat frequency response. Conversely, when either of these two conditions is not met, time/frequency orthogonality cannot be achieved, and FBMC/OQAM inherent distortion appears. Focusing our attention on the channel effect, it is straightforward to see how a frequency selective response causes a frequency-dependent phase rotation in the received signal. In other words, without

Orthogonal Waveforms and Filter Banks for Future Communication Systems. DOI: 10.1016/B978-0-12-810384-5.00009-8

proper equalization, the sharp differentiation between purely real and purely imaginary symbols at the transmit side is lost at the receive side.

The purpose of this chapter is to offer some insight into the effect of the channel response on the received signal of any FBMC/OQAM link. Building on the polyphase implementation, we develop a system model that takes the channel into account, and we show how the dynamics of the channel frequency response (i.e., its derivatives) play a fundamental role in the signal distortion.

## 9.2 A POLYPHASE-BASED FBMC SIGNAL DESCRIPTION AND PERFECT RECONSTRUCTION PROTOTYPE PULSE CONDITIONS

The analysis carried out in the following sections is built on the polyphase representation of FBMC/OQAM systems. As we will see next, this choice allows a compact expression for the signal model, which helps in maintaining the exposition neat and clear. Nevertheless, our results also extend to other implementations (complex-valued prototype pulses, frequency spreading, and, to some extent, fast convolution), all equivalent to the polyphasic one, as explained in Appendix 9.A (see also Chapter 8 or, e.g., [1]). To share a common understanding of the polyphase implementation, this section summarizes its salient points.

### 9.2.1 TRANSMITTER

Let us begin with the transmit side of an FBMC/OQAM system with $M$ subcarriers and prototype pulse $g[l]$, $l = 0, 1, \ldots, KM - 1$, with overlapping factor $K \in \mathbb{N}_+$. As represented in Fig. 9.1, complex subcarrier symbols $\{a_{m,k}\}$ are staggered[1] and then fed into the Synthesis FilterBank (SFB). Here, $m = 0, 1, \ldots, M - 1$ is the subcarrier index, and $k$ is the time index. In the polyphase implementation, the SFB consists of four main steps. First, the subcarrier streams are multiplied by weights[2] $[\Phi^*]_{m,m} = e^{j\pi m \frac{M+2}{2M}}$, $m = 0, 1, \ldots, M - 1$. These coefficients are phase rotations that are needed to achieve the alternation of purely real and purely imaginary symbols in both time and frequency domains, which is a key aspect of FBMC/OQAM modulations. Next, symbols are translated into the time domain by the Inverse Discrete Fourier Transform (IDFT) block, namely a multiplication by the $M \times M$ matrix $\mathcal{F}$ with $[\mathcal{F}]_{m,l} = \frac{1}{\sqrt{M}} e^{j2\pi \frac{ml}{M}}$. The third stage consists of a bank of filters $\{G_m(z^2)\}$,

---

[1]Here, the staggering operation is defined as the mapping from complex symbol $a_{m,k} = b_{m,k} + jc_{m,k}$ to the pair of symbols $(b_{m,k}, jc_{m,k})$. As a result, the output sampling rate is twice the input one. See also Fig. 9.2.

[2]Other choices of $\Phi^*$ are possible (see Appendix 9.A), but we choose this one because it leads to a symmetric representation between AFB and SFB.

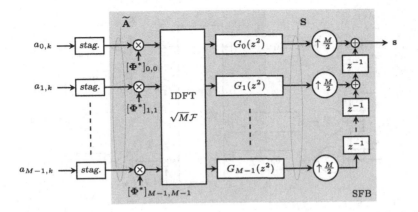

**FIGURE 9.1**

Polyphase implementation of an FBMC/OQAM transmitter (block diagram).

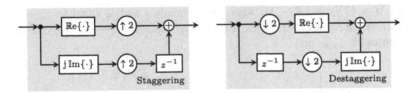

**FIGURE 9.2**

Staggering (left) and destaggering (right) operations.

$m = 0, 1, \ldots, M - 1$, where $G_m(z) = \sum_{r=0}^{K-1} g[m + rM]z^{-r}$ is the $m$th Type-I polyphase component of the transmit prototype pulse $g[l]$. Finally, the upsampling-delay lattice at the right end of Fig. 9.1 transforms the parallel output of the filterbank into a serial stream of symbols. Although the upsampling factor at this point is $M/2$, the total one is $M$ since the staggering operation at the transmit input already introduces an upsampling factor of 2.

Summarizing, let the matrix $\mathbf{A} \in \mathbb{C}^{M \times N_s}$ contain a set of $N_s$ multicarrier symbols, namely, $[\mathbf{A}]_{m,k} = a_{m,k}$. Moreover, let $\widetilde{\mathbf{A}}$ be its staggered version, namely, $\widetilde{\mathbf{A}} = \mathbf{B} \otimes [1 \ 0] + j\mathbf{C} \otimes [0 \ 1]$, with $\mathbf{B}$ and $\mathbf{C}$ the real and imaginary parts of $\mathbf{A} = \mathbf{B} + j\mathbf{C}$, respectively. Then, the corresponding symbol stream at the output of the SFB can be written as (see Fig. 9.1)

$$\mathbf{s} = \mathrm{vec}([\mathbf{S}_1 \ \mathbf{0}] + [\mathbf{0} \ \mathbf{S}_2]), \tag{9.1}$$

where $\mathbf{0}$ denotes a properly sized vector filled with zeros, and where $\mathbf{S}_1$ (resp., $\mathbf{S}_2$) contains the top half (resp., bottom half) rows of the complex matrix

$$\mathbf{S} = \sqrt{M}\mathcal{F}\mathbf{\Phi}^*\widetilde{\mathbf{A}} \star (\mathbf{G} \otimes [1 \ 0]) \tag{9.2}$$

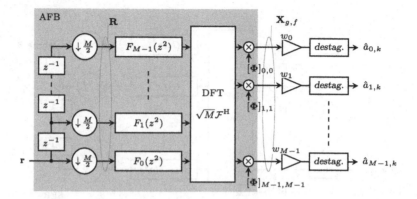

**FIGURE 9.3**

Polyphase implementation of an FBMC/OQAM receiver (block diagram).

of size $M \times (2N_s + 2K - 1)$, where the symbols $\star$ and $\otimes$ stand for (row-wise) convolution and Kronecker product, respectively. In addition, $\mathbf{\Phi}^*$ is a diagonal matrix with entries $[\mathbf{\Phi}^*]_{m,m}$, and

$$
\mathbf{G} = \begin{bmatrix} g[0] & g[M] & \cdots & g[M(K-1)] \\ \vdots & \vdots & \ddots & \vdots \\ g[M-1] & g[2M-1] & \cdots & g[MK-1] \end{bmatrix} \tag{9.3}
$$

(note that the $m$th row corresponds to the coefficients of $G_m(z)$).

## 9.2.2 RECEIVER

At the receive side of the FBMC transceiver, the steps above are reversed (compare also the scheme in Fig. 9.3 to the previous one in Fig. 9.1). First, a delay–downsampling lattice is used to reorganize the received symbols into $M$ parallel streams, forming the matrix

$$
\mathbf{R} = \begin{bmatrix} \mathbf{0} & \mathbf{r}(0) & \mathbf{r}(1) & \cdots & \mathbf{r}(N_{\mathrm{rec}} - 1) \\ \mathbf{r}(0) & \mathbf{r}(1) & \mathbf{r}(2) & \cdots & \mathbf{0} \end{bmatrix}, \tag{9.4}
$$

where $N_{\mathrm{rec}} = 2N_s + 2K - 1$ and $\mathbf{r}(i) = \begin{bmatrix} r[iM/2] & \cdots & r[(i+1)M/2 - 1] \end{bmatrix}^{\mathrm{T}}$. Then, the rows of $\mathbf{R}$ are used as inputs to the bank of filters $\{F_m(z^2)\}$, $m = M-1, M-2, \ldots, 0$, where $F_m(z) = \sum_{r=0}^{K-1} f[m + rM]z^{-r}$ is the $m$th Type-I polyphase component of the receive prototype pulse $f[l]$, $l = 0, 1, \ldots, KM - 1$. Note that these polyphase branches are disposed inversely to their counterparts at the transmit side. The following steps consist of a Discrete Fourier Transform (DFT) (represented in the scheme by the complex matrix $\mathcal{F}^{\mathrm{H}}$) and the set of weights $[\mathbf{\Phi}]_{m,m} = \mathrm{e}^{-\mathrm{j}\pi m \frac{M+2}{2M}}$,

$m = 0, 1, \ldots, M - 1$, applied to the symbols in the frequency domain. As a result, the matrix $\mathbf{X}_{g,f} \in \mathbb{C}^{M \times (2N_s + 4K)}$, built from the symbols at the output of the Analysis FilterBank (AFB) of a Single-Input Single-Output (SISO) point-to-point FBMC/OQAM system with perfect (noiseless and identity) channel, can be written as $\mathbf{X}_{g,f} = \mathbf{X}_{g,f}^{\text{even}} \otimes [1\ 0] + \mathbf{X}_{g,f}^{\text{odd}} \otimes [0\ 1]$, where

$$\mathbf{X}_{g,f}^{\sharp} = \sqrt{M}\,\boldsymbol{\Phi}\,\mathcal{F}^{\text{H}}\big(\mathbf{J}_M \mathbf{F} \star \mathbf{R}^{\sharp}\big), \qquad \sharp \in \{\text{even, odd}\}, \tag{9.5}$$

represents the even or odd columns of $\mathbf{X}_{g,f}$. In the above equation, $\mathbf{J}_M$ is the exchange (anti-identity) matrix of size $M$, and $\mathbf{F}$ is built from $f[l]$ as $\mathbf{G}$ is from $g[l]$ [cf. (9.3)]. Because of the delays introduced by the polyphase filters (i.e., $F_m(z^2)$ as opposed to $F_m(z)$), note that odd and even columns are best treated separately.

Now, it is easy to show that the end-to-end polyphase representation of the FBMC/OQAM signal model can be written as

$$\mathbf{X}_{g,f}^{\text{odd}} = M\boldsymbol{\Phi}\mathcal{F}^{\text{H}}\left([\mathcal{F}\boldsymbol{\Phi}^*\mathbf{B}\ \mathbf{0}\ \mathbf{0}] \star \mathcal{R}_{g,f} + \begin{bmatrix} \mathbf{0}\ \mathrm{j}\mathcal{F}_2\boldsymbol{\Phi}^*\mathbf{C}\ \mathbf{0} \\ \mathrm{j}\mathcal{F}_1\boldsymbol{\Phi}^*\mathbf{C}\ \mathbf{0}\ \mathbf{0} \end{bmatrix} \star \mathcal{S}_{g,f}\right), \tag{9.6a}$$

$$\mathbf{X}_{g,f}^{\text{even}} = M\boldsymbol{\Phi}\mathcal{F}^{\text{H}}\left([\mathbf{0}\ \mathrm{j}\mathcal{F}\boldsymbol{\Phi}^*\mathbf{C}\ \mathbf{0}] \star \mathcal{R}_{g,f} + \begin{bmatrix} \mathbf{0}\ \mathcal{F}_2\boldsymbol{\Phi}^*\mathbf{B}\ \mathbf{0} \\ \mathcal{F}_1\boldsymbol{\Phi}^*\mathbf{B}\ \mathbf{0}\ \mathbf{0} \end{bmatrix} \star \mathcal{S}_{g,f}\right), \tag{9.6b}$$

where we have introduced the matrices

$$\mathcal{R}_{g,f} = \begin{bmatrix} \mathbf{G}_1 \star \mathbf{J}_{M/2}\mathbf{F}_2 \\ \mathbf{G}_2 \star \mathbf{J}_{M/2}\mathbf{F}_1 \end{bmatrix} \qquad \text{and} \qquad \mathcal{S}_{g,f} = \begin{bmatrix} \mathbf{G}_2 \star \mathbf{J}_{M/2}\mathbf{F}_2 \\ \mathbf{G}_1 \star \mathbf{J}_{M/2}\mathbf{F}_1 \end{bmatrix} \tag{9.7}$$

with the matrix $\boldsymbol{\Psi}_1$ (resp. $\boldsymbol{\Psi}_2$) gathering the top half (resp. bottom half) rows of $\boldsymbol{\Psi}$, $\boldsymbol{\Psi} \in \{\mathbf{G}, \mathbf{F}\}$. Similarly, $\mathcal{F}_1$ and $\mathcal{F}_2$ denote the top half rows and bottom half rows of the DFT matrix, $\mathcal{F}$, respectively.

The proof can be sketched as follows: Since $\mathbf{r} = \mathbf{s}$ because of the perfect-channel assumption, we can use (9.1), (9.2), and (9.4) to write

$$\mathbf{R} = \sqrt{M}\begin{bmatrix} \mathbf{0}\ \mathcal{F}_1\boldsymbol{\Phi}^*\widetilde{\mathbf{A}} \star (\mathbf{G}_1 \otimes [1\ 0])\ \mathbf{0} \\ \mathcal{F}_1\boldsymbol{\Phi}^*\widetilde{\mathbf{A}} \star (\mathbf{G}_1 \otimes [1\ 0])\ \mathbf{0}\ \mathbf{0} \end{bmatrix} + \sqrt{M}\begin{bmatrix} \mathbf{0}\ \mathbf{0}\ \mathcal{F}_2\boldsymbol{\Phi}^*\widetilde{\mathbf{A}} \star (\mathbf{G}_2 \otimes [1\ 0]) \\ \mathbf{0}\ \mathcal{F}_2\boldsymbol{\Phi}^*\widetilde{\mathbf{A}} \star (\mathbf{G}_2 \otimes [1\ 0])\ \mathbf{0} \end{bmatrix}. \tag{9.8}$$

Then, inserting (9.8) into (9.5), we can obtain an expression for $\mathbf{X}_{g,f}^{\sharp}$ as a function of $\mathbf{A} = \mathbf{B} + \mathrm{j}\mathbf{C}$. The final form in (9.6) is a consequence of the identities

$$(\mathbf{V}_1 \otimes [1\ 0]) \star (\mathbf{V}_2 \otimes [1\ 0]) = [(\mathbf{V}_1 \star \mathbf{V}_2) \otimes [1\ 0]\ \mathbf{0}]$$

and

$$(\mathbf{V}_1 \otimes [1\ 0]) \star (\mathbf{V}_2 \otimes [0\ 1]) = [\mathbf{0}\ (\mathbf{V}_1 \star \mathbf{V}_2) \otimes [1\ 0]]$$

for any two matrices $\mathbf{V}_1$ and $\mathbf{V}_2$ with the same number of rows.

## 9.2.3 PERFECT RECONSTRUCTION CONSTRAINTS

At this point, it is interesting to verify whether the matrix $\mathbf{A}$ can be recovered free of errors under the perfect-channel assumption. With the compact representation introduced in this chapter, it is straightforward to show that the Perfect Reconstruction (PR) conditions (derived in [1] in terms of Z-transform) read as follows:

$$\mathbf{U}^{+}\mathcal{R}_{g,f} = \frac{2}{M}[\mathbf{0}_{M\times(K-1)} \ \mathbf{1} \ \mathbf{0}_{M\times(K-1)}], \tag{9.9a}$$

$$\mathbf{U}^{-}\mathcal{S}_{g,f} = \mathbf{0}_{M\times(2K-1)}, \tag{9.9b}$$

where

$$\mathbf{U}^{\pm} = \mathbf{I}_2 \otimes (\mathbf{I}_{M/2} \pm \mathbf{J}_{M/2}), \tag{9.10}$$

and where $\mathbf{1}$ is an $M$-element vector filled with ones. Indeed, recalling that $\mathbf{B}$ and $\mathbf{C}$ are real valued and that $\boldsymbol{\Phi}^{*}\mathcal{F} = \boldsymbol{\Phi}\mathcal{F}^{H}(\mathbf{I}_2 \otimes \mathbf{J}_{M/2})$ together with $\mathcal{F} = \mathcal{F}^{T}$, we have

$$\mathbb{Re}\{\mathbf{X}_{g,f}^{\text{odd}}\} = \frac{1}{2}[\mathbf{X}_{g,f}^{\text{odd}} + (\mathbf{X}_{g,f}^{\text{odd}})^{*}]$$

$$= \frac{M}{2}\boldsymbol{\Phi}\mathcal{F}^{H}\left([\mathcal{F}\boldsymbol{\Phi}^{*}\mathbf{B} \ \mathbf{0} \ \mathbf{0}] \star (\mathbf{U}^{+}\mathcal{R}_{g,f}) + \begin{bmatrix} \mathbf{0} \ \mathrm{j}\mathcal{F}_2\boldsymbol{\Phi}^{*}\mathbf{C} \ \mathbf{0} \\ \mathrm{j}\mathcal{F}_1\boldsymbol{\Phi}^{*}\mathbf{C} \ \mathbf{0} \ \mathbf{0} \end{bmatrix} \star (\mathbf{U}^{-}\mathcal{S}_{g,f})\right).$$

After a similar reasoning for $\mathbb{Im}\{\mathbf{X}_{g,f}^{\text{even}}\}$, we readily realize that

$$\mathbb{Re}\{\mathbf{X}_{g,f}^{\text{odd}}\} = [\mathbf{0}_{M\times(K-1)} \ \mathbf{B} \ \mathbf{0}_{M\times(K+1)}], \tag{9.11}$$

$$\mathbb{Im}\{\mathbf{X}_{g,f}^{\text{even}}\} = [\mathbf{0}_{M\times K} \ \mathbf{C} \ \mathbf{0}_{M\times K}] \tag{9.12}$$

whenever (9.9) holds. As a result, this implies that

$$a_{m,k} = \mathbb{Re}\{[\mathbf{X}_{g,f}^{\text{odd}}]_{m,k+K-1}\} + \mathrm{j}\,\mathbb{Im}\{[\mathbf{X}_{g,f}^{\text{even}}]_{m,k+K}\} \tag{9.13}$$

or, in plain words, the symbols $\{a_{m,k}\}$ are perfectly recovered.

By comparing (9.11)–(9.12) with (9.6) we see that the interference generated by other symbols (from both adjacent time slots and subcarriers) is confined to the imaginary (resp. real) part of the odd (resp. even) columns of $\mathbf{X}_{g,f}$ when the channel is ideal and the prototype pulses meet conditions (9.9). As a result, under these assumptions, (9.13) is exact, and no distortion is introduced. Conversely, with frequency selective channels and a poor choice of the prototype pulses, a perfect separation between data signal and interference cannot be achieved. The resulting distortion, inherent in FBMC/OQAM systems, is characterized further.

## 9.3 APPROXIMATIONS FOR LARGE NUMBER OF SUBCARRIERS UNDER STRONG FREQUENCY SELECTIVITY

In the previous section, we have characterized the matrix $\mathbf{X}^{\sharp}_{g,f}$, $\sharp \in \{\text{even, odd}\}$, which contains the samples at the output of the AFB (see Fig. 9.3) under the perfect-channel assumption. Hereafter, we remove this assumption, and we study the case where the signal undergoes a frequency selective channel. Assume that the frequency response of the channel is given by $H(e^{j\omega})$, where $\omega \in [0, 2\pi)$. In what follows, we establish an asymptotic approximation of the signal matrix that will prove to be very useful in several other parts of the book and will provide a very simple method to evaluate the effect of the channel frequency selectivity on the FBMC/OQAM signal.

The primary idea behind the main result in this section consists in approximating the channel frequency response $H(e^{j\omega})$ using a Taylor series around each of the subcarrier frequencies. More specifically, let us consider the Taylor series of $H(e^{j\omega})$ up to order $R \in \mathbb{N}_+$ around a specific subcarrier frequency $\omega_m = \frac{2\pi m}{M}$, that is,

$$H\left(e^{j\omega}\right) \simeq \sum_{r=0}^{R} \frac{j^r}{r!} H^{(r)}\left(e^{j\omega_m}\right)(\omega - \omega_m)^r$$

plus an error term of order $\mathcal{O}((\omega - \omega_m)^{R+1})$, where $H^{(r)}(e^{j\omega_m})$ is defined as $H^{(r)}(z)|_{z=e^{j\omega_m}}$, $H^{(r)}(z)$ being the $r$th complex derivative of $H(z)$ with respect to $z$.

Next, assume that the prototype pulses used at the transmit and receive sides (i.e., $f[l]$ and $g[l]$) are sampled versions of two original analog waveforms $f(t)$ and $g(t)$, so that a total number of $M$ samples are taken for each multicarrier symbol period. It is worth remarking that this assumption reflects common practical choices: For example, the PHYDYAS pulse [2] employed in the simulations below is obtained according to

$$g[l] = \frac{1}{K\sqrt{M}}\left[1 + 2\sum_{i=1}^{K-1}(-1)^i \bar{G}_i \cos\left(\frac{2\pi i}{KM}\left(l + \frac{1}{2}\right)\right)\right]$$

for $l = 0, 1, \ldots, KM - 1$, where the coefficients $\{\bar{G}_i, i = 1, 2, \ldots, K - 1\}$ are reported in Table 9.1 (they are independent of $M$). We denote by $f^{(r)}[l]$ and $g^{(r)}[l]$ the sampled versions of the $r$th derivatives of the analog waveforms. Then, it is not difficult to see that we can approximate the discrete-time Fourier transform of $f^{(r)}[l]$ as

$$F^{(r)}\left(e^{j\omega}\right) \simeq (Mj\omega)^r F\left(e^{j\omega}\right)$$

and equivalently for $G^{(r)}(e^{j\omega})$. According to this observation, we can express the cascade of the channel $H(e^{j\omega})$ and a frequency-translated version of the receive pro-

**Table 9.1** Building coefficients of the PHYDYAS pulse

| $K$ | $\bar{G}_1$ | $\bar{G}_2$ | $\bar{G}_3$ |
|---|---|---|---|
| 2 | 0.70710678 | – | – |
| 3 | 0.91143783 | 0.41143783 | – |
| 4 | 0.97195983 | 0.70710678 | 0.23514695 |

totype pulse, namely, $F\left(e^{j(\omega - \omega_m)}\right)$, as

$$F\left(e^{j(\omega - \omega_m)}\right) H\left(e^{j\omega}\right) \simeq \sum_{r=0}^{R} \frac{1}{M^r} \frac{1}{r!} F^{(r)}\left(e^{j(\omega - \omega_m)}\right) H^{(r)}\left(e^{j\omega_m}\right).$$

In plain words, the above equation suggests that we can approximate the cascade of the channel and the receive filterbank around each subcarrier frequency as a linear combination of $R$ parallel filterbanks, each constructed from a sequential derivative of the original prototype pulse and associated with a derivative of the channel frequency response of increasing order. Of course, the same reasoning is valid for the transmit prototype pulses.

In practice, we need to impose some regularity conditions to guarantee that the above approximations are solid from the mathematical point of view, as specified below (see [3] for further details):

(**As1**)   The transmit and receive pulses are sampled versions of analog waveforms that are $R + 1$ times continuously differentiable. The analog waveforms and their $R + 1$ first derivatives vanish at the end points of their support.

(**As2**)   The channel response $H(z)$ is analytic in an open set containing the boundary of the unit circle.

(**As3**)   The transmitted symbols **A** are independent and identically distributed (i.i.d.), circularly symmetric and bounded zero-mean random variables.

The following proposition, proved in [3], provides a formal framework to approximate the output of the analysis filterbank under strong frequency selectivity as a linear combination of the outputs of $R$ parallel filterbanks, each constructed using a derivative of the original prototype pulse. It strongly relies on assumptions (**As1**)–(**As3**). To present the result, in the notation, it is useful to introduce the dependence on the matrix $\widetilde{\mathbf{A}}$ of staggered symbols. Hence, we will assume that $\mathbf{Y}_{g,f}^{\sharp}\left(\widetilde{\mathbf{A}}\right)$, $\sharp \in \{\text{odd, even}\}$, represents the received symbols at the output of the analysis filterbank when the staggered transmitted symbols in $\widetilde{\mathbf{A}}$ have been transmitted under a frequency selective channel. Likewise, we denote by $\mathbf{X}_{g,f}^{\sharp}\left(\widetilde{\mathbf{A}}\right)$ the same quantity but received under ideal channel conditions (see (9.6)). The following proposition establishes that $\mathbf{Y}_{g,f}^{\sharp}\left(\widetilde{\mathbf{A}}\right)$ can be expressed as a linear combination of matrices of the form $\mathbf{X}_{g,f}^{\sharp}\left(\widetilde{\mathbf{A}}\right)$.

**Proposition 9.1.** *Under* (**As1**)–(**As3**), *for any* $R \in \mathbb{N}_+$, *we can write*

$$\mathbf{Y}^{\sharp}_{g,f}\left(\widetilde{\mathbf{A}}\right) = \sum_{r=0}^{R} \frac{1}{r!M^r} \mathbf{\Delta}\left(H^{(r)}\right)\mathbf{X}^{\sharp}_{g,f^{(r)}}\left(\widetilde{\mathbf{A}}\right) + o\left(M^{-R}\right) \tag{9.14}$$

$$= \sum_{r=0}^{R} \frac{1}{r!M^r}\mathbf{X}^{\sharp}_{g^{(r)},f}\left(\mathbf{\Delta}\left(H^{(r)}\right)\widetilde{\mathbf{A}}\right) + o\left(M^{-R}\right) \tag{9.15}$$

*as* $M \to \infty$, *where* $\mathbf{\Delta}\left(H^{(r)}\right)$ *is an* $M \times M$ *diagonal matrix containing the values* $\{H^{(r)}(e^{j\omega_m}), m = 0, 1, \ldots, M-1\}$, *and where* $o\left(M^{-R}\right)$ *denotes a matrix of dimensions increasing with* $M$ *whose entries decay to zero faster than* $M^{-R}$.

*Proof.* See Proposition 1 in [3]. □

Proposition 9.1 is basically pointing out that we can approximate the received FBMC/OQAM signal under a frequency selective channel as a linear combination of received signals for ideal channel conditions. This linear combination corresponds to the action of multiple AFBs in (9.14), each constructed from a sequential derivative of the original prototype pulse. A very similar approximation holds in (9.15) by considering the combination of multiple SFBs, each corresponding to a different derivative of the corresponding prototype. Note also that, according to the AFB-based approximation in (9.14), the effect of the channel frequency selectivity generates a per-subcarrier weighting at the output of the AFB, whereas in (9.15) the weighting is produced at the input of the SFB. Either way, the weighting corresponds to the multiplication of each subcarrier by a complex coefficient given by the $r$th derivative of the frequency response of the channel.

Fig. 9.4 provides a graphical representation of the physical meaning behind Proposition 9.1. The original model, which consists of an SFB, the frequency selective channel and the AFB, is represented in Fig. 9.4A. Proposition 9.1 is basically stating that the original model is asymptotically equivalent to the model in Fig. 9.4B, which consists of an SFB followed by an ideal channel plus a collection of $R + 1$ AFBs, the outputs of which are weighted and combined according to the sequential derivatives of the channel frequency response. Alternatively, we can formulate an asymptotically equivalent model by considering multiple parallel stages at the transmitter, see Fig. 9.4C, where now multiple SFBs are used.

The main advantage of the result given by Proposition 9.1 is the fact that it allows us to study the behavior of the receive signal matrix $\mathbf{Y}^{\sharp}_{g,f}\left(\widetilde{\mathbf{A}}\right)$, which is very difficult to analyze, in terms of simpler matrices of the form $\mathbf{X}^{\sharp}_{g,f}\left(\widetilde{\mathbf{A}}\right)$, which correspond to the signal model in (9.6). This decomposition will allow us to construct complex equalization structures in Chapters 12 and 14. Furthermore, this asymptotic description will allow us to characterize the residual distortion when more conventional equalization structures are employed, as it is illustrated next.

Consider now the equalization of the FBMC/OQAM signal by single-tap per-subcarrier weighting. More specifically, assume that the $m$th output of the AFB is

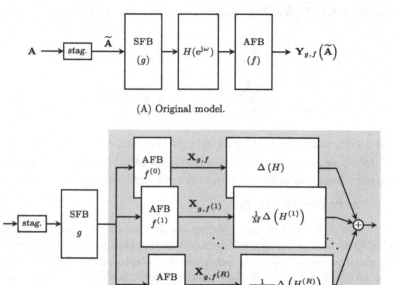

(A) Original model.

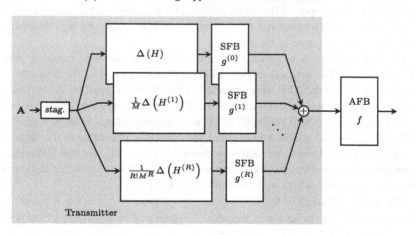

(B) Parallel multistage approximation at the receiver.

(C) Parallel multistage approximation at the transmitter.

**FIGURE 9.4**

Asymptotic representation of the frequency selective channel in FBMC as represented in (A), according to two multistage frequency-flat models associated to the receiver (B) or the transmitter (C).

multiplied by a complex weight $w_m$, and let $\mathbf{W}$ denote the diagonal matrix containing the weights $w_m$, $m = 0, 1, \ldots, M - 1$. By particularizing (9.14) to the case $R = 1$, we see that we can express the matrix of samples at the output of the equalizers as

$$\mathbf{WY}_{g,f}^{\sharp}\left(\widetilde{\mathbf{A}}\right) = \mathbf{W}\boldsymbol{\Delta}\left(H\right)\mathbf{X}_{g,f}^{\sharp}\left(\widetilde{\mathbf{A}}\right) - \frac{\mathrm{j}}{M}\mathbf{W}\boldsymbol{\Delta}\left(H^{(1)}\right)\mathbf{X}_{g,f^{(1)}}^{\sharp}\left(\widetilde{\mathbf{A}}\right) + o\left(M^{-1}\right).$$

Next, assuming that $\mathbf{W} = \boldsymbol{\Delta}(H^{-1})$, that is, the equalizer weights simply invert the channel responses at each specific subcarrier, the last equation can be simplified to

$$\mathbf{WY}_{g,f}^{\sharp}\left(\widetilde{\mathbf{A}}\right) = \mathbf{X}_{g,f}^{\sharp}\left(\widetilde{\mathbf{A}}\right) - \frac{\mathrm{j}}{M}\boldsymbol{\Delta}\left(\frac{H^{(1)}}{H}\right)\mathbf{X}_{g,f^{(1)}}^{\sharp}\left(\widetilde{\mathbf{A}}\right) + o\left(M^{-1}\right).$$

Now, assuming that the pulses $g$, $f$ are constructed to fulfill the perfect-reconstruction conditions presented in Section 9.2.3, we observe that we can recover the original symbols taking real and imaginary parts of the terms $\mathbf{X}_{g,f}^{\mathrm{odd}}\left(\widetilde{\mathbf{A}}\right)$ and $\mathbf{X}_{g,f}^{\mathrm{even}}\left(\widetilde{\mathbf{A}}\right)$, respectively; see further (9.11)–(9.12). Hence, it seems reasonable to estimate the complex-valued transmitted symbols as

$$\hat{a}_{m,k} = \mathrm{Re}\left\{\left[\mathbf{WY}_{g,f}^{\mathrm{odd}}\left(\widetilde{\mathbf{A}}\right)\right]_{m,k+K-1}\right\} + \mathrm{j}\,\mathrm{Im}\left\{\left[\mathbf{WY}_{g,f}^{\mathrm{even}}\left(\widetilde{\mathbf{A}}\right)\right]_{m,k+K}\right\} \qquad (9.16)$$

$$= a_{m,k} + \xi_{m,k} + o\left(M^{-1}\right), \qquad (9.17)$$

where we have defined $\xi_{m,k}$ as the asymptotic residual distortion caused by the channel frequency selectivity, namely,

$$\xi_{m,k} = \mathrm{Re}\left\{\left[\boldsymbol{\Delta}_{\xi}^{\mathrm{odd}}\left(\widetilde{\mathbf{A}}\right)\right]_{m,k+K-1}\right\} + \mathrm{Im}\left\{\left[\boldsymbol{\Delta}_{\xi}^{\mathrm{even}}\left(\widetilde{\mathbf{A}}\right)\right]_{m,k+K}\right\}, \qquad (9.18)$$

where

$$\boldsymbol{\Delta}_{\xi}^{\sharp} = -\frac{\mathrm{j}}{M}\boldsymbol{\Delta}\left(\frac{H^{(1)}}{H}\right)\mathbf{X}_{g,f^{(1)}}^{\sharp}\left(\widetilde{\mathbf{A}}\right).$$

Hence, we can trivially characterize the asymptotic distortion caused by the channel frequency selectivity on estimated symbols $\hat{a}_{m,k}$ by simply analyzing the statistical properties of matrices of the form $\mathbf{X}_{g,f^{(1)}}^{\sharp}\left(\widetilde{\mathbf{A}}\right)$. This will be the objective of the following section.

## 9.4 RESIDUAL INTERFERENCE CHARACTERIZATION

In the previous section, we have seen that the distortion associated with the channel frequency selectivity is proportional to the output of an AFB under ideal channel conditions, where the prototype pulse of this AFB has been replaced by the derivative of the original one. This is more specifically expressed in (9.17), which assumes that a channel inversion per-subcarrier equalizer is employed. The objective of this section is to provide a closed-form expression of the power of this residual interference $\xi_{m,k}$ taking into account the randomness of the transmitted symbols $\mathbf{A}$.

It is clear from the expression of $\xi_{m,k}$ in (9.18) that the statistical properties of this random variable are completely characterized by matrices of the form $\mathbf{X}_{p,q}^{\sharp}(\tilde{\mathbf{A}})$, $\sharp \in \{\text{even, odd}\}$, where $p, q$ are two generic pulses of length $KM$. It is not difficult to see that since the original complex symbols $\mathbf{A}$ have zero mean, we will have $\mathbb{E}\{\mathbf{X}_{p,q}^{\sharp}(\tilde{\mathbf{A}})\} = \mathbf{0}$. Next, we establish the second-order behavior of the entries of these matrices.

To present the main result in this section, we need to introduce some quantities that will ultimately depend on the prototype pulses and their derivatives. To that effect, consider four different pulses $p_1, q_1, p_2, q_2$, all having length $KM$. We define

$$v^{(+,-)}(p_1, q_1, p_2, q_2) = \frac{M}{4} \text{tr}\left[\mathcal{R}_{p_1,q_1}\mathcal{R}_{p_2,q_2}^{\mathsf{T}}\mathbf{U}^+ + \mathcal{S}_{p_1,q_1}\mathcal{S}_{p_2,q_2}^{\mathsf{T}}\mathbf{U}^-\right], \quad (9.19)$$

where $\mathcal{R}_{p,q}$ and $\mathcal{S}_{p,q}$ are as defined in (9.7), and $\mathbf{U}^{\pm}$ as defined in (9.10). Likewise, we define $v^{(-,+)}(p_1, q_1, p_2, q_2)$ as before, interchanging all instances of $+$ and $-$. With these definitions, we are now in the position of presenting the main result of this section.

**Lemma 9.1.** *Consider four generic pulses of length $KM$, namely $p_1, q_1, p_2, q_2$. Let $k \in \{K, K+1, \ldots, N_s - K\}$ (so that tail effects are disregarded), and let $P_s$ denote the power of the transmitted symbols, that is, $P_s = \mathbb{E}\{|[\mathbf{A}]_{m,k}|^2\}$. Then, under (**As3**), we can write*

$$\mathbb{E}\left[\text{Re}\{[\mathbf{X}_{p_1,q_1}^{\text{odd}}(\tilde{\mathbf{A}})]_{m,k}\}\,\text{Re}\{[\mathbf{X}_{p_2,q_2}^{\text{odd}}(\tilde{\mathbf{A}})]_{m,k}\}\right]$$
$$= \mathbb{E}\left[\text{Im}\{[\mathbf{X}_{p_1,q_1}^{\text{even}}(\tilde{\mathbf{A}})]_{m,k}\}\,\text{Im}\{[\mathbf{X}_{p_2,q_2}^{\text{even}}(\tilde{\mathbf{A}})]_{m,k}\}\right] = P_s v^{(+,-)}(p_1, q_1, p_2, q_2),$$

*whereas*

$$\mathbb{E}\left[\text{Im}\{[\mathbf{X}_{p_1,q_1}^{\text{odd}}(\tilde{\mathbf{A}})]_{m,k}\}\,\text{Im}\{[\mathbf{X}_{p_2,q_2}^{\text{odd}}(\tilde{\mathbf{A}})]_{m,k}\}\right]$$
$$= \mathbb{E}\left[\text{Re}\{[\mathbf{X}_{p_1,q_1}^{\text{even}}(\tilde{\mathbf{A}})]_{m,k}\}\,\text{Re}\{[\mathbf{X}_{p_2,q_2}^{\text{even}}(\tilde{\mathbf{A}})]_{m,k}\}\right] = P_s v^{(-,+)}(p_1, q_1, p_2, q_2).$$

*On the other hand, we have*

$$\mathbb{E}\left[\text{Re}\{[\mathbf{X}_{p_1,q_1}^{\text{odd}}(\tilde{\mathbf{A}})]_{m,k}\}\,\text{Im}\{[\mathbf{X}_{p_2,q_2}^{\text{odd}}(\tilde{\mathbf{A}})]_{m,k}\}\right]$$
$$= \mathbb{E}\left[\text{Re}\{[\mathbf{X}_{p_1,q_1}^{\text{even}}(\tilde{\mathbf{A}})]_{m,k}\}\,\text{Im}\{[\mathbf{X}_{p_2,q_2}^{\text{even}}(\tilde{\mathbf{A}})]_{m,k}\}\right] = 0.$$

*Proof.* See Appendix B in [4]. □

**Table 9.2** Value of $\nu^{(+,-)}\left(g, f^{(1)}, g, f^{(1)}\right)$ in decibels for the PHYDYAS prototype pulse

| K | M = 16 | M = 32 | M = 64 | M = 128 | M = 256 | M = 512 |
|---|--------|--------|--------|---------|---------|---------|
| 2 | 3.9224 | 3.9224 | 3.9224 | 3.9224 | 3.9224 | 3.9224 |
| 3 | 3.1782 | 3.1784 | 3.1784 | 3.1784 | 3.1784 | 3.1784 |
| 4 | 3.0054 | 3.0057 | 3.0058 | 3.0058 | 3.0058 | 3.0058 |

Recalling the form of the asymptotic distortion power in (9.18), we can trivially use Lemma 9.1 to obtain a closed-form expression of the power $\mathbb{E}\{|\xi_{m,k}|^2\}$, namely,

$$\mathbb{E}\{|\xi_{m,k}|^2\} = \frac{2P_s}{M^2}\operatorname{Re}^2\left[\frac{H^{(1)}(e^{j\omega_m})}{H(e^{j\omega_m})}\right]\nu^{(+,-)}\left(g, f^{(1)}, g, f^{(1)}\right)$$

$$+ \frac{2P_s}{M^2}\operatorname{Im}^2\left[\frac{H^{(1)}(e^{j\omega_m})}{H(e^{j\omega_m})}\right]\nu^{(-,+)}\left(g, f^{(1)}, g, f^{(1)}\right).$$

This expression can be further simplified under the assumption that the prototype pulses are either symmetric or antisymmetric in the time domain. It can be shown that, under these conditions, we have $\nu^{(+,-)}(p, q, p, q) = \nu^{(-,+)}(p, q, p, q)$ (see Corollary 1 in [4]), and therefore the power of the residual interference takes the form

$$P_e(m) \triangleq \mathbb{E}\{|\xi_{m,k}|^2\} = \frac{2P_s}{M^2}\left|\frac{H^{(1)}(e^{j\omega_m})}{H(e^{j\omega_m})}\right|^2 \nu^{(+,-)}\left(g, f^{(1)}, g, f^{(1)}\right).$$

From this expression we can conclude that the residual distortion experienced by the $m$th subcarrier essentially depends on the quotient of the derivative of the channel frequency response and the response itself. Hence, rapid variations of $H(e^{j\omega_m})$ in the frequency domain will cause peaks in the distortion generated by the channel frequency selectivity. On the other hand, the asymptotic effect of the transmit/receive prototype pulses is the same in all the spectrum, and it is established through the quantity $\nu^{(+,-)}\left(g, f^{(1)}, g, f^{(1)}\right)$. To give an idea about the weight of this factor on the total distortion, Table 9.2 reports its value (in decibels) for different choices of the PHYDYAS prototype pulse.

It turns out that, despite its asymptotic nature, the above formula provides a very accurate approximation of the real distortion power in the whole spectrum. To illustrate this point further, consider an FBMC/OQAM system with $M = 256$ subcarriers constructed using the PHYDYAS pulse [2] with overlapping factor equal to $K = 4$. Fig. 9.5 represents the asymptotic Signal-to-Noise-plus-Distortion Ratio (SNDR) obtained according to the above formula (solid line) versus the simulated one (markers), averaged over $N_s = 5000$ multicarrier symbols. Two different channel realizations were considered, both taken from the Extended Typical Urban (ETU) channel model [5] with different values of the intercarrier separation (denoted as $\Delta_f$ in Fig. 9.5).

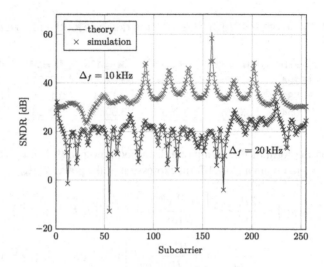

**FIGURE 9.5**

Simulated versus asymptotic signal-to-distortion power ratio for two different realizations of an ETU channel model with different intercarrier separation.

Observe that there is a perfect match between simulated and asymptotic values, even for a relatively low number of subcarriers ($M = 256$).

## 9.5 CHARACTERIZATION OF FBMA

In this section, we present the first example of the insightful applications of the analysis above. In the uplink FilterBank Multiple Access (FBMA) channel, multiple users are assigned different sets of subcarriers to communicate to the same base station. The analysis above can be easily extended to evaluate the amount of interference affecting the users. As we will see next, the linearity of our model allows us to treat the signal at the receive side as the superposition of users' contributions. However, (9.17) suggests that the energy corresponding to each user's signal is not confined to that specific user's subcarriers but leaks into adjacent portions of the spectrum. As a result, subcarrier distortion increases with the number of users, especially if the prototype pulses are not chosen properly.

Let $\mathcal{M}_u^i$ be the set of subcarriers assigned to user $i$. Since all users transmit on different subcarriers, we have $\mathcal{M}_u^i \cap \mathcal{M}_u^{i'} = \emptyset$ for all $i \neq i'$ and $\bigcup_{i=0}^{U-1} \mathcal{M}_u^i \subseteq \{0, 1, \ldots, M-1\}$, with $U$ the total number of users. Then, the signal received at the base station can be written as $\mathbf{Y}_{g,f} = \sum_{i=0}^{U-1} \mathbf{Y}_{g,f}^i$, where $\mathbf{Y}_{g,f}^i$ takes the form in (9.14) after replacing $\mathbf{A}$ by $\mathbf{A}^i = \mathbf{B}^i + j\mathbf{C}^i$, the matrix of symbols transmitted by user $i$. FBMA subcarrier allocation is easily represented by forcing $[\mathbf{A}^i]_{m,k} = a_{m,k}^i = 0$

when subcarrier $m$ is not assigned to user $i$, that is, when $m \notin \mathcal{M}_u^i$, whereas $a_{m,k}^i \in \mathbb{C}$ otherwise.

Now, let us consider a low-complexity base station where the FBMC receiver is complemented by a single-tap per-subcarrier equalizer as that analyzed in the previous section. This choice, which represents cheap real-life devices, also has the didactic purpose of simplifying mathematical expressions hereafter. The corresponding expression for the symbol estimates is

$$\hat{a}_{m,k}^i = \sum_{u=0}^{U-1} \mathrm{Re}\left\{\left[\mathbf{WY}_{g,f}^{u,\mathrm{odd}}\right]_{m,k+K-1}\right\} + j \sum_{u=0}^{U-1} \mathrm{Im}\left\{\left[\mathbf{WY}_{g,f}^{u,\mathrm{even}}\right]_{m,k+K}\right\}, \quad (9.20)$$

where the diagonal matrix $\mathbf{W} = \mathrm{diag}\{w_m\}$ collects the equalizer coefficients.

Following the steps explained before for the single-user case, as a direct consequence of linearity and superposition principles, we readily see that all users can be received free of errors when the prototype pulses meet the PR constraints and all channels are ideal. Indeed, under these assumptions, there is no intercarrier interference, and, as a result, $[\mathbf{Y}_{g,f}^u]_{m,k} = 0$ for all $u \neq i$ and $m \in \mathcal{M}_u^i$. Conversely, the analysis of Sections 9.3 and 9.4 (see (9.17) and (9.18)) establishes that frequency selective channels, and non-PR prototype pulses cause all subcarriers to be received with a distortion whose mean-square value can be written as

$$P_e(m) = \mathbb{E}\left\{\left|\hat{a}_{m,k}^i - w_m \sum_{u=0}^{U-1} H^u\left(e^{j\omega_m}\right) a_{m,k}^u\right|^2\right\}$$

$$= 2\mathbb{E}\left\{\sum_{u=0}^{U-1}\left|\mathrm{Re}\left\{\left[\mathbf{WY}_{g,f}^{u,\mathrm{odd}}\right]_{m,k+K-1}\right\} - w_m H^u\left(e^{j\omega_m}\right) b_{m,k}^u\right|^2\right\}, \quad (9.21)$$

where $H^u(e^{j\omega_m})$ is the nominal value of the channel frequency response of user $u$ at subcarrier $m$. The second step above has been obtained with the help of (9.20) under the standard assumption of symbols $a_{m,k}^i$ being circularly symmetric random variables independent across user index $i$, subcarrier index $m$, and time index $k$.

For subcarrier $m$ assigned to user $i$, (9.21) clearly shows that the corresponding distortion power $P_e(m)$ is the sum of contributions of all users, not just of user $i$. This fact can cause unbearable levels of distortion and impair communications. Fortunately, interference can be controlled by a proper choice of the prototype pulses. Briefly,[3] the distortion due to each user can be made to decay quickly outside the user's assigned subcarriers, as shown in Fig. 9.6, and a single silent subcarrier is sufficient to isolate contiguous users when typical prototype pulses are employed.

---

[3]Details are out of the scope of this text and can be found in [6].

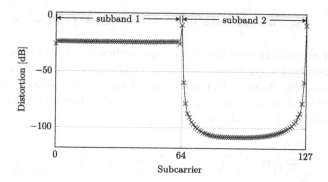

**FIGURE 9.6**

Example of distortion generated by a user transmitting on subcarriers 0–63 (subband 1) of a system with 128 subcarriers. The distortion is negligible in the entire subband 2 except for its boundary subcarriers (i.e., subcarriers 64 and 127).

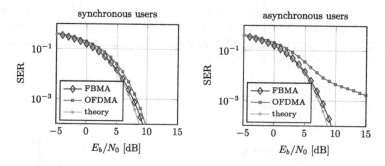

**FIGURE 9.7**

Symbol error rate for FBMA and CP-OFDMA with a 4-QAM constellation: synchronous users case (left) and asynchronous users case (right).

As a final remark, it is worth comparing FBMA with Cyclic Prefix (CP)-Ortho-gonal Frequency-Division Multiple Access (OFDMA), which is the common bench-mark for frequency-division multiple access channels (see, e.g., [7,8] and references therein). Fig. 9.7 compares the symbol error rate obtained by the two schemes with a 4-QAM constellation. We can see that performances are similar when users are perfectly synchronized (leftmost graph). However, FBMA proves itself supe-rior when users are not synchronized (rightmost graph). As explained with more details in [6], filterbank multiple access schemes are less sensitive to delays be-tween users as compared to CP-OFDMA schemes. Indeed, user orthogonality is lost in the latter approach when the relative user delay is longer than the cyclic prefix.

## 9.6 CONCLUDING REMARKS

In this chapter, we have shown how the output of an FBMC/OQAM AFB with a sufficiently large number of subcarriers can be approximated by a series of terms that depend on the derivatives of progressively higher orders of the channel response. Simulations confirm that, thanks to this approximation, the distortion caused by highly frequency selective channels can be estimated with great accuracy.

Besides distortion characterization, this result has several other applications. As an example, Section 9.5 showed how it is useful to analyze the interactions between different users in an FBMA uplink channel and to compare the latter with a more classic solution based on cyclic prefix. Similarly, the approximation obtained in Section 9.3 leads to a novel equalization scheme in Section 12.4, whereas in Sections 13.3.2 and 14.3.4, we extend it to the analysis and design of linear precoders and receivers in Multiple-Input Multiple-Output (MIMO) FBMC/OQAM systems.

---

## APPENDIX 9.A  RELATIONSHIP BETWEEN POLYPHASE AND TRADITIONAL FBMC/OQAM REPRESENTATION

In this appendix, we provide a more formal proof of the equivalence between the polyphase and the traditional representation of the FBMC/OQAM signal waveform.

### 9.A.1 POLYPHASE EQUIVALENT OF THE TRANSMITTER

Consider the traditional representation of the transmitted waveform corresponding to the transmission of $N_s$ multicarrier symbols, namely,

$$s[l] = \sum_{m=0}^{M-1} \sum_{n=0}^{2N_s-1} e^{j\varphi_{m,n}} d_{m,n} g\left[l - nM/2\right] e^{j\frac{2\pi}{M}m\left(l-\frac{L_g-1}{2}\right)}, \tag{9.A.1}$$

where $L_g = MK$, $d_{m,n} \in \mathbb{R}$ represents the real-valued transmitted sequence, and $\varphi_{m,n}$ is some phase sequence that accepts multiple choices [1] such as $\varphi_{m,n} = (m+n)\pi/2 - mn\pi$. Here, we will consider a different choice, which guarantees the transmitter/receiver symmetry in the polyphase representation:

$$\varphi_{m,n} = \begin{cases} \dfrac{\pi}{2}\left(2K+1\right)m, & n \text{ even,} \\[2mm] \dfrac{\pi}{2}\left(2K-1\right)m + \dfrac{\pi}{2}, & n \text{ odd.} \end{cases} \tag{9.A.2}$$

Observe that the transmitted samples $s[l]$ are only different from zero in the index interval going from $l = 0$ to $l = (2N_s + 2K - 1)M/2 - 1$. We gather these transmitted samples into a column vector $\mathbf{s} = \left[\mathbf{s}^{\mathrm{T}}(0) \ \cdots \ \mathbf{s}^{\mathrm{T}}(2N_s + 2K - 2)\right]^{\mathrm{T}}$, where each $\mathbf{s}(i)$

has length $M/2$ and is defined as

$$\mathbf{s}(i) = \begin{bmatrix} s[iM/2] & \cdots & s[(i+1)M/2 - 1] \end{bmatrix}^{\mathsf{T}}.$$

The objective is to show that $\mathbf{s}$ can be expressed as in (9.1). We begin by separating the sum in $n$ in (9.A.1) into two terms, corresponding to odd $n$ and even $n$, and using the expression of $\varphi_{m,n}$ above, which allows us to write

$$s[l] = \sum_{\substack{m=0 \\ }}^{M-1} \sum_{\substack{n=0 \\ n \text{ even}}}^{2N_s-1} [\boldsymbol{\Phi}^*]_{m,m} d_{m,n} g\left[l - nM/2\right] e^{j\frac{2\pi}{M}ml}$$

$$+ j \sum_{\substack{m=0 \\ }}^{M-1} \sum_{\substack{n=0 \\ n \text{ odd}}}^{2N_s-1} [\boldsymbol{\Phi}^*]_{m,m} d_{m,n} g\left[l - nM/2\right] e^{j\frac{2\pi}{M}m\left(l+\frac{M}{2}\right)}.$$

To introduce the polyphase representation, let $((n))_M$ denote the residue modulo $M$, and $\lfloor x \rfloor$ the largest integer not greater than $x$. Let $g_r[k]$, $r = 0, 1, \ldots, M-1$, denote the $r$th polyphase component of $g[l]$ and observe that $g[l] = g_{((l))_M}[\lfloor l/M \rfloor]$, so that

$$g\left[l - nM/2\right] = \begin{cases} g_{((l))_M}\left[\lfloor l/M \rfloor - n/2\right], & n \text{ even}, \\ g_{((\tilde{l}))_M}\left[\lfloor \tilde{l}/M \rfloor - (n+1)/2\right], & n \text{ odd}, \end{cases} \tag{9.A.3}$$

where $\tilde{l} = l + M/2$. Using this, we may rewrite the transmitted signal $s[l]$ as

$$s[l] = \sqrt{M} \sum_{\substack{n=0 \\ n \text{ even}}}^{2N_s-1} [\mathcal{F}\boldsymbol{\Phi}^*\mathbf{D}]_{((l))_M,n} g_{((l))_M}\left[\lfloor l/M \rfloor - n/2\right]$$

$$+ \sqrt{M} \sum_{\substack{n=0 \\ n \text{ odd}}}^{2N_s-1} [\mathcal{F}\boldsymbol{\Phi}^*j\mathbf{D}]_{((\tilde{l}))_M,n} g_{((\tilde{l}))_M}\left[\lfloor \tilde{l}/M \rfloor - (n+1)/2\right],$$

where $\mathbf{D} \in \mathbb{R}^{M \times 2N_s}$ gathers the transmitted real symbols, namely, $[\mathbf{D}]_{m,n} = d_{m,n}$.

To simplify the notation, we denote $\mathbf{E} = \sqrt{M}\mathcal{F}\boldsymbol{\Phi}^*\tilde{\mathbf{A}}$, where

$$\tilde{\mathbf{A}} = \mathbf{D}\left(\mathbf{I}_{N_s} \otimes \begin{bmatrix} 1 & 0 \\ 0 & j \end{bmatrix}\right),$$

so that writing $e_{m,n} = [\mathbf{E}]_{m,n}$, we have

$$\sum_{\substack{n=0 \\ n \text{ even}}}^{2N_s-1} e_{((l))_M,n} g_{((l))_M}\left[\lfloor l/M \rfloor - n/2\right] = [\mathbf{E} \star (\mathbf{G} \otimes [1\ 0])]_{((l))_M, 2\lfloor l/M \rfloor}$$

$$= [\mathbf{S}]_{((l))_M, 2\lfloor l/M \rfloor},$$

$$\sum_{\substack{n=0 \\ n \text{ odd}}}^{2N_s-1} e_{(\!(\tilde{i})\!)_M,n} g_{(\!(\tilde{i})\!)_M} \left[ \lfloor \tilde{l}/M \rfloor - (n+1)/2 \right] = \left[ \mathbf{E} \star (\mathbf{G} \otimes [1\ 0]) \right]_{(\!(\tilde{i})\!)_M, 2\lfloor l/M \rfloor -1}$$

$$= [\mathbf{S}]_{(\!(\tilde{i})\!)_M, 2\lfloor l/M \rfloor -1},$$

where $\mathbf{S}$ is defined in (9.2). Using these identities, we readily obtain (9.1).

## 9.A.2 POLYPHASE EQUIVALENT OF THE RECEIVER

Let $r[l]$ denote the samples at the input of the AFB. The conventional receiver recovers transmitted symbol $d_{m,n}$ as[4]

$$d_{m,n} = \begin{cases} \mathrm{Re}\{[\mathbf{X}_{g,f}]_{m,n+2K-1}\}, & \text{even } n, \\ \mathrm{Im}\{[\mathbf{X}_{g,f}]_{m,n+2K-1}\}, & \text{odd } n, \end{cases}$$

where $\mathbf{X}_{g,f}$ is the matrix of the samples at the output of the AFB. The entries of this matrix are conventionally expressed as[5]

$$[\mathbf{X}_{g,f}]_{m,n+2K-1} = (\delta_{n \text{ even}} + j\delta_{n \text{ odd}}) \sum_{l \in \mathbb{Z}} r[l] \left( \tilde{f}_{n,m}[l] \right)^*,$$

where $\delta_n$ is the Kronecker delta (e.g., $\delta_{n \text{ even}} = 1$ if and only if $n$ is even and zero otherwise), and

$$\tilde{f}_{m,n}[l] = e^{j\varphi_{m,n}} \tilde{f}[l - nM/2] e^{j\frac{2\pi}{M}m\left(l - \frac{L_f-1}{2}\right)}$$

with $\varphi_{m,n}$ equal to the phase used at the transmit prototype, $L_f = MK$, and $\tilde{f}[l] = f[MK - l - 1]$ denoting the time-reversed version of the receive prototype $f[l]$. The objective of this subsection is to show that the matrix $\mathbf{X}_{g,f}$ can be expressed as in (9.5).

Using the expression of $\varphi_{m,n}$ given in (9.A.2), we are able to write

$$[\mathbf{X}_{g,f}]_{m,n+2K-1} = \sum_{l \in \mathbb{Z}} r[l] \left( [\mathbf{\Phi}]_{m,m} \tilde{f}[l - nM/2] e^{-j\frac{2\pi}{M}ml} \right) \delta_{n \text{ even}}$$

$$+ \sum_{l \in \mathbb{Z}} r[l] \left( [\mathbf{\Phi}]_{m,m} \tilde{f}[l - nM/2] e^{-j\frac{2\pi}{M}m\left(l + \frac{M}{2}\right)} \right) \delta_{n \text{ odd}}.$$

---

[4]The offset of $2K - 1$ symbols is a consequence of the memory of the AFB.
[5]Observe that we introduce the factor $(\delta_{n \text{ even}} + j\delta_{n \text{ odd}})$ because we estimate the symbols taking real parts of even samples and imaginary parts of odd samples.

Now, applying identity (9.A.3) with $g$ replaced by $\tilde{f}$, we see that

$$
\begin{aligned}
\left[\mathbf{X}_{g,f}\right]_{m,n+2K-1} \\
= \sqrt{M} \sum_{i=0}^{M-1} \left[\mathbf{\Phi}\mathcal{F}^{\mathrm{H}}\right]_{m,i} \sum_{k\in\mathbb{Z}} r\left[kM+i\right] \tilde{f}_i\left[k-n/2\right] \delta_{n \text{ even}} \\
+ \sqrt{M} \sum_{i=0}^{M/2-1} \left[\mathbf{\Phi}\mathcal{F}^{\mathrm{H}}\right]_{m,i+M/2} \sum_{k\in\mathbb{Z}} r\left[kM+i\right] \tilde{f}_{i+M/2}\left[k-\frac{n+1}{2}\right] \delta_{n \text{ odd}} \\
+ \sqrt{M} \sum_{i=0}^{M/2-1} \left[\mathbf{\Phi}\mathcal{F}^{\mathrm{H}}\right]_{m,i} \sum_{k\in\mathbb{Z}} r\left[kM+i+M/2\right] \tilde{f}_i\left[k-\frac{n-1}{2}\right] \delta_{n \text{ odd}},
\end{aligned}
$$

where we have applied the change of variables $l = kM + i$, namely, $k = \lfloor l/M \rfloor$ and $i = ((l))_M$. Now, using the fact that $\tilde{f}_i[k] = f_{M-i-1}[K-k-1]$ and the definition of the matrix $\mathbf{R}$ in (9.4), we can rewrite this identity as

$$
\begin{aligned}
\left[\mathbf{X}_{g,f}\right]_{m,n+2K-1} \\
= \sqrt{M} \sum_{i=0}^{M-1} \left[\mathbf{\Phi}\mathcal{F}^{\mathrm{H}}\right]_{m,i} \sum_{k \text{ odd}} [\mathbf{R}]_{i,k}\left[\mathbf{J}_M\mathbf{F}\otimes[1\ 0]\right]_{i,2K-1-k+n} \delta_{n \text{ even}} \\
+ \sqrt{M} \sum_{i=0}^{M-1} \left[\mathbf{\Phi}\mathcal{F}^{\mathrm{H}}\right]_{m,i} \sum_{k \text{ even}} [\mathbf{R}]_{i,k}\left[\mathbf{J}_M\mathbf{F}\otimes[1\ 0]\right]_{i,2K-1-k+n} \delta_{n \text{ odd}}.
\end{aligned}
$$

Now, observe that in the sum with respect to $k$, only the even columns of the matrix $\mathbf{J}_M\mathbf{F}\otimes[1\ 0]$ are selected. However, since the odd columns of this matrix are all zero, we expand both sums to all $k \in \mathbb{Z}$, which directly leads to

$$
\mathbf{X}_{g,f} = \sqrt{M}\mathbf{\Phi}\mathcal{F}^{\mathrm{H}}\left(\mathbf{R}\star(\mathbf{J}_M\mathbf{F}\otimes[1\ 0])\right),
$$

as we wanted to show.

## REFERENCES

[1] P. Siohan, C. Siclet, and N. Lacaille, "Analysis design of OFDMA/OQAM systems based on filterbank theory," *IEEE Trans. Signal Process.*, vol. 50, pp. 1170–1183, May 2002.

[2] M. G. Bellanger, "Specification and design of a prototype filter for filter bank based multi-carrier transmission," in *Proc. IEEE ICASSP 2001*, Salt Lake City, UT, May 2001.

[3] X. Mestre and D. Gregoratti, "Parallelized structures for MIMO FBMC under strong channel frequency selectivity," *IEEE Trans. Signal Process.*, vol. 64, pp. 1200–1215, Mar. 2016.

[4] X. Mestre, M. Majoral, and S. Pfletschinger, "An asymptotic approach to parallel equalization of filter bank based multicarrier signals," *IEEE Trans. Signal Process.*, vol. 61, pp. 3592–3606, Jul. 2013.

[5] G. T. 36.101, "User equipment (UE) radio transmission and reception." 3rd Generation Partnership Project; Technical Specification Group Radio Access Network; Evolved Universal Terrestrial Radio Access (E-UTRA), Tech. Rep., 2013. http://www.3gpp.org.

[6] D. Gregoratti and X. Mestre, "Uplink FBMC/OQAM-based multiple access channel: Distortion analysis under strong frequency selectivity," *IEEE Trans. Signal Process.*, vol. 64, pp. 4260–4272, Aug. 2016.

[7] H. Saeedi-Sourck, Y. Wu, J. W. M. Bergmans, S. Sadri, and B. Farhang-Boroujeny, "Complexity and performance comparison of filter bank multicarrier and OFDM in uplink of multicarrier multiple access networks," *IEEE Trans. Signal Process.*, vol. 59, pp. 1907–1912, Apr. 2011.

[8] V. Dalakas and E. Kofidis, "Filter bank-based multicarrier modulation for multiple access in next generation satellite uplinks: A DVB-RCS2-based experimental study." HAL–archives-ouvertes, HAL Id: hal-01381072, Oct. 13, 2016. [Online]. Available: https://hal.archives-ouvertes.fr/hal-01381072.

# FBMC Synchronization Techniques

# 10

**Jérôme Louveaux\*, Martin Fuhrwerk†, Davide Mattera‡, Markku Renfors§,
Mario Tanda‡**

*Université catholique de Louvain, Louvain-la-Neuve, Belgium\**
*Leibniz Universität Hannover, Hannover, Germany†*
*Università degli Studi di Napoli Federico II, Napoli, Italy‡*
*Tampere University of Technology, Tampere, Finland§*

## CONTENTS

## 10.1 INTRODUCTION

Synchronization is a crucial operation in multicarrier systems. In particular, Orthogonal Frequency-Division Multiplexing (OFDM) is known to be very sensitive to Carrier Frequency Offset (CFO). Frame (or symbol timing) synchronization also has to be achieved in order to locate the Cyclic Prefix (CP) precisely enough to avoid

*Orthogonal Waveforms and Filter Banks for Future Communication Systems.* DOI: 10.1016/B978-0-12-810384-5.00010-4

intersymbol and intercarrier interference. Likewise, systems employing FilterBank MultiCarrier with Offset-QAM subcarrier modulation (FBMC/OQAM) also require good frequency offset and symbol timing synchronization but usually have slightly relaxed synchronization requirements with respect to OFDM as the residual errors can be efficiently corrected by a multitap equalizer if it is present. The study of the sensitivity of FBMC/OQAM to residual timing and frequency offsets is the first topic of this chapter and is analyzed in Section 10.2. The focus is on the case where only a single-tap equalizer is used. These results allow us to assess the accuracy required from the timing and frequency offset estimators to only result in negligible effects at the receiver. Then, the chapter focuses on the synchronization algorithms themselves. As is common in burst transmission nowadays, two types of training information can be used for the purpose of synchronization, preamble training sequences and scattered pilots. Both cases are covered in this chapter.

However, the FBMC/OQAM context presents a number of challenges. First and foremost, the intrinsic interference of the Offset Quadrature Amplitude Modulation (OQAM) has to be taken into account in the estimation process. Secondly, for preamble-based synchronization, the size of the preamble has to be kept to a minimum to reduce the overhead. This is slightly more challenging in FilterBank MultiCarrier (FBMC), where the filters are longer, and hence longer guard periods are needed to avoid interference from the payload data sequence. Finally, the choice has to be made between two families of methods. Time Domain (TD) methods operate before the Analysis FilterBank (AFB) at the receiver. They have access to all the information and are able to compensate perfectly for the synchronization errors, but tend to have higher complexity since they do not take advantage of the particular structure of the signal. On the other hand, Frequency Domain (FD) techniques perform the estimation (and often also the compensation) after the AFB, combining the information from the different subcarriers. Less complex estimation methods can usually be designed in this way, and they are less sensitive to interference from other users, but the compensation is less accurate. Both types of methods are described in this chapter.

In Section 10.3, the requirements for preamble design and different options for the structure of the preamble are discussed. In Section 10.3.1, the problem of estimating the Symbol Timing Offset (STO) affecting the received signal is addressed. Several algorithms in the TD and FD are presented, and a system design that reduces the influence of a misaligned AFB is proposed. In Section 10.3.2, CFO estimation is considered. Both TD and FD algorithms are described. Blind synchronization is briefly surveyed in Section 10.4, where the most well-known techniques and results from the literature are summarized. In Section 10.5, synchronization based on scattered pilots (in that case, obviously, in the FD) is discussed.

Finally, one important advantage of FBMC/OQAM systems is their increased ability to separate the users in frequency thanks to the very low sidelobes of the modulation filters, thus requiring no (or at most relaxed) synchronization among the users. However, it is still necessary for the base station to synchronize with all the different users individually. This scenario is considered in Section 10.6 using Fast-Convolution FilterBank (FC-FB) processing.

## 10.2 SENSITIVITY ANALYSIS OF FBMC/OQAM TO SYNCHRONIZATION ERRORS

In this section, the effect of residual synchronization errors on the performance of an FBMC/OQAM system is analyzed. The system is studied in the presence of a frequency selective channel and using single-tap equalization at the receiver. Both residual CFO and STO are considered.

The discrete-time transmitted sequence $s[l]$ can be written as

$$s[l] = \sum_{n=-\infty}^{\infty} \sum_{m \in \mathcal{M}_{\mathrm{u}}} j^{n+m} d_{m,n} e^{j\frac{2\pi}{M}m\left(l-n\frac{M}{2}\right)} \theta^m g\left[l - nM/2\right], \qquad (10.1)$$

where

$$\theta = \exp\left(-j\pi \frac{L_g - 1}{M}\right), \qquad (10.2)$$

$M$ is the total number of subcarriers, $\mathcal{M}_{\mathrm{u}}$ denotes the set of active subcarriers, and $g$ is the (unit energy) prototype filter impulse response, of length $L_g = KM$, with $K$ being the overlapping factor. The real-valued information-symbols $d_{m,n}$ are assumed to be independent and identically distributed (i.i.d.) symbols of zero-mean and variance $\sigma_d^2$. The discrete-time low-pass signal received after a multipath channel with additive noise and possibly after the time domain stage aimed at compensating for the synchronization mismatches can be written as

$$r[l] = e^{j2\pi \varepsilon l/M} \sum_{\rho=0}^{P-1} c_\rho s\left[l - \tau_\rho\right] + v[l], \qquad (10.3)$$

where the noise $v[l]$ is assumed to be Additive White Gaussian Noise (AWGN) of power spectral density $2N_0/T_{\mathrm{s}}$, where $T_{\mathrm{s}}$ is the sampling period, the parameter $\varepsilon$ measures the normalized residual CFO still present after rough frequency synchronization, and the (discrete) time delay $\tau_0$, that is, the delay of the first multipath component, denotes the residual STO after rough timing compensation. The $m$th complex-valued output $y_{m,n}$ of the AFB for symbol $n$ can be written as

$$y_{m,n} = \sum_{l=-\infty}^{\infty} r[l] \left\{ j^{n+m} e^{j\frac{2\pi}{M}m\left(l-n\frac{M}{2}\right)} \theta^m g\left[l - n\frac{M}{2}\right] \right\}^*. \qquad (10.4)$$

Using relations (10.4), (10.3), and (10.1), the output $y_{m,n}$ can be written as the sum of a useful term, an interference term, and an additive noise term:

$$y_{m,n} = y_{m,n}^{(\text{U})} + y_{m,n}^{(\text{I})} + y_{m,n}^{v}.$$ (10.5)

The first two terms are given by

$$y_{m,n}^{(\text{U})} = e^{j\pi\varepsilon n}\mathcal{I}_{0,0}[m, \varepsilon, h]d_{m,n},$$ (10.6)

$$y_{m,n}^{(\text{I})} = e^{j\pi\varepsilon n}\sum_{\delta=n_{\min}}^{n_{\max}}\sum_{q\in\mathcal{A}^m,\,(\delta,q)\neq(0,0)}\mathcal{I}_{\delta,q}[m, \varepsilon, h]d_{((m-q))_M, n-\delta},$$ (10.7)

where $((\ell))_M$ denotes $\ell$ modulo $M$, and

$$\mathcal{I}_{\delta,q}[m, \varepsilon, h] = j^{-\delta-q}\theta^{-q}(-1)^{\delta(m-q)}\sum_{\rho=0}^{P-1}c_\rho r_g^{q-\varepsilon}\left[\tau_\rho - \delta M/2\right]e^{-j\frac{2\pi}{M}(m-q)\tau_\rho}.$$

(10.8)

with

$$r_g^{\alpha}[p] = \sum_{l=-\infty}^{\infty} g[l]g[l-p]e^{-j\frac{2\pi\alpha}{M}l}.$$ (10.9)

Note that $r_g^0[0] = 1$ since we have assumed that the prototype filter is normalized to unit energy. The coefficients $\mathcal{I}$ in (10.8) depend on the channel response $h$ through the set of coefficients $\{c_\rho\}_{\rho=0,1,...,P-1}$ and delays $\{\tau_\rho\}_{\rho=0,1,...,P-1}$. Moreover, the set $\mathcal{A}^m$ is defined as follows:

$$\mathcal{A}^m = \{q \in \{0, 1, \ldots, M-1\} : ((m-q))_M \in \mathcal{M}_{\text{u}}\}.$$ (10.10)

The finite length of the prototype filter implies that $r_g^{\alpha}[p] = 0$ for $|p| \geq L_g$ and, consequently,

$$n_{\min} = \left\lfloor \frac{-L_g + \min_\rho \tau_\rho}{M/2} + 1 \right\rfloor, \qquad n_{\max} = \left\lceil \frac{L_g + \max_\rho \tau_\rho}{M/2} - 1 \right\rceil.$$ (10.11)

Furthermore, the variance of the noise term is given by

$$\mathbb{E}\left\{|y_{m,n}^{v}|^2\right\} = 2\sigma_d^2/\gamma,$$ (10.12)

where $\gamma = E_s/N_0$ with $E_s$ equal to the energy of the useful term of the received signal in a multicarrier symbol period dedicated to each active subcarrier.

It is then assumed that a single-tap equalizer is used at each subcarrier to compensate for all remaining effects, yielding the decision variable $y_{m,n}^{(\text{E})}$ as follows:

$$y_{m,n}^{(\text{E})} = \mathbb{Re}\{W_m e^{-j\pi\varepsilon n}y_{m,n}\}$$ (10.13)

with

$$W_m = \left( r_g^{-\varepsilon}[0] \sum_{\rho=0}^{P-1} c_\rho \exp\left(-j2\pi\tau_\rho m/M\right) \right)^{-1}, \qquad (10.14)$$

based on a common approximation of $\mathcal{I}_{0,0}[m, \varepsilon, h]$:

$$\mathcal{I}_{0,0}[m, \varepsilon, h] = \sum_{\rho=0}^{P-1} c_\rho r_g^{-\varepsilon}[\tau_\rho] e^{-j\frac{2\pi}{M}m\tau_\rho} \simeq r_g^{-\varepsilon}[0] \sum_{\rho=0}^{P-1} c_\rho e^{-j\frac{2\pi}{M}m\tau_\rho}. \qquad (10.15)$$

The term on the right-hand side of (10.15) is easier to calculate since it only involves the channel response at normalized frequency $m/M$ and the term $r_g^{-\varepsilon}[0] = \sum_{l=-\infty}^{\infty} g^2[l] e^{j\frac{2\pi\varepsilon}{M}l}$, which only depends on the residual CFO. The use of the coefficients on the left-hand side of (10.15) may offer performance improvements in some scenarios, as discussed in [1]. We assume that $W_m$ can be computed exactly at the receiver. From (10.13), (10.5), and (10.7) it easily follows that

$$y_{m,n}^{(E)} = \sum_{\delta=n_{\min}}^{n_{\max}} \sum_{q\in\mathcal{A}^m} \mathcal{J}_{\delta,q}[m, \varepsilon, h] d_{(m-q)_M, n-\delta} + \text{Re}\{e^{-j\pi\varepsilon n} W_m y_{m,n}^v\} \qquad (10.16)$$

with

$$\mathcal{J}_{\delta,q}[m, \varepsilon, h] = \text{Re}\left\{W_m \mathcal{I}_{\delta,q}[m, \varepsilon, h]\right\}. \qquad (10.17)$$

For a given symbol constellation, model (10.16) allows a simple evaluation of the Bit Error Rate (BER) using a sufficiently large number of independent Monte Carlo trials. Moreover, (10.16) also enables a simple evaluation of the Signal-to-Interference-plus-Noise Ratio (SINR) of the decision variable on the $m$th subcarrier, in the presence of residual mismatches due to the synchronization procedure, as follows:

$$\text{SINR}_m(\varepsilon, h) = \frac{\text{Re}^2\left\{W_m \mathcal{I}_{0,0}[m, \varepsilon, h]\right\}}{\sum_{\delta=n_{\min}}^{n_{\max}} \sum_{q\in\mathcal{A}^m\ (\delta,q)\neq(0,0)} \text{Re}^2\left\{W_m \mathcal{I}_{\delta,q}[m, \varepsilon, h]\right\} + |W_m|^2/\gamma}. \qquad (10.18)$$

When the synchronization is ideal and the frequency selectivity of the channel is low, the SINR should approach $\text{SINR}_m(\varepsilon, h) \simeq \gamma$. The degradation due to the residual timing and frequency offsets can then be evaluated by computing the loss of $\text{SINR}_m(\varepsilon, h)$ with respect to $\gamma$.

The numerical results presented below consider the PHYDYAS prototype filter [2]. In Fig. 10.1, we report $\text{SINR}_m(\varepsilon, h)$ versus $\gamma$ for different values of $\varepsilon$ assuming a null residual STO. In Fig. 10.2, we report $\text{SINR}_m(\varepsilon, h)$ versus $\gamma$ for different values of the STO $\tau_0$ in the absence of CFO. The plots were obtained for $M = 1024$, $K = 4$, and $m = 38$, but we have verified that they are independent of these values when all the subcarriers are active. Obviously, the sensitivity of the

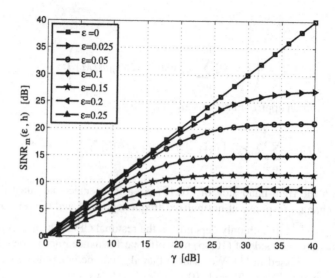

**FIGURE 10.1**

$\mathrm{SINR}_m(\varepsilon, h)$ versus $\gamma$ on an ideally flat channel without STO for different values of $\varepsilon$.

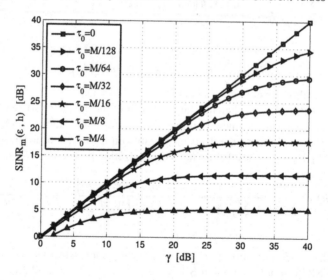

**FIGURE 10.2**

$\mathrm{SINR}_m(\varepsilon, h)$ versus $\gamma$ on an ideally flat channel without frequency offset for different values of $\tau_0$.

system to synchronization mismatches increases with $\gamma$, and, consequently, the accuracy of the synchronization procedures becomes more important when higher-order constellations are adopted. For instance, at the values of $\gamma$ that are relevant for a

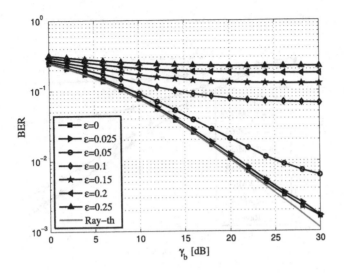

**FIGURE 10.3**

BER with 64-QAM constellation in the ITU Veh-A channel versus $\gamma_b$ for different values of $\varepsilon$ in the absence of STO ($\tau_0 = 0$).

4-Quadrature Amplitude Modulation (QAM) constellation, a CFO of $\varepsilon \le 0.1$ and an STO of $\tau_0 \le M/16$ can easily be tolerated. When the 64-QAM constellation is adopted, only $\varepsilon \le 0.025$ and $\tau_0 \le M/64$ can be tolerated. It is interesting to compare the results obtained for FBMC systems with those previously obtained for OFDM systems. Under reasonable approximations, it is shown in [3] that the sensitivity of the FBMC system to the presence of CFO is lower than that of the classical OFDM system.

Some BER results have also been obtained by simulation experiments in two kinds of subchannel frequency response. In both cases, the parameters were set to $(M, 1/T_s) = (1024, 11.2 \text{ MHz})$. Multipath International Telecommunication Union (ITU) Vehicular-A (Veh-A) or ITU Vehicular-B (Veh-B) channels were considered. The first channel is mildly dispersive and can easily be equalized by a single-tap equalizer, whereas the second one has significant frequency selectivity. The simulation experiments have been performed by exploiting the exact model in (10.16) and using $10^5$ (for 64-QAM constellation) or $10^6$ (for 4-QAM constellation) independent trials. In each trial, a different channel realization has been considered. The case of a unit-energy single-path Rayleigh-fading channel has been included in the figures for the sake of the comparison and is labeled Ray-th. For the Veh-A channel, the sensitivity to CFO is illustrated in Fig. 10.3, reporting the BER versus $\gamma_b$ (defined as $\gamma/k$ for $2^k$-QAM constellation) in the absence of STO for a 64-QAM constellation. The sensitivity to STO is shown in Fig. 10.4 (again for the case of 64-QAM constellation) in the absence of CFO. The results confirm that a negligible performance loss

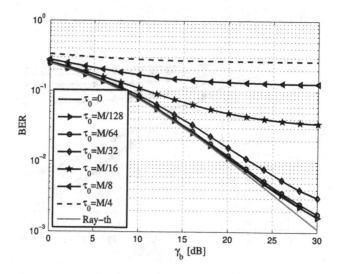

**FIGURE 10.4**

BER with 64-QAM constellation in channel ITU Veh-A versus $\gamma_b$ for different values of $\tau_0$ in the absence of CFO ($\varepsilon = 0$).

is observed if the previously reported conditions on the synchronization errors are satisfied.

In the Veh-B case instead, the presence of the highly dispersive channel makes the single-tap equalizer unable to reach the optimal (Ray-th) performance. Fig. 10.5 illustrates the sensitivity to the CFO (reporting the BER versus $\gamma_b$ in the absence of STO for 4-QAM constellation). Fig. 10.6 illustrates the sensitivity to STO (reporting the BER in the absence of CFO for 4-QAM constellation). The sensitivity to synchronization errors themselves (observed by comparing to the curve with $\varepsilon = \tau_0 = 0$) is lower since most of the performance loss is due to the imperfect equalization. The range of acceptable residual synchronization errors is thus wider in this case, but it is likely to get smaller if superior equalizers are employed.

## 10.3 PREAMBLE-BASED SYNCHRONIZATION

In a standard transceiver, the preamble is a sequence of $N_b$ complex-valued (or $2N_b$ real-valued) known symbols inserted in the transmitted burst before the remaining $N_s$ complex-valued payload symbols. In this section, real transmitted symbols corresponding to the known preamble are denoted by $S_{m,n}$. Hence, the presence of the preamble is characterized by

$$d_{m,n} = S_{m,n} \quad \text{for } n \in \{0, 1, \ldots, 2N_b - 1\}. \tag{10.19}$$

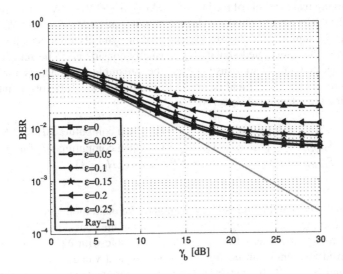

**FIGURE 10.5**

BER with 4-QAM constellation in channel ITU Veh-B versus $\gamma_b$ for different values of $\varepsilon$ in the absence of STO ($\tau_0 = 0$).

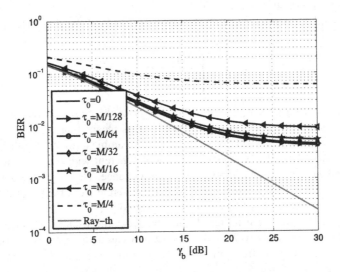

**FIGURE 10.6**

The BER with 4-QAM constellation in channel ITU Veh-B versus $\gamma_b$ for different values of $\tau_0$ in the absence of the CFO ($\varepsilon = 0$).

The remaining real symbols of the burst for $2N_b \leq n \leq 2N_b + 2N_s - 1$ correspond to payload data. The case of blind synchronization is equivalent to setting $N_b = 0$ and, therefore, to assuming that only the $N_s$ information symbols are transmitted.

Although the transmitted signal $s[l]$ is usually considered to be zero for $l < 0$, it was suggested in [4] to consider that the preamble sequence is nonzero for $n \in \{-2(K-1), \ldots, -1\}$ and is set deterministically, but only for the purpose of constructing the transmitted signal for $l \geq 0$:

$$d_{m,n} = \begin{cases} 0, & n \in \{-2(K-1), \ldots, -1\} \text{ when computing } s[l] \text{ for } l < 0, \\ S_{m,n}, & n \in \{-2(K-1), \ldots, -1\} \text{ when computing } s[l] \text{ for } l \geq 0, \\ S_{m,n}, & n \in \{0, 1, \ldots, 2N_b - 1\}. \end{cases}$$

(10.20)

This is usually named the preloading technique. The main motivation for adopting it lies in the fact that the resulting transmitted signal in the considered interval is equal to a standard signal generated using a longer preamble with $N_b + K - 1$ symbols. The achieved efficiency can be significant. For typical values, such as $L_g = 4M$, $N_b = 2$, and $N_s = 50$, the transmitted packet is shortened, and the fraction of time dedicated to preamble transmission changes its value from $\dfrac{N_b + 4 - 1}{N_b + 4 - 1 + N_s} \simeq 9\%$ to $\dfrac{N_b}{N_b + N_s} \simeq 3.85\%$ without significantly affecting the spectral containment of the transmitted signal. It is worth emphasizing that the advantage of the preloading technique is important when $K > N_b$.

A standard approach for setting the known symbols of the preamble is proposed in [4]:

$$S_{m,n} = \begin{cases} d_m^{(T)} & \text{for even } n, \\ 0 & \text{for odd } n, \end{cases}$$

(10.21)

where $d_m^{(T)} = \pm E_d$ with $E_d$ equal to the energy of the constellation used for generating the complex-valued payload symbols. This means that the same symbol is repeated $N_b$ times on each subcarrier during the preamble (with every other subsymbol being zero). This choice of preamble is not coherent with the usual assumption of independent transmitted symbols with fixed variance. Therefore, the power spectral density of the transmitted signal is slightly different from the usual expression. However, the introduced difference can be neglected in practice.

The symbol values transmitted during the preamble usually belong to a finite alphabet (as assumed in (10.21)). This constraint can also be removed. In [5], the authors assume complete freedom in determining the values of $S_{m,n}$. They are able to determine a data-dependent expression for these values so that a Constant Amplitude Zero Auto-Correlation (CAZAC) preamble is transmitted. A data-independent approximation of this expression is also developed. Consequently, the synchronization task is greatly simplified when such a preamble structure is adopted. Although this contribution may pave the way to important advances in FBMC synchronization, we do not introduce such a degree of freedom here and consider the more difficult task of performing synchronization under the usual constraint that the symbol values transmitted during the preamble belong to a finite alphabet. The advantage of the considered approach lies in the fact that only minimal variations are introduced in the spectral behavior of the overall transmitted signal.

## 10.3.1 SYMBOL TIMING SYNCHRONIZATION

In this subsection, we consider the problem of estimating the time delay affecting the received signal. Such an estimation is performed so as to compensate this delay at the receiver stage before the AFB. The effect of the resulting estimation error, denoted by $\tau_0$, was studied in Section 10.2, where a useful reference was determined for the required performance of estimation methods. The goal of this chapter is not to provide an extensive description of all existing synchronization methods in the literature. Rather a few methods that perform well are presented. These methods can operate in the TD or FD. Their description is split into two subsections.

### 10.3.1.1 *Time Domain Methods*

We first need to recall the Conjugate Symmetry Property (CSP) of a vector. If an $M \times 1$ vector $\mathbf{u} = [u_0 \ \mathbf{u}_1^T \ u_{M/2} \ \mathbf{u}_2^T]^T$ (where each of the vectors $\mathbf{u}_1$ and $\mathbf{u}_2$ has $M/2-1$ components) is the Discrete Fourier Transform (DFT) of a real-valued vector, then $\mathbf{u}_1 = \mathbf{u}_2^{\#}$, where $\mathbf{u}_2^{\#}$ denotes the flipped and conjugate version of $\mathbf{u}_2$.

Suppose that the entries of the vector $\mathbf{u}$ represent $M$ samples of the transmitted signal in one symbol period and let us assume that the training period is constructed in such a way that the CSP holds, as it is the case in OFDM with a properly chosen training symbol (see [6] for details). Then, we can identify the beginning of this single symbol period by determining the time instant where the CSP is strongest, following the procedure suggested in [7]:

$$\hat{\tau}_0 = \arg\max_{\chi} \Psi(\chi) = \arg\max_{\chi} \frac{2|(\mathbf{v}_1(\chi))^H \mathbf{v}_2^{\#}(\chi)|}{\|\mathbf{v}_1(\chi)\|^2 + \|\mathbf{v}_2(\chi)\|^2}, \qquad (10.22)$$

where $\hat{\tau}_0$ denotes the estimate of the delay $\tau_0$, and the $M$-vector $[v_0(\chi) \ \mathbf{v}_1(\chi) \ v_{M/2}(\chi) \ \mathbf{v}_2(\chi)]$ has been collected from the received signal at the candidate delay $\chi$. It can be easily shown that such a procedure is robust to the presence of a phase offset.

The use of the preamble structure described in (10.21) implies that the vector $\mathbf{s}_n$ containing the $M$ samples to be transmitted in the interval $\{nM, \ldots, nM + M - 1\}$ is the same in the first two intervals: $\mathbf{s}_0 = \mathbf{s}_1$. Moreover, the presence of the factor $j^m$ in (10.1) implies that the vector $\mathbf{s}_0$ is obtained by applying a left cyclic shift of $M/4$ samples on a vector having the CSP. Furthermore, the fact that the energy of the prototype filter is mainly concentrated on a short interval centered around $(L_g - 1)/2$ implies that $\mathbf{s}_2 \simeq \mathbf{s}_0$ and the first half of $\mathbf{s}_3$ is well approximated by the first half of $\mathbf{s}_0$.

These approximations, whose accuracy depends on the chosen prototype filter, are described in detail in [8], where an additional important property is shown: the search in (10.22) can be equivalently replaced by the procedure

$$\hat{\tau}_0 = \arg\max_{\chi} \Psi_L(\chi), \qquad \Psi_L(\chi) = \frac{2\left|\sum_{l=0}^{\frac{L}{2}-1} r[\chi - l]r[\chi + l]\right|}{\sum_{l=0}^{\frac{L}{2}-1}|r[\chi - l]|^2 + \sum_{l=0}^{\frac{L}{2}-1}|r[\chi + l]|^2}$$

(10.23)

with $L = M$. Based on the particular structure of the preamble, large values of $\Psi_M(\chi)$ can be found for five different positions $\chi$. The location of such positions depends on the particular signal structure; the presence of the CSP can be detected if (and only if) a proper position $\chi$ for evaluating $\Psi_L(\cdot)$ is chosen. This makes the detector of CSP an important tool for estimating the STO. The first of such five positions is located $3M/4$ samples after the beginning of the first multicarrier symbol period of the burst, and the other four ones are regularly spaced and separated by $M/2$ samples. Moreover, at $M/2$ samples before the first position and at $M/2$ samples after the fifth position, a weak CSP is detected. Finally, as described in [8], by filtering the sequence $\Psi_M(\chi)$ with the filter matched to the ideal value expected for $\Psi_M(\chi)$ we can estimate the STO.

The experimental results reported in [8] show that the performance of the above STO estimator is able to meet the requirements imposed by the sensitivity analysis carried out in Section 10.2. The simulations refer to a multipath scenario described by the ITU Veh-A and ITU Veh-B channel models with a bandwidth of 11.2 MHz and show that the performance of the algorithm is acceptable, provided that the channel frequency response is approximately flat on each subchannel, i.e., provided that the conditions needed to correctly operate with a single-tap equalizer are satisfied. This is strictly related to the derivation of the synchronization procedure, which is based on such an assumption, i.e., it is based on the assumption that the CSP exhibited by the transmitted signal is not destroyed by the multipath channel.

When the blind approach is followed (no preamble is sent, and hence $N_b = 0$), the useful signal $s[l]$ can be shown to still exhibit an approximate CSP, which can be detected at the receiver using the function $\Psi_{M/2}(\chi)$, i.e. as defined in (10.23) with $L = M/2$. The proof provided in [9] is based on the exploitation of the PHYDYAS prototype filter [2]. More specifically, the behavior of $\Psi_{M/2}(\chi)$ is characterized by a single maximum value located $7M/4$ samples after the beginning of the burst, at least in the absence of noise and channel distortion. Therefore, the blind timing estimator proposed in [9] determines the point $\chi_{top}$ where $\Psi_{M/2}(\chi)$ assumes its maximum, and, then, it estimates the beginning of the burst as $\chi_{top} - 7M/4$. The statistics of $\Psi_{M/2}(\chi)$ in (10.23) can be shown [9] to be independent of the frequency offset $\varepsilon$.

Since the test (10.23) has a significant computational complexity, it is best to reduce the search range of the positions $\chi$ by using a coarse estimate of the beginning of the burst. The procedure for coarse estimation proposed in [9] exploits the instantaneous power of the received signal. First, the squared amplitude $|r[l]|^2$ of the received signal $r[l]$ is filtered with a series of two (simple to implement) causal moving-average filters of $M$ samples by generating the following two sequences:

$$
\begin{cases}
P_s[l] &= P_s[l-1] + |r[l]|^2 - |r[l-M]|^2, \\
P[l] &= P[l-1] + P_s[l] - P_s[l-M].
\end{cases}
\tag{10.24}
$$

These sequences are initialized with null values. The procedure then looks for the value closest to the average between the minimal and maximal values of $P[l]$, denoted as $P_{min}$ and $P_{max}$, respectively. The coarse timing estimate proposed in [9] is equal to $N_p - \frac{11}{4}M$, where

$$
N_p = \arg\min_l \left| P[l] - \frac{P_{max} + P_{min}}{2} \right|.
\tag{10.25}
$$

The coarse estimate is used to define the interval of $M/4$ samples within which the maximum of $\Psi_{M/2}(\chi)$ is searched.

The simulation results are detailed in [9]. They show that the performance of the proposed methods is satisfactory, provided that $M \geq 1024$. More specifically, the values $M \in \{1024, 2048, 4096\}$ are considered in the experiments. The coarse procedure for $M = 1024$ with the ITU Veh-B channel is able to correctly set the interval for fine TD search in about 99% of the experiments. In all the other cases and in both channel models considered, the coarse procedure is able to work correctly in 100% of the cases, provided that $\gamma_b \geq 10$ dB. The fine procedure using the statistics in (10.23) is able to greatly improve the coarse procedure with the ITU Veh-A channel. With the ITU Veh-B channel, a performance that is superior or close to that obtained by the coarse procedure in ideal (flat) channel is obtained, achieving an RMS Error (RMSE) of the blind timing estimator below (or close to) 0.02. Taking into account the results in Section 10.2, such a performance is compatible with a negligible BER degradation in comparison with the case of perfect time synchronization ($\tau_0 = 0$).

### 10.3.1.2 *Frequency Domain Methods*

In general, FD processing enhances the synchronization performance in interference scenarios, where interferers with high power may significantly deteriorate the performance of TD synchronization algorithms. Processing in the FD provides the ability to neglect interfered parts of the spectrum and can thus provide a considerable Signal-to-Interference Ratio (SIR) gain. Due to the excellent spectral containment of FBMC modulation, especially in comparison to Cyclic Prefix Orthogonal Frequency-Division Multiplexing (CP-OFDM), the spread of the interference power is limited to a few neighboring subcarriers, which reduces the required amount of guard bands in Non-Contiguous OFDM (NC-OFDM) scenarios [10,11].

### Preamble Design for FD Synchronization

The key aspects in preamble design for synchronization in FD are the intrinsic interference of FBMC systems induced by the AFB, the required estimation range of the STO (and CFO) estimator(s), and the amount of resources to be spent for the preamble. In [12], the STO is estimated based on a modified preamble structure, which was

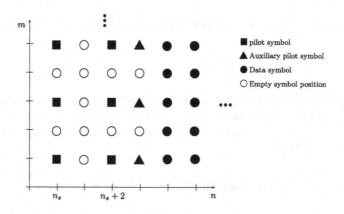

**FIGURE 10.7**

Frame structure of the pilot, auxiliary and data symbol positions [17].

originally proposed in [13] and is a variant of the preamble defined in the 802.11a standard of the Institute of Electrical and Electronics Engineers (IEEE). Here, the effect of self-interference is mitigated by occupying only every second subchannel. The STO is estimated based on the received energy in FD as also mentioned in [13]. However, it requires a sample-wise demodulation of the output of the AFB and a sufficiently large gap between the preamble and the payload data to find the maximum of the metric. Based on a reduced version of this preamble, the same authors propose an iterative approach for CFO and STO estimation that comes close to the maximum likelihood estimator [14]. These proposals are not considered here since they face the drawback of needing a sample-wise demodulation leading to a high computational complexity.

In [15–17], a more compact preamble was proposed, which is composed of four consecutive FBMC symbols as depicted in Fig. 10.7, where only every second subcarrier and FBMC symbol is allocated. To reduce or even remove the gap between the preamble and the data, the intrinsic interference needs to be combated, which is realized by inserting auxiliary pilots next to the pilot symbols [18]. The synchronization sequence $S_{m,n}$ within the frame symbols $d_{m,n}$ is defined, from a pseudo-random binary sequence $R[m]$, as

$$S_{m,n_s} = S_{m,n_s+2} = \begin{cases} R\left[\frac{m-1}{2}\right] & \text{for odd } m, \\ 0 & \text{otherwise,} \end{cases} \qquad (10.26)$$

where $n_s$ indicates the position of the first synchronization symbol within the symbol stream. If used as a preamble, $n_s = 0$ is the time of the first symbol within a frame. For all simulation results presented in this section, a Prototype Filter Function (PFF) with an overlapping factor $K = 3$ and design parameters according to [19] are applied. Although the suggested preamble structure utilizes guard symbols in the time

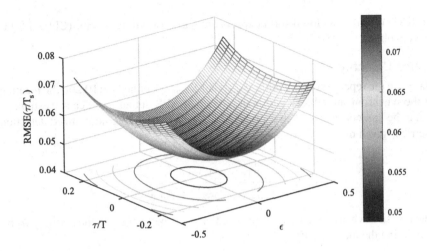

**FIGURE 10.8**

CRLB-based RMSE for the normalized STO $\tau/T_s$ plotted over the $\tau$–$\epsilon$ offset plane for SNR = 0 dB in the AWGN channel [17].

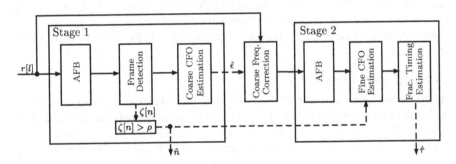

**FIGURE 10.9**

Block diagram of the parts of a two-step FD synchronization applied for STO estimation. Dashed lines represent the forwarding of estimated parameters, and solid lines indicate the signal flow.

and frequency domains, the RMSE of the STO estimation is mainly influenced by the CFO as shown in [17]. To illustrate this, Fig. 10.8 represents the RMSE derived from the FD CRLB according to RMSE = $\sqrt{\text{CRLB}}$ [17]. Therefore, the timing estimation method presented here uses a two-stage approach as sketched in Fig. 10.9. In the first stage, the signal processed by the AFB $y_{m,n}$ is autocorrelated to obtain a coarse estimate $\hat{n}$ of the symbol position of the synchronization sequence. Then, a coarse CFO estimate $\hat{\epsilon}$ is calculated, which is used to correct the TD signal before passing it to the second stage where a cross-correlation-based metric (Cross-Correlation Estima-

tor (CCE)) [16] or a closed-form expression (Closed-Form Estimator (CFE)) [20] is
used to obtain the STO.

### Frame Detection

Based on the repetition of the pilots $S_{m,n_s}$ and $S_{m,n_s+2}$ defined in (10.26), the location
of the synchronization block within the stream of AFB outputs $y_{m,n}$ can be estimated
using the autocorrelation metric in the FD. The corresponding normalized detection
metric is given by

$$\zeta[n] = \frac{2\left|\sum_{\mu \in \mathcal{M}_u} y_{\mu,n} y_{\mu,n+2}^*\right|}{\sum_{\mu \in \mathcal{M}_u}\left(|y_{\mu,n}|^2 + |y_{\mu,n+2}|^2\right)}.$$

The decision whether the frame is detected or not is performed on the metric $\zeta[n]$ by
comparing the metric value at the estimate

$$\hat{n} = \arg\max_n(\zeta[n])$$

with a predefined threshold value $\rho$. If $\zeta[\hat{n}] > \rho$, then $\hat{n}$ provides an estimate of
the frame start in integer multiples of the OQAM symbol spacing, $T/2$. An example
of comparison between analytical and simulation-based results for the probability of
misdetection and false alarm for the frame detection in AWGN channels can be found
in [17].

### Symbol Timing Offset

Based on the coarse estimate of the position of the synchronization sequence $S_{m,n}$
within the symbol stream, the CFO of the received pilots symbols can be coarsely estimated and corrected (see Section 10.3.2.2). Then, a more accurate estimation of the
fractional symbol timing $\tau = \tau_0$ can be performed. Utilizing the time shift property
of the Fourier transform, i.e., $x(lT_s - \tau) \circ\!\!-\!\!\bullet X(j\omega)e^{-j\omega\tau}$, the STO estimate $\hat{\tau}_{CCE}$
calculated by the CCE is given by

$$\hat{\tau}_{CCE} = \arg\max_{\hat{\tau}}\left(\frac{2\left|\sum_{\mu \in \mathcal{M}_u}\sum_{q \in \{0,2\}} y'_{\mu,\hat{n}+q} S_{\mu,q}^* e^{j2\pi \frac{\mu}{M}\hat{\tau}}\right|}{\sum_{\mu \in \mathcal{M}_u}\sum_{q \in \{0,2\}}\left(|y'_{\mu,\hat{n}+q}|^2 + |S_{\mu,q}|^2\right)}\right)$$

with the frequency corrected symbol $y'_{\mu,\hat{n}+q}$. The STO estimate $\hat{\tau}_{CFE}$ obtained by
using the CFE is

$$\hat{\tau}_{CFE} = \frac{T}{4\pi}\angle\left(\sum_{q \in \{0,2\}}\sum_{\mu \in \mathcal{M}'_u}\frac{y'_{\mu,\hat{n}+q} y'^*_{\mu+2,\hat{n}+q}}{S_{\mu,n_s+q,\mu} S_{\mu+2,n_s+q}}\right)$$

with the set $\mathcal{M}'_u$ containing all $\mu$ such that both $\mu$ and $\mu + 2$ belong to $\mathcal{M}_u$. This
estimator determines the STO by calculating the average phase shift between two

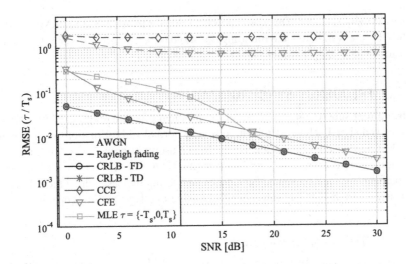

**FIGURE 10.10**

RMSE of the STO for the CCE, CFE, and MLE in AWGN and multitap Rayleigh fading channels ($\mathbb{E}\{|h[l]|^2\} \propto e^{-0.5l}$, $l \in \{0, 1, \dots, M/4 - 1\}$, $M = 32$) [17].

neighboring pilots ($S_{\mu,n}$ and $S_{\mu+2,n}$). A third estimator with a low practicality is the Maximum Likelihood Estimator (MLE) derived in [17], which has a limited estimation range of $\tau \in \{-T_s, 0, T_s\}$ and requires a symmetric allocation of the subchannels. This estimator $\hat{\tau}_{\text{MLE}}$ is given by

$$\hat{\tau}_{\text{MLE}} = \frac{T}{4\pi} \sum_{q \in \{0,2\}} \frac{\sum_{\mu \in \mathcal{M}_u} \mu |y'_{\mu,\hat{n}+q}| (\angle y'_{\mu,\hat{n}+q} - \angle S_{\mu,\hat{n}+q})}{\sum_{\mu \in \mathcal{M}_u} \mu^2 |y'_{\mu,\hat{n}+q}|}.$$

Fig. 10.10 depicts the RMSE for the presented STO estimation algorithms and compares their performance with the TD and FD CRLB derived in [21] and [17] for AWGN channels, respectively. For the limited estimation range of $-T_s \leq \tau \leq T_s$ in AWGN conditions, the MLE can achieve the CRLB for SNR values higher than 20 dB. In contrast to the MLE, the CFE is treating all subchannels equally and is not considering the subchannel index $m$ in its expression. Hence, it does not utilize the complete received information. Therefore, the CFE approaches the CRLB but is suboptimal since a gap between the RMSE of the estimation and the theoretical bound persists even for high SNR values in the case of an AWGN channel. The performance of the CFE is significantly degraded in the presence of a multitap Rayleigh fading channel, as a result of the signal spread over multiple channel taps and the resulting interaction between different paths at the pilot positions in the frequency domain. The CCE achieves a similar performance as the CFE in the multitap channel for the same reasons. In the case of AWGN, the CCE, based on the integer nature of the estimate

$\tilde{\tau}$, has virtually no error. This corresponds to the observation that the RMSE values of the CFE are well below the rounding threshold of 0.5, where rounding the residual error to the next integer would yield zero as well [17].

## 10.3.2 FREQUENCY SYNCHRONIZATION

In the current subsection, we consider the problem of estimating the CFO affecting the received signal. The effect of the resulting normalized estimation error, denoted by $\varepsilon$, has been studied in Section 10.2, where a useful reference was determined for the required performance of estimation methods. A few CFO estimation methods that perform well are presented here, first considering the case where the estimation is carried out in the TD and subsequently considering those methods operating in the FD.

### 10.3.2.1 *Time Domain Methods*

The use of the preamble structure described in (10.21) for $N_b$ identical multicarrier symbols can be used for CFO estimation. Specifically, if $N_b \geq 2$, then a CFO estimate can be obtained by using the fact that the received signal in the $n$th ($0 \leq n \leq N_b - 2$) multicarrier symbol period, collected in the vector $\mathbf{r}_1$, and in the $(n+1)$th one, collected in the vector $\mathbf{r}_2$, satisfy the following property in the absence of noise and timing synchronization errors: $\mathbf{r}_2 = \mathbf{r}_1 e^{j2\pi\varepsilon}$. Therefore, the following CFO estimator can be used:

$$\hat{\varepsilon} = \frac{1}{2\pi} \angle \{\mathbf{r}_1^H \mathbf{r}_2\}. \tag{10.27}$$

The experimental results reported in [8] show that the performance of this CFO estimator is able to meet the requirements imposed by the sensitivity analysis carried out in Section 10.2.

When the blind approach is followed and a burst of $N_s$ data symbols is transmitted without preamble, the useful signal $s[l]$ satisfies the following approximate property (see [9] for the details):

$$\frac{g\left[M/2+l\right]s\left[3M/2+l\right]}{g\left[3M/2+l\right]s\left[M/2+l\right]} \simeq \exp(j2\pi\varepsilon), \quad l = 0, 1, \ldots, M/2 - 1. \tag{10.28}$$

When the correct timing synchronization has been obtained and assuming that the channel distortion and noise can be neglected, property (10.28) can be exploited to obtain $M/2$ estimates of $\exp(j2\pi\varepsilon)$ by processing the initial burst of the received signal $r[\cdot]$. Such estimates can be averaged to better estimate the normalized CFO before entering in the AFB stage. The use of a censured mean (rejecting the estimates of $\exp(j2\pi\varepsilon)$ with magnitude larger than 2) is proposed in [9].

Another property of the transmitted signal can also be exploited for the purpose of CFO estimation [9]. Defining the function $Q(\chi) = \sum_{l=0}^{\frac{M}{4}-1} r[\chi - l]r[\chi + l]$ at the numerator of (10.23), we have the following approximate property [9]:

$$\angle Q(\hat{\tau}_0) - \angle Q(\hat{\tau}_0 - M/2) \simeq 2\pi\varepsilon + \pi. \tag{10.29}$$

The proof in [9] also relies on some properties of the PHYDYAS prototype filter [2] regarding the sign of some of its samples. An extension to other prototype filters has not yet been considered in the literature.

The results of the simulations, performed under the same scenario as the one previously considered, are discussed in [9]. They show that:

- the performance of the CFO estimator is nearly the same in both ITU Veh-A and ITU Veh-B channels;
- on ITU Veh-A and ITU Veh-B channels, the performance of the estimator based on (10.28) is better than that of the estimator based on (10.29), provided that the value of $\gamma_b$ is not too small. Similar performance, instead, is achieved for smaller values of $\gamma_b$;
- on an ideally flat channel, the performance of the estimator based on (10.29) is superior to that based on (10.28) for sufficiently small $\gamma_b$ and is nearly the same for larger values of $\gamma_b$;
- reasonable values of $\gamma_b$ and a number of subcarriers $M \geq 1024$ are needed to achieve an RMSE of the blind CFO estimator lower than or equal to 0.02. Note that the value of 0.02 for the normalized CFO $\varepsilon$ has been shown in Section 10.2 to be practically equivalent to perfect synchronization on the considered channels.

### 10.3.2.2 *Frequency Domain Methods*

#### Maximum Likelihood Estimator

Similarly to the STO estimation, the estimation of the CFO in FBMC systems has to address the intrinsic interference effect. Fig. 10.11 depicts the RMSE derived from the FD CRLB [17]. Once again, the CFO has a higher impact on the estimation error. In [22], an MLE has been derived for CP-OFDM systems and small CFO values, so that the influence of interference and amplitude degradation can be neglected. This estimator has also been suggested for application in FBMC systems [10,12,14,15,17, 20] and is given by

$$\hat{\epsilon} = \frac{1}{2\pi} \angle \left( \sum_{\mu \in \mathcal{M}_u} y_{\hat{n}+2,\mu} y_{\hat{n},\mu}^* \right).$$

As mentioned before, the estimator is influenced not only by noise but also by interference from neighboring subchannels and the misalignment of the transmit and receive filters in case of larger offsets as shown in the results. Fig. 10.12 compares the performance of the MLE for CFO in AWGN and Rayleigh fading channels to the TD and FD CRLBs derived in [21] and [17]. The STO and CFO are uniformly distributed in the ranges $-T/4 + T_s \leq \tau \leq T/4 + T_s$ and $-0.5 \leq \epsilon \leq 0.5$, respectively. The results show that the intrinsic interference, induced by the large offsets

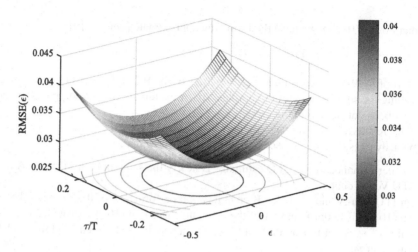

**FIGURE 10.11**

CRLB based RMSE for the normalized CFO $\epsilon$ plotted over the $\tau-\epsilon$ offset plane for SNR = 0 dB in the AWGN channel [17].

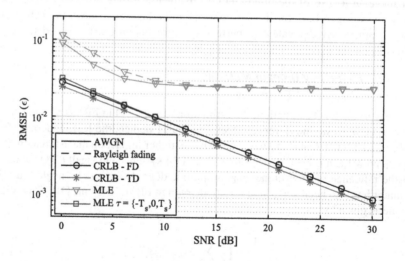

**FIGURE 10.12**

RMSE of the CFO for the CCE, CFE and MLE in AWGN and Rayleigh fading channels $(\mathbb{E}\{|h[l]|^2\} \propto e^{-0.5l}$, $l \in \{0, 1, \ldots, M/4 - 1\}$, $M = 32)$ [17].

to be estimated, results in a high residual CFO causing a saturation for SNR values higher than 5 dB.

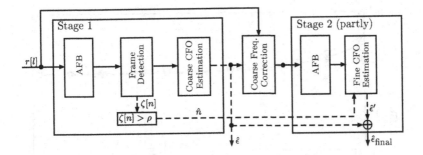

**FIGURE 10.13**

Block diagram of the parts of a two-step FD synchronization applied for CFO estimation.
Dashed lines represent the forwarding of estimated parameters, and solid lines indicate the
signal flow.

### Two-Step Approach

Therefore, in [17], a two-step approach is proposed, which applies two instances
of the MLE as sketched in Fig. 10.13. In this approach, a coarse CFO estimate $\hat{\epsilon}$
is calculated after the frame acquisition. Then, the TD signal is corrected by the
first CFO estimate, and afterwards, the residual CFO $\hat{\epsilon}'$ is estimated. The total CFO
estimation $\hat{\epsilon}_{\text{final}}$ is calculated by

$$\hat{\epsilon}_{\text{final}} = \hat{\epsilon} + \hat{\epsilon}',$$

which may be used to correct the data part of the signal in TD or tune the receiver
Phased-Locked Loop (PLL). The estimation error for a one-time and two-time CFO
estimation is plotted in Fig. 10.14. It can be seen that a second CFO estimator significantly reduces the RMSE of the CFO estimation by up to one order of magnitude
for an SNR higher than 20 dB. Hence, using the two-time approach and considering
the simplicity and the optimality in case of small CFO values, the MLE is deemed to
be a practical solution for FD estimation.

## 10.4 BLIND SYNCHRONIZATION

In this section, blind and semiblind synchronization schemes for FBMC/OQAM systems are briefly reviewed. The main advantages and disadvantages of these various
schemes are summarized. For a more detailed description of the algorithms themselves, we refer the readers to the corresponding references.

The first family of methods is based on the cyclostationarity property of
FBMC/OQAM signals. These methods compute well-chosen second-order statistics of the received signals, which, under certain conditions, exhibit a dependence on
the synchronization parameters. With sufficient data (and hence sufficient accuracy

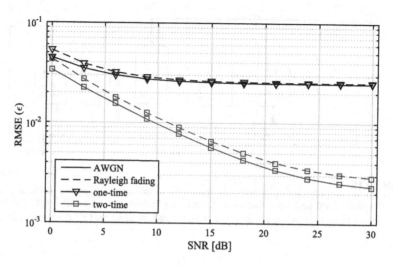

**FIGURE 10.14**

Performance comparison of the one-time and two-time CFO estimation approaches in AWGN and Rayleigh fading channels ($\mathbb{E}\{|h[l]|^2\} \propto e^{-0.5l}$, $l \in \{0, 1, \ldots, M/4 - 1\}$, $M = 32$) [17].

on the statistics), the estimation of the synchronization parameters can then be performed. The first cyclostationarity-based method for FBMC systems was presented in [23]. It uses the correlation (also called unconjugate correlation) as the second-order statistic of choice. Both CFO and symbol timing are estimated. However, to obtain a sufficient dependence on these parameters, it is necessary to apply a subcarrier weighting across the frequencies, which may impose unnecessary constraints. To circumvent that problem, it is proposed in [24] to use the *conjugate* correlation instead of the usual unconjugate one. The obtained cyclostationarity properties appear to be better suited to the estimation of synchronization parameters. Only the CFO estimation was studied in [24]. This method was then slightly improved in [25] (again for the estimation of CFO) by using FD processing. In that formulation however, the need for weighting across the subcarriers (or for the presence of null subcarriers) reappears. All the methods in this family are based on the maximization of an objective function, and the search of this maximum is performed in two steps: a coarse acquisition and a refinement. Because of the coarse step, the methods are sensitive to the outlier effect. They also typically require a relatively large number of samples (on the order of 1000 FBMC symbols), and their complexity is so high that they have only really be simulated for small numbers of subcarriers (less than 100).

The second family of methods is based on various approximations of the Maximum Likelihood (ML) estimator. In [26], it is assumed that the signal is approximately Gaussian (which is acceptable for a large number of subcarriers). Based on this approximation, an ML CFO estimation method (still requiring a maximum search) is derived, as well as a closed-form Best Linear Unbiased Estimator (BLUE). This closed-form estimator is shown to outperform the cyclostationarity-based methods, especially for a small number of samples, and to be less sensitive to the outlier effect. In [27], a low SNR approximation is used to derive an ML estimator of the timing offset (similarly to what is done in [28] for Filtered MultiTone (FMT)). The relationship between this expression and the conjugate and unconjugate correlations is also emphasized. The method requires a search of the maximum of the cost function but seems to provide good performance for short signals. In the presence of uncorrected CFO, the performance of this timing estimator degrades, especially when a large number of observations is used (as the impact of CFO gets larger).

The third family of methods is the set of tracking algorithms, usually based on decision-directed estimators. In [29], a semiblind method for CFO tracking is presented using both pilots and a differential detection procedure. Reference [13] also includes some simple tracking methods for both CFO, and symbol timing based on decision directed error estimations in a closed loop. The phase tracking estimator is improved in [30] by adding some interference suppression. A low complexity version is also presented in [30] based on sign decisions, still providing a good performance.

Finally, the method presented in [9] was already discussed in Sections 10.3.1.1 for STO and 10.3.2.1 for CFO. It is based on some symmetry properties of the beginning of an FBMC/OQAM burst (without the knowledge of the transmitted data). The symbol timing estimation is performed in two steps, starting from a coarse estimation and then finely searching for the position where the symmetry property is best achieved. Then a closed-form expression of the CFO can be obtained based on the structure of the initial burst. The method provides a good performance for a small number of symbols (as it works on the initial burst) but requires a reasonably high SNR (on the order of 10 dB). The method is also able to perform reasonably well in the presence of mild frequency selectivity.

## 10.5 SCATTERED PILOT-BASED SYNCHRONIZATION

In this section, we introduce data-aided STO and CFO estimation methods, which utilize scattered pilot symbols, i.e., relatively sparse patterns of pilot symbols placed among data symbols. Scattered pilots are commonly used for channel estimation, as discussed in Chapter 11, but they can be used also for synchronization, for refining the STO and CFO estimates after coarse synchronization and tracking the variations of the synchronization parameters. To avoid the effects of intrinsic interference, the auxiliary (or help) pilot scheme, illustrated in Fig. 10.7 for the case of the preamble-based estimation, is commonly applied for all scattered pilots. In this scheme, an OQAM subsymbol is used as the actual pilot, and its preceding or following subsymbol is chosen in such a way that the intrinsic interference on the pilot due to adjacent data symbols is canceled [18,20].

A straightforward approach is to utilize the phase rotation between two pilots closely located in the frequency (resp. time) direction for STO (resp. CFO) estimation [20]. Assuming that the pilots take arbitrary binary values, i.e., $d_{m,n} = \pm 1$, when the index pair $(m, n)$ corresponds to a pilot, the pilot modulation can be removed by using $\tilde{y}_{m,n} = d_{m,n} y_{m,n}$ for STO and CFO estimation. The basic estimation schemes are as follows.

- The delay induced by the STO is seen as a phase difference between two pilots at the same time at two different subcarriers, $m$ and $m + \Delta m$. Then the STO can be estimated as

$$\hat{\tau}_0 = \frac{\angle\{\tilde{y}_{m+\Delta m,n}\} - \angle\{\tilde{y}_{m,n}\} - \Delta m \pi/2}{2\pi \Delta m} T. \tag{10.30}$$

- Likewise, the CFO is seen as a phase difference between two pilots at the same subcarrier at two different subsymbol instants, $n$ and $n + \Delta n$. Then the CFO can be estimated as

$$\hat{\epsilon} = \frac{\angle\{\tilde{y}_{m,n+\Delta n}\} - \angle\{\tilde{y}_{m,n}\} - \Delta n \pi/2}{\pi \Delta n} F. \tag{10.31}$$

Naturally, estimates based on a single pilot pair would be very noisy. However, these basic estimates can be averaged over multiple pilot pairs within the transmission frame over which the channel response and synchronization offsets are expected to remain constant.

One fundamental issue in phase rotation-based estimation is the $2\pi$ ambiguity, which depends on the pilot distance. The ambiguity is encountered when the phase rotation exceeds the range of $\pm\pi$. Consequently, the unambiguous estimation ranges are given by $\pm T/(2\Delta m)$ for STO and $\pm F/\Delta n$ for CFO.

It is also possible to compensate for STO and CFO in the FD with a per-subcarrier processing [20]. The STO compensation is basically a special case of channel equalization to compensate the delay-dependent linear phase response. The basic approach for CFO compensation is through complex exponential multiplication to cancel the frequency shift. An improved performance can be achieved by also compensating the mismatched subchannel filter frequency response in subchannel processing [20]. The performance of such STO and CFO compensation schemes using 3-tap subcarrier equalizers was evaluated in [20]. An iterative scheme to enhance the performance was also proposed, based on the idea that after compensating the estimated offsets (possibly together with channel equalization), the performance can be improved by reestimating and recompensating the offsets. Fig. 10.15 shows an example of the resulting performance in a Worldwide Interoperability for Microwave ACcesS (WiMAX) (802-16e)-like FBMC/OQAM system.

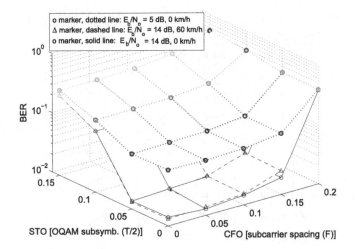

**FIGURE 10.15**

BER performance versus STO and CFO for WiMAX-like uplink transmission of $4 \times 2$ AMC23 slots (each with 18 subcarriers by 3 symbols, 6 pilots per slot). Veh-A channel with $E_b/N_0 = 5$ dB and $E_b/N_0 = 14$ dB using 4-QAM. Static user and user with 60 km/h mobility. Synchronization with STO and CFO compensations and frequency response correction with one iteration.

## 10.6 SYNCHRONIZATION IN ASYNCHRONOUS MULTIUSER SCENARIOS

When considering scenarios using fragmented (noncontiguous) spectrum, there is great interest to move the time and frequency synchronization functions from the TD processing side to the FD, to be implemented through subchannel processing. Otherwise, it is difficult to manage the strong interfering spectral components, which may be allocated between the used portions of the received spectrum. Similar problems are encountered in asynchronous cellular uplink or ad hoc scenarios, in which case the other users' signals appear as interference in the synchronization process for a specific user. Considering applications with low data rate and sporadic transmission of data packets, continuous synchronization of devices would introduce significant overhead in energy consumption. Therefore, immediate synchronization to each transmission burst would be favored [31–33].

Filterbank signal processing provides powerful tools for separating different users' signals from each other. However, for TD synchronization algorithms, the filtering capability of the AFB cannot be effectively utilized, and the channelization should be done separately, greatly increasing the complexity. In traditional FBMC waveform processing, the time and frequency resolution of subchannel signals is not sufficient to carry out the synchronization tasks directly at the AFB output at the receiver, especially regarding the fine CFO and STO compensation. As seen in the

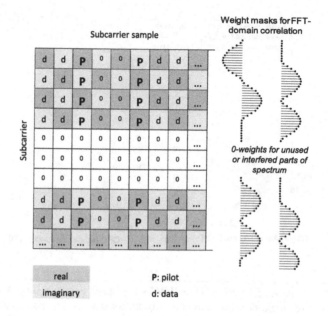

**FIGURE 10.16**

Left: Pilot and data allocation in the used frame structure. Right: FC weight masks for FFT-domain correlation. If the channelization weights are real, then the weights used for correlation are real in even subcarriers and imaginary in odd subcarriers.

previous section, multitap subcarrier processing helps to improve the CFO and STO compensation quality, but with 3-tap subcarrier processing, the feasible CFO and STO compensation ranges cannot cover the case where all users have completely different STOs and CFOs. On the other hand, the FC-FB processing structure (see Section 8.3) effectively supports simultaneous processing of nonsynchronized signals from different users, providing tools for individual CFO and STO compensation for different subchannel multiplexes. Similar ideas have been recently proposed also in the Frequency Spreading FBMC/OQAM (FS-FBMC) context [33,34].

In this section, we first introduce an FC-FB-based CFO and STO estimation and frame detection scheme. Then offset compensation techniques are introduced. We focus here on the FBMC/OQAM case, but the schemes can also be used for FMT and Single-Carrier (SC) waveforms.

## 10.6.1 STO AND CFO ESTIMATION BASED ON FREQUENCY DOMAIN CORRELATION

Here we use a preamble structure consisting of two FBMC/OQAM symbols, i.e., four OQAM subsymbols, as shown in Fig. 10.16. We assume that the two OQAM subsymbols in the middle are zero-valued and that the first and fourth symbols consist of ±1 in even subcarriers and ±j in odd subcarriers. The pilots are loaded on the active data subcarriers only. This scheme is used independently for each (possibly noncontiguous) multiplex of subchannels that are not synchronized with each other. The idea can be straightforwardly extended to other preamble structures, and it is also possible to load data in the two OQAM subsymbols in the middle with reasonable degradation in the synchronization performance.

The STO estimation scheme is based on the cross-correlation between the preamble sequence and the corresponding received signal. The reference is the transmitted baseband sequence with the data surrounding the preamble set to zero. This reference sequence is correlated with the corresponding received baseband data sequence. With sliding correlation, the beginning of the frame can be identified. Instead of implementing the TD correlation, we utilize the fact that the correlation becomes a multiplication in the FD. The correlation is realized in the receiver's FC-based AFB structure through multiplication of the Fast Fourier Transform (FFT)-domain signal by the FFT sequence appearing in the transmitter's Synthesis FilterBank (SFB) due to the preamble.

For the numerical results to be reported later, the selected preamble was found by searching over a large number of random sequences, targeting to maximize the synchronization performance and minimize the Peak-to-Average Power Ratio (PAPR) of the transmitted time domain signal. On the receiver side, the sharp bandpass filtering provided by the AFB helps to suppress interference leakage from the unused spectral slots. The timing estimation process is based on the CCE approach presented in Section 10.3.1.2 and includes the following steps:

- Long FFT of the FC AFB at the receiver, jointly for all received multiplexes. The same FFT output is used later for data detection within the same FC processing block.
- Multiplication of the FFT-domain vector corresponding to the active subcarriers by the precomputed FFT-domain reference sequence. This step is repeated for both subsymbols used in synchronization.
- Long Inverse Fast Fourier Transform (IFFT) to obtain the TD correlation. This step is repeated for both subsymbols used.
- The IFFT outputs are combined noncoherently (squared magnitude sum) because the delay difference of the subsymbols introduces phase rotation to the correlation peaks.
- Finding the peak of the combined IFFT output.

Depending on the number of active subcarriers in the target multiplex, it may be possible to use reduced IFFT length to get the TD correlation, with reduced resolution in the STO estimation.

CFO estimation is combined with this structure by estimating the phase rotation between the two OQAM subsymbols at the correlation peak. For the used pilot structure, a CFO of one FFT bin corresponds to a phase rotation of $6\pi/L$, where $L$ denotes here the short transform (IFFT) length used in the FC-based receiver. Due to phase ambiguity, the CFO estimation range is around $\pm L/6$ FFT bins or $\pm F/3$, where $F$ is the subcarrier spacing.

## 10.6.2 TIMING OFFSET COMPENSATION

Considering effective implementation of the base station receiver, it would be desirable to use a single FC-FB engine to process multiple uplink signals emerging from different nonsynchronized mobiles. These signals may use different or differently parameterized waveforms, but they are assumed to be nonoverlapping, i.e., a narrow guardband is inserted between different users' spectra. The relative frequency offsets are assumed to be small enough such that sufficient spectral isolation is maintained. As an example, with the 15 kHz subcarrier spacing of Long-Term Evolution (LTE), a few kHz frequency offsets can be tolerated without introducing significant cross-talk even when using a single subcarrier guardband. Naturally, this depends on the FilterBank (FB) design and expected power level differences between uplink signals. Our target is to be able to compensate STOs in a range exceeding the OQAM sub-symbol duration.

In the FC-FB structure, STO compensation can be achieved by using the same idea as in Section 10.5, i.e., by adjusting the filtering delay through the phase response of each AFB subchannel. A timing offset (delay) of $\tau$ can be compensated by introducing an additional linear phase response to the FFT-domain weights [35]:

$$\hat{w}_{m,\ell} = w_{m,\ell} e^{2j\pi\tau(m-1+2\ell/L)/M}, \tag{10.32}$$

where $w_{m,\ell}$, $\ell = 0, 1, \ldots, L-1$, are the original FFT-domain filtering weights for subchannel $m$ (the used "FFT-shifted" indexing scheme for FFT bins is indicated in Fig. 10.17). With STO compensation, it is necessary to use complex FFT-domain weights in the AFB, even if the basic FC-FB system is designed for real weights. STO compensation can be seen as a special case of embedded channel equalization, to be introduced in Section 12.5.2.

## 10.6.3 CFO COMPENSATION

Effective CFO compensation methods can also be implemented through subcarrier processing utilizing the FC-FB structure [36]. Shifting the FFT-domain subchannel weight mask by an integer number of FFT bins allows us to compensate CFO with precision of $\pm 0.5$ times the FFT bin spacing. With the typical subcarrier spacing of 8 FFT bins, the residual CFO is within $\pm 0.0625$ times the subcarrier spacing. Clearly, this is too much for obtaining reasonable link performance.

Our approach for fine CFO compensation is to shift the subchannel weight mask by a fraction of the FFT bin spacing. The weight mask shift can be achieved, e.g., by sampling the square-root-raised cosine function with the desired frequency offset or by optimizing different sets of weight masks for a sufficient number of fractional frequency shifts using the methods presented in Section 8.3. Another approach is to interpolate the weight mask for the target frequency shift from the basic weight mask.

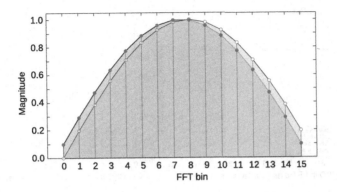

**FIGURE 10.17**

FFT weight mask shifting for CFO compensation.

Actually, it will be seen that linear interpolation is sufficient for most practical needs. The general idea is illustrated in Fig. 10.17. The mask is first shifted by an integer number of FFT bins, after which the residual CFO, normalized to FFT bin spacing, satisfies $\delta_{CFO} \in (-1, 0]$. Then, for linear interpolation, the new weight values are computed as follows:

$$\hat{w}_{m,\ell} = (1 - \delta_{CFO})w_{m,\ell} + \delta_{CFO}w_{m,((\ell+1))_L}. \tag{10.33}$$

Interpolation methods, and linear interpolation in particular, can be applied in flexible ways. After recalculating the weights, they can be used for processing longer sequences of data samples, as long as the CFO value remains constant. This is equivalent to weighting the FFT bins by the original weight factors and linearly interpolating the data samples between FFT bins afterwards.

In addition to the weight mask shifting, it is necessary to use In-phase/Quadrature (I/Q) mixing after the IFFT of the receiver analysis bank to shift the subcarrier spectrum to the right position. The shifted weight mask implements the subchannel filtering in a proper way, and the mixing brings the subchannel to the correct spectral position. The needed mixing sequence is

$$c_{mix}[n] = e^{-j2\pi n\delta_{CFO}/L}. \tag{10.34}$$

## 10.6.4 NUMERICAL RESULTS

We first consider the STO and CFO estimation performance in an LTE-like case with about 1 MHz bandwidth, 72 active subcarriers, and 15 kHz subcarrier spacing. For a larger number of subcarriers in the multiplex, the performance is expected to be better. The FC-FB parameters are $N = 8M = 1024$ and $L = 16$. As mentioned earlier, the Binary Phase-Shift Keying (BPSK) model is used for the pilot symbols, and their power level is the same as that of the Offset Quadrature Phase-Shift Keying

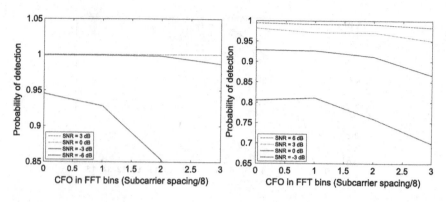

**FIGURE 10.18**

Probability of correct timing estimation for Veh-A channel (left) and HT channel (right).

(OQPSK) data symbols. Fig. 10.18 shows the correct timing offset estimation probability for different SNR levels as a function of CFO for ITU Veh-A and Hilly Terrain (HT) channels. For the Veh-A channel, the criterion for correct timing offset is that the delay exactly matches the channel time delay in high-rate samples, whereas in the HT case, ±2 sample offsets are tolerated. This is because the HT channel exhibits such variations in effective delays. Over 95% detection probability is reached with Veh-A channel for −3 dB SNR within a CFO range of ±3 FFT bins, corresponding to ±5.6 kHz with the considered parameters. With HT channel, 92% detection probability is reached with 0 dB SNR for a CFO range of ±2 FFT bins. It can be noted that the feasible CFO range in timing offset estimation exceeds the range of unambiguous CFO estimation. Furthermore, the STO estimation performance is independent of the actual STO value in a range significantly exceeding OQAM symbol duration.

Fig. 10.19 shows CFO estimation results for both channels, only including the cases where the correct timing estimate was found. It can be seen that the bias of the estimate is very small. The Root Mean Squared (RMS) error is 0.2 to 0.3 times the FFT bin spacing for small CFOs and 0.3 to 0.6 times the FFT bin spacing for CFOs approaching 2 FFT bins. When the CFO exceeds ±2 FFT bins, the estimation performance degrades rapidly. It was observed that increasing the distance of the two pilot subsymbols improves the CFO estimation accuracy with small CFOs and reduces the estimation range. The proposed preamble design is considered to make a feasible trade-off between CFO estimation range and accuracy.

An important consideration in the proposed way of timing offset compensation is the inevitable cyclic distortion in the FC implementation of multirate filters. In optimized design, the effect is symmetric within the processing block. With timing offset compensation, the cyclic truncation effect becomes more severe on one side of the effective linear impulse response. This appears as reduced attenuation of out-of-band spectral components and increased inband interference. The inband and out-of-band effects can be seen in Fig. 10.20 for an optimized FC-FB design with

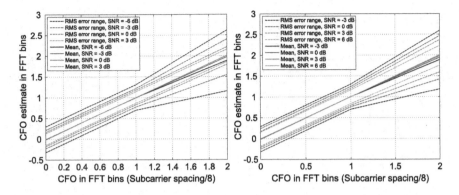

**FIGURE 10.19**

RMS and mean errors in CFO estimation for Veh-A channel (left) and HT channel (right).

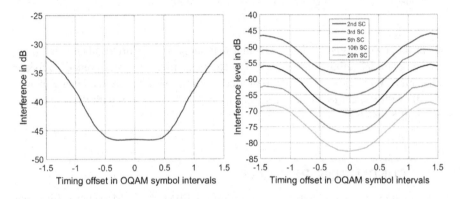

**FIGURE 10.20**

Inband effects (left) and out-of-band effects (right) of timing offset control in the FC-FB implementation of FBMC/OQAM.

complex weights. The inband performance variations are very small well beyond the target range of timing offset control, namely ±0.5 OQAM subsymbol intervals (i.e., ±0.25 OQAM symbols). The out-of-band test situation includes just one active subcarrier, which is not synchronized to the receiver processing. We measure the interference power level at the 2nd, 3rd, 5th, 10th, and 20th SC from the active one. It can be seen that the variations in the out-of-band performance are also relatively small within the target range of timing offset control. In conclusion, both inband and out-of-band interference performances are not significantly compromised, even within much wider tuning range.

Fig. 10.21 shows the CFO compensation results with and without weight mask interpolation for a noncontiguous multicarrier scenario with two deactivated Physical Resource Blocks (PRBs) of 12 subcarriers. The spectra are measured through the

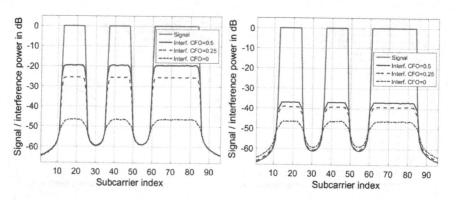

**FIGURE 10.21**

Interference levels with CFO compensation for FBMC/OQAM waveform. Left: Subchannel weight mask not shifted. Right: Weight mask shifted through linear interpolation.

subcarrier power levels. It can be seen that without properly shifted subchannel filtering (i.e., plain I/Q-modulation-based CFO compensation), the interference levels are rather high. With linear interpolation-based shifting, the interference level is below −37 dB with respect to the active subcarrier power levels, even with the worst-case CFO of 0.5 FFT bins.

## 10.7 CONCLUDING REMARKS

In this chapter, we discussed synchronization for FBMC/OQAM systems. First, we analyzed the effect of residual synchronization errors (CFO and STO), both analytically and by simulations, for a receiver based on single-tap equalization. This enables to estimate the required precision of the synchronization procedures, depending on the operating SNR and on the constellation. Then, we studied preamble-based synchronization, considered several TD and FD methods and presented the corresponding preamble designs. Regarding STO synchronization, the methods perform well in mildly selective channels, but their performance decreases when the channel has higher frequency selectivity. In practice, however, those highly selective channels require multitap equalization, which is able to handle some large residual timing offset by itself. Hence, the methods appear to be adequate in the highly frequency selective case as well. Regarding CFO synchronization, the suggested TD method is able to satisfy the requirements. In the frequency domain, we considered a two-step approach to improve the performance and to reach an accuracy close to the CRLB. We also presented blind TD methods for both STO and CFO estimation. These methods provide sufficient accuracy if the number of subcarriers is large enough and if the available SNR is around 10 dB or higher.

We also presented methods based on scattered pilots. They are based on the principle of auxiliary pilots and naturally operate in the FD. These methods can be used for refining the STO and CFO estimates after coarse synchronization and tracking the variations of the synchronization parameters.

Finally, one advantage of FBMC/OQAM systems is their ability to allow unsynchronized users to coexist by separating them very efficiently in the FD, thanks to a good selectivity of the filters. In the last section, we considered this multiuser scenario. We showed how the FC-FB scheme makes it possible for the base station to simultaneously synchronize with the different users, effectively implementing estimation and compensation of the different CFOs and STOs.

# REFERENCES

[1] D. Mattera and M. Tanda, "Optimum single-tap per-subcarrier equalization for OFDM/OQAM systems," *Digit. Signal Process.*, vol. 49, pp. 148–161, 2016.

[2] M. Bellanger, "Specification and design of a prototype filter for filter bank based multicarrier transmissions," in *Proc. ICASSP-2001*, Salt Lake City, UT, May 2001.

[3] Q. Bai and J. Nossek, "On the effects of carrier frequency offset on cyclic prefix based OFDM and filter bank based multicarrier systems," in *Proc. SPAWC-2010*, Marrakesh, Morocco, Jul. 2010.

[4] M. G. Bellanger, "Efficiency of filter bank multicarrier techniques in burst radio transmission," in *Proc. GLOBECOM-2010*, Miami, FL, Dec. 2010.

[5] W. Chung, C. Kim, S. Choi, and D. Hong, "Synchronization sequence design for FBMC/OQAM systems," *IEEE Trans. Wireless Commun.*, vol. 15, no. 10, pp. 7199–7211, Oct. 2016.

[6] M. Tanda, "Blind symbol-timing and frequency-offset estimation in OFDM systems with real data symbols," *IEEE Trans. Commun.*, vol. 52, no. 10, pp. 1609–1612, Oct. 2004.

[7] H. Minn, V. Bhargava, and K. Letaief, "A robust timing and frequency synchronization for OFDM systems," *IEEE Trans. Wireless Commun.*, vol. 2, no. 4, pp. 822–839, Jul. 2003.

[8] D. Mattera and M. Tanda, "Data-aided synchronization for OFDM/OQAM systems," *Signal Process.*, vol. 92, no. 9, pp. 2284–2292, Sep. 2012.

[9] D. Mattera and M. Tanda, "Blind symbol timing and CFO estimation for OFDM/OQAM systems," *IEEE Trans. Wireless Commun.*, vol. 12, no. 1, pp. 268–276, Jan. 2013.

[10] C. Thein, M. Fuhrwerk, and J. Peissig, "Practical evaluation of NC-OFDM system designs in dynamic spectrum access with narrow-band interference," in *Proc. ISWCS-2013*, Ilmenau, Germany, Aug 2013.

[11] T. A. Weiss and F. K. Jondral, "Spectrum pooling: An innovative strategy for the enhancement of spectrum efficiency," *IEEE Commun. Mag.*, vol. 42, no. 3, pp. S8–S14, Mar. 2004.

[12] H. Saeedi-Sourck and S. Sadri, "Frequency-domain carrier frequency and symbol timing offsets estimation for offset QAM filter bank multicarrier systems in uplink of multiple access networks," *Wirel. Pers. Commun.*, vol. 70, no. 2, pp. 601–615, 2013.

[13] P. Amini and B. Farhang-Boroujeny, "Packet format design and decision directed tracking methods for filter bank multicarrier systems," *EURASIP J. Adv. Signal Process.*, vol. 2010, p. 307983, 2010.

[14] H. Saeedi-Sourck, S. Sadri, Y. Wu, and B. Farhang-Boroujeny, "Near maximum likelihood synchronization for filter bank multicarrier systems," *IEEE Wireless Commun. Lett.*, vol. 2, no. 2, pp. 235–238, 2013.

[15] C. Thein, M. Fuhrwerk, and J. Peissig, "About the use of different processing domains for synchronization in non-contiguous FBMC systems," in *Proc. PIMRC-2013*, London, UK, Sep. 2013.

[16] C. Thein, M. Fuhrwerk, and J. Peissig, "Frequency-domain processing for synchronization and channel estimation in OQAM-OFDM systems," in *SPAWC 2013*, Darmstadt, Germany, Jun. 2013.

[17] C. Thein, M. Schellmann, and J. Peissig, "Analysis of frequency domain frame detection and synchronization in OQAM-OFDM systems," *EURASIP J. Adv. Signal Process.*, vol. 2014, p. 83, 2014.

[18] C. Lélé, R. Legouable, and P. Siohan, "Channel estimation with scattered pilots in OFDM/OQAM," in *Proc. SPAWC-2008*, Recife, Brazil, Jul. 2008.

[19] A. Viholainen, T. Ihalainen, T. H. Stitz, M. Renfors, and M. Bellanger, "Prototype filter design for filter bank based multicarrier transmission," in *Proc. EUSIPCO-2009*, Glasgow, Scotland, Aug. 2009.

[20] T. H. Stitz, T. Ihalainen, A. Viholainen, and M. Renfors, "Pilot-based synchronization and equalization in filter bank multicarrier communications," *EURASIP J. Adv. Signal Process.*, vol. 2010, p. 741429, 2010.

[21] T. Fusco, A. Petrella, and M. Tanda, "A data-aided symbol timing estimation algorithm for OFDM/OQAM systems," in *ICC 2009*, Dresden, Germany, Jun. 2009.

[22] P. H. Moose, "A technique for orthogonal frequency division multiplexing frequency offset correction," *IEEE Trans. Commun.*, vol. 42, no. 10, pp. 2908–2914, 1994.

[23] H. Bölcskei, "Blind estimation of symbol timing and carrier frequency offset in wireless OFDM systems," *IEEE Trans. Commun.*, vol. 49, no. 6, pp. 988–999, Jun. 2001.

[24] P. Ciblat and E. Serpedin, "A fine blind frequency offset estimator for OFDM/OQAM systems," *IEEE Trans. Signal Process.*, vol. 52, no. 1, pp. 291–296, Jan. 2004.

[25] G. Lin, L. Lundheim, and N. Holte, "New methods for blind fine estimation of carrier frequency offset in OFDM/OQAM systems," in *Proc. SPAWC-2006*, Cannes, France, Jul. 2006.

[26] T. Fusco and M. Tanda, "Blind frequency-offset estimation for OFDM/OQAM systems," *IEEE Trans. Signal Process.*, vol. 55, no. 5, pp. 1828–1838, May 2007.

[27] T. Fusco, L. Izzo, A. Petrella, and M. Tanda, "Blind symbol timing estimation for OFDM/OQAM systems," *IEEE Trans. Signal Process.*, vol. 57, no. 12, pp. 4952–4958, Dec. 2009.

[28] V. Lottici, M. Luise, C. Saccomando, and F. Spalla, "Non-data-aided timing recovery for filter-bank multicarrier wireless communications," *IEEE Trans. Signal Process.*, vol. 54, no. 11, pp. 4365–4375, Nov. 2006.

[29] G. Dainelli, V. Lottici, M. Moretti, and R. Reggiannini, "Efficient carrier recovery for FBMC systems over time-frequency selective channels," in *Proc. 2011 8th International Workshop on Multi-Carrier Systems Solutions (MC-SS)*, Herrsching, Germany, May 2011.

[30] J. B. Doré and V. Berg, "Blind phase tracking algorithm for FBMC receivers," in *Proc. ISWCS-2015*, Brussels, Belgium, Aug. 2015.

[31] F. Boccardi, R. W. Heath, A. Lozano, T. L. Marzetta, and P. Popovski, "Five disruptive technology directions for 5G," *IEEE Commun. Mag.*, vol. 52, no. 2, pp. 74–80, Feb. 2014.

[32] C. Bockelmann, N. Pratas, H. Nikopour, K. Au, T. Svensson, C. Stefanovic, P. Popovski, and A. Dekorsy, "Massive machine-type communications in 5G: Physical and MAC-layer solutions," *IEEE Commun. Mag.*, vol. 54, no. 9, pp. 59–65, Sep. 2016.

[33] J. Doré, V. Berg, N. Cassiau, and D. Kténas, "FBMC receiver for multi-user asynchronous transmissions on fragmented spectrum," *EURASIP J. Adv. Signal Process.*, p. 41, 2014.

[34] D. Mattera, M. Tanda, and M. Bellanger, "Frequency domain CFO compensation for FBMC systems," *Signal Process.*, vol. 114, pp. 183–197, Sep. 2015.

[35] M. Renfors and J. Yli-Kaakinen, "Timing offset compensation in fast-convolution filter bank based waveform processing," in *Proc. ISWCS-2013*, Ilmenau, Germany, Aug. 2013.

[36] J. Yli-Kaakinen and M. Renfors, "Multi-mode filter bank solution for broadband PMR coexistence with TETRA," in *Proc. EuCNC-2014*, Bologna, Italy, Jun. 2014.

# CHAPTER

# FBMC Channel Estimation Techniques

# 11

**Eleftherios Kofidis**\*,†, **Leonardo Gomes Baltar**‡, **Xavier Mestre**§, **Faouzi Bader**¶,
**Vincent Savaux**‖

*University of Piraeus, Piraeus, Greece\**
*Computer Technology Institute & Press "Diophantus" (CTI), Patras, Greece†*
*Intel Deutschland GmbH, Neubiberg, Germany‡*
*Centre Tecnològic de Telecomunicacions de Catalunya (CTTC/CERCA), Barcelona, Spain§*
*CentraleSupélec, Rennes, France¶*
*b<>com, Cesson-Sévigné, France‖*

## CONTENTS

## 11.1 INTRODUCTION

Numerous gains are expected from replacing Orthogonal Frequency-Division Multiplexing (OFDM) by FilterBank MultiCarrier (FBMC) in future communication systems [1], but they may come at the cost of losing the simplicity of common receiver functions, including channel synchronization, estimation, and equalization. These can be quite challenging tasks in systems employing FilterBank MultiCarrier with Offset-QAM subcarrier modulation (FBMC/OQAM) due to the *intrinsic* interference effect, particularly, under realistic propagation conditions (including those envisaged in next generation networks) that may involve severe frequency and/or time selectivity for the physical channel [2]. For example, high mobile speeds necessitate the use of a relatively small number of subcarriers in an FBMC system, in order to cope with the Doppler spread effects [3]. This in turn implies that the subchannels are no longer well described by a frequency flat model, in contrast to what is assumed in most of the existing FBMC/OQAM channel estimation schemes [4]. Although such a model (along with more restricting requirements on the coherence bandwidth of the channel) greatly simplifies the channel estimation task in such systems, effectively translating the problem to an OFDM-like one, it typically leads to severe error floors at medium to high Signal-to-Noise Ratio (SNR) values when it is not sufficiently accurate [4]. The reason for this is that the subchannel model inaccuracy manifests itself at such weak noise regimes as the residual intrinsic interference prevails over noise.

It is thus a necessity to study channel estimation in FBMC systems through the development of more accurate (and less OFDM-like) schemes that will not rely on invalid simplifying assumptions. Moreover, additional information, such as the color of the noise at the demodulator output, must be incorporated in the estimator to optimize its performance. In addition to adopting more powerful estimation techniques, the training design problem needs to be studied in the light of these increased requirements. Coping with frequency selective subchannels may require the use of longer training sequences [5]. On the other hand, shorter ones will be more suitable in fast fading scenarios. An additional problem in designing training and algorithms for FBMC/OQAM channel estimation stems from the undesired interference that the training part of the frame (preamble or isolated pilot) suffers from neighboring control and/or data symbols. This has to be appropriately addressed so as not to significantly compromise estimation performance while spending as few resources (concerning both training bandwidth and computational time) as possible for estimating the channel. (Semi)blind methods provide one possible way to meet these requirements.

This chapter aims at providing a comprehensive overview of past and current research in these topics. First, in Section 11.2, the FBMC/OQAM system model is recalled, and basic notation is introduced. Section 11.3 addresses FBMC/OQAM channel estimation based on a known preamble. Both types of training preamble are considered, namely, block-type (also called full), where all subcarriers carry pilots, and comb-type (also called sparse), whose pilot tones are isolated in frequency. The most well-studied case of channels of low frequency selectivity is covered first, and estimators based on both Least Squares (LS) and Minimum Mean-Squared Error (MMSE) criteria are described for this scenario. For the case of highly frequency selective channels, impulse response estimators are studied. Notably, the estimators developed can work for any given training sequence of any given length. One of the methods reported permits the transmission of data at the nonpilot subcarriers while training is performed, thus offering the possibility for increased bandwidth efficiency. Emphasis is given to designing optimal short preambles in view of their importance in practical fast fading channels. To be able to track channel variations in time, pilots that are scattered throughout the payload part of the frame are also needed. Designing and placing such pilots in FBMC/OQAM turn out to be rather tricky. This problem and possible solution approaches are reviewed in Section 11.4, where recent developments concerning the highly frequency selective case are given more attention. (Semi-)blind FBMC/OQAM channel estimation is discussed in Section 11.5. Concluding remarks, including a discussion of possible future research directions, are made in Section 11.6.

## 11.2 SYSTEM MODEL

The discrete-time signal at the output of the Synthesis FilterBank (SFB) is given by [6]

$$s[l] = \sum_{m=0}^{M-1} \sum_{n} d_{m,n} g_{m,n}[l], \tag{11.1}$$

where $d_{m,n}$ are (*real*-valued) Pulse Amplitude Modulation (PAM) symbols (Offset Quadrature Amplitude Modulation (OQAM) half-symbols), and

$$g_{m,n}[l] = g\left[l - n\frac{M}{2}\right] e^{j\frac{2\pi}{M}m\left(l - \frac{L_g-1}{2}\right)} e^{j\varphi_{m,n}}$$

with $g$ being the *real symmetric* prototype filter impulse response (assumed here of unit energy) of length $L_g$, $M$ being the *even* number of subcarriers, and $\varphi_{m,n} = \frac{\pi}{2}(m+n) - mn\pi$ [6]. The filter $g$ is usually designed to have length $L_g = KM$ with $K$ being the overlapping factor. The double subscript $(\cdot)_{m,n}$ denotes the $(m,n)$th Frequency-Time (F-T) point. Thus, $m$ is the subcarrier index, and $n$ is the FBMC/OQAM symbol time index. The modulator output is transmitted through a channel of length $L_h$, which is, as usual in block transmissions, assumed to be invariant in the duration of a MultiCarrier (MC) symbol [7]. At the receiver front-end, noise $v$ is added, which is assumed white Gaussian with zero mean and variance $\sigma_v^2$. The noisy channel output is then given by

$$r[l] = \sum_{k=0}^{L_h-1} s[k-l]h[k] + v[l], \tag{11.2}$$

where

$$\mathbf{h} = \begin{bmatrix} h[0] & h[1] & \cdots & h[L_h-1] \end{bmatrix}^{\mathsf{T}}$$

is the Channel Impulse Response (CIR).

The pulse $g$ is designed so that the associated subcarrier functions $g_{m,n}$ are orthogonal in the *real* field, that is,

$$\text{Re}\left\{\sum_l g_{m,n}[l]g_{p,q}^*[l]\right\} = \delta_{m,p}\delta_{n,q}, \tag{11.3}$$

where $\delta_{i,j}$ is the Kronecker delta. This implies that even in the absence of channel distortion and noise and with perfect time and frequency synchronization, there will be some intercarrier (and/or intersymbol) interference at the output of the Analysis FilterBank (AFB), which is purely real or imaginary (depending on the parity of the F-T point) and is known as *intrinsic* interference [8]. Assuming, for simplicity of the presentation, that the AFB processing includes a multiplication by $e^{-j\varphi_{m,n}}$, the interference can be always seen as being imaginary, and hence we can write the response of the FBMC/OQAM TransMUltipleXer (TMUX) from the F-T point $(m, n)$ to $(p, q)$ as

$$\Gamma_{p,m}^{q,n} = \sum_l g_{m,n}[l]g_{p,q}^*[l] = j\langle g\rangle_{m,n}^{p,q} \tag{11.4}$$

for $(p, q) \neq (m, n)$, where (using the notation of [9]) $\langle g\rangle_{m,n}^{p,q}$ is real-valued. Moreover, $\Gamma_{p,p}^{q,q} = 1$. An example (for even $p$) is shown in Table 11.1, where the PHYDYAS [10] prototype filter with $K = 4$ is employed. Notice that this is a Near Perfect Reconstruction (NPR) TMUX, as one can see in the gray-shaded cells, which would contain zeros for a Perfect Reconstruction (PR) system. It is more important to remark the fact that (as shown in the tabulated example) there is no interference from non-adjacent subcarriers at the same symbol time (Inter-Carrier Interference (ICI)). The interference from preceding and following time instants and next to adjacent subcarriers, i.e., from F-T points $(m \pm 2, n \pm 1)$, is, in general, very small [4]. For the PHYDYAS TMUX, it is zero.

The study of FBMC/OQAM systems can be greatly simplified through approximations relying on the channel coherence bandwidth and time and on a good time-frequency localization enjoyed by the prototype filter $g$. Thus, for channels of length $L_h$ relatively small compared to the size $(M)$ of the FilterBank (FB), it can be shown that the AFB output at the F-T point $(p, q)$ can be approximated by [9]

$$y_{p,q} = H_{p,q}d_{p,q} + j\underbrace{\sum_{m=0}^{M-1}\sum_{\substack{n \\ (m,n)\neq(p,q)}} H_{m,n}d_{m,n}\langle g\rangle_{m,n}^{p,q}}_{I_{p,q}} + \eta_{p,q}, \tag{11.5}$$

where $H_{m,n}$ is the Channel Frequency Response (CFR) at $(m, n)$, and $\eta_{p,q}$ is the noise at the AFB output, which is also Gaussian with zero mean and variance $\sigma_\eta^2 = \sigma_v^2$, but correlated in both time and frequency directions, due to the AFB filtering [4].

**Table 11.1** Response of an FBMC/OQAM TMUX employing the PHYDYAS prototype filter [10] with $K = 4$. The case of even subcarrier index $p$ is depicted

| $m$ | | | | | $n$ | | | | |
| --- | --- | --- | --- | --- | --- | --- | --- | --- | --- |
| | $q-4$ | $q-3$ | $q-2$ | $q-1$ | $q$ | $q+1$ | $q+2$ | $q+3$ | $q+4$ |
| $p-2$ | 0 | j0.006 | −0.0001 | 0 | 0 | 0 | −0.0001 | −j0.006 | 0 |
| $p-1$ | j0.0054 | −j0.0429 | j0.125 | −j0.2058 | j0.2393 | −j0.2058 | j0.125 | −j0.0429 | j0.0054 |
| $p$ | 0 | j0.0668 | −0.0002 | j0.5644 | 1 | −j0.5644 | −0.0002 | −j0.0668 | 0 |
| $p+1$ | −j0.0054 | −j0.0429 | −j0.125 | −j0.2058 | −j0.2393 | −j0.2058 | −j0.125 | −j0.0429 | −j0.0054 |
| $p+2$ | 0 | j0.006 | −0.0001 | 0 | 0 | 0 | −0.0001 | −j0.006 | 0 |

Even under this *flat subchannel* model, channel estimation has to cope with the interference term $I_{p,q}$, which is in general complex-valued and not purely imaginary due to the complex CFR gains [8,9]. A further step to come closer to the simplicity of Cyclic Prefix Orthogonal Frequency-Division Multiplexing (CP-OFDM) is to exploit the fact that the interference from F-T points outside a neighborhood $\Omega_{p,q}$ of $(p,q)$ is negligible while assuming that the CFR is (almost) constant over this time-frequency area. The latter, of course, implies even a shorter channel delay spread than that required to validate (11.5), which can then be written as

$$y_{p,q} \approx H_{p,q} x_{p,q} + \eta_{p,q}, \tag{11.6}$$

namely, in a CP-OFDM fashion, where

$$x_{p,q} = d_{p,q} + \mathrm{j} \underbrace{\sum_{(m,n)\in\bar{\Omega}_{p,q}} d_{m,n} \langle g \rangle_{m,n}^{p,q}}_{u_{p,q}} = d_{p,q} + \mathrm{j} u_{p,q}, \tag{11.7}$$

with $\bar{\Omega}_{p,q} = \Omega_{p,q} \setminus \{(p,q)\}$, is the *virtual* transmitted symbol at $(p,q)$, consisting of the transmitted symbol itself *plus* the interference coming from its Time-Frequency (T-F) neighborhood,

$$u_{p,q} = \sum_{(m,n)\in\bar{\Omega}_{p,q}} d_{m,n} \langle g \rangle_{m,n}^{p,q}. \tag{11.8}$$

$\bar{\Omega}_{p,q}$ is most commonly taken to be the first-order T-F neighborhood of $(p,q)$ consisting of its immediate neighboring F-T points (shown hatched in the example of Table 11.1). The corresponding interference weights $\mathrm{j}\langle g \rangle_{m,n}^{p,q}$ can be shown to satisfy the following symmetries [4]:

$$\begin{matrix} \mathrm{j}(-1)^p\delta & -\mathrm{j}\beta & \mathrm{j}(-1)^p\delta \\ -\mathrm{j}(-1)^p\gamma & d_{p,q} & \mathrm{j}(-1)^p\gamma \\ \mathrm{j}(-1)^p\delta & \mathrm{j}\beta & \mathrm{j}(-1)^p\delta \end{matrix} \tag{11.9}$$

with the horizontal direction corresponding to time and the vertical one to frequency. Thus, the interference from $(p-1,q)$ is $-\mathrm{j}\beta d_{p-1,q}$, etc. The quantities $\beta, \gamma, \delta$ can be a priori computed from $g$ (see [4] for detailed expressions) and are positive and smaller than one. Generally, $\gamma > \beta > \delta$. In the example of Table 11.1, $\beta = 0.2393$, $\gamma = 0.5644$, and $\delta = 0.2058$. The above symmetries play a central role in designing training input and associated channel estimation methods, as it will be seen in the sequel.

## 11.3 PREAMBLE-BASED CHANNEL ESTIMATION

Known (training) symbols for the purpose of channel estimation can be gathered in the beginning of a frame and/or placed at isolated F-T points throughout the frame. These are two principal pilot configurations, known as *preamble* and *scattered pilots*, respectively. This section is devoted to FBMC/OQAM channel estimation methods based on an appropriately designed preamble.[1] As it is common in preamble-based estimation, we assume the channel to be time invariant during the preamble period. Furthermore, we will consider no synchronization errors. Recent works taking this and other kinds of inaccuracies (e.g., In-phase/Quadrature (I/Q) imbalance) into account include [11–13].

Difficulties in FBMC/OQAM preamble design come from the intrinsic interference effect, as explained in the previous section. Guard FBMC symbols of all zeros (almost invariably in the related literature) precede and follow the pilot symbol(s) to *protect* them from being interfered by the unknown (control/data) part of the previous and current frame, respectively [9]. Often, the inter-frame gaps can replace the guards preceding the pilots [4]. Hence, in view of the commonly made assumption of the interference being negligible outside the first-order T-F neighborhood, one guard symbol suffices, resulting in a preamble overhead comparable with that of a single OFDM symbol. It must be noted, however, that despite its popularity in the FBMC/OQAM literature, this assumption is only approximately (more or less) met in practice, as, for example, Table 11.1 shows.

### 11.3.1 CHANNELS OF LOW FREQUENCY SELECTIVITY

Research on FBMC/OQAM channel estimation has been almost exclusively focused on scenarios of (relatively to the selectivity of the FB) low frequency selectivity, namely, channels for which (11.6) holds with sufficient accuracy. This is because the task then becomes less challenging since the input–output model resembles that of CP-OFDM, and hence existing OFDM techniques can be (more easily) adapted to this problem. This subsection aims at briefly reviewing some of the most well-known methods of this category. For more information on this subject, we refer the reader to [4] and the references therein.

#### 11.3.1.1 The Pair of Pilots (POP) Method

The POP method, proposed in [9], aims at computing an estimate of $H_{p,q}$ by using (11.6) at two different (in practice consecutive) time instants $q_1, q_2$ to construct a system of equations for its real and imaginary parts. The method is described here in a different, yet equivalent way, based, instead, on the Zero Forcing (ZF) equalizer $W_{p,q} = 1/H_{p,q}$ [4]. Dropping the time index from $W$ and $H$ (in view of the time

---

[1]The same principles apply if the pilot blocks are placed in the middle of the frame (midambles).

invariance assumption) and *neglecting* the noise allow us to write

$$y_{p,q_1} W_p = d_{p,q_1} + ju_{p,q_1},$$
$$y_{p,q_2} W_p = d_{p,q_2} + ju_{p,q_2}.$$

Taking the real parts yields a system of two equations in two unknowns, namely, the real $(W_p^R)$ and imaginary $(W_p^I)$ parts of $W_p$, whose solution can be (compactly) written as

$$W_p = W_p^R + jW_p^I = j\frac{d_{p,q_1} y_{p,q_2}^* - d_{p,q_2} y_{p,q_1}^*}{\mathbb{Im}(y_{p,q_1}^* y_{p,q_2})}. \qquad (11.10)$$

In practice, the preamble would comprise the first two FBMC symbols, that is, $q_1 = 0$ and $q_2 = 1$. In that case, with the preamble suggested in [9], where $d_{p,0} = (-1)^p$, and the second symbol is all zeros, the above would simplify to

$$W_p = j\frac{(-1)^p y_{p,1}^*}{\mathbb{Im}(y_{p,0}^* y_{p,1})}. \qquad (11.11)$$

The CFR would then be computed as $H_p = 1/W_p$.

An advantage of the POP scheme, besides its simplicity, is that it does not explicitly depend on the employed prototype filter (provided, of course, that (11.6) is satisfied). However, it must be emphasized that the above derivation only holds when the noise is negligible. We can see that (see [9]), in the presence of both noise and interference from the unknown part of the frame, the method can have unpredictable performance. This is because, in such a case, the degree of the noise enhancement also depends on unknown (hence uncontrollable) data.

### 11.3.1.2 *The Interference Cancellation Method (ICM)*

A more straightforward approach consists of avoiding or canceling the intrinsic interference effect. This can be achieved, for example, by relying on the zero guard FBMC symbol to avoid interference from the unknown part of the frame while appropriately choosing the entries $d_{p,0}$ of the pilot FBMC symbol so as to avoid/cancel the interference among them. A number of ways that have been proposed are reviewed in [4]. One of them, which relies on the common assumption of the first-order T-F interference neighborhood and the symmetries in (11.9), was proposed in [14] and here will be referred to as ICM. It is easy to verify from (11.9) that if $d_{p-1,0} = d_{p+1,0}$, then there is null interference to the $(p, 0)$ point. Thus, the corresponding virtual transmitted symbol will be equal to $x_{p,0} = d_{p,0}$, and the LS estimate of the CFR directly results from (11.6) as

$$\hat{H}_p = \frac{y_{p,0}}{d_{p,0}}.$$

Fig. 11.1D shows an example of such a preamble for the simple but illustrative case of $M = 8$ and Binary Phase-Shift Keying (BPSK) symbols $d_{p,0}$.

| (A) | | (B) | | (C) | | (D) | |
|---|---|---|---|---|---|---|---|
| 1 | 0 | $d_0$ | 0 | 1 | 0 | 1 | 0 |
| $-1$ | 0 | $-jd_0$ | 0 | $-j$ | 0 | $-1$ | 0 |
| $-1$ | 0 | $-d_0$ | 0 | $-1$ | 0 | 1 | 0 |
| 1 | 0 | $-d_1$ | 0 | $j$ | 0 | $-1$ | 0 |
| 1 | 0 | $jd_1$ | 0 | 1 | 0 | 1 | 0 |
| $-1$ | 0 | $d_1$ | 0 | $-j$ | 0 | $-1$ | 0 |
| $-1$ | 0 | $-d_0$ | 0 | $-1$ | 0 | 1 | 0 |
| 1 | 0 | $jd_0$ | 0 | $j$ | 0 | $-1$ | 0 |

**FIGURE 11.1**

Preamble structures for (A) Interference Approximation Method (IAM)-R, (B) IAM-I, (C) IAM-C, and (D) ICM (and POP) methods. $M = 8$. OQPSK modulation is assumed. $d_0, d_1$ are randomly chosen BPSK symbols.

### 11.3.1.3 The Interference Approximation Method (IAM)

An alternative to avoiding/canceling the intrinsic interference is to use it in favor of the estimation accuracy. If the immediate F-T neighbors of $(p, q)$ in (11.7) carry training (known) symbols, then an approximation of the interference term can be computed. This in turn allows us to construct the so-called *pseudo-pilots* $x_{p,q}$ and use them to get a channel estimate in exactly the same way as in CP-OFDM. This method is called (for obvious reasons) the IAM method and requires all input symbols in the immediate neighborhood of $(p, q)$ to be known. This is the case for the entries of the pilot FBMC symbol in the preamble structure discussed here. In a matrix–vector form, (11.6) is written as

$$\mathbf{y} = \mathbf{X}\mathbf{H} + \boldsymbol{\eta}, \tag{11.12}$$

where $\mathbf{X} = \mathrm{diag}(x_{0,0}, x_{1,0}, \ldots, x_{M-1,0})$, $\mathbf{y} = \begin{bmatrix} y_{0,0} & y_{1,0} & \cdots & y_{M-1,0} \end{bmatrix}^{\mathrm{T}}$, and the $M \times 1$ vectors $\mathbf{H}$, $\boldsymbol{\eta}$ are similarly defined. Clearly, the LS estimate of $\mathbf{H}$ can be computed as

$$\hat{\mathbf{H}}_{\mathrm{LS}} = \mathbf{X}^{-1}\mathbf{y}, \tag{11.13}$$

that is,

$$\hat{H}_p = H_p + \frac{\eta_{p,0}}{x_{p,0}} \tag{11.14}$$

for $p = 0, 1, \ldots, M - 1$. The latter relation suggests that the pilots should be so chosen as to result in pseudo-pilots of maximum magnitude [9]. A number of IAM preambles have been reported in the literature [4,9,15], each with its own pros and cons. The most well known are briefly described here, starting from the most straightforward choice of pseudo-randomly chosen pilots (referred to as IAM1

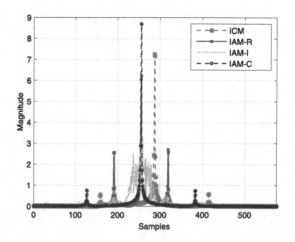

**FIGURE 11.2**

Magnitude of the modulated (SFB output) preambles considered for low frequency selective channels. The PHYDYAS FB with $M = 128$ and $K = 4$ was employed.

in [15]). Obviously, such a preamble is far from being optimal in terms of minimum estimation Mean-Squared Error (MSE); however, it enjoys the property of a low Peak-to-Average Power Ratio (PAPR) because of its pseudo-random nature, which is important in practical implementations involving a High Power Amplifier (HPA) operating in its nonlinear region [16]. The pattern (11.9) suggests that we can obtain pseudo-pilots of maximum magnitude using *Real*-valued (PAM) pilot symbols if the latter are chosen so as to satisfy the condition $d_{p+1,0} = -d_{p-1,0}$ for all $p$. Such is the case of the so-called IAM-R preamble construction, an example of which is given in Fig. 11.1A. We can do better than that if *Complex*-valued pilot symbols are also allowed and chosen so that pseudo-pilots of maximum magnitude (that are either purely real or imaginary) result. The idea in this so-called IAM-C scheme [4,17,18] is to simply multiply the entries at the odd-indexed subcarriers of the IAM-R pilot symbol by j, as shown in the example of Fig. 11.1C. IAM-R and IAM-C enjoy MSE-optimality (over real- and complex(imaginary)-valued pilots, respectively), however, at the expense of a high PAPR, as shown in Fig. 11.2. This is due to their periodicity, as we can see in Fig. 11.2 for the ICM preamble as well. A compromise between estimation accuracy and PAPR is achieved in the IAM-I scheme [4,9], where the idea of constructively adding interference is only applied in one third of the subcarriers, with the selection of the pilot symbols being random otherwise. This way, *I*maginary pseudo-pilots of maximum magnitude result in only one third of the subcarriers, whereas the rest of them deliver complex pseudo-pilots of smaller magnitude. An example of this IAM variant is shown in Fig. 11.1B. Fig. 11.2 demonstrates its considerably better PAPR behavior. The estimation performance of these preamble-based methods is evaluated in the example of Fig. 11.3, where the

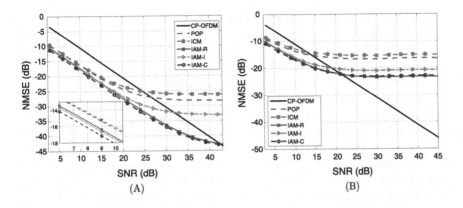

**FIGURE 11.3**

Performance comparison of preamble-based channel estimation methods made for low frequency selective channels. The PHYDYAS FB with $M = 128$ and $K = 4$ was employed. (A) ITU Ped-A ($L_h = 9$) and (B) ITU Veh-B ($L_h = 47$) channel models [19].

Normalized Mean-Squared Error (NMSE) $\mathbb{E}\{\frac{\|\mathbf{H}-\hat{\mathbf{H}}\|^2}{\|\mathbf{H}\|^2}\}$ is plotted versus the *receive* SNR for two scenarios, one of low and one of high frequency selectivity. To more clearly show the effect of the modeling assumptions on the estimation accuracy, no data are transmitted before or after the preamble. The CP-OFDM curve is also included as a reference; the LS estimator was used, with a preamble of all equal pilots, proved to be optimal in [20]. Note how the constructive use of the intrinsic interference in the IAM schemes succeeds to outperform the ICM and POP methods. These results confirm the optimality of IAM-C and demonstrate a near optimal performance for IAM-R at all the SNR values considered. On the other hand, IAM-I only outperforms IAM-R at low SNR. At higher SNR values, it exhibits an error floor higher than that of IAM-R. This is the price paid for its good PAPR property. Error floors such as those observed in this example are typical for FBMC techniques that rely on approximations like (11.6). They are due to the interference residing from the model inaccuracy, which is hidden by the noise at low SNR values and shows up at the weak noise regime. As a result, the advantage over CP-OFDM is lost at higher values of the SNR, and, as exemplified in this figure, this effect is strengthened as the channel frequency selectivity becomes stronger and hence the approximation in (11.6) becomes cruder.

Could an alternative preamble structure yield pseudo-pilots that are stronger than those of IAM-C? It has been shown in [21, Section 5.4] and [22,23] that such a preamble can indeed be formed in an analogous way as previously by also employing the left- and right-hand sides of the T-F neighborhood (11.9) in a structure of three instead of one pilot FBMC symbols and no guard ones. Such an IAM scheme, called Extended IAM-C (E-IAM-C), was designed and successfully tested in [4,22].

Additional improvements can be achieved by further exploiting the frequency-domain smoothness of the channel response. Ways of doing this include taking the finite length (Finite Impulse Response (FIR)) of the channel into account through Discrete Fourier Transform (DFT) interpolation [20,24,25] and/or directly smoothing (averaging) in the frequency domain [23,24,26–28].

A number of works have also targeted the spectral inefficiency resulting from the use of guard FBMC symbols. Thus, preambles comprising one reference FBMC symbol and no guards have been proposed and studied (see [4] for a review). The *unknown* interference to the pilots from surrounding symbols is invariably dealt with in such schemes via an iterative procedure, where the channel and the surrounding data are alternately estimated (see, e.g., [24]). The channel and data estimates are improved from one iteration to the other. Such preamble-based methods are, however, more computationally demanding and generally unable to attain the performance obtained with preamble structures that utilize guard symbols [4,24]. More recently, a modification of the IAM-C method, allowing the transmission of information symbols on half of the F-T points, which would otherwise be used to carry nulls, was developed in [29]. It is not of an iterative nature (albeit involving again considerably more computations than IAM-C), whereas it still achieves a performance comparable with that of IAM-C.

In addition to this kind of preambles, namely with all frequencies occupied by (nonzero) pilots, referred to as *full* (block-type), *sparse* (comb-type) preamble structures, where only $L_h$ isolated subcarriers carry pilots, with the rest of them being nulled, have also been considered. Preambles of the latter type have been proposed as a straightforward means of avoiding interference among pilots (see [4] for a review). Optimality of preambles of this kind, in the sense of attaining the minimum possible MSE with a given transmit energy budget, was studied in [30]. Similarly with the corresponding optimality condition for CP-OFDM sparse preambles (with the energy budget not including the energy spent for the Cyclic Prefix (CP)$^2$), in [30], it was proved that the pilot-carrying subcarriers in an optimal sparse FBMC/OQAM preamble are equidistant, with the pilots being equipowered.

### 11.3.1.4 *Linear Minimum Mean Squared Error (LMMSE) Estimation*

It was recently shown that LMMSE channel estimation can also be performed in FBMC/OQAM systems, in a similar way with CP-OFDM, under the assumption of a channel satisfying Eq. (11.6) [31,32]. The LMMSE estimate of the $M$-point CFR can be expressed as

$$\hat{\mathbf{H}}_{\text{LMMSE}} = \underbrace{\mathbf{R}_{\mathbf{H}}\left(\mathbf{R}_{\mathbf{H}} + \mathbf{R}_{\eta}\left(\mathbf{X}\mathbf{X}^{\text{H}}\right)^{-1}\right)^{-1}}_{\mathbf{K}} \hat{\mathbf{H}}_{\text{LS}}, \tag{11.15}$$

---

$^2$As shown in [20], when the CP energy is also taken into account, the optimal pilots should be equal.

where $\mathbf{X}$ is defined as previously, namely as the $M \times M$ diagonal matrix containing the pseudo-pilots on its main diagonal, and $\hat{\mathbf{H}}_{LS} = \mathbf{X}^{-1}\mathbf{y}$ denotes the LS estimate (cf. (11.13)). The $M \times M$ channel covariance matrix is defined as $\mathbf{R_H} = \mathbb{E}\{\mathbf{HH}^H\}$, and similarly for the covariance matrix of the noise component, $\mathbf{R}_\eta$.

It must be emphasized that the direct computation of (11.15) is a heavy task, requiring $\mathcal{O}(M^3)$ complex operations. Furthermore, $\mathbf{R}_\eta$ and $\mathbf{R_H}$ are usually unknown at the receiver. To overcome this and reduce the complexity of the LMMSE estimator, two approximate solutions were proposed in [33], analogous to those followed in CP-OFDM [34]. Thus, following [34], the matrix $\mathbf{XX}^H$ is approximated by $\mathbb{E}\{\mathbf{XX}^H\} \propto \mathbf{I}_M$. Furthermore, the exact but unknown channel covariance matrix $\mathbf{R_H}$ is replaced by a matrix $\tilde{\mathbf{R}}_\mathbf{H}$ whose eigenvalues are assumed to be a priori known. The corresponding estimator matrix in (11.15) is denoted by $\tilde{\mathbf{K}}$. Consider its eigenvalue decomposition

$$\tilde{\mathbf{K}} = \tilde{\mathbf{U}}\tilde{\mathbf{\Delta}}\tilde{\mathbf{U}}^H, \tag{11.16}$$

where $\tilde{\mathbf{\Delta}}$ is diagonal and $\tilde{\mathbf{U}}$ is unitary. We summarize the two approximations as follows:

- *Approximation 1*: The MMSE estimate of the noise variance $\hat{\sigma}_\eta^2$ [32] is used, and the noise correlation is neglected, that is, $\mathbf{R}_\eta \approx \hat{\sigma}_\eta^2 \mathbf{I}_M$. This then implies that $\tilde{\mathbf{R}}_\mathbf{H}$ is diagonalizable by $\tilde{\mathbf{U}}$. In that case, the complexity is reduced to $\mathcal{O}(M^2)$ operations since the matrix inversion now is for a diagonal matrix.
- *Approximation 2*: The unknown channel covariance matrix is approximated by a circulant one, which has the same eigenvalues as in Approximation 1. Then $\tilde{\mathbf{U}}$ is the $M$th-order DFT matrix with entries $\frac{1}{\sqrt{M}} \exp(-j\frac{2\pi ij}{M})$, $i, j = 0, 1, \ldots, M - 1$, and the computational cost reduces to the order of the Fast Fourier Transform (FFT), namely, $\mathcal{O}(M \log_2 M)$ operations.

Fig. 11.4 shows the MSE performances of these two approximate solutions versus SNR. The PHYDYAS FB with $M = 256$ and $K = 4$ was employed. The 3rd Generation Partnership Project (3GPP) TU channel model was considered [35], with a subcarrier spacing of 15 kHz. The noise variance $\sigma_\eta^2$ was assumed known (or accurately estimated, e.g., as in [32]). The result of the LMMSE estimator in (11.15) with the exact channel covariance matrix is also plotted as a reference. In all cases, the ICM preamble (see Fig. 11.1) was adopted. We can observe that Approximation 2 achieves a lower MSE than Approximation 1 for any considered SNR value. It is worth noticing that the performance reaches a lower bound (error floor), which was analyzed in [31]. We can conclude that Approximation 2 fits the FBMC/OQAM problem better than Approximation 1, with the latter being an effective adaptation of a technique originally proposed for CP-OFDM. As expected, the exact LMMSE estimator of (11.15) achieves superior performance since it relies on exact knowledge of the channel and noise statistics, however, at the expense of a much higher computational effort.

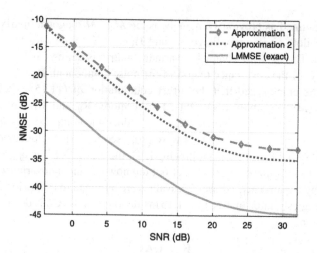

**FIGURE 11.4**

Performance comparison of the exact and approximate LMMSE estimators. The PHYDYAS FB with $M = 256$ and $K = 4$ was employed in TU channels. The subcarrier spacing was 15 kHz. (Figure adapted from [33].)

## 11.3.2 CHANNELS OF HIGH FREQUENCY SELECTIVITY

The assumption of a large coherence bandwidth (relatively to the subcarrier spacing) and large coherent time (relatively to the symbol duration) for the channel was seen in the previous subsection to greatly simplify the channel estimation task in FBMC systems, effectively translating the problem to an OFDM-like one. However, as the typical example of Fig. 11.3 (particularly (B)) demonstrates, this assumption may lead to severe error floors at medium to high SNR values when it is not sufficiently accurate. A consequence of this is that the advantage that FBMC has over CP-OFDM at lower SNR values is lost in the higher SNR regime, also preventing the adoption of input constellations of higher order in such systems. Such is the case in scenarios involving severe frequency [36] and/or time selectivity [3] for the physical channel, for example, those encountered in Professional (or Private) Mobile Radio (PMR) systems [37]. For example, high mobile speeds necessitate the use of a relatively small number of subcarriers in an FBMC system to cope with the Doppler spread effects [3]. It is thus a necessity to revisit FBMC channel estimation through the development of more accurate schemes that will not rely on invalid simplifying assumptions. Moreover, additional information such as the color of the noise at the demodulator output must be incorporated in the estimator to optimize its performance. In addition to adopting more powerful estimation techniques, the training design problem needs to be revisited as well in the light of these increased requirements.

Two preamble-based channel estimation approaches that explicitly take into account the above considerations are briefly described in this subsection. Both were developed and/or studied in detail within the EMPhAtiC project [37].

### 11.3.2.1 *Parametric Methods*

Channel estimation techniques of this type rely on the path-delay parameterization of the channel and have been previously applied in OFDM (see, e.g., [38]). Their value is found at the explicit incorporation of the channel parameterization in the estimation algorithm, hence their name. One technique of this kind was developed for FBMC/OQAM systems in [39] (and in other works of the same authors) and further studied in [27]. The underlying idea is to model the frequency selective subchannel responses as Taylor polynomials around the central subcarrier frequencies (see also Chapter 9). Varying the order of the polynomials provides a more or less good approximation of the subchannel responses. This, along with the standard path-delay model for the channel, results in a formulation of the problem as one of array processing with the delays playing the roles of directions of arrivals. A subspace-based algorithm, based on MUltiple SIgnal Classification (MUSIC), is applied for estimating the delays. The estimation of the path gains is then made easy and can be performed either in the LS or the MMSE (if statistical information is available) sense. Estimating the number of paths is an important problem in this type of algorithms. A solution based on a model order selection criterion, which is free of assumptions on either the problem statistics or the data set size, was successfully applied in [27]. The simulation results demonstrated a very important performance improvement over both IAM-based methods and corresponding CP-OFDM MUSIC-LS schemes. A notable limitation of this technique is the limited freedom in selecting the preamble pilots. Furthermore, its computational load is significantly increased by the need to perform a MUSIC step.

### 11.3.2.2 *Time Domain (TD) Methods*

A computationally much simpler approach, applicable with any type of preamble, was developed in [27] and publicized in [40–43]. It simply relies on writing out the AFB output directly in terms of the CIR and with no assumption made on its length, that is,

$$
y_{p,q} = \sum_{k=0}^{L_h-1} h[k] \sum_{m=0}^{M-1} e^{-j\frac{2\pi}{M}mk} \sum_{n} d_{m,n} j^{m+n-p-q} (-1)^{mn-pq}
$$
$$
\times \sum_{l} g\left[l - k - n\frac{M}{2}\right] g\left[l - q\frac{M}{2}\right] e^{j\frac{2\pi}{M}(m-p)\left(l-\frac{L_g-1}{2}\right)} + \eta_{p,q}.
\tag{11.17}
$$

In [44], such a formulation is also independently adopted, however, without delving into the analysis of the problem structure and the preamble optimization reported here. The value of (11.17) is found in the fact that it is *exact*, regardless of the channel length $L_h$ (only assumed not to exceed $M$), and can therefore be used to address the channel estimation problem for the general case, namely, when the subchannels are not well approximated by the frequency flat model (11.5).

Adopting a preamble of the commonly used type, namely consisting of one pilot FBMC symbol followed by one or more guards, Eq. (11.17) yields

$$y_{p,0} = \sum_{k=0}^{L_h-1} \Gamma_{p,k} h[k] + \eta_{p,0},$$ (11.18)

where

$$\Gamma_{p,k} = \sum_{m=0}^{M-1} e^{-j\frac{2\pi}{M}mk} d_{m,0} j^{m-p} e^{-j\frac{2\pi}{M}(m-p)\left(\frac{L_g-1}{2}\right)} \sum_{l=k}^{L_g-1} g[l-k]g[l]e^{j\frac{2\pi}{M}(m-p)l}$$ (11.19)

can be seen as the response of the TMUX to this particular input for a channel equal to a $k$-samples delay and can thus be easily computed. The AFB output sample is the sum of those responses multiplied by the channel gains; in matrix–vector notation,

$$\mathbf{y} = \mathbf{\Gamma}\mathbf{h} + \boldsymbol{\eta}$$ (11.20)

with $[\mathbf{\Gamma}]_{p,k} = \Gamma_{p,k}$. The matrix $\mathbf{\Gamma}$ is of dimensions $M \times L_h$, i.e., tall, and hence the equation can be solved for $\mathbf{h}$ using LS.

However, recall that the noise $\eta$ is not uncorrelated among subcarriers [4]. Its covariance matrix $M \times M$ is known to be given by

$$\mathbf{R}_\eta = \sigma^2 \begin{bmatrix} 1 & j\beta & 0 & \cdots & 0 & \mp j\beta \\ -j\beta & 1 & j\beta & \cdots & 0 & 0 \\ \vdots & \ddots & \ddots & \ddots & \ddots & \vdots \\ \pm j\beta & 0 & 0 & \cdots & -j\beta & 1 \end{bmatrix} \triangleq \sigma_\eta^2 \mathbf{B}$$ (11.21)

with the signs of its lower left and upper right corner entries depending on the prototype filter used [43], where $\mathbf{B}$ is *tridiagonal* in view of the negligible ICI between nonadjacent subcarriers and (almost) *circulant*. The latter implies that it is DFT-diagonalizable [45], i.e.,

$$\mathbf{B} = \mathcal{F}\mathbf{\Lambda}_0\mathcal{F}^H,$$ (11.22)

where $\mathcal{F}$ is the *unitary $M$-point DFT matrix*, and $\mathbf{\Lambda}_0 = \text{diag}(\lambda_{0,0}, \lambda_{0,1}, \ldots, \lambda_{0,M-1})$ is diagonal with the (real and positive) eigenvalues of $\mathbf{B}$ on its main diagonal. These eigenvalues were studied in [46] and can be expressed as

$$\lambda_0[m] = 1 + 2\beta \sin\left(\frac{2\pi m}{M}\right), \quad m = 0, 1, \ldots, M-1.$$

Note that $\lambda_0[0] = \lambda_0[\frac{M}{2}] = 1$.

With colored noise, Gauss–Markov estimation is better suited than LS, resulting in the following estimate for $\mathbf{h}$ [47]:

$$\hat{\mathbf{h}} = \left( \boldsymbol{\Gamma}^H \mathbf{R}_\eta^{-1} \boldsymbol{\Gamma} \right)^{-1} \boldsymbol{\Gamma}^H \mathbf{R}_\eta^{-1} \mathbf{y}$$

or, equivalently,

$$\hat{\mathbf{h}} = \left( \boldsymbol{\Gamma}^H \mathbf{B}^{-1} \boldsymbol{\Gamma} \right)^{-1} \boldsymbol{\Gamma}^H \mathbf{B}^{-1} \mathbf{y}, \tag{11.23}$$

which shows that no knowledge of the noise power $\sigma_\eta^2$ is required. Because of the Gaussianity of the noise, this is also the Maximum Likelihood (ML) estimate [47]. Since this estimation technique stems from expressing the problem in terms of the TD channel (i.e., CIR), it will henceforth be referred to simply as the TD method. The above estimator can also be written as

$$\hat{\mathbf{h}} = \tilde{\boldsymbol{\Gamma}}^\dagger \tilde{\mathbf{y}}, \tag{11.24}$$

with

$$\tilde{\boldsymbol{\Gamma}} = \boldsymbol{\Lambda}_0^{-1/2} \mathcal{F}^H \boldsymbol{\Gamma}, \tag{11.25}$$

$$\tilde{\mathbf{y}} = \boldsymbol{\Lambda}_0^{-1/2} \mathcal{F}^H \mathbf{y}, \tag{11.26}$$

where $\boldsymbol{\Lambda}_0^{-1/2} \mathcal{F}^H$ is an (a priori known) square root of the matrix $\mathbf{B}$, and $\tilde{\boldsymbol{\Gamma}}^\dagger$ is the Moore–Penrose pseudo-inverse of $\tilde{\boldsymbol{\Gamma}}$.

Consider the design of the pilot FBMC symbol $\mathbf{d} = \begin{bmatrix} d_{0,0} & d_{1,0} & \cdots & d_{M-1,0} \end{bmatrix}^T$ so as to minimize the channel estimation MSE subject to a constraint on the transmit energy (this problem was addressed in [20,30] for low frequency selective channels). In view of (11.25) and (11.24), this problem can be stated as follows [41,47]:

$$\min_{\mathbf{d}} \ \mathrm{tr} \left\{ \left( \tilde{\boldsymbol{\Gamma}}^H \tilde{\boldsymbol{\Gamma}} \right)^{-1} \right\} \tag{11.27}$$

$$\text{such that} \quad \mathbf{d}^H \mathbf{B} \mathbf{d} \leq \mathcal{E}, \tag{11.28}$$

where the energy is considered at the SFB output (i.e., channel input). For the case of a full preamble, the latter is not trivially related to the energy at the SFB input [20]. The results of this optimization are outlined in the sequel for both full and sparse preambles. Details can be found in [40–43].

Rewrite (11.19) as $\Gamma_{p,k} = \bar{\mathbf{g}}_{p,k}^H \mathbf{d}$ with the obvious definition for the $M \times 1$ vectors $\bar{\mathbf{g}}_{p,k}$ and $\mathbf{d}$. We can then express the matrix $\boldsymbol{\Gamma}$ in the form

$$\boldsymbol{\Gamma} = \mathcal{G} \mathbf{D}, \tag{11.29}$$

where

$$\mathcal{G} = \begin{bmatrix} \bar{\mathbf{g}}_{0,0}^{\mathrm{H}} & \bar{\mathbf{g}}_{0,1}^{\mathrm{H}} & \cdots & \bar{\mathbf{g}}_{0,L_h-1}^{\mathrm{H}} \\ \bar{\mathbf{g}}_{1,0}^{\mathrm{H}} & \bar{\mathbf{g}}_{1,1}^{\mathrm{H}} & \cdots & \bar{\mathbf{g}}_{1,L_h-1}^{\mathrm{H}} \\ \vdots & \vdots & \ddots & \vdots \\ \bar{\mathbf{g}}_{M-1,0}^{\mathrm{H}} & \bar{\mathbf{g}}_{M-1,1}^{\mathrm{H}} & \cdots & \bar{\mathbf{g}}_{M-1,L_h-1}^{\mathrm{H}} \end{bmatrix}$$

and

$$\mathbf{D} = \mathbf{I}_{L_h} \otimes \mathbf{d}.$$

The matrix $\mathcal{G}$ has a special structure, which can be used to better understand and solve the training design problem. Indeed, $\mathcal{G}$ can be written as

$$\mathcal{G} = \begin{bmatrix} \mathcal{G}_0 & \mathcal{G}_1 & \cdots & \mathcal{G}_{L_h-1} \end{bmatrix}, \tag{11.30}$$

where, for $k = 0, 1, \ldots, L_h-1$, the $M \times M$ matrix $\mathcal{G}_k = \begin{bmatrix} \bar{\mathbf{g}}_{0,k} & \bar{\mathbf{g}}_{1,k} & \cdots & \bar{\mathbf{g}}_{M-1,k} \end{bmatrix}^{\mathrm{H}}$ satisfies [41,43]

$$\mathcal{G}_k = \mathbf{E}^k \mathbf{G}_k, \tag{11.31}$$

where

$$\mathbf{E} = \mathrm{diag}\left(1, e^{-j\frac{2\pi}{M}}, e^{-j2\frac{2\pi}{M}}, \ldots, e^{-j(M-1)\frac{2\pi}{M}}\right),$$

and the matrix $\mathbf{G}_k$ is Hermitian and (almost) tridiagonal. The latter stems from the fact that ICI only exists among adjacent subcarriers. In fact, we can readily verify that $\mathcal{G}_0 = \mathbf{B}$.[3] Moreover, $\mathbf{G}_k$ is (almost) circulant and hence can be diagonalized with the DFT matrix [45], that is,

$$\mathbf{G}_k = \mathcal{F}\mathbf{\Lambda}_k\mathcal{F}^{\mathrm{H}}. \tag{11.32}$$

Its eigenvalues are real valued and can be easily computed a priori (since they only depend on the employed FB) as follows [43]:

$$\lambda_k[m] \triangleq [\mathbf{\Lambda}_k]_{m,m} = r_g[k] + 2c_g[k] \sin\left[\frac{2\pi}{M}\left(\frac{k}{2}+m\right)\right] \tag{11.33}$$

---

[3]Indeed, the vector of pseudo-pilots generated by this preamble can be expressed as $\mathbf{x} = \mathbf{Bd}$ [20]. The model in (11.6), valid for low frequency selective channels, can be also expressed through the above formulation, namely by taking $\mathbf{G}_k = \mathbf{B}$, $k = 0, 1, \ldots, L_h - 1$. IAM-C can then result as a special case of the optimal TD preamble (see [43] for details). Furthermore, $\mathbf{B} = \mathbf{I}_M$ corresponds to the CP-OFDM system, and existing optimality results for CP-OFDM channel estimation can result as special cases of the above analysis (see [20,48]).

for $m = 0, 1, \ldots, M - 1$, where

$$r_g[k] = \sum_{l=k}^{L_g-1} g[l-k]g[l] \tag{11.34}$$

is the lag-$k$ autocorrelation of $g$, and

$$c_g[k] = \sum_{l=k}^{L_g-1} g[l-k]g[l] \cos\left[\frac{2\pi}{M}\left(l - \frac{L_g-1+k}{2}\right)\right] \tag{11.35}$$

for $k = 0, 1, \ldots, L_h - 1$.

It is known that a necessary condition for a cost function of the form (11.27) to be minimized is that the matrix $\tilde{\boldsymbol{\Gamma}}$ have orthogonal columns [47]. Using the above analysis, it can be verified that this condition suggests that an optimal $\mathbf{d}$ is built as a linear combination of $\frac{M}{2^l}$ columns of the DFT matrix, which are equispaced at intervals of length $\frac{M}{2^l}$ if $2^{l-1} < L_h \leq 2^l$, $l = 1, 2, \ldots, \log_2 M$. Thus, for channels of length $L_h > \frac{M}{2}$, the pilot vector will be a scaled version of *one* of the DFT columns, that is,

$$\mathbf{d} = \pm\sqrt{\frac{\mathcal{E}}{\lambda_0[m]}}\mathcal{F}(:, m), \tag{11.36}$$

where Matlab indexing notation has been used. It can be shown that the index $m$ that minimizes (11.27) can be determined as[4] [41,43]

$$m_{\text{opt}} = \arg\min_{0 \leq m \leq M-1} \lambda_0[m] \sum_{k=1}^{L_h-1} \frac{\lambda_0[((k+m))_M]}{\lambda_k^2[m]}, \tag{11.37}$$

with $((\cdot))_M$ denoting modulo $M$. This (offline) search can be facilitated if the symmetries enjoyed by the eigenvalues (see [40,41,43]) are utilized. The quantity in (11.37) is plotted as a function of $m$ in Fig. 11.5 for the PHYDYAS prototype filter with $M = 128$ and $K = 4$ and for several possible channel lengths. We can observe that the appropriate selection of $m$ in (11.37) is more critical for longer channels. Moreover, with increasing $L_h$, the cost tends to be symmetric around $m = \frac{M}{2}$, attaining its maximum at that index.

The pilots in (11.36) are complex valued in general. Constraining them to be real valued (i.e., PAM symbols), so as to strictly conform with the OQAM modulation, would reduce the search set for $m_{\text{opt}}$ to $\{0, \frac{M}{2}\}$, which in turn corresponds to all equal $d_{m,0}$ or equal with alternating signs. As the example of Fig. 11.5 shows, the former

---

[4]The term for $k = 0$ is always equal to one and hence is omitted from the sum.

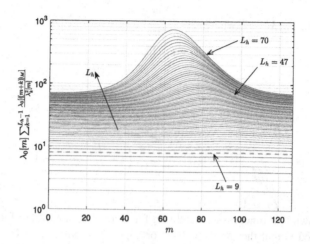

**FIGURE 11.5**

The cost (normalized MSE) of (11.37) as a function of $m$. The direction of increasing $L_h$ is shown. The PHYDYAS FB was used with $M = 128$ and $K = 4$.

choice ($m = 0$) is the best, that is,

$$\mathbf{d} = \pm\sqrt{\frac{\mathcal{E}}{M}}\mathbf{1}_M, \tag{11.38}$$

where $\mathbf{1}_M$ denotes the $M \times 1$ vector of all ones, and the equality $\lambda_0[0] = 1$ has been used. Note that $\mathbf{d}$ in (11.38) is an eigenvector of all $\mathbf{G}_k$ matrices, that is,

$$\mathbf{G}_k\mathbf{d} = \lambda_k[0]\mathbf{d}$$

with $\lambda_0[0] = 1$. It can then be readily verified that

$$\tilde{\mathbf{\Gamma}} = \pm\sqrt{\mathcal{E}}\mathbf{\Lambda}_0^{-1/2}\begin{bmatrix} \mathbf{L}_0 \\ \mathbf{0}_{(M-L_h)\times L_h} \end{bmatrix},$$

where $\mathbf{L}_0 = \mathrm{diag}(\lambda_0[0], \lambda_1[0], \dots, \lambda_{L_h-1}[0])$, whereby the estimator in (11.24) becomes

$$\hat{\mathbf{h}} = \pm\frac{1}{\sqrt{\mathcal{E}}}\mathbf{L}_0^{-1}\begin{bmatrix} \mathbf{I}_{L_h} & \mathbf{0}_{L_h\times(M-L_h)} \end{bmatrix}\mathcal{F}^H\mathbf{y}. \tag{11.39}$$

The latter expression leads to the following simple computational procedure for realizing the TD channel estimator:

**i)** Take the *first $L_h$* terms of the Inverse Discrete Fourier Transform (IDFT) of $\mathbf{y}$:

$$\mathbf{z} = \begin{bmatrix} \mathbf{I}_{L_h} & \mathbf{0}_{L_h\times(M-L_h)} \end{bmatrix}\mathcal{F}^H\mathbf{y}.$$

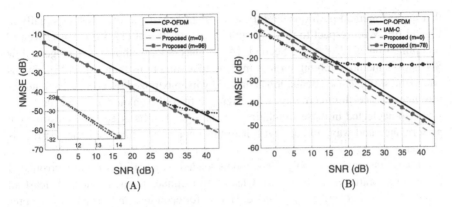

**FIGURE 11.6**

Performance evaluation of full preamble-based channel estimation methods made for high frequency selective channels. The PHYDYAS FB with $M = 128$ and $K = 4$ was employed. (A) ITU Ped-A ($L_h = 9$) and (B) ITU Veh-B ($L_h = 47$) channel models [19].

**ii)** Divide by the diagonal entries of $\mathbf{L}_0$ and scale:

$$\hat{\mathbf{h}} = \pm \frac{1}{\sqrt{\mathcal{E}}} \mathbf{z} \oslash \text{diag}(\mathbf{L}_0).$$

It is of interest to observe that IDFT($\mathbf{y}$) may also be directly found from the (polyphase) FB structure (see, e.g., [6, Fig. 4]), and hence its first $L_h$ entries can be made available at practically no extra cost. The computational cost of the estimator is then proportional to $L_h$. Observe that $\begin{bmatrix} \mathbf{I}_{L_h} & \mathbf{0}_{L_h \times (M-L_h)} \end{bmatrix} \mathcal{F}^H = \mathcal{F}(:, 0 : L_h - 1)^H$ and note that an estimate of the $M$-point CFR can then be computed as

$$\hat{\mathbf{H}} = \sqrt{M} \mathcal{F}(:, 0 : L_h - 1)\hat{\mathbf{h}}. \tag{11.40}$$

We can come up with a procedure analogous to the above for the general (possibly complex-valued) optimal preamble in (11.36). Then, instead of the eigenvalues $\lambda_k[0]$, the $m_{\text{opt}}$th eigenvalues of the $\mathbf{G}_k$ matrices would be involved. The reader is referred to [42,43] for details.

The results of applying this method in the examples of Fig. 11.3 are depicted in Fig. 11.6. Real pilots, as in (11.38), were employed for the TD estimator. Examples illustrating the performance loss from adopting (11.36) with suboptimal values of $m$ (dictated by Fig. 11.5) were also included. Note that this loss is negligible in the low frequency selective case (A) as one would expect from Fig. 11.5. CP-OFDM channel estimation and the IAM-C method were also tested using DFT-interpolation for the sake of fairness of the comparison. The TD method generally outperforms IAM-C in both examples, and, as expected, it does not exhibit error floors, which are typical for IAM at medium to high SNR values. A more careful look, however, shows that the TD method performs slightly worse than IAM-C at low SNR, especially in Fig. 11.6B. With respect to this, we should first recall that (11.36) is not proven to be optimal unless $L_h > \frac{M}{2}$. Second, IAM-C has already shown advantage at low SNRs over methods relying on less inaccurate assumptions than itself [46]. This is due to the strong noise attenuation achieved by its maximized pseudo-pilots in noise-(not interference-)limited regimes. Interference from the unknown part of the frame was also neglected in these simulations, as in Fig. 11.3. Its effect on the estimation performance and ways to cope with it in this context are discussed later on in this chapter.

In a sparse preamble, only the symbols loaded on a set of isolated (surrounded by nulls) subcarriers are nonzeros. Clearly, to estimate $L_h$ parameters, at least an equal number of equations are needed. Hence, for economy and to simplify the presentation, we will assume that there are no more than $L_h$ pilot tones, indexed as $\mathcal{P} = \{p_0, p_1, \ldots, p_{L_h-1}\} \subset \{0, 1, \ldots, M-1\}$; $L_h$ is either the channel length or an upper bound thereof. Moreover, for simplicity and without loss of generality, we will assume that $\frac{M}{L_h}$ is an integer. The preamble design problem then consists of (a) finding the right places for the pilot tones and (b) choosing the values for the pilot symbols. Let $\mathbf{d}_{\mathcal{P}} = \begin{bmatrix} d_{p_0,0} & d_{p_1,0} & \cdots & d_{p_{L_h-1},0} \end{bmatrix}^{\mathrm{T}}$ be the vector of pilots loaded on the pilot subcarriers. Specializing the previous analysis to this preamble structure leads to the conclusion that the optimal solution is given by *equipowered and equispaced* pilot tones [42,43] with

$$|d_{p_i,0}| = \sqrt{\frac{\mathcal{E}}{L_h}} \tag{11.41}$$

and

$$p_i = p_0 + i\frac{M}{L_h} \tag{11.42}$$

for $i = 0, 1, \ldots, L_h - 1$ with $p_0$ freely chosen. For example, $p_0 = 0$ if subcarrier number 0 is to be a pilot tone. The noise covariance matrix for the active subcarriers is then equal to $\mathbf{R}_{\eta|\mathcal{P}} = \mathbb{E}\{\boldsymbol{\eta}_{\mathcal{P}}\boldsymbol{\eta}_{\mathcal{P}}^{\mathrm{H}}\} = \sigma_\eta^2 \mathbf{I}_{L_h}$, and Eq. (11.24) takes the form

$$\hat{\mathbf{h}} = \left(\boldsymbol{\Gamma}_{\mathcal{P}}^{\mathrm{H}}\boldsymbol{\Gamma}_{\mathcal{P}}\right)^{-1}\boldsymbol{\Gamma}_{\mathcal{P}}^{\mathrm{H}}\mathbf{y}_{\mathcal{P}} = \boldsymbol{\Gamma}_{\mathcal{P}}^{-1}\mathbf{y}_{\mathcal{P}}, \tag{11.43}$$

where $\boldsymbol{\Gamma}_{\mathcal{P}}$ is the $L_h \times L_h$ submatrix of $\boldsymbol{\Gamma}$ consisting of its rows and columns corresponding to the pilot subcarrier indexes, and similarly for $\mathbf{y}_{\mathcal{P}}$. Obviously, its inverse, $\boldsymbol{\Gamma}_{\mathcal{P}}^{-1}$, can be computed *offline*. The above computation then requires $L_h^2$ complex multiplications. An alternative computational procedure, which resembles more the one usually employed in CP-OFDM, can be easily devised [42,43] and described below. For simplicity and without loss of generality, let $p_0 = 0$. Define $\mathbf{H}_{\mathbf{r}_g} = \begin{bmatrix} H_{\mathbf{r}_g}[0] & H_{\mathbf{r}_g}[1] & \cdots & H_{\mathbf{r}_g}[L_h-1] \end{bmatrix}^{\mathrm{T}}$ as the $L_h$-point DFT of the impulse

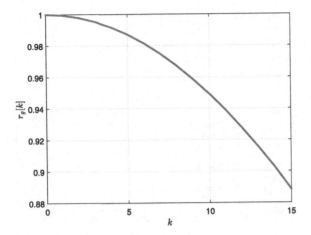

$r_g[k]$ for the PHYDYAS prototype filter ($M = 64$, $K = 4$) and $L_h = 16$.

response weighted (or windowed) by the $r_g[k]$:

$$H_{\mathbf{r}_g}[l] = \sum_{k=0}^{L_h-1} r_g[k]h[k]e^{-jkl\frac{2\pi}{L_h}}, \quad l = 0, 1, \ldots, L_h - 1,$$

with $\mathbf{r}_g = \begin{bmatrix} r_g[0] & r_g[1] & \cdots & r_g[L_h - 1] \end{bmatrix}^{\mathsf{T}}$ (see (11.34)). Then we can implement the estimator as follows:

1. Compute the "weighted" CFR vector:

$$\hat{\mathbf{H}}_{\mathbf{r}_g} = \mathbf{y}_{\mathcal{P}} \oslash \mathbf{d}_{\mathcal{P}}. \tag{11.44}$$

2. Compute the "weighted" impulse response via $L_h$-point Inverse Fast Fourier Transform (IFFT) and divide by the weights $r_g[k]$ to arrive at the CIR estimate:

$$\hat{\mathbf{h}} = \mathrm{IDFT}(\hat{\mathbf{H}}_{\mathbf{r}_g}) \oslash \mathbf{r}_g. \tag{11.45}$$

This alternative procedure costs $\frac{L_h}{2} \log_2 L_h$ complex multiplications for the IFFT, $L_h$ complex divisions (nevertheless, amenable to simplification) for (11.44), and $2(L_h - 1)$ real divisions for (11.45).

Recall that equispaced and equipowered pilot tones are also optimal with low frequency selective channels [30]. What makes the difference with the present scenario is that the "weights" $r_g[\cdot]$ are not all equal in the estimation algorithm above (see Fig. 11.7). With $r_g[k] = 1$, for all $k$, the algorithm above reduces to the well-known time-domain estimation scheme for sparse preambles [4].

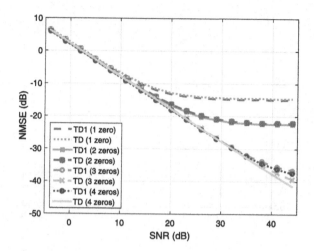

**FIGURE 11.8**

Performance evaluation of the TD method with $\mathbf{r}_g = \mathbf{1}$ ("TD1") and $\mathbf{r}_g$ given by (11.34) ("TD") for ITU Veh-A channels of length $L_h = 11$, using $L = 16$ pilot tones and 1–4 guard symbols. The PHYDYAS FB with $M = 64$ and $K = 4$ was employed.

An example of the use of sparse preambles is shown in Fig. 11.8 for a PHY-DYAS FBMC system with $M = 64$ subcarriers spaced 15 kHz apart and ITU Veh-A channels of length $L_h = 11$. The CFR estimation NMSE (with the CFR computed from the CIR estimate as in (11.40)) is plotted versus the receive SNR. For the number of pilot tones to divide $M$, $L = 16$ equispaced and equipowered tones were employed. The classical method, namely (11.44) and (11.45) with $\mathbf{r}_g = \mathbf{1}_L$, is referred to as "TD1". A realistic scenario involving nonnegligible interference from the (Quadrature Phase-Shift Keying (QPSK)-modulated) data was considered. The results of using one to four guard (null) FBMC symbols between the pilot symbol and the payload are shown. Note how the TD estimator with the optimal preamble manages to mitigate the error floor effect.

The previous examples show that one guard symbol may not be sufficient to protect the pilots from being interfered by the unknown (control or information) symbols, and we would in general require more than one to attain an interference-free performance. An alternative would be to boost the pilots over the payload, thus trading spectral for power efficiency. Iteratively estimating the channel and the unknown symbols affecting the pilots in an alternating manner was recently proposed as a bandwidth efficient approach to TD channel estimation [49]. An iterative scheme of the Expectation Maximization (EM) type was proposed in [50] in connection with the subcarrier model outlined below.

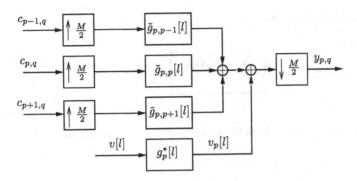

**FIGURE 11.9**

FBMC/OQAM subcarrier model.

### 11.3.2.3 *The Structured Approach*

An alternative way to formulate the problem is through the so-called *structured* approach [51], where the knowledge of that part of the overall (end-to-end) channel that depends on the FB is elegantly modeled and exploited to arrive at an efficient model for the system. The efficiency is meant in terms of both facilitating the development of (sub)channel estimation algorithms and significantly reducing the required training sequence length [51]. This model is schematically shown in Fig. 11.9, where the fact that only adjacent subcarriers overlap is made explicit. In this formulation, (11.17) is written as

$$y_{p,q} = \sum_{m=p-1}^{p+1} \tilde{g}_{p,m}[q] \star c_{m,q} + \eta_{p,q}, \tag{11.46}$$

where $c_{m,q} \triangleq d_{m,q}e^{j\varphi_{m,q}}$ is the sequence of input OQAM symbols, $\star$ stands for linear convolution, and $\tilde{g}_{p,m}[q]$ is the impulse response of length $L_{\tilde{g}} = \left\lceil \frac{2L_g+L_h-2}{M/2} \right\rceil$, which results from the downsampling by $\frac{M}{2}$ of the convolution between the transmit filter $g_m$, the receive filter $g_p$, and the frequency selective channel $h$. The corresponding $L_{\tilde{g}} \times 1$ vector can be expressed as

$$\tilde{\mathbf{g}}_{p,m} = \bar{\mathbf{G}}_{p,m}\mathbf{h}, \tag{11.47}$$

where $\bar{\mathbf{G}}_{p,m}$ results from keeping only every $\frac{M}{2}$th row of the $(2L_g + L_h - 2) \times L_h$ convolution matrix of $g_p^* \star g_m$. We can then write the sequence of $L_o$ samples received at the $p$th AFB output as

$$\mathbf{y}_p = \underbrace{\left( \sum_{m=p-1}^{p+1} \mathbf{C}_m \bar{\mathbf{G}}_{p,m} \right)}_{\mathbf{S}_p} \mathbf{h} + \boldsymbol{\eta}_p = \mathbf{S}_p \mathbf{h} + \boldsymbol{\eta}_p, \tag{11.48}$$

where $\boldsymbol{\eta}_p$ is accordingly built from the sequence of noise samples, and

$$\mathbf{C}_m = \begin{bmatrix} c_{m,0} & c_{m,1} & c_{m,2} & \cdots & c_{m,L_{\tilde{g}}-1} \\ c_{m,1} & c_{m,2} & c_{m,3} & \cdots & c_{m,L_{\tilde{g}}} \\ \vdots & \vdots & \vdots & \ddots & \vdots \\ c_{m,L_o-1} & c_{m,L_o} & c_{m,L_o+1} & \cdots & c_{m,L_c-1} \end{bmatrix}$$

is an $L_o \times L_{\tilde{g}}$ Hankel matrix built out of the sequence of $L_c = L_o + L_{\tilde{g}} - 1$ input OQAM symbols. Consider $M_{tr}$ *nonadjacent* subcarriers $p_i$, $i = 0, 1, \ldots, M_{tr} - 1$, carrying training sequences of length $L_c$.[5] Then we can write[6]

$$\underbrace{\begin{bmatrix} \mathbf{y}_{p_0} \\ \mathbf{y}_{p_1} \\ \vdots \\ \mathbf{y}_{p_{M_{tr}-1}} \end{bmatrix}}_{\bar{\mathbf{y}}} = \underbrace{\begin{bmatrix} \mathbf{S}_{p_0} \\ \mathbf{S}_{p_1} \\ \vdots \\ \mathbf{S}_{p_{M_{tr}-1}} \end{bmatrix}}_{\mathbf{S}} \mathbf{h} + \underbrace{\begin{bmatrix} \boldsymbol{\eta}_{p_0} \\ \boldsymbol{\eta}_{p_1} \\ \vdots \\ \boldsymbol{\eta}_{p_{M_{tr}-1}} \end{bmatrix}}_{\bar{\boldsymbol{\eta}}}. \tag{11.49}$$

Then, similarly to (11.23), we can compute an estimate of $\mathbf{h}$ from the above as follows:

$$\hat{\mathbf{h}} = \left( \mathbf{S}^H \mathbf{R}_{\bar{\eta}}^{-1} \mathbf{S} \right)^{-1} \mathbf{S}^H \mathbf{R}_{\bar{\eta}}^{-1} \bar{\mathbf{y}}, \tag{11.50}$$

where the covariance matrix of $\bar{\boldsymbol{\eta}}$ can be approximately written as

$$\mathbf{R}_{\bar{\eta}} = \mathrm{diag}(\mathbf{R}_{\boldsymbol{\eta}_{p_0}}, \mathbf{R}_{\boldsymbol{\eta}_{p_1}}, \ldots, \mathbf{R}_{\boldsymbol{\eta}_{p_{M_{tr}-1}}}) = \mathbf{I}_{M_{tr}} \otimes \mathbf{R}_{\eta_-}$$

with $\mathbf{R}_{\eta_-}$ being the covariance matrix of the noise sequence at an AFB output. A detailed analysis of the correlation of the FBMC/OQAM noise in both frequency and time can be found in [53].

---

[5]Note that this extends the short preambles considered previously to longer ones. This in turn restricts the applicability of this approach to slowly varying channels. A study of the estimation performance as a function of the training sequence length was reported in [52]. The TD method with optimally designed preambles that consist of more than one consecutive pilot FBMC symbols was addressed in [53].
[6]Since the training subcarriers are not contiguous, it is possible to have the training symbols be frequency multiplexed with data symbols [51].

The discussion so far ignored a constraint on the preamble structure, which is very common in practical implementations and presents an additional challenge to the preamble design and the realization of the estimation algorithm. In a realistic system, only a subset of $M_u < M$ subcarriers are modulated. This may be due to the insertion of null (*virtual*) subcarriers at the extremities of the spectrum in order to meet the Radio Frequency (RF) mask requirement [54]. Another possible scenario involving inactive subcarriers in a large fraction of the spectrum occurs in a multiuser environment [55]. In such cases, unless special care is taken to take the null subcarriers into account, a considerable loss in the channel estimation performance may be experienced, known as "border effect" [56]. Existing means of coping with this problem include using robust versions of the pseudoinverse of the DFT submatrix in (11.40), computed via truncated Singular Value Decomposition (SVD) [54] or LMMSE regularization [55]. Another possibility, considered for FBMC/OQAM systems in [57], is to take a multirate view at the CIR estimation problem. First, a *downsampled* version of the CIR, $\mathbf{h}_\downarrow$, of length $L_{h_\downarrow}$ is estimated from the active (available) part of the spectrum. The CIR corresponding to the entire spectrum is then approximated with the aid of a linear transformation $\mathbf{h} = \mathbf{A}\mathbf{h}_\downarrow$, equivalent to a fractional *upsampling* of $\mathbf{h}_\downarrow$ by a factor of $\frac{L_h}{L_{h_\downarrow}}$, where $L_{h_\downarrow} = \left\lfloor \frac{M_u}{M} L_h \right\rfloor$. This is performed in three steps, namely, upsampling by a factor of $L_h$, low-pass filtering, and downsampling by a factor of $L_{h_\downarrow}$. Substituting $\mathbf{h} = \mathbf{A}\mathbf{h}_\downarrow$ into (11.49) yields a more robust estimator, where the matrix to be inverted is now well conditioned. The corresponding MSE is given by [57]

$$\text{MSE}_\downarrow = \sigma_\eta^2 \frac{M_u}{M} \text{tr}\left\{ \left(\mathbf{A}^H\mathbf{S}^H\mathbf{R}_{\bar{\eta}}^{-1}\mathbf{S}\mathbf{A}\right)^{-1}\right\}. \tag{11.51}$$

## 11.4 SCATTERED PILOTS-BASED CHANNEL ESTIMATION

When *scattered* pilots are employed, as it is the case for estimating and tracking time-varying channels, the neighbors of the pilot F-T point $(m, n)$ carry unknown (data) symbols, implying that the pilot will be contaminated by unknown interference. Hence, we cannot (directly) approximate the interference in (11.7). However, by properly choosing one (or more) of the neighboring symbols, this interference can be considerably reduced or (theoretically) forced to zero, which renders the pseudo-pilot in (11.7) real-valued and equal to $d_{m,n}$. This is the idea that lies behind most of the scattered pilots-based FBMC/OQAM channel estimation methods. The interference may be canceled by applying an appropriate transmission pattern around the pilot (exploiting the symmetry in (11.9)) and/or transmitting zeros at the F-T points that contribute most to the interference (e.g., [58]). A more spectrally efficient solution is to set a neighboring symbol, known as *help* [8] or *auxiliary* [59] pilot, so as to achieve interference reduction/nulling. Combinations of this idea with that of IAM have also been tested [58,60–62]. Generalizations of the Help Pilot (HP) idea

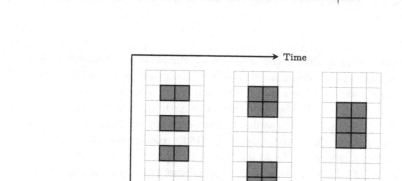

**FIGURE 11.10**

Three different ways of distributing $2(R + 1)K_p = 12$ pilots in six, three, and two clusters of two, four, and six pilots each, respectively. The total number of pilots remains constant. The first case ($R = 0$) corresponds to the way HPs are used.

that rely on a coding-like transformation of the neighboring symbols have been recently reported [63–65]. Iterative joint estimation/detection schemes that overcome the need for HPs have been also studied [66,67]. More recently, a Kalman filter-based pilot-assisted approach for time-varying channels appeared in [68]. With the exception of only a few works [69,70], the common assumption underlying the literature on scattered pilot-based FBMC/OQAM channel estimation is that of low (or mild) frequency selectivity, which validates the model in (11.6). In the rest of this section, the HP scheme and its several variants are briefly reviewed, followed by a more detailed presentation of recent results [70] about the strongly frequency selective case.

## 11.4.1 CHANNELS OF LOW FREQUENCY SELECTIVITY

Constructing the HP is made possible through the assumption of the CFR being constant in both frequency and time over the neighborhood of the pilot that contributes most to the interference [8]. This is usually (but not necessarily [59]) taken to be the first-order T-F neighborhood (see Table 11.1). The idea is to set the value of one of the T-F neighbors of a pilot $d_{m,n}$ in a way that (most of) the interference to the pilot is nulled (so as to have $x_{m,n} \approx d_{m,n}$ in (11.7)) (see, for example, the first configuration in Fig. 11.10). Thus, in view of (11.9), the value of an HP positioned at $(m, n + 1)$

would be chosen as

$$d_{m,n+1} = \frac{1}{\gamma}\big[\gamma d_{m,n-1} - (-1)^m \beta (d_{m+1,n} - d_{m-1,n})$$
$$- \delta(d_{m-1,n-1} + d_{m-1,n+1} + d_{m+1,n-1} + d_{m+1,n+1})\big]. \qquad (11.52)$$

In fact, this or the $(m, n-1)$ position is commonly preferred in order to avoid resulting in an HP of a large amplitude, which would generate a high PAPR (recall that $\gamma$ is generally the largest of the weights in (11.9)).

It must be emphasized that the total training overhead per pilot position in the HP scheme, i.e., two real-valued symbols, is equal to that for OFDM [71]. However, as explained before, it is a rather power-inefficient method as compared to pilot-aided OFDM channel estimation in view of the fact that the HP amplitude is not perfectly controlled and hence may suffer from a high PAPR effect. A number of variants of the HP idea have thus been proposed with the aim of mitigating this drawback. The minimization of the amplitude of the HP ("precoding symbol" [72]), at the cost of accepting a nonzero value (out of a set of possible values) for the interference to the main pilot, was proposed in [72]. The search in the set of (discrete but many) interference values entails a considerable computational cost, which is traded for the improvement of the estimation accuracy due to the IAM-like effect of the resulting pseudo-pilot. Such a boosting of the virtual pilot is also achieved in [62] through appropriately setting the values of the pair of pilots, computed with the aid of an interference-aware extension of the POP method. The PAPR problem is further mitigated with the aid of an approach common in MC systems [48], namely multiplying the pilots with PseudoNoise (PN) sequences. The use of a Wiener (MMSE) estimator with iterative interference cancellation and Low-Density Parity-Check (LDPC) decoding was recently proposed in [73]. An approach analogous to spreading and despreading in direct sequence Code-Division Multiple Access (CDMA) systems was devised in [63], where all the neighbors of the pilot symbol are freely chosen but first *coded* (through a linear transformation) before being transmitted. The coding matrix is built so as to reduce the interference to the pilot while also ensuring a low computational cost for the forward and the inverse transformation. Moreover, since it is chosen to be orthogonal, it also addresses the power inefficiency of the classical HP scheme. The coding approach was also studied in [67] and evaluated against the straightforward iterative joint estimation/detection approach of [66]. Along with the HP approach, it was also recently considered in a channel adaptive modulation context [74]. Analogous schemes, based on coding (or "scrambling" [75]) of the symbols in the pilot neighborhood and aiming at improving power efficiency, were reported in [64,75] and [65], where a more systematic construction of the coding matrix was conceived. In the latter work, an alternative, more computationally efficient way to address the power inefficiency problem was also investigated, which involves more than one HP symbol. Their values are found as the minimum-norm (hence low PAPR) LS solution of an underdetermined system of linear equations expressing the requirement of null interference in a more general setting than (11.52). The effect of the

HPs on the throughput of the system was also studied, demonstrating gains from the adoption of multiple HPs at medium to low SNR values. The resulting Bit Error Probability (BEP) and its minimization with respect to the HP to data power ratio were studied in [76]. The performance of this scheme in a real-world (testbed) scenario was successfully evaluated in [77]. Distributing the pilot information in two neighboring symbols, which are set so as to effectively address the intrinsic interference effect, was proposed in [78] as an alternative to the use of a data-independent pilot with auxiliary F-T neighbors. This so-called *dual-dependent pilots (DDP)* scheme was shown to achieve only a slight performance improvement over that of [75] at a comparable computational cost.

## 11.4.2 CHANNELS OF HIGH FREQUENCY SELECTIVITY

In the presence of less smooth channel responses, the previous schemes fail to provide good channel estimates and exhibit performance characterized by severe error floors at medium to high SNR [42]. Recent efforts in extending the above ideas to more frequency selective channels include [65,75], which, however, still rely on the assumption of a per-subcarrier frequency flat response, and [69], which is based on a principal component analysis of the intrinsic interference and relies on a priori knowledge of the channel second-order statistics to compute the generalized auxiliary pilots. An iterative approach, analogous to that of [66], was recently proposed in [79], where the channel frequency selectivity is modeled through a Frequency Spread FBMC/OQAM (FS-FBMC) implementation.

It was shown in [70] that the HP approach can be generalized to the highly frequency selective scenario by using the technical tools established in Chapter 9. The subchannel frequency selectivity is modeled with the aid of a Taylor polynomial of order $R \in \mathbb{N} \cup \{0\}$, and clusters of more than two pilot symbols are considered for channels of (relatively) significant frequency selectivity ($R > 0$). To describe the idea and the associated method, assume, first, a set of $K_p$ disjoint clusters of pilots in the time-frequency plane, each of them composed of $2(R+1)$ real-valued pilot symbols (see Fig. 11.10 for a few examples). This amounts to a total number of $2(R+1)K_p$ pilot symbols. The traditional HP configuration is obtained by forcing $R = 0$, so that each cluster consists of two pilots, the original and the auxiliary one. When $R$ increases, pilots tend to be grouped into a smaller number of highly populated clusters, as illustrated in Fig. 11.10.

According to Proposition 9.1, under some regularity conditions, we can write (as $M \to \infty$)

$$y_{m,n} = \sum_{r=0}^{R} \frac{1}{r! M^r} H^{(r)}\left(e^{j\omega_m}\right) x_{m,n}^{(r)} + o\left(M^{-R}\right),$$

where $H^{(r)}(z)$ is the $r$th derivative of the channel transfer function $H(z)$, and $x_{m,n}^{(r)}$ is the signal received at the $m$th subcarrier output of an AFB under noiseless perfect channel conditions when the receive prototype pulse is replaced by its $r$th derivative in the time domain (see Chapter 9). The above equation is basically stating that we can describe the output of the AFB in the presence of a highly frequency selective channel as a linear combination of contributions obtained under frequency flat conditions, simply by filtering the received signal with sequential derivatives of the prototype pulse.

Assume now that the received signal is processed by an AFB constructed from the $s$th derivative of the original prototype pulse and denote by $y_{m,n}^{(s)}$ the corresponding output (so that, in particular, $y_{m,n}^{(0)} = y_{m,n}$). Here again, a direct application of Proposition 9.1 allows us to write

$$\frac{y_{m,n}^{(s)}}{M^s} = \sum_{r=0}^{R-s} \frac{1}{r!} H^{(r)}\left(e^{j\omega_m}\right) \frac{x_{m,n}^{(r)}}{M^{r+s}} + o\left(M^{-R}\right)$$

for $s = 0, 1, \ldots, R$. Now, assuming that the quantities $x_{m,n}^{(r)}$ are known, the above identities can be expressed as a system of $R+1$ equations with unknowns $H^{(r)}(e^{j\omega_m})$, $r = 0, 1, \ldots, R$. It is easy to see that there is a direct relationship between these unknowns and the CIR $\mathbf{h}$ because

$$H^{(r)}\left(e^{j\omega_m}\right) = \omega_m^{\mathrm{H}}(-1)^r \mathbf{\Delta}^r \mathbf{h},$$

where $\mathbf{\Delta} = \mathrm{diag}(0, 1, \ldots, L_h - 1)$ and $\omega_m = \begin{bmatrix} 1 & e^{j\omega_m} & \cdots & e^{j\omega_m(L_h-1)} \end{bmatrix}^{\mathrm{T}}$. Using this and obviating the error term, we can express the above system of $R + 1$ linear equations in matrix form as

$$\mathbf{y}_{m,n} = \left(\mathbf{\Psi}_{m,n} \otimes \omega_m^{\mathrm{H}}\right) \mathbf{\Upsilon}_R \mathbf{h}, \tag{11.53}$$

where $\mathbf{y}_{m,n} \in \mathbb{C}^{(R+1)\times 1}$ has the entries $\frac{y_{m,n}^{(s)}}{M^s}$, $s = 0, 1, \ldots, R$,

$$\mathbf{\Upsilon}_R = \begin{bmatrix} \mathbf{I}_{L_h} & (-1)\mathbf{\Delta} & \cdots & \frac{(-1)^R}{R!}\mathbf{\Delta}^R \end{bmatrix}^{\mathrm{T}},$$

and $\mathbf{\Psi}_{m,n}$ is a Hankel upper triangular matrix with the $r$th entry of its first row equal to $\frac{x_{m,n}^{(r)}}{M^r}$, $r = 0, 1, \ldots, R$.

At this point, all the ingredients are there to formulate the corresponding channel estimation method. For each of the $K_p$ available clusters of pilots, we can write a system of linear equations such as the one in (11.53), where $\mathbf{y}_{m,n}$ gathers the observations at the output of a set of $R + 1$ parallel FBs, constructed using the sequential derivatives of the original prototype filter. Stacking the $K_p$ systems on top of one another results in

$$\mathbf{y} = \mathbf{\Omega}\mathbf{h},$$

where

$$\Omega = \begin{bmatrix} \Psi_1 \otimes \omega_1^H \\ \vdots \\ \Psi_{K_p} \otimes \omega_{K_p}^H \end{bmatrix},$$

with the obvious definition of $y \in \mathbb{C}^{(R+1)K_p \times 1}$ and the notation for the $\Psi$ matrices and the $\omega$ vectors being slightly abused to make their correspondence with the $K_p$ clusters explicit. It can be shown that the above system of equations is full column rank if $L_h \leq K_p$, provided that the matrices $\Psi_k$ are designed to be full rank. We can then compute the LS CIR estimate as

$$\hat{h} = \left(\Omega^H \Omega\right)^{-1} \Omega^H y.$$

A crucial point in this development is the fact that the values of $x_{m,n}^{(r)}$, $r = 0, 1, \ldots, R$, namely, the output values of the $R + 1$ parallel FBs at the F-T position $(m, n)$ under noiseless perfect channel conditions, are assumed to be known. Observe that, according to (11.7), these values will uniquely depend on the data $d_{k,\ell}$ transmitted on the support set $\Omega_{m,n}$. It can readily be shown that, by fixing the value of a subset of $2(R + 1)$ symbols in $\Omega_{m,n}$, we can perfectly determine the value of $x_{m,n}^{(r)}$, $r = 0, 1, \ldots, R$ (further details are provided in [70]). It should be pointed out, however, that the power of the inserted pilots cannot be easily controlled, and just as in the conventional HP approach ($R = 0$), these pilots will typically have a power higher than the average power of the information symbols.

An example of the estimation performance of the above method is given here for an FBMC/OQAM system with $M = 64$ subcarriers employing the PHYDYAS prototype filter [10] with overlapping factor $K = 4$. The pilots were distributed in 27, 14, and 9 equispaced clusters of 2, 4, and 6 pilots each (see Fig. 11.10) for $R = 0, 1, 2$ respectively, so that the total number of pilots was fixed to be approximately constant regardless of $R$. The 3GPP Extended Typical Urban (ETU) [35] was considered, and the intercarrier separation was assumed to be relatively large (equal to 25 kHz), hence leading to subchannels of significant frequency selectivity. Fig. 11.11 plots the resulting MSE as a function of the SNR for different values of $R$, with the SNR being computed with respect to the information symbols. It must be pointed out that, as $R$ increases, the proposed method saturates at a lower MSE floor due to the increased order in the channel selectivity model. At lower values of the SNR, the estimator is penalized due to the fact that the number of pilot clusters is reduced, and therefore the channel is sounded at sparser points with increasing $R$. It was shown in [70] that we can obtain a theoretical characterization of the MSE for each value of the SNR, so that we can determine the optimum value of $R$ depending on the SNR operational point.[7]

---

[7] An analogous remark, concerning the optimum number of HPs, was also made in [77].

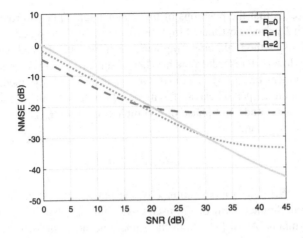

**FIGURE 11.11**

NMSE as a function of the SNR for the pilot-based channel estimation method of [70] with different values of $R$. The ETU channel model with 25 kHz subcarrier spacing was considered.

## 11.5 (SEMI)BLIND TECHNIQUES

In applications where only little or no training information is available, a semi-blind or blind estimation approach can be relevant [80]. Such techniques have the potential of improving the bandwidth efficiency of training-based schemes, albeit at the cost of an increase in computational complexity and/or the size of the required data set. In contrast to the large extent of the (semi)blind OFDM channel estimation literature, only a few related works for FBMC/OQAM systems have been reported.

The problem of blind FBMC/OQAM channel estimation was first studied in [81] through a Second-Order Statistics (SOS) approach, i.e., based on the properties and estimates of the correlation function. It was shown that the overlapping of the FBMC/OQAM symbols in time induces a cyclostationarity of period $M$ to the transmitted signal, although no CP (as in related OFDM works) is used. This property was used to arrive at a homogeneous system of linear equations involving products of the cyclic spectra of the transmitted and received signals computed at two different cycles $k_1$, $k_2$, which have to be different from zero in order to (at least theoretically) cancel noise.[8] Solving this system (in practice, in the LS sense) yields the CIR vector up to a scalar ambiguity if $L_h \leq \frac{M}{|k_1 - k_2|}$. The latter identifiability condition is always satisfied for (relatively) low frequency selective channels. The method works with the AFB input signal (i.e., in the time domain) and generally needs very long data records to obtain good estimates, particularly, for long prototype filters [81]. Among

---

[8]Of course, the noise affects the accuracy of the SOS estimates [81].

its nice features is that the positions of the channel zeros play no role and that the channel length $L_h$ can be estimated from the dimension of the signal subspace.

The previous method, although implicitly requiring the channel delay spread to be sufficiently low with respect to the number of subcarriers, does not explicitly rely for its development on (11.5) or (11.6). On the other hand, the method proposed in [82] is based on the per-subcarrier flat model (11.6) to view the problem as one of estimating the magnitude and phase of a complex rotation $H_{m,n}$. The magnitude is estimated from the relation

$$\sigma_{y_{m,n}}^2 = |H_{m,n}|^2 2\sigma_d^2 + \sigma_\eta^2,$$

where $\sigma_{y_{m,n}}^2 = \mathbb{E}(|y_{m,n}|^2)$ is an estimate of the power of $y_{m,n}$, using the fact that the intrinsic interference term of the pseudo-symbol $x_{m,n}$ has power equal to that of the data symbols. As in [81], both the signal and noise powers are assumed to be known. The phase of $H_{m,n}$ is estimated from the dominant eigenvector of the so-called spatial-sign covariance matrix, defined as $\mathbb{E}(\underline{\mathbf{y}}_{m,n}\underline{\mathbf{y}}_{m,n}^T)$ with the spatial-sign vector $\underline{\mathbf{y}}_{m,n}$ being the unit-norm version of $\left[\mathrm{Re}(y_{m,n}) \quad \mathrm{Im}(y_{m,n})\right]^T$. Nevertheless, a preamble (like those employed by the IAM method in Section 11.3.1) is still needed to resolve the remaining sign ambiguity. Adaptive versions of this technique, based on constant- and multi-modulus algorithms, were reported in [83].

The latter requirement of [82], namely, of a training preamble, can be criticized as compromising the advantage of the blind approach over purely training-based estimation. Moreover, similarly to [81], this method requires long data records to arrive at accurate estimates. Both of these drawbacks are mitigated in [84], whose basic idea is to smooth in frequency (exploiting the smoothness of the CFR as the frequency smoothing techniques do; see, e.g., [26]) instead of in time. This way, significant savings in the required data length are shown to be possible. To further improve the estimation performance, scattered zero-valued pilots may be also employed, as they are shown to facilitate the estimation of the phase. The CFR smoothness assumption also helps in resolving the sign ambiguity, a task now requiring only one pilot tone.

Finally, a High-Order Statistics (HOS) approach was followed in [85] to develop a blind estimation method applicable in both OFDM and FBMC modulations. Like the previous techniques, it applies in channels of low frequency selectivity. Based on the knowledge of the 4th-order moment of the input data and an estimate of the 4th-order moments of the demodulated signal, an estimate of the 4th power of the CFR is readily computed. When transformed in the time domain, the problem then translates to a 4th-order autodeconvolution to determine the CIR. Both OFDM and FBMC modulations are addressed although for the FBMC/OQAM system, it is the composite CIR (i.e., including both the transmit-receive pulse shaping and the physical channel) that is computed. A notable advantage of using HOS is that the Gaussian noise is nulled (largely suppressed in practice). Despite the method being based on HOS, a good performance is demonstrated in the simulation results with a reasonable dataset size.

## 11.6 CONCLUDING REMARKS

FBMC/OQAM channel estimation has been a subject of extensive research for the last decade or so. Active research in this area is fueled by both the potential and interesting challenges underlying the application of this type of FBMC in a number of communication systems ranging from wireless cellular and ad hoc [86] to wired (see, e.g., [87]) networks. This chapter presented an overview of this fascinating subject with clear emphasis given to recent advances in the more challenging problem of high frequency selective channel estimation. Both training (preamble or pilot)-based and (semi)blind techniques were considered. In spite of the effort made by the authors to provide an extensive account of the related literature, some important aspects or techniques had to be omitted or described in less detail to make the chapter better accessible to nonexperts and keep its length within reasonable limits. Such topics include, for example, channel estimation for CP-based FBMC/OQAM systems [88], optimal pilot allocation in the presence of virtual subcarriers [89], multisymbol preambles [22,90] and their optimization [53], channel estimation with Transmitter (TX)/Receiver (RX) impairments taken into account [11], and systems using Biorthogonal Frequency-Division Multiplexing with Offset-QAM subcarrier modulation (BFDM-OQAM) [4].

Despite being an already large and fast growing subject, more research is needed in FBMC/OQAM channel estimation before it reaches the extent and maturity found in the CP-OFDM channel estimation literature. Open problems, with promises for both theoretical and practical interest, include channel estimation with joint consideration of all realistic constraints and impairments (e.g., Carrier Frequency Offset (CFO) [12,91], I/Q imbalance [13], HPA and limited latency [16], virtual subcarriers [89]), and bandwidth-efficient and high-performance training and techniques for doubly (i.e., time and frequency) dispersive channels [18,83,92]. Further improvements in the estimation performance and the bandwidth efficiency of these techniques can be sought in the exploitation of the sparse/compressible nature of wireless CIRs [93]. Related work for FBMC/OQAM systems has so far been constrained to rather simple scenarios with a straightforward (and OFDM-mimicking) application of compressive sensing ideas and techniques [27,94–96]. We hope that this chapter will inspire further research advances in this hot and fast growing subject.

## REFERENCES

[1] P. Banelli, S. Buzzi, G. Colavolpe, A. Modenini, F. Rusek, and A. Ugolini, "Modulation formats and waveforms for 5G networks: Who will be the heir of OFDM?," *IEEE Signal Process. Mag.*, vol. 31, no. 6, pp. 80–93, Nov. 2014.

[2] J. G. Andrews, S. Buzzi, W. Choi, S. Hanly, A. Lozano, A. C. K. Soong, and J. C. Zhang, "What will 5G be?," *IEEE J. Sel. Areas Commun.*, vol. 32, no. 6, pp. 1065–1082, Jun. 2014.

[3] G. Garbo, S. Mangione, and V. Maniscalco, "Wireless OFDM-OQAM with a small number of subcarriers," in *Proc. WCNC-2008*, Las Vegas, NV, Mar.–Apr. 2008.

[4] E. Kofidis, D. Katselis, A. Rontogiannis, and S. Theodoridis, "Preamble-based channel estimation in OFDM/OQAM systems: A review," *Signal Process.*, vol. 93, no. 7, pp. 2038–2054, Jul. 2013.

[5] T. Hidalgo Stitz, T. Ihalainen, and M. Renfors, "Practical issues in frequency domain synchronization for filter bank based multicarrier transmission," in *Proc. ISCCSP-2008*, Malta, Mar. 2008.

[6] P. Siohan, C. Siclet, and N. Lacaille, "Analysis and design of OFDM/OQAM systems based on filterbank theory," *IEEE Trans. Signal Process.*, vol. 50, no. 5, pp. 1170–1183, May 2002.

[7] P. Banelli and L. Rugini, "OFDM and multicarrier signal processing," in *Academic Press Library in Signal Processing, vol. 2*, R. Chellappa and S. Theodoridis, Eds., Elsevier Ltd., 2014, pp. 187–293.

[8] J.-P. Javaudin, D. Lacroix, and A. Rouxel, "Pilot-aided channel estimation for OFDM/OQAM," in *Proc. VTC-2003 (Spring)*, Jeju island, Korea, Apr. 2003.

[9] C. Lélé, J.-P. Javaudin, R. Legouable, A. Skrzypczak, and P. Siohan, "Channel estimation methods for preamble-based OFDM/OQAM modulations," *Trans. Emerg. Telecommun. Technol.*, vol. 19, no. 7, pp. 741–750, Nov. 2008.

[10] M. G. Bellanger, "Specification and design of a prototype filter for filter bank based multicarrier transmission," in *Proc. ICASSP-2001*, Salt Lake City, UT, May 2001.

[11] A. Ishaque and G. Ascheid, "I/Q imbalance and CFO in OFDM/OQAM systems: Interference analysis and compensation," in *Proc. PIMRC-2013*, London, UK, Sep. 2013.

[12] S. Van Caekenberghe, A. Bourdoux, L. Van der Perre, and J. Louveaux, "Preamble-based frequency-domain joint CFO and STO estimation for OQAM-based filter bank multicarrier," *EURASIP J. Adv. Signal Process.*, 2014. http://dx.doi.org/10.1186/1687-6180-2014-118.

[13] M. Sakai, H. Lin, and K. Yamashita, "Joint estimation of channel and I/Q imbalance in OFDM/OQAM systems," *IEEE Commun. Lett.*, vol. 20, no. 2, pp. 284–287, Feb. 2016.

[14] S.-W. Kang and K.-H. Chang, "A novel channel estimation scheme for OFDM/OQAM-IOTA system," *ETRI J.*, vol. 29, no. 4, pp. 430–436, Aug. 2007.

[15] C. M. Lélé, "OFDM/OQAM: Méthodes d'estimation de canal, et combinaison avec l'accès multiple CDMA ou les systèmes multi-antennes." Ph.D. dissertation. Conservatoire National des Arts et Métiers (CNAM), 2008.

[16] T. Levanen, M. Renfors, T. Ihalainen, E. Lähetkangas, V. Syrjälä, and M. Valkama, "On the performance of time constrained OQAM-OFDM waveforms with preamble based channel estimation," in *Proc. WCNC-2016 Workshop on Novel Waveform and MAC Design for 5G (NWM5G-2016)*, Doha, Qatar, Apr. 2016.

[17] J. Du and S. Signell, "Novel preamble-based channel estimation for OFDM/OQAM systems," in *Proc. ICC-2009*, Dresden, Germany, Jun. 2009.

[18] M. Nájar, *et al.*, "MIMO channel estimation and tracking." PHYDYAS, Deliverable 4.1, Jan. 2009. [Online]. Available: http://www.ict-phydyas.org/delivrables/PHYDYAS-D4.1.pdf/view.

[19] Recommendation ITU-R M.1225. "Guidelines for evaluation of radio transmission technologies for IMT-2000," ITU, 1997. [Online]. Available: https://www.itu.int/dms_pubrec/itu-r/rec/m/R-REC-M.1225-0-199702-I!!PDF-E.pdf.

[20] D. Katselis, E. Kofidis, A. Rontogiannis, and S. Theodoridis, "Preamble-based channel estimation for CP-OFDM and OFDM/OQAM systems: A comparative study," 2009. arXiv:0910.3928v1 [cs.IT] (extended version of [30]).

[21] J. Du, "Pulse shape adaptation and channel estimation in generalized frequency division multiplexing systems." Licentiate thesis. Stockholm, Sweden: KTH, Dec. 2008.

[22] E. Kofidis and D. Katselis, "Improved interference approximation method for preamble-based channel estimation in FBMC/OQAM," in *Proc. EUSIPCO-2011*, Barcelona, Spain, Aug.–Sep. 2011.

[23] D. Kong, D. Qu, P. Gao, C. Wang, and T. Jiang, "Frequency domain averaging for channel estimation in OQAM-OFDM systems," in *Proc. WCNC-2013*, Shanghai, China, Apr. 2013.

[24] X. Chen, M. Zhao, and C. Xu, "Preamble-based channel estimation methods with high spectral efficiency for pulse shaping OFDM/OQAM systems," in *Proc. 6th Int'l Conf. Wireless Communications and Signal Processing (WCSP-2014)*, Hefei, China, Oct. 2014.

[25] X. Lin, X. Pingping, and L. Ying, "Improved DFT-based channel estimation for FBMC/OQAM wireless communication systems," in *Proc. ISCIT-2016*, Qingdao, China, Sep. 2016.

[26] D. Katselis, C. R. Rojas, M. Bengtsson, and H. Hjalmarsson, "Frequency smoothing gains in preamble-based channel estimation for multicarrier systems," *Signal Process.*, vol. 93, no. 9, pp. 2777–2782, Sep. 2013.

[27] E. Kofidis, *et al.*, "Training design and algorithms for channel estimation." EMPhAtiC, Deliverable 3.1, Dec. 2013. [Online]. Available: http://www.ict-emphatic.eu/images/deliverables/deliverable_d3.1_final.pdf.

[28] D. Kong, D. Chen, D. Qu, and T. Jiang, "Preamble optimization based on frequency domain averaging for channel estimation in OQAM-OFDM systems," in *Proc. WCSP-2015*, Nanjing, China, Oct. 2015.

[29] W. Liu, D. Chen, D. Kong, and T. Jiang, "Preamble overhead reduction with IAM-C for channel estimation in OQAM-OFDM systems," in *Proc. ChinaSIP-2015*, Chengdu, China, Jul. 2015.

[30] D. Katselis, E. Kofidis, A. Rontogiannis, and S. Theodoridis, "Preamble-based channel estimation for CP-OFDM and OFDM/OQAM systems: A comparative study," *IEEE Trans. Signal Process.*, vol. 58, no. 5, pp. 2911–2916, May 2010.

[31] V. Savaux and F. Bader, "Mean square error analysis and linear minimum mean square error application for preamble-based channel estimation in orthogonal frequency division multiplexing/offset quadrature amplitude modulation systems," *IET Commun.*, vol. 9, no. 14, pp. 1763–1773, Sep. 2015. http://dx.doi.org/10.1049/iet-com.2014.1181.

[32] V. Savaux, F. Bader, and Y. Louët, "A joint MMSE channel and noise variance estimation for OFDM/OQAM modulation," *IEEE Trans. Commun.*, vol. 63, no. 11, pp. 4254–4266, Nov. 2015.

[33] V. Savaux, Y. Louët, and F. Bader, "Low-complexity approximations for LMMSE channel estimation in OFDM/OQAM," in *Proc. 23rd Int. Conf. Telecommunications (ICT-2016)*, Thessaloniki, Greece, May 2016.

[34] O. Edfors, M. Sandell, J.-J. van de Beek, S. K. Wilson, and P. O. Börjesson, "OFDM channel estimation by singular value decomposition," *IEEE Trans. Commun.*, vol. 46, no. 7, pp. 931–939, Jul. 1998.

[35] TS 36.104 V8.2. "3rd Generation Partnership Project; Technical Specification Group Radio Access Network; Evolved Universal Terrestrial Radio Access (E-UTRA), Base

Station (BS) radio transmission and reception (Release 8), 3GPP, 2008. [Online]. Available: http://www.arib.or.jp/IMT-2000/V700Sep08/5_Appendix/Rel8/36/36104-820.pdf.

[36] X. Mestre, M. Sánchez-Fernández, and A. Pascual-Iserte, "Characterization of the distortion of OFDM/OQAM modulations under frequency selective channels," in *Proc. EUSIPCO-2012*, Bucharest, Romania, Sep. 2012.

[37] "Enhanced Multicarrier techniques for Professional Ad-hoc and cell-based Communications (EMPhAtiC)." FP7 ICT project. [Online]. Available: http://www.ict-emphatic.eu/.

[38] B. Yang, K. B. Letaief, R. S. Cheng, and Z. Cao, "Channel estimation for OFDM transmission in multipath fading channels based on parametric channel modeling," *IEEE Trans. Commun.*, vol. 49, no. 3, pp. 467–479, Mar. 2001.

[39] G. Garbo, S. Mangione, and V. Maniscalco, "Orthogonal multicarrier transmission with modal channel estimation," in *Proc. ICT-2009*, Marrakesh, Morocco, May 2009.

[40] E. Kofidis, "Preamble-based channel estimation in OFDM/OQAM systems: A time-domain approach," Jan. 2014. arXiv:1306.2581v2 [cs.IT].

[41] E. Kofidis, "Short preamble-based estimation of highly frequency selective channels in FBMC/OQAM," in *Proc. ICASSP-2014*, Florence, Italy, May 2014.

[42] E. Kofidis, "Channel estimation in filter bank-based multicarrier systems: Challenges and solutions," in *Proc. ISCCSP-2014*, Athens, Greece, May 2014.

[43] E. Kofidis, "Preamble-based estimation of highly frequency selective channels in FBMC/OQAM systems," *IEEE Trans. Signal Process.*, vol. 65, no. 7, pp. 1855–1868, Apr. 2017.

[44] D. Kong, D. Qu, and T. Jiang, "Time domain channel estimation for OQAM-OFDM systems: Algorithms and performance bounds," *IEEE Trans. Signal Process.*, vol. 62, no. 2, pp. 322–330, Jan. 2014.

[45] R. M. Gray, *Toeplitz and Circulant Matrices: A Review*, Foundations and Trends in Communications and Information Theory, 2005 (now publ.).

[46] D. Katselis, M. Bengtsson, C. R. Rojas, H. Hjalmarsson, and E. Kofidis, "On preamble-based channel estimation in OFDM/OQAM systems," in *Proc. EUSIPCO-2011*, Barcelona, Spain, Sep. 2011.

[47] S. M. Kay, *Fundamentals of Statistical Signal Processing, Vol. I: Estimation Theory*. Prentice-Hall, 1993.

[48] D. Katselis, "Some preamble design aspects in CP-OFDM systems," *IEEE Commun. Lett.*, vol. 16, no. 3, pp. 356–359, Mar. 2012.

[49] Y. Zhao, X. Chen, L. Xue, J. Liu, and Z. Xie, "Iterative preamble-based time domain channel estimation for OFDM/OQAM systems," *IEICE Trans. Commun.*, vol. E99-B, no. 10, pp. 2221–2227, Oct. 2016.

[50] L. G. Baltar, A. Mezghani, and J. A. Nossek, "EM based per-subcarrier ML channel estimation for filter bank multicarrier systems," in *Proc. 10th Int'l Symp. Wireless Communication Systems (ISWCS-2013)*, Ilmenau, Germany, Aug. 2013.

[51] L. G. Baltar, M. Newinger, and J. A. Nossek, "Structured subchannel impulse response estimation for filter bank based multicarrier systems," in *Proc. ISWCS-2012*, Paris, France, Aug. 2012.

[52] L. G. Baltar, T. Laas, M. Newinger, A. Mezghani, and J. A. Nossek, "Enhancing spectral efficiency in advanced multicarrier techniques: A challenge," in *Proc. EUSIPCO-2014*, Lisbon, Portugal, Sep. 2014.

[53] E. Kofidis, "On optimal multi-symbol preambles for highly frequency selective FBMC/OQAM channel estimation," in *Proc. ISWCS-2015*, Brussels, Belgium, Aug. 2015.

[54] M. Diallo and M. Hélard, "Channel estimation methods with low complexity for 3GPP/LTE," *Rev. Afr. Rech. Inform. Math. Appl. (ARIMA)*, vol. 18, pp. 93–116, 2014. [Online]. Available: http://arima.inria.fr/018/pdf/vol.18.pp.93-113.pdf.

[55] F. Rottenberg, Y. Medjahdi, E. Kofidis, and J. Louveaux, "Preamble-based channel estimation in asynchronous FBMC-OQAM distributed MIMO systems," in *Proc. ISWCS-2015*, Brussels, Belgium, Aug. 2015.

[56] M. Diallo, M. Hélard, L. Cariou, and R. Rabineau, "DFT based channel estimation methods for MIMO-OFDM systems," in *Vehicular Technologies: Increasing Connectivity*, M. Almeida, Ed., InTech, 2011. [Online]. Available: http://cdn.intechweb.org/pdfs/15300.pdf.

[57] L. G. Baltar, A. Mezghani, and J. A. Nossek, "Spectral efficient channel estimation algorithms for FBMC/OQAM systems," in *Proc. 11th Int. Symp. Wireless Communication Systems (ISWCS-2014)*, Barcelona, Spain, Aug. 2014.

[58] X. He, Z. Zhao, and H. Zhang, "A pilot-aided channel estimation method for FBMC/OQAM communications system," in *Proc. Int. Symp. Communications and Information Technologies (ISCIT-2012)*, Gold Coast, Australia, Oct. 2012.

[59] T. H. Stitz, T. Ihalainen, A. Viholainen, and M. Renfors, "Pilot-based synchronization and equalization in filter bank multicarrier communications," *EURASIP J. Appl. Signal Process.*, 2010. http://dx.doi.org/10.1155/2010/741429.

[60] F. Bader and M. Shaat, "Pilot pattern adaptation and channel estimation in MIMO WiMAX-like FBMC system," in *Proc. 6th Int. Conf. Wireless and Mobile Communications (ICWMC-2010)*, Valencia, Spain, Sep. 2010.

[61] F. Bader and M. Shaat, "Pilot pattern design for PUSC MIMO WiMAX-like filter banks multicarrier system," *Int. J. Adv. Telecommun.*, vol. 4, no. 1–2, pp. 156–165, 2011.

[62] J.-M. Choi, Y. Oh, H. Lee, and J.-S. Seo, "Interference-dependent pair of pilots for channel estimation in FBMC systems," in *11th Int. Symp. Broadband Multi-media Systems and Broadcasting (BMSB-2016)*, Nara, Japan, Jun. 2016.

[63] C. Lélé, R. Legouable, and P. Siohan, "Channel estimation with scattered pilots in OFDM/OQAM," in *Proc. SPAWC-2008*, Recife, Brazil, Jul. 2008.

[64] W. Cui, D. Qu, T. Jiang, and B. Farhang-Boroujeny, "Coded auxiliary pilots for channel estimation in FBMC-OQAM systems," *IEEE Trans. Veh. Technol.*, vol. 65, no. 5, pp. 2936–2948, May 2016.

[65] R. Nissel and M. Rupp, "On pilot-symbol aided channel estimation in FBMC-OQAM," in *Proc. ICASSP-2016*, Shanghai, China, Mar. 2016.

[66] C. Lélé, R. Legouable, and P. Siohan, "Iterative scattered pilot channel estimation in OFDM/OQAM," in *Proc. SPAWC-2009*, Perugia, Italy, Jun. 2009.

[67] C. Lélé, "Iterative scattered-based channel estimation method for OFDM/OQAM," *EURASIP J. Adv. Signal Process.*, vol. 42, 2012.

[68] M. Aldababseh and A. Jamoos, "Estimation of FBMC/OQAM fading channels using dual Kalman filters," *Sci. World J.*, 2014. http://dx.doi.org/10.1155/2014/586403 [Online].

[69] B. Lahami, M. Siala, and I. Kammoun, "Optimization of training-based channel estimation for OFDM/OQAM systems operating on highly time-frequency dispersive channels," in *Proc. 2nd Int. Conf. Communications and Networking (ComNet-2010)*, Tozeur, Tunisia, Nov. 2010.

[70] X. Mestre and E. Kofidis, "Pilot-based channel estimation for FBMC/OQAM systems under strong frequency selectivity," in *Proc. ICASSP-2016*, Shanghai, China, Mar. 2016.

[71] S. Coleri, M. Ergen, A. Puri, and A. Bahai, "Channel estimation techniques based on pilot arrangement in OFDM systems," *IEEE Trans. Broadcast.*, vol. 48, no. 3, pp. 223–229, Sep. 2002.

[72] J. Bazzi, P. Weitkemper, and K. Kusume, "Power efficient scattered pilot channel estimation for FBMC/OQAM," in *Proc. 10th Int. ITG Conf. Systems, Communications and Coding (SCC-2015)*, Hamburg, Germany, Feb. 2015.

[73] Y. J. Harbi and A. G. Burr, "Wiener filter channel estimation for OFDM/OQAM with iterative interference cancellation in LTE channel," in *Proc. 19th Int'l Conf. OFDM and Frequency Domain Techniques (ICOF-2016)*, Essen, Germany, Aug. 2016.

[74] M. Fuhrwerk, S. Moghaddamnia, and J. Peissig, "Scattered pilot-based channel estimation for channel adaptive FBMC-OQAM systems," *IEEE Trans. Wireless Commun.*, vol. 16, no. 3, pp. 1687–1702, Mar. 2017.

[75] Z. Zhao, N. Vucic, and M. Schellmann, "A simplified scattered pilot for FBMC/OQAM in highly frequency selective channels," in *Proc. ISCWS-2014*, Barcelona, Spain, Aug. 2014.

[76] R. Nissel and M. Rupp, "Bit error probability for pilot-symbol aided channel estimation in FBMC-OQAM," in *Proc. ICC-2016*, Kuala Lumpur, Malaysia, May 2016.

[77] R. Nissel, S. Caban, and M. Rupp, "Experimental evaluation of FBMC-OQAM channel estimation based on multiple auxiliary symbols," in *Proc. SAM-2016*, Rio de Janeiro, Brazil, Jul. 2016.

[78] B. Yu, S. Hu, P. Sun, S. Chai, C. Qian, and C. Sun, "Channel estimation using dual-dependent pilots in FBMC/OQAM systems," *IEEE Commun. Lett.*, vol. 20, no. 11, pp. 2157–2160, Nov. 2016. https://doi.org/10.1155/2017/6784142.

[79] Y.-J. Won, J.-G. Oh, J.-S. Lee, and J.-T. Kim, "A study of an iterative channel estimation scheme of FS-FBMC system," *Wirel. Commun. Mob. Comput.*, 2017.

[80] L. Tong and S. Perreau, "Multichannel blind identification: From subspace to maximum likelihood methods," *Proc. IEEE*, vol. 86, no. 10, pp. 1951–1968, Oct. 1998.

[81] H. Bölcskei, P. Duhamel, and R. Hleiss, "A subspace-based approach to blind channel identification in pulse shaping OFDM/OQAM systems," *IEEE Trans. Signal Process.*, vol. 49, no. 7, pp. 1594–1598, Jul. 2001.

[82] W. Hou and B. Champagne, "Semiblind channel estimation for OFDM/OQAM systems," *IEEE Signal Process. Lett.*, vol. 22, no. 4, pp. 400–403, Apr. 2015.

[83] T. Su, "Adaptive semiblind channel estimation for OFDM/OQAM systems." MEng thesis. Montreal, Canada: McGill University, Sep. 2015. [Online]. Available: http://www.ece.mcgill.ca/~bchamp/Theses/MEngTheses/Su2015.pdf.

[84] B. Su, "Semiblind channel estimation for OFDM/OQAM systems assisted by zero-valued pilots," in *Proc. DSP-2015*, Singapore, Jul. 2015.

[85] R. Maoudji, I. Ahriz, A. Savarit, L. Féty, and M. Terré, "4th order statistics based blind channel estimation for multicarrier transmission," in *Proc. ICT-2014*, Lisbon, Portugal, May 2014.

[86] M. Renfors, F. Bader, L. Baltar, D. Le Ruyet, D. Roviras, P. Mège, M. Haardt, and T. H. Stitz, "On the use of filter bank based multicarrier modulation for professional mobile radio," in *Proc. VTC-2013 (Spring)*, Dresden, Germany, Jun. 2013.

[87] X. Fang, Y. Xu, Z. Chen, and F. Zhang, "Frequency-domain channel estimation for polarization-division-multiplexed CO-OFDM/OQAM systems," *J. Lightw. Technol.*, vol. 33, no. 13, pp. 2743–2750, Jul. 2015.

[88] D. Kong, X.-G. Xia, T. Jiang, and X. Gao, "Channel estimation in CP-OQAM-OFDM systems," *IEEE Trans. Signal Process.*, vol. 62, no. 21, pp. 5775–5786, Nov. 2014.

[89] F. Rottenberg, F. Horlin, E. Kofidis, and J. Louveaux, "Generalized optimal pilot allocation for channel estimation in multicarrier systems," in *Proc. SPAWC-2016*, Edinburgh,
UK, Jul. 2016.

[90] J. Wang, H. Zhao, Z. Yuyan, F. Li, and L. Zhao, "Intrinsic interference elimination for
preamble-based channel estimation in FBMC systems," in *Globecom Workshops 2016*,
Washington, DC, Dec. 2016.

[91] Y. Zeng and M. W. Chia, "Joint time-frequency synchronization and channel estimation
for FBMC," in *Proc. PIMRC-2013*, Washington, DC, Sep. 2014.

[92] S. Nedic, *et al.*, "Adaptive equalization and Successive self-Interference Cancellation
(SIC) methods." EMPhAtiC, Deliverable 3.2, May 2014. [Online]. Available: http://www.
ict-emphatic.eu/images/deliverables/deliverable_d3.2_final.pdf.

[93] G. Taubock, F. Hlawatsch, D. Eiwen, and H. Rauhut, "Compressive estimation of doubly selective channels in multicarrier systems: Leakage effects and sparsity-enhancing
processing," *IEEE J. Sel. Topics Signal Process.*, vol. 4, no. 2, pp. 255–271, Apr. 2010.

[94] X. Liu, Z. Cai, A. Jia, and Z. Ou, "A novel channel estimation method based on compressive sensing for OFDM/OQAM systems," *J. Comput. Inf. Syst.*, vol. 9, no. 15,
pp. 5955–5963, Aug. 2013.

[95] H. Wang, W. Du, and L. Xu, "A new sparse adaptive channel estimation method
based on compressive sensing for FBMC/OQAM transmission network," *Sensors*, 2016.
http://dx.doi.org/10.3390/s16070966.

[96] X. Liu, X. Chen, L. Xue, and Z. Xie, "Channel estimation of OQAM/OFDM based on
compressed sensing," *IEICE Trans. Commun.*, 2016, article ID: 2016EBP3280.

# FBMC Channel Equalization Techniques

# 12

Leonardo Gomes Baltar*, Pascal Chevalier[†], Markku Renfors[‡],
Juha Yli-Kaakinen[‡], Jérôme Louveaux[§], Xavier Mestre[¶], Faouzi Bader[‖],
Vincent Savaux[**]

*Intel Deutschland GmbH, Neubiberg, Germany*
*CEDRIC Laboratory, CNAM, Paris, France[†]*
*Tampere University of Technology, Tampere, Finland[‡]*
*Université catholique de Louvain, Louvain-la-Neuve, Belgium[§]*
*Centre Tecnològic de Telecomunicacions de Catalunya (CTTC/CERCA), Barcelona, Spain[¶]*
*CentraleSupélec, Rennes, France[‖]*
*b<>com, Cesson-Sévigné, France[**]*

## CONTENTS

Orthogonal Waveforms and Filter Banks for Future Communication Systems. DOI: 10.1016/B978-0-12-810384-5.00012-8

**299**

## 12.1 INTRODUCTION

One of the main advantages of MultiCarrier Modulation (MCM) schemes for broadband wireless communications is their robustness to multipath propagation[1] channels, stemming from the fact that they divide the channel spectrum into very narrow subbands, and, in the extreme case, no frequency selectivity, i.e., only flat fading, is observed in each of them. For an increased spectral efficiency, most practical MCM schemes have their subbands overlapped in frequency. In Cyclic Prefix Orthogonal Frequency-Division Multiplexing (CP-OFDM), Inter-Symbol Interference (ISI) and Inter-Carrier Interference (ICI) can be completely removed if the Cyclic Prefix (CP) is at least as long as the channel delay spread. Thus, at the expense of reducing the spectral efficiency due to the CP redundancy, the subchannels corresponding to the different subbands are completely decoupled. The equalization in CP-OFDM then becomes trivial and can be performed by a single complex multiplication per subcarrier, giving rise to the so-called single-tap equalizer. Usually, this is of the Zero Forcing (ZF) type, which directly inverts the frequency response of the channel at each subcarrier. More sophisticated equalizers, such as the Minimum Mean-Squared Error (MMSE) one, are adopted if the noise and channel statistics are known or estimated.

FilterBank MultiCarrier with Offset-QAM subcarrier modulation (FBMC/OQAM) systems do not have to employ a CP, and they enjoy (real-field) orthogonality in ideal propagation scenarios. This also means that full orthogonality exists by considering the Quadrature Amplitude Modulation (QAM) symbols before the Offset Quadrature Amplitude Modulation (OQAM) staggering at the Synthesis FilterBank (SFB) and after OQAM destaggering at the Analysis FilterBank (AFB), as originally proved for Perfect Reconstruction (PR) Modified Discrete Fourier Transform (MDFT) FilterBank (FB) [1] and later for FBMC/OQAM systems [2]. In other words, the so-called self-interference can easily be removed by the OQAM destaggering. For realistic propagation scenarios, where channel distortions are present, the symbols received at the AFB output are contaminated by both ISI and ICI, which are channel induced. With mildly frequency selective channels, a single-tap equalizer, like that presented in Section 12.2, should be sufficient to compensate for the channel effects and minimize both kinds of interference. However, with moderate to highly frequency selective channels, more elaborate equalizers have to be used, which will

---

[1] We mainly refer to wireless channels here although MCM and most of the solutions presented in this chapter could also be employed in broadband wireline systems, e.g., in Digital Subscriber Line (DSL), power lines, and fiber optics communications, where frequency selective channels are also involved.

also increase the receiver complexity. Such equalizers can be designed and implemented in the time or frequency domain, and they have the ability to compensate also for time and phase shifts.

Another advantage of FBMC/OQAM systems, apart from the CP-free transmission, is that they provide a flexibility in choosing the subcarrier spacing. In the cases where a higher symbol rate per subcarrier is necessary to reduce latency or frequency offset/phase noise are relevant impairments, a higher subcarrier bandwidth (large subcarrier spacing) can be considered. The consequences of this include an increase in the complexity of the per-subcarrier equalization due to the larger number of taps required. In other cases where a higher granularity in frequency domain is desired and a higher Peak-to-Average Power Ratio (PAPR) can be tolerated,[2] a longer symbol duration can be accepted. This results in narrower subbands (smaller subcarrier spacing) and the possibility to employ low complexity equalizers, the single-tap one being the simplest possible.

The compensation of the effects of multipath propagation in FBMC/OQAM systems was first presented in [3], where it was shown that it is possible to completely eliminate ISI and ICI, and to compensate for time and carrier phase deviations, if a per-subcarrier $T/2$-spaced[3] equalizer with a sufficient number of taps is employed. Here $T$ is the symbol period. The equalizer coefficients are computed using an MMSE steepest descent adaptive algorithm. The analytical solution for the multi-tap equalizer presented in Section 12.3.1 shares many objectives and properties with that in [3]. It is worth mentioning that in [3] two structures for the implementation of the fractionally spaced equalizer are introduced, which correspond to the equalizer operating at the $2/T$ or $1/T$ sampling rate. Much later, in [4], both fractionally and symbol-spaced adaptive steepest descent equalizers were proposed. In the nonfractionally spaced case, three equalizers per subcarrier are employed to remove ISI and ICI. These equalizers are placed after the OQAM destaggering and combine the output of the subcarrier of interest with its neighbors. In the fractionally spaced case, also three equalizers per subcarrier are employed, and two variants are provided: With and without the OQAM destaggering in the adaptation loop. In [5], a combined equalization and echo cancellation solution was presented, where a fractionally spaced Finite Impulse Response (FIR) filter for the equalization and another for the echo cancellation part are employed. For the latter, a pre-processing before the FIR filter is included to emulate the SFB- and AFB-equivalent response. The per-subcarrier equalization for odd-stacked FBMC/OQAM systems was revisited in [6], where specific equalizer structures were presented to compensate different levels of frequency selectivity. In [7], an equalizer similar to that presented in Section 12.3.1 was de-

---

[2] It should be noted that, while the subcarrier spacing directly affects the number of active subcarriers in the used frequency band, also the PAPR properties (see Chapter 18) become an important concern in the FBMC/OQAM system dimensioning.

[3] We refer here to a fractionally spaced equalizer because its inputs are the signals before the OQAM destaggering or complex symbols before the Pulse Amplitude Modulation (PAM) demodulation, depending on which subcarrier model is employed.

signed so as to cope with ICI from all subcarriers and channel time selectivity. An evaluation of the spectral efficiency as a function of the time and frequency spread was performed and showed that the MMSE multitap equalizer significantly increases spectral efficiency. To improve the robustness to ISI and ICI, a combination of the Walsh–Hadamard transform with FBMC/OQAM was proposed in [8] (see also [20] and [16]). The effect of the transform is to spread the symbols over all subcarriers, resulting in frequency diversity. An MMSE equalizer was employed at the receiver. It is worth mentioning here that frequency diversity can also be achieved by combining bit-interleaved channel coding and the equalization schemes presented in this chapter. The authors in [9] performed an analysis of ISI and ICI and proposed a new equalizer structure that uses the interference effect in a positive way. A single-tap ZF equalizer before OQAM destaggering is combined with an interference estimation and cancellation scheme applied after the destaggering on a per-subcarrier basis.

For mildly frequency selective channels, the classical single-tap ZF equalizer, applied before the OQAM destaggering, was compared to two alternative equalizers in [10]: a dispersion receiver, where the AFB is designed to match the combination of SFB and channel, and ZF equalization is then applied, and an interference-free receiver, which preprocesses the received signal before the AFB, to transform the equivalent channel to one with purely real or imaginary Channel Frequency Response (CFR), thus allowing the interference to be completely eliminated. It was shown that all three receivers behave similarly for mildly frequency selective channels. If the channel dispersion increases, the first one presents better performance than the other two. In [11], also for mildly frequency selective channels and considering channel coding, a method for calculating the Log-Likelihood Ratio (LLR) values was derived for the specific OQAM signaling in FilterBank MultiCarrier (FBMC) systems when single-tap ZF equalizers are employed. More recently, a single-tap equalizer that maximizes the Signal-to-Interference Ratio (SIR) was derived in [12], again for mildly frequency selective channels. It was shown that the maximum SIR criterion leads to improved performance compared to the ZF one. The authors of [13] propose to jointly design the per-subchannel equalizer and the AFB prototype filter, based on maximizing Signal-to-Interference-plus-Noise Ratio (SINR). An iterative two-step approach is followed, where in the first step the equalizer and in the second step the AFB prototype filter are optimized in an alternating manner. In [14], for mildly doubly selective channels and single-tap equalizers, the authors derive Bit Error Probability (BEP) expressions to compare FBMC/OQAM with OFDM. For channels with strong frequency selectivity, [15] extends previous results on MMSE decision-feedback equalization by including two ICI-suppressing filters at each subcarrier.

It should be noted that many of the equalizers discussed in this chapter, along with some other solutions, were studied within the Framework Programme 7 (FP7) Information and Communications Technology (ICT) projects PHYDYAS [17,18] and EMPhAtiC [19,20] (see also the references therein). With the exception of Sections 12.2 and 12.6, the focus is this chapter will be on equalizers for highly frequency selective channels.

Widely Linear Processing (WLP) results, in general, in better performance for OQAM-based systems than strictly linear processing. For this reason, this chapter includes an introduction to WLP and a literature review of how it can be applied to wireless communications and, more specifically, to FBMC/OQAM systems (Section 12.3.2). In Section 12.4, similarly as in Sections 12.3.1 and 12.3.3, another receiver structure specially made for highly frequency selective channels is presented. It involves multiple AFBs and equalizers operating in parallel. In Section 12.5, equalizer solutions that are specific for Fast-Convolution FilterBank (FC-FB) are presented and evaluated. The FC-FB scheme allows us to efficiently realize flexible MultiCarrier (MC) systems, where different subcarriers can have different bandwidths and data rates. Finally, Section 12.6 discusses blind equalizer solutions for FBMC/OQAM systems, where no channel knowledge or training sequences are involved. Section 12.7 concludes the chapter, outlining some possible directions for future research in this area.

A large part of the definitions, notations, and system models employed in this chapter are described in detail in Chapters 9 and 11. Hence, we recommend the reader first to consult those chapters, especially, Section 11.2.

## 12.2 SINGLE-TAP EQUALIZERS

For mildly frequency selective channels or, in other words, transmission scenarios where the subcarrier bandwidth is small compared to the coherence bandwidth of the channel, a single-tap complex equalizer per subcarrier can be employed, similarly to CP-OFDM systems. Moreover, it can be assumed, as it is common in MC systems, that the coherence time of the channel covers at least one MC symbol. With these assumptions, FBMC/OQAM systems have a similar complexity for their equalization with conventional CP-OFDM with the difference that the pulse shaping and the OQAM scheme modify the system model. This needs to be taken into account in order not to have a poor equalization performance. ISI and ICI limit the performance of the ZF equalizer. Hence, in the following, a single-tap MMSE equalizer is developed so as to take into account the effect of the pulse shaping and the OQAM modulation properties.

The pulse shaping at the $m$th subcarrier, $m = 0, 1, \ldots, M - 1$, is given by (see also Section 11.2)

$$g_m[l] = g[l] e^{j\frac{2\pi}{M}m\left(l - \frac{L_g-1}{2}\right)}, \quad l = 0, 1, \ldots, L_g - 1, \quad (12.1)$$

where $g[l]$ is the (symmetric and unit energy) prototype filter impulse response of typical length $L_g = KM$, $M$ is the (even) number of subcarriers, and $K$ is the overlapping factor. The latter should be kept as small as possible so as not only to limit the complexity but also to reduce the time-domain spreading of the symbols and the transceiver latency. A typical value is $K = 4$, and the roll-off factor of the prototype filter is usually set to one. Thus, $g_m[l]$ has a non-negligible overlap in frequency only with its two adjacent filters (see also Table 11.1). Observe that the subcarrier filter response in (12.1) is the paraconjugate of itself, namely, $g_m^*[L_g - 1 - l] = g_m[l]$.

The subcarrier model commonly adopted for a single-tap equalizer design is described in Section 11.2 and relies on the assumption that the channel is sufficiently slowly varying in frequency and time that its CFR is (almost) invariant over the first-order Time-Frequency (T-F) neighborhood of the given Frequency-Time (F-T) point $(p, q)$. Then Eq. (11.6) applies, and the AFB output at subcarrier $p$ and time instant $q$ can be written as

$$y_{p,q} \approx H_{p,q}(d_{p,q} + j u_{p,q}) + \eta_{p,q},$$

where the PAM symbol $d_{p,q}$ contains the information of interest, $j u_{p,q}$ stands for the interference contributed by its T-F neighbors, and $\eta_{p,q}$ is the corresponding noise component. Consider the complex-valued single-tap equalizer $w_{p,q}$. The real part of its output is taken to yield an estimate of $d_{p,q}$[4]:

$$\tilde{d}_{p,q} = \mathrm{Re}\{w_{p,q} y_{p,q}\} = w_{p,q}^{(R)} y_{p,q}^{(R)} - w_{p,q}^{(I)} y_{p,q}^{(I)}$$
$$= \bar{\mathbf{w}}_{p,q}^{T} \left( \bar{\mathbf{H}}_{p,q} \check{\mathbf{c}}_{p,q} + \bar{\boldsymbol{\eta}}_{p,q} \right), \qquad (12.2)$$

where

$$\check{\mathbf{c}}_{p,q} = \begin{bmatrix} d_{p,q} & u_{p,q} \end{bmatrix}^{T},$$

$$\bar{\mathbf{H}}_{p,q} = \begin{bmatrix} H_{p,q}^{(R)} & -H_{p,q}^{(I)} \\ H_{p,q}^{(I)} & H_{p,q}^{(R)} \end{bmatrix},$$

$$\bar{\mathbf{w}}_{p,q} = \begin{bmatrix} w_{p,q}^{(R)} & w_{p,q}^{(I)} \end{bmatrix}^{T},$$

$$\bar{\boldsymbol{\eta}}_{p,q} = \begin{bmatrix} \eta_{p,q}^{(R)} & \eta_{p,q}^{(I)} \end{bmatrix}^{T}$$
$$= \begin{bmatrix} \mathbf{g}_p^{(R)} & \mathbf{g}_p^{(I)} \\ -\mathbf{g}_p^{(I)} & \mathbf{g}_p^{(R)} \end{bmatrix}^{T} \begin{bmatrix} \mathbf{v}^{(R)} \\ \mathbf{v}^{(I)} \end{bmatrix} = \bar{\mathcal{G}}_p \bar{\mathbf{v}},$$

with $\mathbf{g}_p \in \mathbb{C}^{L_g \times 1}$ containing the coefficients of the analysis filter impulse response of the $p$th subcarrier and $\mathbf{v} \in \mathbb{C}^{L_g \times 1}$ containing samples of the noise at the channel output. The latter is assumed to be zero-mean Gaussian with variance $\sigma_v^2$. The AFB output noise $\eta$ is then also Gaussian with zero mean and variance $\sigma_\eta^2 = \sigma_v^2$. See Section 11.2 for more details on the system model.

The ZF equalizer is a first and straightforward solution for single-tap equalization, which has been extensively employed in the literature [10]. Similarly to CP-OFDM,

---

[4]The decomposition into real-valued vectors and matrices shown here was first introduced in the context of OQAM equalizers in [21] for Single-Carrier (SC) systems.

the ZF equalizer is given by $w_{p,q} = 1/H_{p,q}$. But it may not be the best possible option to be employed here because of the existing interference, which strongly limits its performance. Another solution frequently used in practice[5] is the Linear Minimum Mean Squared Error (LMMSE) equalizer, which must take into account the statistics of the interference and of the noise, in addition to the properties of the OQAM signaling. The latter solution is given by

$$\bar{w}_{p,q}^{\text{opt}} = \arg\min_{\bar{w}_{p,q}} \mathbb{E}\left\{|\tilde{d}_{p,q} - d_{p,q}|^2\right\}$$

$$= \left(\bar{H}_{p,q}\begin{bmatrix} 1 & 0 \\ 0 & \frac{\sigma_u^2}{\sigma_d^2} \end{bmatrix}\bar{H}_{p,q}^{\text{T}} + \frac{\sigma_v^2}{2\sigma_d^2}\bar{\mathcal{G}}_p\bar{\mathcal{G}}_p^{\text{T}}\right)^{-1}\begin{bmatrix} H_{p,q}^{(\text{R})} \\ H_{p,q}^{(\text{I})} \end{bmatrix},$$

where $\sigma_d^2$ and $\sigma_u^2$ are the variances of the input symbols and the interference term, respectively. Recalling the pattern of interference from Eq. (11.9) and assuming the input symbols to be independent and identically distributed (i.i.d.), we can write $\sigma_u^2 \approx 2\sigma_d^2(\gamma^2 + \beta^2 + 2\delta^2)$, where $\gamma$, $\beta$, and $\delta$ only depend on the prototype filter. The complex single-tap equalizer is then obtained by destacking the two components of $\bar{w}_{p,q}$ into its real and imaginary parts.

## 12.3 MULTITAP EQUALIZERS

### 12.3.1 CLASSICAL TECHNIQUES

Multitap equalizers can be divided into two basic categories depending on how they are computed and realized. On a per-subcarrier basis, equalizer design and implementation can be performed both in the time and frequency domains. Moreover, the computation of the equalizer coefficients does not impose a specific structure to be adopted. This also means that, during the design step, a specific structure can be assumed, which is not exactly the implemented one. For example, in [6], the authors base the design of the equalizers on the frequency responses of the channel and the FB prototype, but the implementation is primarily based on time domain real- and complex-valued FIR filters. On the other hand, in the Frequency Spreading FBMC/OQAM (FS-FBMC) structure [22] or in the equalizers presented in Section 12.5, multiple frequency bins per subcarrier are assumed, and the equalizer is designed in the frequency domain and implemented as one complex multiplier per frequency bin. Clearly, one multiplier per frequency bin is not the same as a single tap per subcarrier presented previously and can also address frequency selectivity at the subchannel level. The frequency domain design can also follow ZF or MMSE criteria.

---

[5] Although the ZF is the mostly commonly encountered solution in the CP-OFDM literature, in practical implementations, such as in Long-Term Evolution (LTE), it is the MMSE that is more often adopted.

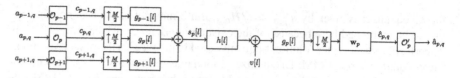

**FIGURE 12.1**

Subcarrier model for the multitap equalizer design.

This section focuses on classical Tap Delay Line (TDL) structure-based design. One or more equalizers per subcarrier are represented as FIR filters, and their coefficients are then computed. It is worth noting that a TDL or any other specific structure necessarily needs to be implemented, that is, other FIR filter structures, such as polyphase, nonrecursive lattice, CORDIC-based, etc., can be employed in the equalizer operation. Moreover, the design that follows can be also effectively implemented in the frequency domain, as long as the number of frequency bins is the same as the number of equalizer taps in the time domain.

In an FBMC/OQAM system, the SFB combines the $M_u$ complex-valued QAM input signals $a_{m,n} = d_{m,2n} + jd_{m,2n-1}$, $m = 0, 1, \ldots, M_u - 1$, generated at a rate of $1/T$, into a single complex-valued signal $s[l]$ of a higher sampling rate $1/T_s = M/T$. The signal is transmitted to the receiver through a frequency selective channel, and white Gaussian noise is added. In our system, $M$ corresponds to the total number of subcarriers available, and $M_u$ is the number of subcarriers used for transmission. The AFB separates the received signal back into its $M_u$ components at the lower rate $1/T$ per subcarrier. A subcarrier model that can be used to design multitap equalizers is given in Fig. 12.1. $\mathcal{O}_p$ and $\mathcal{O}'_p$ represent the staggering and destaggering operations, respectively.

The first operation in the SFB is the OQAM staggering of each input $a_{m,n}$ that generated the output sequence $\mathbf{c}_{m,n}$ as

$$
\mathbf{c}_{m,n} = \begin{cases} \begin{bmatrix} d_{m,n} & jd_{m,n-1} & d_{m,n-2} & \cdots \end{bmatrix}^{\mathrm{T}}, & m+n \text{ is odd}, \\ \begin{bmatrix} jd_{m,n} & d_{m,n-1} & jd_{m,n-2} & \cdots \end{bmatrix}^{\mathrm{T}}, & m+n \text{ is even}. \end{cases}
$$

The input symbol $a_{m,n}$ is split into its real $a_{m,n}^{(\mathrm{R})} = d_{m,2n}$ and j times its imaginary $ja_{m,n}^{(\mathrm{I})} = jd_{m,2n-1}$ parts and up-sampled by a factor of two. Then, depending on the parity of the subcarrier index that we observe, either $d_{m,n}$ or $jd_{m,n}$ is delayed by exactly $T/2$ samples, and finally these components are added together. At the receiver, the AFB applies OQAM destaggering to reconstruct the complex QAM symbols $\hat{a}_{p,q}$ from the equalizer outputs $\hat{c}_{p,2q}$ and $\hat{c}_{p,2q-1}$ at the observed subcarrier $p$ and time slot $q$.

To have a simple notation for the system in Fig. 12.1, we define the following filtering and downsampling operation $\tilde{g}_{p,m}[q] = (g_p \star h \star g_m)[l] \,|_{l=q\frac{M}{2}}$. This represents the overall impulse response from subcarrier $m$ at the Transmitter (TX) to subcarrier $p$ at the Receiver (RX) with $m \in \{p-1, p, p+1\}$. The resulting filter has $L_{\tilde{g}} = \left\lceil \frac{2L_g + L_h - 2}{M/2} \right\rceil$ coefficients. Moreover, we define the transposed convolution matrix $\mathbf{G}_{p,m} \in \mathbb{R}^{L_{eq} \times (L_{\tilde{g}} + L_{eq} - 1)}$ generated from the impulse response $\tilde{g}_{p,m}[q]$, where $L_{eq}$ is the number of taps of the equalizer. Furthermore, we assume the channel impulse response $h[l]$ or an estimate of it to be available.

Define the purely real input sequence $\mathbf{d}_{m,n} = \begin{bmatrix} d_{m,n} & d_{m,n-1} & \cdots \end{bmatrix}^{\mathrm{T}}$ such that $\mathbf{c}_{m,n} = \mathbf{J}_{m,n} \mathbf{d}_{m,n}$, with

$$
\mathbf{J}_{m,n} = \begin{cases} \mathrm{diag}\begin{bmatrix} 1 & j & 1 & j & \cdots \end{bmatrix}, & m+n \text{ is odd,} \\ \mathrm{diag}\begin{bmatrix} j & 1 & j & 1 & \cdots \end{bmatrix}, & m+n \text{ is even.} \end{cases}
$$

This extracts the imaginary js from the input signal. We then right multiply the transposed convolution matrix of $\tilde{g}_{p,m}[q]$ by $\mathbf{J}_{m,n}$ and obtain $\mathbf{G}'_{p,m} = \mathbf{G}_{p,m} \mathbf{J}_{m,n}$. Hence,

$$
\hat{c}_{p,q} = \bar{\mathbf{w}}_p^{\mathrm{T}} \left( \bar{\mathbf{G}}'_{p,p} \mathbf{d}_{p,q} + \bar{\mathbf{G}}'_{p,p-1} \mathbf{d}_{p-1,q} + \bar{\mathbf{G}}'_{p,p+1} \mathbf{d}_{p+1,q} + \mathcal{G}_p \bar{\mathbf{v}} \right),
$$

where we define

$$
\mathbf{d}_{m,q} \in \mathbb{R}^{(L_{\tilde{g}} + L_{eq} - 1) \times 1},
$$

$$
\bar{\mathbf{G}}'_{p,m} = \begin{bmatrix} (\mathbf{G}'_{p,m})^{(\mathrm{R})} \\ (\mathbf{G}'_{p,m})^{(\mathrm{I})} \end{bmatrix} \in \mathbb{R}^{(2L_{eq}) \times (L_{\tilde{g}} + L_{eq} - 1)}
$$

$$
\text{with } m \in \{p-1, \ p, \ p+1\},
$$

$$
\bar{\mathbf{w}}_p = \begin{bmatrix} \mathbf{w}_p^{(\mathrm{R})} \\ \mathbf{w}_p^{(\mathrm{I})} \end{bmatrix} \in \mathbb{R}^{(2L_{eq}) \times 1},
$$

$$
\mathcal{G}_p = \begin{bmatrix} (\check{\mathbf{G}}_p)^{(\mathrm{R})} & -(\check{\mathbf{G}}_p)^{(\mathrm{I})} \\ (\check{\mathbf{G}}_p)^{(\mathrm{I})} & (\check{\mathbf{G}}_p)^{(\mathrm{R})} \end{bmatrix} \in \mathbb{R}^{(2L_{eq}) \times 2(L_{\tilde{g}} + L_{eq} - 1)}, \text{ and}
$$

$$
\bar{\mathbf{v}} = \begin{bmatrix} \mathbf{v}^{(\mathrm{R})} \\ \mathbf{v}^{(\mathrm{I})} \end{bmatrix} \in \mathbb{R}^{2(L_{\tilde{g}} + L_{eq} - 1) \times 1}.
$$

The matrix $\check{\mathbf{G}}_p$ is obtained by taking every $\frac{M}{2}$th row of the convolution matrix generated from the RX subcarrier impulse response $g_p[l]$.

We make here the usual assumption that the input symbols are i.i.d. and Gaussian distributed although they belong in reality to a discrete alphabet. The covariance matrix of $\mathbf{d}_{m,n}$ is defined as $\mathbb{E}\left[\mathbf{d}_{m,n} \mathbf{d}_{m,n}^{\mathrm{T}}\right] = \sigma_d^2 \mathbf{I}$, and, again, the noise vector $\mathbf{v}$ at the AFB input has covariance $\sigma_v^2 \mathbf{I}$.

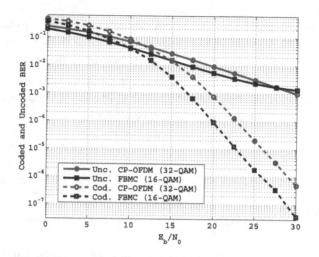

**FIGURE 12.2**

Uncoded and coded BER comparison between FBMC and CP-OFDM for the same throughput, ITU Veh-B static channel model, and the parameters in Table 12.1.

The multitap MMSE equalizer is given by [23]

$$
\bar{\mathbf{w}}_p^{\text{opt}} = \arg\min_{\bar{\mathbf{w}}_p} \mathbb{E}\left\{ |\hat{c}_{p,q} - c_{p,q-\nu}|^2 \right\}
$$

$$
= \left( \sum_{m=p-1}^{p+1} \bar{\mathbf{G}}'_{p,m} (\bar{\mathbf{G}}'_{p,m})^{\text{T}} + \frac{\sigma_v^2}{2\sigma_d^2} \mathbf{\mathcal{G}}_p \mathbf{\mathcal{G}}_p^{\text{T}} \right)^{-1} \bar{\mathbf{G}}'_{p,p} \mathbf{e}_{\nu+1}. \qquad (12.3)
$$

The vector $\mathbf{e}_{\nu+1} \in \{0, 1\}^{(L_{\tilde{g}}+L_{\text{eq}}-1)\times 1}$ contains a unity at the $(\nu+1)$th position, and $\nu$ is the equalizer delay typically chosen as $\nu = 2K + \lceil L_{\text{eq}}/2 \rceil$. It can be easily proved that separately solving for real or imaginary symbols gives the same result. Moreover, there is a number of solutions in the literature to simplify the matrix inversion above in order to reduce the costs of its implementation.

An example of the per-subcarrier multitap MMSE equalizer performance when perfect Channel State Information (CSI) is available at the RX side is depicted in Fig. 12.2. It shows a comparison of the uncoded and coded Bit Error Rate (BER) for FBMC and CP-OFDM with the parameters in Table 12.1, and both systems have the same data throughput. This example was taken from [24], where more details of the simulation setup can be found. It can be observed an $E_b/N_0$ advantage of 2.5 dB for FBMC/OQAM compared to CP-OFDM in the coded BER. It is worth noting that in FBMC, the subcarrier spacing can be increased and the same BER performance will be observed. The drawback is then higher complexity due to longer equalizers.

The per-subcarrier linear fractionally spaced analytical MMSE solution presented above was initially proposed in [23], and it was later extended to a Decision Feed-

**Table 12.1** Parameters for coded and uncoded BER comparison between FBMC and CP-OFDM in Fig. 12.2

| Parameter | Value |
| --- | --- |
| Total number of subcarriers | $M = 1024$ |
| Used subcarriers | $M_u = 768$ |
| Subcarrier spacing | $\Delta f = 10.9$ kHz |
| Total bandwidth | BW = 10 MHz |
| Sampling period | $T/M = 89.28$ ns |
| CP duration | $T_{CP} = 22.85$ μs (1/4 symbol length) |
| Equalizer length | $L_{eq} = 5$ |
| Channel model | ITU Veh-B static |
| Root Mean Squared (RMS) delay spread | $\tau_{RMS} = 4$ μs |
| Symbols per subcarrier | 1000 |
| Channel realizations | 200 |
| Channel code | Convolutional |
| Code rate | $\mathcal{R} = 1/2$ |
| Code polynomials | $1 + D^1 + D^2 + D^3 + D^6$ |
|  | $1 + D^2 + D^3 + D^5 + D^6$ |
| Decoder | Soft Max-log-MAP |

back Equalizer (DFE) in [25]. The DFE is only necessary in extreme cases, where the frequency selectivity in each subcarrier is very high. In [26] and [27], adaptive solutions using the set-membership Normalized Least Mean Squares (NLMS) algorithm were presented for linear and decision feedback equalizers. The adaptive solutions are especially important when we need to track time variations of the channel in high-mobility scenarios. Moreover, in [24], the unbiased MMSE solution is presented and compared with the Maximum Likelihood Sequence Estimation (MLSE) scheme. Furthermore, [28] shows how to transform linear equalizers and DFEs into linear precoders and Tomlinson–Harashima Precoder (THP) with the help of the Mean-Squared Error (MSE)-duality for Single-Input Single-Output (SISO) systems. More recently, in [29], the equalizer presented above was extended for doubly selective channels. Finally, the multitap MMSE solution above can be further extended to consider imperfect CSI if the statistics of the channel or its estimate are available.

## 12.3.2 WIDELY LINEAR MULTITAP EQUALIZERS

### 12.3.2.1 Widely Linear Processing (WLP) Basics

Before dealing with Widely Linear (WL) equalizers, let us first introduce some WLP basics. It is well known that the MMSE estimate $\hat{d}$ of a scalar random variable $d$ given a random vector $\mathbf{y}$ is the conditional expectation $\hat{d} = \mathbb{E}\{d/\mathbf{y}\}$. If $\mathbf{z} = [d, \mathbf{y}^T]^T$ is a real-valued zero-mean Gaussian vector, then the regression is linear and takes the form $\hat{d} = \mathbf{w}^T\mathbf{y}$, where $\mathbf{w}$ is a real-valued filter. However, for Gaussian zero-mean complex data, it is shown in [30] that the regression remains linear, i.e., of the form $\hat{d} = \mathbf{w}^H\mathbf{y}$, where $\mathbf{w}$ is a complex-valued filter, only if $\mathbf{z}$ is Second-Order (SO) circular (or proper) [31], i.e., if $\mathbb{E}\left\{\mathbf{z}\mathbf{z}^T\right\} = \mathbf{0}$. If $\mathbf{z}$ is SO noncircular (or improper), i.e., if $\mathbb{E}\left\{\mathbf{z}\mathbf{z}^T\right\} \neq \mathbf{0}$, then regression is linear only if both $\mathbf{y}$ and $\mathbf{y}^*$ are considered. This extended notion of linearity is called WL [30], and WLP can be expressed as

$$\hat{d} = \mathbf{w}_1^H \mathbf{y} + \mathbf{w}_2^H \mathbf{y}^*,$$

where $\mathbf{w}_1$ and $\mathbf{w}_2$ are complex-valued filters. It can be shown that, for SO noncircular (or improper) complex data, even non-Gaussian, using WL instead of linear filters should yield improvements in MMSE estimation problems [32]. In fact, this potential performance improvement in noncircular contexts is not limited to such problems but concerns many more applications as it will be discussed in the following subsections. Note that the usefulness of WL filtering was already pointed out before in [33] and [34] but in very restricted contexts.

### 12.3.2.2 *Usefulness of WLP for Wireless Communications*

For SISO wireless systems, the complex observation $y(t)$ at time $t$ corresponds to the complex envelope of a real signal. The signal $y(t)$ is SO noncircular if there is $(t, \tau)$ such that $c_y(t, \tau) \triangleq \mathbb{E}\{y(t)y(t - \tau)\} \neq 0$. Many signals encountered in the wireless communications domain are SO noncircular. The so-called Rectilinear (R) and Quasi-Rectilinear (QR) signals are particularly important examples. An R signal uses a mono-dimensional modulation, such as a Binary Phase-Shift Keying (BPSK) or an Amplitude Shift Keying (ASK) modulation, whereas a QR signal is a filtered version of an R signal after a simple derotation operation [35]. Examples of QR signals are Minimum-Shift Keying (MSK) and OQAM signals, whereas a Gaussian Minimum-Shift Keying (GMSK) signal is an approximated version of a QR signal. For such signals, WL filters improve the performance of linear ones. For this reason, WLP is worth considering for FBMC/OQAM waveforms, as discussed further.

More precisely, to within a complex constant factor, the complex envelope of R and QR signals is given by

$$s(t) = \sum_k x_k g(t - kT_d), \qquad (12.4)$$

where $T_d$ is the symbol duration for R, MSK, and GMSK signals and half the symbol duration for OQAM signals, $g(t)$ is a real-valued pulse shaping filter, and $x_k$ are i.i.d. random variables corresponding to the transmitted symbols for R signals and a function of the latter for QR signals. For R signals, $x_k$ is real-valued and denoted by $d_k$, whereas for QR signals, $x_k = j^k d_k$, where $d_k$ is real valued. In this latter case, denoting by $s_d(t) \triangleq j^{-t/T_d} s(t)$ the derotated signal $s(t)$, it is straightforward to verify that

$$s_d(t) = \sum_k d_k g_d(t - kT_d), \qquad (12.5)$$

where $g_d(t) = j^{-t/T_d} g(t)$ is the complex-valued derotated pulse-shaping filter. Eq. (12.5) shows that a derotated QR signal is nothing else but a filtered R signal.

Note that in the particular case of a QAM signal transmitting the complex symbols $c_k = c_k^{(R)} + jc_k^{(I)}$ of duration $T$, where $(c_k^{(R)}, c_k^{(I)})$ are real-valued symbols, the complex envelope $s(t)$ of the associated OQAM signal, obtained through staggering $c_k^{(R)}$ and $c_k^{(I)}$ by $T/2$, is given by

$$s(t) = \sum_k c_k^{(R)} g(t - kT) + jc_k^{(I)} g(t - kT - T/2) \qquad (12.6)$$

and can be written as in (12.4), where $x_k = j^k d_k$, $d_{2k} = (-1)^k c_k^{(R)}$, $d_{2k+1} = (-1)^k c_k^{(I)}$, and $T_d = T/2$.

For an R signal, it is easy to verify that $\mathbb{E}\left\{s(t)^2\right\} = \mathbb{E}\left\{|s(t)|^2\right\} \neq 0$, which generates an instantaneous SO noncircularity coefficient $\gamma_s = \mathbb{E}\left\{s(t)^2\right\}/\mathbb{E}\left\{|s(t)|^2\right\}$, whose modulus is maximal and equal to one. For this reason, the best improvements offered by a WLP with respect to a linear one are obtained for R signals. Moreover, since a derotated QR signal is a filtered R signal, we may expect, for QR signals in general and for OQAM signals in particular, improvements by WLP not far from the ones obtained for R signals.

### 12.3.2.3 *Applications of WLP in Wireless Communications*

As SO noncircular signals are present in many domains of interest, since the pioneering works [30–34], there has been an increasing interest for optimal WL filters in SO noncircular contexts. General overviews of complex processing for noncircular signals can be found in [36–38], whereas many applications of WLP in wireless communications are presented in the references contained in [39,40]. Among these applications, two of them, which concern FBMC/OQAM waveforms, have received a particular interest and respectively correspond to channel equalization and Co-Channel Interference (CCI) mitigation for R or QR signals. We briefly discuss these applications before considering specificities of WLP for FBMC/OQAM waveforms.

### 12.3.2.4 *WLP for Equalization and CCI Mitigation*

WL equalization in the temporal domain has been considered in several papers among which we can cite [41–44] for R signals and [41,45] for QR ones. All these papers consider ZF and/or MMSE approaches. [41] well synthesizes the main advantages of WL equalization with respect to the linear one. WL equalization exploits the phase of the multiple paths in addition to the temporal or frequency dimension. Its interest increases with the phase discrimination between the multiple paths and with the frequency selectivity of the propagation channel. Besides, the WLP generally reduces the required size of equalizers.

SISO WLP for CCI mitigation, also called Single Antenna Interference Cancellation (SAIC), has been considered in several works, among which we can cite [39], [46–48] for SC and [49–51] for Orthogonal Frequency-Division Multiplexing (OFDM) modulation. In all these papers, except for [49] and [50], the CCI has the same waveform, the same constellation, and the same carrier frequency as the Signal-of-Interest (SOI), whereas in [49] and [50], the SOI is an FBMC/OQAM signal with the CCI being a SO circular narrow-band interference. [47], [48], and [51] consider R constellations, [46] considers GMSK constellations, whereas both R and QR constellations are considered in [39]. The propagation channel is assumed to be frequency selective, with the exception of [39], where it is assumed to be flat. [39] well explains both the behavior of SAIC and the advantages of WLP with respect to linear processing for CCI mitigation. WLP exploits a phase discrimination between SOI and CCI and hence the SAIC capability as long as there is a phase discrimination between the sources.

### 12.3.2.5 *WLP for FBMC/OQAM Waveforms: Specificities*
#### i) State-of-the-Art

Papers dealing with WLP for FBMC/OQAM waveforms remain very scarce up to now and mainly include [52–54] for SISO and [55,56] for MIMO systems. [55] considers a MIMO system with no CSI available at the TX, transmitting several spatially multiplexed data streams per subcarrier, under the assumption of the channel being flat over three consecutive subcarriers. It proposes a per-subcarrier processing based on two steps. The first one implements a linear MMSE processing to mitigate ICI, whereas the second one uses MMSE WLP to mitigate Inter-Stream Interference (ITI). In that paper, ICI is not processed by WL filtering, which is suboptimal, whereas ITI is processed by WL filtering, which exploits both phase and spatial dimensions. Moreover, the potential frequency selectivity of the channel is not taken into account. [56] considers the same kind of systems but with CSI assumed to be available at the TX side and for frequency selective propagation channels. It proposes a per-subcarrier processing consisting of jointly optimizing, in terms of the MMSE criterion, the single-tap linear precoder and multitap linear or WL receivers but without comparing them. Again, WL processing exploits both phase and spatial discrimination between the streams. In [53], a SISO FBMC/OQAM system corrupted by FBMC/OQAM CCI is considered under the assumption of a flat fading channel. A per subcarrier processing is proposed based on linear or WL MMSE processing to cancel CCI thanks to phase discrimination between the sources. The processing of ICI does not seem to be explained in that work but should be simplified by the flatness assumption of the channel. Finally, [52] and [54] seem to be the only ones that consider a SISO FBMC/OQAM system under frequency selective fading channels using linear and WL receivers to mitigate both ISI and ICI. A discrete-time MMSE-based approach is considered therein in the time- [52,54] or frequency-domain [54]. However, [54] assumes the presence of TX and RX In-phase/Quadrature (I/Q) imbalance, which does not allow a clear evaluation of the usefulness of WLP, whereas in [52] the advantage of WLP over the linear one is presented in terms of the MMSE expressions without BER results. Only in [21], BER results illustrating the advantage of WLP over the linear one for single-carrier systems based on OQAM are presented.

We can see from the above review of the literature that it remains unclear how much exactly WLP may outperform linear processing for SISO FBMC/OQAM systems under frequency selective fading channels without I/Q imbalance. In fact, we will try to briefly explain in the following that, for frequency selective channels, the FBMC/OQAM waveform has special characteristics, which require special attention when developing powerful WL techniques that outperform linear ones. This can explain the low number of publications dealing with WLP for FBMC/OQAM systems under frequency selective fading.

## ii) Observation Model

Let us consider the reception of an FBMC/OQAM signal corrupted by a frequency selective fading channel and background noise. Under these assumptions and using a discrete-time model for (12.4) and (12.6), the complex envelope of the observation at time index $l$ at the RX after frequency synchronization can be written as

$$
\begin{aligned}
y[l] &= \left( \sum_{m=0}^{M_u-1} \sum_n j^{m+n} d_{m,n} g\left[l - n\frac{M}{2}\right] e^{j\frac{2\pi}{M}m\left(l-\frac{L_g-1}{2}\right)} \right) \star h[l] + v[l] \\
&= \sum_{m=0}^{M_u-1} \sum_n j^{m+n} d_{m,n} g_m^{(h)}\left[l - n\frac{M}{2}\right] e^{j\frac{2\pi}{M}m\left(l-\frac{L_g-1}{2}\right)} + v[l] \\
&\triangleq \sum_{m=0}^{M_u-1} j^m s_m^{(h)}[l] e^{j\frac{2\pi}{M}m\left(l-\frac{L_g-1}{2}\right)} + v[l].
\end{aligned}
$$

Here, $g_m^{(h)}[l] \triangleq g[l] \star h_m[l]$, where $h_m[l] \triangleq e^{-j\frac{2\pi}{M}ml}h[l]$, and $s_m^{(h)}[l]$ is defined as a discrete-time version of (12.4), where $x_k$ and $g(t)$ are replaced by $j^k d_{m,k}$ and $g_m^{(h)}[l]$, respectively. The transmitted symbols $d_{m,n}$ are assumed to be i.i.d.. Let $m_0$ be the subcarrier of interest for a per-subcarrier RX processing and denote by $\mu$ the quantity $\mu \triangleq m - m_0$. To simplify the notation, we simply denote by $d_{\mu,n}$, $h_\mu[l]$, $g_\mu^{(h)}[l]$, and $s_\mu^{(h)}[l]$ the symbol $d_{m_0+\mu,n}$, the channels $h_{m_0+\mu}[l]$, $g_{m_0+\mu}^{(h)}[l]$, and the signal $s_{m_0+\mu}^{(h)}[l]$, respectively. Under these assumptions and after a frequency shift of $e^{-j\frac{2\pi}{M}m_0\left(l-\frac{L_g-1}{2}\right)}$, the new observation $y_{m_0}[l] \triangleq e^{-j\frac{2\pi}{M}m_0\left(l-\frac{L_g-1}{2}\right)} y[l]$ can be written as

$$
\begin{aligned}
y_{m_0}[l] &= \sum_n j^{m_0+n} d_{0,n} g_0^{(h)}\left[l - n\frac{M}{2}\right] \\
&\quad + \sum_{\mu=-m_0, \mu\neq 0}^{M_u-1-m_0} \sum_n j^{m_0+\mu+n} d_{\mu,n} g_\mu^{(h)}\left[l - n\frac{M}{2}\right] e^{j\frac{2\pi}{M}\mu\left(l-\frac{L_g-1}{2}\right)} + v_{m_0}[l] \\
&\triangleq j^{m_0} s_0^{(h)}[l] + \sum_{\mu=-m_0, \mu\neq 0}^{M_u-1-m_0} j^{m_0+\mu} s_\mu^{(h)}[l] e^{j\frac{2\pi}{M}\mu\left(l-\frac{L_g-1}{2}\right)} + v_{m_0}[l], \qquad (12.7)
\end{aligned}
$$

where $v_{m_0}[l] \triangleq e^{-j\frac{2\pi}{M}m_0\left(l-\frac{L_g-1}{2}\right)} v[l]$. For the subcarrier of interest $m_0$, the signal $j^{m_0} s_0[l]$ corresponds to the received SOI, whereas the signal $j^{\mu+m_0} s_\mu^{(h)}[l] \times$

$e^{j\frac{2\pi}{M}\mu\left(l-\frac{L_g-1}{2}\right)}$ corresponds to the received ICI number $\mu$. The OQAM signal, which is transmitted on subcarrier $m_0$, is then filtered by the frequency selective channel $j^{m_0}h_0[l]$ and corrupted, in the worst case, by $M_u - 1$ statistically independent ICI components and background noise. Each ICI component with index $\mu$ can be considered as an OQAM signal filtered by the frequency selective channel $j^{\mu+m_0}h_\mu[l]$ and frequency shifted by the frequency offset $\Delta f_\mu = \mu/T$. Besides, the frequency selective channels are more or less correlated from a source to another, depending on the frequency selectivity of the channel. Finally, as the assumption made previously, only two ICI components ($\mu = \pm 1$) spectrally overlap with the SOI.

### iii) Problem Formulation and Difficulties for WLP

For a per-subcarrier processing, from the observation model (12.7) the problem is to recover the SOI symbols $d_{0,n}$, i.e., the symbols of a QR SOI, with the highest possible accuracy. To this end and for a PHYDYAS prototype filter, a WLP has to jointly equalize the frequency selective channel of the OQAM SOI and to remove the two ICI components ($\mu = \pm 1$), which spectrally overlap with the SOI. These are filtered by correlated frequency selective channels, which are frequency shifted. This situation is very different from those considered in [39,46–48,51] since:

**a)** Two QR ICI components ($\mu = \pm 1$) must be removed instead of one in those works.

**b)** Each ICI component $\mu$ is frequency shifted by $\Delta f_\mu = \mu/T$ with respect to the SOI, whereas this frequency shift is null in the previous references.

**c)** Each ICI component $\mu$ spectrally overlaps with ICI components $\mu - 1$ and $\mu + 1$, which is an additional difficulty.

For these reasons, conventional time-invariant SISO WLP [39,46–48,51], implemented at the symbol rate $1/T_d$, which can remove at most one CCI component with zero frequency shift while mitigating the ISI, cannot perform well as soon as the channels seen by the two ICI components are sufficiently different, i.e., beyond a certain level of frequency selectivity of the channel $h[l]$. In such situations, other solutions must be found.

### 12.3.2.6 *WLP for FBMC/OQAM: Advanced Research Directions*

To remove two ICI components associated with two different propagation channels, we have to create a virtual Single-Input Multiple-Output (SIMO) reception with at least three inputs instead of two.

A first idea is to oversample the observations by a factor $N_{up} \geqslant 2$, i.e., to implement WLP at a rate equal to at least $N_{up}/T_d$, and to build a $2N_{up} \times 1$ augmented observation vector, at time $nT_d$, defined by

$$\bar{y}[n] \triangleq \left[y[n], y[n+1], \ldots, y[n+N_{up}-1], y^*[n], y^*[n+1], \ldots, \right.$$
$$\left. y^*[n+N_{up}-1]\right]^{\mathsf{T}}.$$

Such a choice was made, for example, in [46], in the Global System for Mobile communications (GSM) context. A time-invariant multitap linear processing is then implemented, at the rate $1/T_d$, on each component of $\tilde{\mathbf{y}}[n]$. As $N_{up}$ becomes very large, the processing tends to a time-invariant SISO continuous-time WL processing. Such a continuous time-invariant WLP was proposed in [57], using a pseudo-MLSE approach, to optimize the reception of R or QR SOI corrupted by one R or QR CCI without differential frequency offset in frequency selective channels. Despite its continuous-time structure, this SISO processing does not allow us to process more than one CCI sharing the same bandwidth as the SOI. The oversampling of the data is then not sufficient to process more than one ICI component and to solve the SISO FBMC/OQAM problem for frequency selective fading channels.

The second idea, which aims at removing two CCI components from one antenna, is to exploit the cyclo-stationary property of the QR signals through the implementation of linear or WL frequency-shifted (FRESH) filtering [34], for SO circular or noncircular signals, respectively, from the knowledge of the cyclic frequencies of the received signals. Such a time-variant WL approach was proposed in [58] for QR signals to improve the performance of the time-invariant WL filtering approach proposed in [57]. Such an approach was also proposed in [59] to take into account a differential frequency offset between the sources for frequency selective channels. In the latter case, the approach proposed in [59] discriminates the SOI and CCI spectrally and by phase. Extensions to processing two ICI components from one RX signal have not been published but are straightforward. One may then think that this approach may solve our FBMC/OQAM problem. In fact, this solves the two first difficulties (a) and (b) presented in Section 12.3.2.5. The third one, (c), may be solved by low-pass filtering the observations around the SOI bandwidth before WL FRESH processing to remove the ICI components $\mu$ for $|\mu| > 1$. However, this low-pass filtering destroys the spectral correlation properties of the ICI components $\mu$ for $|\mu| = 1$ and degrades the performance of the WL FRESH processing, as shown recently in [60]. This demonstrates the difficulty of implementing efficient SISO WL receivers for FBMC/OQAM waveforms.

### 12.3.3 TWO-STAGE INTERFERENCE CANCELLATION EQUALIZERS

Although MMSE equalizers (see Section 12.3.1 or [23]) take into account the interferences coming from adjacent symbols and adjacent subcarriers, they are usually not able to remove them completely. Some ICI and ISI still remain, thus limiting the performance at high Signal-to-Noise Ratio (SNR). One way to mitigate this issue, and therefore further improve the performance of equalizers at high SNR and/or in high selectivity scenarios, is to use interference cancellation techniques. This section presents a two-stage interference cancellation equalizer as described in [61].

The idea behind the Two-Stage Interference Cancellation (TS-IC) equalizer is to start with a classical MMSE equalizer in order to provide a first estimate of the transmitted symbols on all subcarriers. Then, for each subcarrier, the interferences coming from the adjacent subcarriers can be canceled based on the knowledge of these symbols. Channel estimation is obviously necessary to perform this operation but should be available anyway to compute the MMSE equalizer in the first step. After removing ICI from the received signal for each subcarrier, only ISI remains. A new MMSE equalizer can then be derived to cope with this remaining ISI without having to take ICI into account. Thanks to its increased degrees of freedom, this new MMSE equalizer is able to offer improved performance. The tentative decisions obtained in the first stage are only used to cancel the interference from adjacent subcarriers, and not to cancel the interference inside the subcarrier of interest, so that error propagation is limited. The two-stage MMSE equalizer is briefly described in the following two paragraphs.

*Stage 1*: At this stage, initial decisions are obtained on the transmitted symbols. The equalizer that provides these decisions is the one given by (12.3) in Section 12.3.1. Assuming that the channel is known and that the tentative decisions are correct, it is possible to remodulate them and compute the convolution with the cascade of synthesis filter, channel and analysis filter in order to obtain the contributions of each of the symbols on the received samples $\hat{c}_{p,q}$. Therefore, these contributions can be completely removed from the received signal:

$$
\begin{aligned}
\mathbf{y}^-_{p,q} &= \mathbf{y}_{p,q} - \mathbf{G}'_{p,p-1}\hat{\mathbf{d}}_{p-1,q} - \mathbf{G}'_{p,p+1}\hat{\mathbf{d}}_{p+1,q} \\
&\approx \mathbf{G}'_{p,p}\mathbf{d}_{p,q} + \check{\mathbf{G}}_p\mathbf{v},
\end{aligned}
\tag{12.8}
$$

where $\mathbf{y}^-_{p,q} = \begin{bmatrix} y^-_{p,q} & y^-_{p,q-1} & \cdots \end{bmatrix}^{\mathrm{T}}$ is the received vector (at the AFB output) after removing the contributions of the adjacent subcarriers. It is clear that ICI has indeed been removed (up to decision errors and channel estimation errors) and that only ISI and noise remain.

*Stage 2*: Now to mitigate ISI, a new MMSE-type equalizer is considered that is based on the new received samples (12.8). Following a similar reasoning as for the MMSE equalizer in Section 12.3.1, it can be easily shown that the MMSE equalizer for received signal (12.8) is given by

$$
\bar{\mathbf{w}}_{p,\mathrm{TS}} = \left( \bar{\mathbf{G}}'_{p,p}(\bar{\mathbf{G}}'_{p,p})^{\mathrm{T}} + \frac{\sigma_v^2}{2\sigma_d^2}\mathcal{G}_p\mathcal{G}_p^{\mathrm{T}} \right)^{-1} \bar{\mathbf{G}}'_{p,p}\mathbf{e}_{v+1},
\tag{12.9}
$$

where the same definitions as in (12.3) hold. This new MMSE equalizer only copes with ISI and does not have to take ICI into account. The additional degrees of freedom allow us to obtain better performance. As can be seen from (12.9), this new equalizer can be computed at the same cost with the classical MMSE equalizer because all the terms appearing in (12.9) also appear in (12.3). So the additional computation of the second equalizer is only a small increase of complexity.

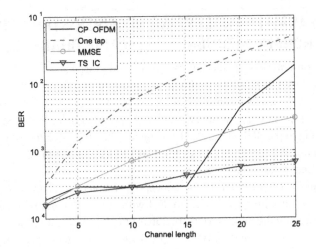

**FIGURE 12.3**

Effect of the channel length on the performance of the equalizers: BER vs. channel length. $E_b/N_0 = 30$ dB, MMSE equalizers of length 3.

In order to assess the performance of the TS-IC equalizer, the obtained BER is compared to those of other equalizers. A PHYDYAS FB is employed with $M = 64$ subcarriers, an overlapping factor of $K = 4$, and 16-QAM modulated input. The CP length for the CP-OFDM system is $\frac{M}{4}$. The FIR channels are chosen randomly, with flat power delay profiles. In all simulations, the channel is assumed to be perfectly known. Due to the use of a CP in OFDM, it should be noted that its throughput rate is lower than that of the FBMC system and the energy wasted in the CP is taken into account in $E_b/N_0$.

Fig. 12.3 shows the effect of the channel length on the performance of the equalizers. Different channel lengths are considered: 2, 5, 10, 15, 20, and 25. The BER performance of various equalizers is compared for a high SNR scenario, $E_b/N_0 = 30$ dB. For short channels of 2 to 5 taps, it can be seen that all the algorithms approximately have the same performance. In these cases, the channel can be considered flat per subcarrier, and hence the classical single-tap equalizer gives a satisfactory performance with low complexity. For longer (more frequency selective) channels, the TS-IC equalizer provides the best performance thanks to its ability to remove ICI. From these results it can be seen that the proposed scheme is particularly interesting in the case of highly frequency selective transmission channels and at high SNR values. As expected, in CP-OFDM, the performance is good as long as the channel length is smaller than the CP, but it degrades rapidly for longer channels.

## 12.4 A PARALLEL PROCESSING RECEIVER STRUCTURE FOR STRONGLY FREQUENCY SELECTIVE CHANNELS

In past sections, we have explored several methods for channel equalization of the FBMC/OQAM signal based on multitap processing at the output of the AFB. In what follows, we will introduce an alternative equalization method based on a parallel multifilterbank processing architecture. The proposed equalizer consists of a set of AFBs working in parallel on the received signal. The outputs of these AFBs are then properly combined at the subcarrier level to produce an estimate of the perfectly equalized signal prior to destaggering.

The main idea is to build on the concepts presented in Chapter 9 and propose an equalization structure that is able to deal with strong channel frequency selectivity while exploiting the multicarrier structure of the signal. To introduce the proposed strategy, let us assume that the FBMC/OQAM signal undergoes a frequency selective channel with frequency response $H\left(e^{j\omega}\right)$. Fig. 12.4A shows the most natural way of equalizing the received signal. Let **A** denote an $M \times N_s$ matrix containing the complex multicarrier symbols to be transmitted. After the staggering operation, these symbols go through an SFB with prototype pulse $g$, which generates the FBMC/OQAM transmitted signal. At the receiver side, the frequency selective channel is equalized with a filter with frequency response $W\left(e^{j\omega}\right)$. The result is then passed through an AFB with prototype pulse $f$ followed by a destaggering operation, from which an estimation of the symbols **A** is generated.

Now, in practice, the implementation in Fig. 12.4A is too costly from the computational point of view, mainly because the equalizer must process the whole signal bandwidth without exploiting the multicarrier nature of the signal. For this reason, it is customary to assume that the equalizer is approximately flat within each subcarrier bandwidth, so that we can approximately change the position of equalizer and the AFB, as illustrated in Fig. 12.4B. In this figure, we have denoted by $\boldsymbol{\Delta}(W)$ an $M \times M$ diagonal matrix with its $m$th diagonal entry equal to $W\left(e^{j\omega_m}\right)$, where $\omega_m = 2\pi m/M$, $m = 0, 1, \ldots, M - 1$, is the $m$th subcarrier frequency. Hence, multiplication by the matrix $\boldsymbol{\Delta}(W)$ is equivalent to equalization of the $m$th subcarrier signal via multiplication by the weight $W\left(e^{j\omega_m}\right)$ (single tap per-subcarrier equalization).

In this section, we show that we can further refine this process and consider more accurate approximations of the ideal scheme in Fig. 12.4A while still exploiting the nature of the multicarrier signal. To this end, we consider the development presented in Chapter 9 and more specifically the result presented in Proposition 9.1. The rationale is as follows. Consider the cascade of the equalizer $W\left(e^{j\omega}\right)$ in Fig. 12.4A and the $m$th subcarrier branch of the AFB. If $F\left(e^{j\omega}\right)$ denotes the Fourier transform of the receive prototype pulse, then we know that the frequency response associated with the $m$th subcarrier of the AFB is proportional to $F\left(e^{j(\omega-\omega_m)}\right)$. Considering the Taylor series development of $W\left(e^{j\omega}\right)$ around $\omega = \omega_m$ up to the $R$th order, we are able to

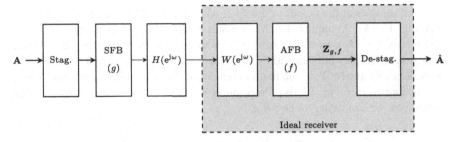

(A) Ideal implementation of the equalizer.

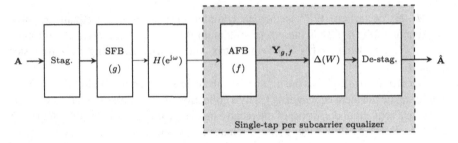

(B) First order approximation of (A).

**FIGURE 12.4**

Ideal implementation of the equalizer at the receiver (A) and baseline simplification with single-tap per-subcarrier weighting (B). Thin arrows represent time domain signals, whereas thick ones should be understood as frequency domain $M$-dimensional connections.

approximate

$$W\left(e^{j\omega}\right)F\left(e^{j(\omega-\omega_m)}\right) \approx \sum_{r=0}^{R} \frac{j^r}{r!} W^{(r)}\left(e^{j\omega_m}\right)(\omega-\omega_m)^r F\left(e^{j(\omega-\omega_m)}\right),$$

where $W^{(r)}(z)$ is the $r$th derivative of the transfer function $W(z)$. The above expression is pointing out that the cascade of the $m$th branch of the AFB and the ideal equalizer can be approximated by the sum of $R+1$ parallel systems, each one consisting of a single per-subcarrier weight $W^{(r)}\left(e^{j\omega_m}\right)/r!$ plus the $m$th branch of an AFB with prototype $(j\omega)^r F\left(e^{j\omega}\right)$ in the frequency domain. Now, it turns out that when the number of subcarriers grows without bound, we can see $(j\omega)^r F\left(e^{j\omega}\right)$ as the Fourier transform of the $r$th derivative of the analog waveforms associated to the original prototype. This means that we can effectively approximate the ideal response of the equalizer by using a set of $R+1$ parallel AFBs.

Let us formulate the above line of reasoning in more formal mathematical terms, building on the results presented in Chapter 9. Let us denote by $\mathbf{Z}_{g,f}$ the matrix of received samples at the output of the AFB when the ideal equalizer is implemented (see Fig. 12.4A). Assuming that $N_s$ complex multicarrier symbols are transmitted, this matrix has $M$ rows and at least $2N_s$ columns (corresponding to the $2N_s$ real-valued symbols plus tails caused by the prototypes, the channel, and the equalizer). Now, let us denote by $\mathbf{Y}_{g,f}$ the matrix containing the same samples when no equalizer is applied (see Fig. 12.4B). Proposition 9.1 ensures that, under some standard conditions, we can write, for any $R \in \mathbb{N}_+$,

$$\mathbf{Z}_{g,f} = \sum_{r=0}^{R} \frac{1}{r!M^r} \Delta\left(W^{(r)}\right) \mathbf{Y}_{g,f^{(r)}} + o\left(M^{-R}\right) \tag{12.10}$$

as $M \to \infty$, where $W^{(r)}$ is the $r$th complex derivative of the equalizer transfer function $W(z)$, $f^{(r)}$ is the $r$th time-domain derivative of the original prototype pulse[6] $f$, and where $o\left(M^{-R}\right)$ is a matrix with entries decaying to zero faster than $M^{-R}$.

According to (12.10), as the number of subcarriers $M$ increases to infinity, we may approximate the ideal equalizer $W(z)$ by a collection of $R+1$ parallel AFBs, each employing a sequential derivative of the receive prototype pulse. From the output of each of these $R + 1$ AFBs we can produce a matrix of samples $\mathbf{Y}_{g,f^{(r)}}$, $r = 0, 1, \ldots, R$. Now, by properly combining these matrices we can obtain an approximation of $\mathbf{Z}_{g,f}$ as accurate as desired, simply by adding more parallel AFBs. Furthermore, observe that the weighting applied to each of these outputs corresponds to a single per-subcarrier multiplication by a complex weight. Indeed, note that $\Delta\left(W^{(r)}\right)$ is a diagonal matrix, so the $m$th subcarrier output associated to the $r$th parallel AFB is multiplied by the complex coefficient $\frac{1}{r!M^r} W^{(r)}\left(e^{j\omega_m}\right)$, which is proportional to the $r$th derivative of the equalizer transfer function $W(z)$ evaluated at $z = e^{j\omega_m}$.

The corresponding multistage architecture is outlined in Fig. 12.5. As explained before, the proposed equalization structure is composed of $R + 1$ parallel filterbanks, each using a sequential derivative of the original prototype pulse $f$. The outputs of these AFBs are then combined at the subcarrier level to produce an estimate of the intended signal prior to destaggering. Each additional parallel stage can be associated with a new term of the Taylor expansion of the equalizer around each subcarrier frequency. Hence, by increasing the number of parallel stages we can achieve an arbitrarily accurate approximation of the ideal equalizer in Fig. 12.4A. On the other hand, observe that the per-subcarrier single tap equalizer in Fig. 12.4B is a specific case of the proposed multistage architecture when the number of parallel stages is fixed to one (i.e., $R = 0$).

At this point, a natural question arises as to how to determine the number of parallel stages that are needed to guarantee a certain performance of the multistage equalizer. To address this point, it is useful to evaluate the residual interference at the

---

[6]In more formal terms, $f$ is assumed to be a sampled version of a smooth analog waveform, and $f^{(r)}$ is obtained by sampling its $r$th derivative.

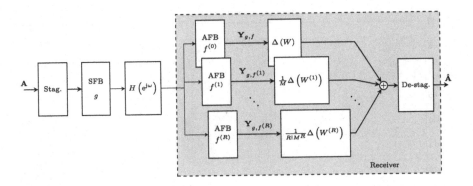

**FIGURE 12.5**

Parallel multistage equalizer consisting of $R+1$ concurrent stages.

output of the destaggering block in Fig. 12.5 following the approach in [62,63]. Let us consider the case where the equalizer simply inverts the channel, so that $W(z) = 1/H(z)$. Assuming that the prototype pulses $g$, $f$ are symmetric or antisymmetric in the time domain and meet the PR conditions, we can express the asymptotic residual error at the output of the equalizer corresponding to the $m$th subcarrier as [62]

$$
\mathbb{E}\left[|\{A - \hat{A}\}_{m,k}|^2\right] = \frac{2\eta_{R+1}^{(+,-)} \left|H\left(e^{j\omega_m}\right)\right|^2}{M^{2(R+1)}((R+1)!)^2} \left.\left|\frac{d^n H^{-1}\left(e^{j\omega}\right)}{d\omega^n}\right|^2\right|_{\omega=\omega_m} + o\left(M^{-2(R+1)}\right),
$$

$$(12.11)$$

where the pulse-specific quantity $\eta_{R+1}^{(+,-)} = \eta^{(+,-)}\left(g, f^{(R+1)}, g, f^{(R+1)}\right)$ is as defined in Chapter 9. This formula allows us to compute the residual distortion for a given channel response $H(z)$ and therefore fix the minimum number of stages, $R+1$, so that a certain minimum quality is achieved across the signal spectrum.

Despite its asymptotic nature, the above formula provides a very accurate description of the actual distortion at the output of the multistage equalizer. To illustrate this point, we simulated an FBMC/OQAM system with $M = 256$ subcarriers employing a PHYDYAS pulse [64] with overlapping factor $K = 4$ at both the transmitter and the receiver. Fig. 12.6 shows a particular realization of an Extended Typical Urban (ETU) channel model [65] (subcarrier separation equal to 15 kHz) and the corresponding performance in terms of signal to residual distortion power ratio when one ($R = 0$) and two ($R = 1$) parallel stages are implemented. The lower plot in this figure shows both the simulated performance (averaged over 5000 symbols) and the asymptotic one according to (12.11). Observe that the asymptotic performance is almost indistinguishable from the actual one although the number of subcarriers is relatively modest.

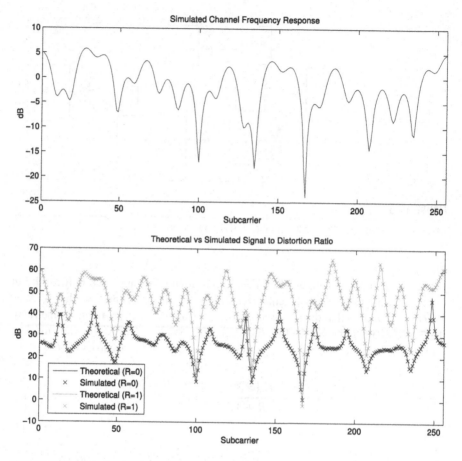

**FIGURE 12.6**

Signal-to-distortion power ratio of the multistage equalizer with $R = 0$ and $R = 1$ (lower subplot) for a specific realization of an ETU channel model (upper subplot).

Fig. 12.7 represents the cumulative distribution function of the subcarrier signal-to-distortion power ratio obtained from 100 realizations of the ETU channel model in a noiseless scenario. Apart from the performance obtained with the multi-stage architecture, we also represent the output signal-to-distortion ratio achieved with a multitap equalizer with 3, 5, and 7 taps per subcarrier, constructed using the frequency sampling approach. The legend of the figure incorporates the percentage of increase in computational complexity (real-valued additions and real-valued multiplications) with respect to the single-tap per-subcarrier equalizer. Observe that the multistage equalizer is a very competitive solution in terms of performance vs. computational complexity as compared to more traditional multitap equalizers.

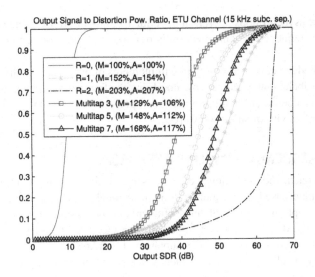

**FIGURE 12.7**

Cumulative distribution function of the signal-to-distortion power ratio for different equalizers. In the legend, M and A respectively represent the relative increase in the number of real-valued multiplications and additions with respect to the single-tap per-subcarrier equalizer.

## 12.5 CHANNEL EQUALIZATION IN FAST CONVOLUTION-BASED RECEIVERS

Fragmented spectrum use scenarios, like the heterogeneous Professional (or Private) Mobile Radio (PMR) case, impose new challenges due to the existence of strong interfering unwanted spectral components between the used spectral slots. In this section, flexible FC-FB signal processing (see Section 8.3) is utilized for effectively suppressing the unwanted spectral components from the channel estimation/equalization processes. The frequency-domain subcarrier processing concept utilizes the frequency-domain weights of the FC-FB structure also for channel equalization, timing, and frequency offset compensation and for timing and frequency offset estimation, as described in Section 10.6.

Integrating the synchronization and channel equalization functions with the FC-FB processing structure leads to efficient overall implementation. Even more importantly, the target is to move these functions from the time-domain processing side to the frequency domain, to be implemented through subcarrier processing. This is important in fragmented spectrum scenarios, i.e., noncontiguous multicarrier systems because otherwise it is difficult to manage the strong interfering spectral components, which may be dynamically allocated between the used portions of the received spectrum. Furthermore, the FC-FB processing structure is able to support simultaneous processing of nonsynchronized signals from different users. In this respect, our goal is to support asynchronous operation in cellular uplink or ad hoc scenarios such that only coarse frequency synchronization is required to be performed in the time domain.

The embedded equalization scheme implements optimal fractionally spaced multitap linear subchannel equalizers by modifying the frequency-domain weights, without additional coefficients, and provides good performance also in case of significant frequency selectivity of subchannels [66].

### 12.5.1 FREQUENCY SAMPLING APPROACH

In this approach, the subchannel equalizers are designed using a frequency sampling technique based on frequency-domain channel estimates. The principal idea is to solve or optimize the equalizer coefficients in such a manner that the frequency response of the designed filter is forced to take the given target values at a set of considered frequency points within a subchannel [6,67,68].

This design scheme enables frequency selective subchannel processing. In the case of time-domain equalizers, the number of frequency grid points used in the design is typically twice the number of subchannels. In this case, subchannel equalizers designed based on one, two, or three target points per subchannel can be derived. Between the given set of frequency points, the equalizer response is approximating the optimal frequency selective response. In general, a larger number of points can be used in the design [67], however, at the expense of somewhat more complex equalizer structures and more demanding derivation of the optimized filter coefficients.

For example, with the ZF criterion, this optimal response is the inverse of the channel transfer function. The level of approximation error depends on the level of frequency selectivity within the subchannel bandwidth. Moreover, it should be noted that this approach gives some degrees of freedom in the system design with respect to the size of the FB, subchannel bandwidth, and the order of the subchannel processing performed.

### 12.5.2 FC-FB-BASED EQUALIZER

Here we consider a subchannel-wise channel equalization structure where the channelization weights of the FC-FB structure (see Section 8.3) are also used for channel equalization purposes. This frequency-domain equalization scheme is referred to as an embedded equalizer. The frequency-sampled subcarrier-wise equalization model of [6] is employed. This model approximates closely a linear fractionally spaced equalizer and the conceptual model of the receiver front-end includes the following elements [69,70]: (i) Pulse-shape matched filter: the FC-FB realizes the square Root Raised Cosine (RRC) type pulse shaping filter through the weights $\mathbf{w}_m$, where $m$ is the subchannel index. (ii) Channel matched filter $\mathbf{H}_m^*$, where $\mathbf{H}_m$ is the subchannel frequency response at the corresponding frequency-domain bins. (iii) Folding effect (aliasing) due to resampling at the symbol rate. (iv) Linear equalizer with the MMSE criterion.

The resulting frequency-domain weights can be determined as

$$\widehat{\mathbf{w}}_m = \operatorname{diag}([\mathbf{D}_{m,0} + \mathbf{D}_{m,1} + \zeta \mathbf{I}_{L_m}])^{-1} \operatorname{diag}(\mathbf{w}_m)\mathbf{H}_m^*,$$

where

$$\mathbf{D}_{m,0} = \operatorname{diag}\left(\mathbf{H}_{m,0}\mathbf{H}_{m,0}^{\mathsf{T}}\right) \qquad \text{and} \qquad \mathbf{D}_{m,1} = \operatorname{diag}\left(\mathbf{H}_{m,1}\mathbf{H}_{m,1}^{\mathsf{T}}\right)$$

with

$$\mathbf{H}_{m,0} = \operatorname{diag}(\mathbf{w}_m)\mathbf{H}_m \qquad \text{and} \qquad \mathbf{H}_{m,1} = \operatorname{diag}(\mathbf{J}_R\mathbf{w}_m)\mathbf{H}_m.$$

Here, $\zeta$ is the noise-to-signal power ratio, $\mathbf{I}_{L_m}$ is the $L_m \times L_m$ identity matrix, and $\mathbf{J}_R$ is the circular permutation matrix obtained by cyclically left shifting the $L_m \times L_m$ reverse identity matrix by $L_m/2 - 1$ positions.

To construct the embedded equalizer, the channel estimate should be available at all the frequency-domain bins of the active subchannels. In scattered pilot-based estimation, the channel estimates at the pilot positions can be evaluated using the techniques reviewed in Chapter 11. Naturally, piecewise constant or piecewise linear interpolation can be used to approximate the channel estimate with a frequency resolution of a fraction of the subcarrier spacing if the channel is not highly frequency selective within the subbands. The same embedded equalizer model can be used for FBMC/OQAM [66,71], Filtered MultiTone (FMT) [71], and SC waveforms [72].

### 12.5.3 NUMERICAL RESULTS

The examples here are for the 1.4 MHz LTE-like case with FBMC/OQAM waveform, 15 kHz subcarrier spacing, and $M_u = 72$ active subcarriers (out of $M = 128$ in total). The equalizer adaptation is based on the assumption of Perfect Channel Information (PCI).

As a benchmark for the comparisons, we consider the polyphase filterbank structure, using the PHYDYAS prototype filter design [73] with overlapping factor of $K = 4$, together with three-tap frequency-sampling-based MMSE-type subcarrier equalizers [68]. In the FC-FB case, the short transform length is $L_m = 16$ for all active subchannels, the overlap factor is 6/16, and the instantaneous channel response in the middle of each 5-symbol Fast Fourier Transform (FFT) processing block is used. In the polyphase case, the channel is updated for each OQAM half-symbol. Figs. 12.8 and 12.9 show the performance of both configurations for Quadrature Phase-Shift Keying (QPSK) modulation in the case of Vehicular-A (Veh-A) channel (about 2.5 μs delay spread) and Hilly Terrain (HT) channel (20 μs delay spread) models, respectively, for both the static user and the user with 200 km/h mobility with the carrier frequency of $f_c = 400$ MHz. As can be seen from these figures, the performance of the embedded FC-FB equalizer is more sensitive to high mobility, but it is able to equalize well channels with long channel delay spread, i.e., high frequency selectivity, like the HT channel.

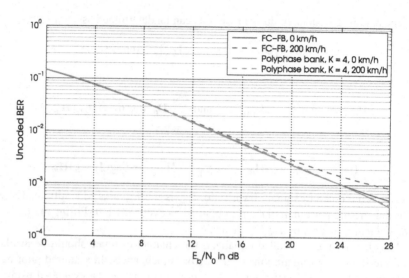

**FIGURE 12.8**

Performance of the polyphase FBMC and FC-FB in the case of Veh-A channel model for a static user and a user with 200 km/h mobility. QPSK modulation and PCI is assumed. The overlapping factor for the polyphase filterbank is $K = 4$.

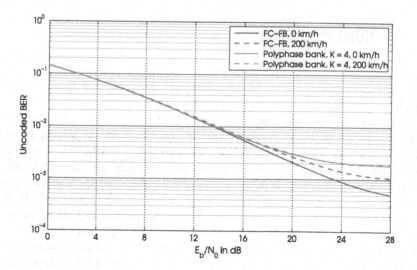

**FIGURE 12.9**

Performance of the polyphase FBMC and FC-FB in the case of HT channel model for a static user and a user with 200 km/h mobility. QPSK modulation and PCI is assumed. The overlapping factor for the polyphase filterbank is $K = 4$.

## 12.6 BLIND FBMC/OQAM EQUALIZERS

### 12.6.1 INTRODUCTION

One of the most widely used blind equalization algorithms is the Constant Modulus Algorithm (CMA) proposed by Godard [74] in 1980 (see also [75]). An overview of blind equalization using CMA can be found in [76]. The CMA converges independently of the phase of the channel, but this feature leads to a misadjustment in the detected symbols, which needs to be corrected. The Multi-Modulus Algorithm (MMA) presented in [77] (originally called modified CMA) has been proposed to solve the phase ambiguity issue by employing the CMA on both the real and imaginary parts of the received signal. More recently, MMA-based methods have been derived, which aim at improving the performance of CMA for higher-order constellations [78] and have been shown to outperform both CMA and MMA.

Although a thoroughly studied topic in OFDM modulation, blind equalization has been studied very little in the FBMC/OQAM context. In [79], the cyclo-stationarity induced by the adoption of the OQAM modulation was exploited to blindly estimate the channel based on Second-Order Statistics (SOS). However, this technique has a high complexity, and very long data records have to be stored to obtain good estimates of the SOS of the signal. The "classical" CMA still requires long data records and was adapted to cosine modulated FBs in [80]. Its convergence behavior was analyzed in [81]. The idea behind CMA applied in FBMC/OQAM is to update the equalizer by comparing the real part of its output with a given real constant that depends on the constellation size. However, the use of real-valued symbols in FBMC/OQAM restricts the choice of the cost function to those that are adapted to real constellations, whereas numerous cost functions adapted to different complex constellations have been proposed in the blind equalization literature [82,83].

In this subsection, we discuss the application in FBMC/OQAM of modified and adapted versions of CMA-based, Constant Norm Algorithm (CNA) [84,85], and $\beta$-MMA [86] blind equalization methods.

### 12.6.2 UPDATE ALGORITHM

We consider *direct* blind equalization techniques, namely not relying on a (blind) channel estimation stage. The output of the (single-tap) equalizer at subcarrier $p$ and at time $q$ can be expressed as

$$\tilde{d}_{p,q} = \mathrm{Re}\{w_{p,q} y_{p,q}\}. \qquad (12.12)$$

To estimate the ZF equalizer $w_{p,q}^{\mathrm{opt}} = \frac{1}{H_{p,q}}$, the receiver has access to only a few features of the input signal, such as the constellation size. Therefore, a conventional way to obtain $w_{p,q}^{\mathrm{opt}}$ consists in successively approximating this limit. The simplest technique to implement this is the Least Mean Squares (LMS) algorithm, which can be written as

$$w_{p,q+1} = w_{p,q} - \kappa \phi(\tilde{d}_{p,q}), \qquad (12.13)$$

where $\kappa$ is the step-size parameter of the algorithm, and the function $\phi(\cdot)$ is defined as $\phi(\tilde{d}_{p,q}) = \frac{\partial J(\tilde{d}_{p,q})}{\partial w_{p,q}^*}$ with $J$ being a given cost function that will be detailed shortly. Notice from (12.12) that $\tilde{d}_{p,q}$ is not a holomorphic function of $w_{p,q}$ due to the presence of $\mathbb{Re}\{\cdot\}$. Therefore, Wirtinger's differentiation [87] should be used instead of the usual complex differentiation to obtain $\phi(\tilde{d}_{p,q})$. An example will be provided in the sequel. Alternatives to LMS can also be used in (12.13), for example, the Recursive Least Squares (RLS) algorithm [88], yet with higher complexity requirements.

## 12.6.3 CONSTANT MODULUS ALGORITHM IN FBMC/OQAM

As previously stated, CMA is undoubtedly the most widely used blind equalization algorithm, owing its popularity to its simplicity. However, the application of CMA in FBMC/OQAM is not as trivial as presented in [80]. In fact, the CMA cost function can be expressed as

$$J(\tilde{d}_{p,q}) = \mathbb{E}\{(|\tilde{d}_{p,q}|^2 - \varsigma)^2\}, \tag{12.14}$$

where $\varsigma = \frac{\mathbb{E}\{|d_m|^4\}}{\mathbb{E}\{|d_m|^2\}}$, as indicated in [74]. Then, substituting (12.14) into (12.13) yields

$$w_{p,q+1} = w_{p,q} - 2\kappa\left(|\tilde{d}_{p,q}|^2 - \varsigma\right)\frac{\partial|\tilde{d}_{p,q}|^2}{\partial w_{p,q}^*}, \tag{12.15}$$

where $\frac{\partial|\tilde{d}_{p,q}|^2}{\partial w_{p,q}^*}$ stands for the Wirtinger derivative. It can be noted that, thanks to (12.12), $|\tilde{d}_{p,q}|^2$ can be rewritten as

$$|\tilde{d}_{p,q}|^2 = \frac{1}{4}(w_{p,q}y_{p,q})^2 + \frac{1}{4}((w_{p,q}y_{p,q})^*)^2 + \frac{1}{2}w_{p,q}y_{p,q}(w_{p,q}y_{p,q})^*.$$

It is known [87, Theorem 2.5] that the Wirtinger derivative yields $\frac{\partial z}{\partial z^*} = \frac{\partial z^*}{\partial z} = 0$ for any $z \in \mathbb{C}$. Therefore, (12.15) can be rewritten as

$$w_{p,q+1} = w_{p,q} - 2\kappa y_{p,q}^* \tilde{d}_{p,q}\left(|\tilde{d}_{p,q}|^2 - \varsigma\right), \tag{12.16}$$

and we actually find the usual expression of the CMA update. It must be emphasized that most of blind equalization algorithms, such as Sato's [89] and MMA [77] algorithms, are strictly equivalent to CMA in the FBMC/OQAM context because $d_{p,q}$ are real valued. However, as shown next, algorithms beyond CMA can be applied through an appropriate transformation to complex-valued symbols.

## 12.6.4 BEYOND CMA

To take advantage of the different blind equalization techniques found in the literature, it was proposed in [85] to redesign the received samples into *equivalent* complex transmitted symbols. This goes as follows:

1. We define $y_{p,q}^c$ as the sum of two consecutive received symbols, namely,

$$
\begin{aligned}
y_{p,q}^c &= j y_{p,q} + y_{p,q+1} \\
&= H_{p,q}\left( j d_{p,q}\left(1 + \langle g \rangle_{p,q}^{p,q+1}\right) + d_{p,q+1}\left(1 - \langle g \rangle_{p,q+1}^{p,q}\right)\right) \\
&\quad - \underbrace{\sum_{\substack{(m,n)\in\tilde{\Omega}_{p,q} \\ n\neq q+1}} H_{m,n}d_{m,n}\langle g \rangle_{m,n}^{p,q}}_{\tilde{I}_{p,q}} + j \underbrace{\sum_{\substack{(m,n)\in\tilde{\Omega}_{p,q+1} \\ n\neq q}} H_{m,n}d_{m,n}\langle g \rangle_{m,n}^{p,q+1}}_{\tilde{I}_{p,q+1}} \\
&\quad + j\eta_{p,q} + \eta_{p,q+1},
\end{aligned}
\tag{12.17}
$$

where $j\langle g\rangle_{m,n}^{p,q}$ is the TransMUltipleXer (TMUX) response (cf. (11.4)). Note that it was tacitly assumed above that $H_{p,q} = H_{p,q+1}$, which means that the channel is slowly varying.

2. We use the symmetry property of $\langle g\rangle_{m,n}^{p,q}$, i.e., $\langle g\rangle_{p,q+1}^{p,q} = -\langle g\rangle_{p,q}^{p,q+1}$ (see Eq. (11.9)) to define $\tilde{y}_{p,q}$ as

$$
\begin{aligned}
\tilde{y}_{p,q} &= \frac{y_{p,q}^c}{1 + \langle g\rangle_{p,q}^{p,q+1}} \\
&= H_{p,q}\underbrace{\left(j d_{p,q} + d_{p,q+1}\right)}_{\tilde{c}_{p,q}'} \\
&\quad + \underbrace{\frac{-\tilde{I}_{p,q} + \tilde{I}_{p,q+1}}{1 + \langle g\rangle_{p,q}^{p,q+1}}}_{\tilde{I}_{p,q}^r} + \underbrace{\frac{j\eta_{p,q} + \eta_{p,q+1}}{1 + \langle g\rangle_{p,q}^{p,q+1}}}_{\tilde{\eta}_{p,q}}.
\end{aligned}
\tag{12.18}
$$

The sample $\tilde{y}_{p,q}$ can be seen as the *equivalent* complex transmitted symbols $\tilde{c}_{p,q}'$ distorted by the channel $H_{p,q}$, the interference $\tilde{I}_{p,q}^r$, and the colored noise $\tilde{\eta}_{p,q}$.

In [85], the symbols $\tilde{y}_{p,q}$ are used to feed the update algorithm using CNA, and the equalizer coefficients $w_{p,q}$ are applied to $y_{p,q}$ before the OQAM destaggering as in (12.12). However, it can be noticed that, alternatively, the equalization stage can be directly applied to the sample $\tilde{y}_{p,q}$. In both cases, the equalizer coefficients using $\tilde{y}_{p,q}$ are updated at the rate $1/T$ in contrast to the update rate of $2/T$ of the usual CMA in (12.16).

It has been noticed in [85] that the energy of the interference plus noise term $\tilde{I}_{p,q}^r + \tilde{\eta}_{p,q}$ in (12.18) is much higher than the one that disturbs the OQAM symbol in $y_{p,q}$ at the output of the AFB. However, the ratio between the energy of the useful signal and that of the interference plus noise is much higher in (12.18) than at the output of the AFB. Furthermore, such a reformation of the received signal allows the receiver to use other algorithms than CMA with better performance, such as the $\beta$-MMA [86] or CNA [84,85] methods. In the following, we actually focus on these two equalization methods. CNA is similar to CMA, but norms $\| \cdot \|_\varpi$ of order $\varpi > 2$ are adopted instead of the Euclidean norm used in (12.14). The basic principle of $\beta$-MMA is to apply the CMA to both the real and imaginary parts of $w_{p,q}$.

## 12.6.5 PERFORMANCE RESULTS

We compare the performance of CMA in (12.16) with CNA and $\beta$-MMA using the redesign of complex symbols described in (12.18). More details on the cost functions and the update algorithms of CNA and $\beta$-MMA are given in [84–86]. The simulation parameters are as follows: $M = 128$ subcarriers, 16-QAM constellation, and the different update algorithms are initialized with $w_{p,0} = 0.1$ for all $p$. Moreover, the channel is the one considered in [80], i.e., $\mathbf{h} = \begin{bmatrix} 1 & -0.2 & 0.3 & 0.2 & 0.1 & 0.2 & 0.35 & -0.2 \end{bmatrix}^T$. The step-size parameter $\kappa$ has been set as follows: $7.34 \times 10^{-5}$ for CMA, $3.67 \times 10^{-4}$ for CNA, and $7.34 \times 10^{-4}$ for $\beta$-MMA, ensuring that all three methods have the same steady-state performance.

Fig. 12.10A depicts the MSE of the blind equalizers, defined as MSE $= \mathbb{E}\{(|\tilde{d}_{p,q}| - |d_{p,q}|)^2\}$, versus the number of FBMC/OQAM symbols used for their computation. It can be observed that $\beta$-MMA and CMA converge at the same rate, whereas CNA achieves a higher convergence speed since it requires 1500 less FBMC/OQAM symbols to reach its steady state. Furthermore, it must be recalled that the update of CNA and $\beta$-MMA using the redesign of complex symbols is performed every two, instead of every one FBMC/OQAM symbols with CMA in (12.16). In Fig. 12.10B, we show the steady-state performances of the three algorithms as functions of the number of FBMC/OQAM symbols, which are required to reach this steady state. The CNA outperforms both CMA and $\beta$-MMA. Together with the observations in Fig. 12.10A, this reveals that the redesign of complex symbols at the receiver can improve the performance of blind equalization in FBMC/OQAM systems.

## 12.7 CONCLUDING REMARKS

This chapter provided an overview of the research in FBMC/OQAM equalization with emphasis given to more recent advances. Solutions that range from the simple, single-tap equalizer to more complex schemes that can also cope with highly frequency selective channels were discussed. For the latter scenarios, multitap equalizers are necessary, such as the classical MMSE solution. It is able to effectively mitigate ISI and ICI because the equalizer is fractionally spaced and the prototype filter and the OQAM signaling structure are taken into account in the design. Further research on multitap equalizers should focus on reducing their complexity and enhancing their ability to cope with doubly dispersive channels. Other optimization objectives for specific scenarios and performance goals, such as a constant MSE value in exchange for lower complexity and latency, should also be considered. Moreover,

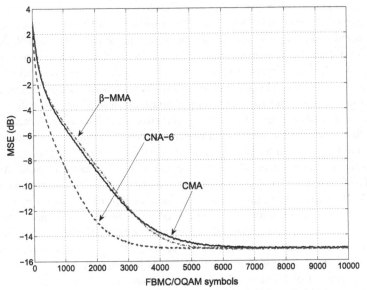

(A)  MSE performance versus FBMC/OQAM symbols.

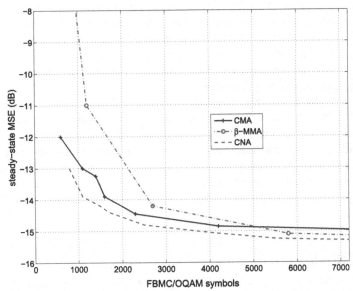

(B) Steady-state MSE performance versus FBMC/OQAM symbols.

**FIGURE 12.10**

MSE performance of CMA, CNA [84,85], and $\beta$-MMA [86].

the effects of the AFB noise color should be further studied, particularly, for longer equalizers and in combination with channel coding.

To achieve a more effective ISI and ICI mitigation, a two-stage multitap equalizer scheme was also presented. It employs interference cancellation with decisions made from a first stage to allow performance improvements in a second stage. It is also applicable for channels of strong frequency selectivity. The parallel multistage architecture is a powerful equalization scheme for such demanding scenarios. It consists of multiple parallel AFBs, each employing a different sequential derivative of the receive prototype filter. This is a very attractive alternative to per-subcarrier multitap equalization, which does not introduce additional latency. The equalizer in each stage may be seen as a trivial single-tap processor, so that a linear combination of the outputs of these stages allows us to equalize highly frequency selective channels. The frequency sampling-based multitap equalization idea and its embedding in the FC-FB structure were also discussed. Such an approach allows a more flexible partition of the spectrum and has shown good performance in channels with strong frequency selectivity. In the light of the potential performance gains from the use of WLP in OQAM-based transmission, the state-of-the-art in this special topic was also discussed in detail along with possible directions for future research.

In the context of blind equalization, the applicability of the well-known CMA algorithm in FBMC/OQAM was studied, and its limitations, coming from the real-valued nature of the PAM symbols, were pointed out. A method of forming complex-valued symbols was developed, which allows the use of more effective algorithms, such as CNA and $\beta$-MMA. Future research should include the extension of this method to constellations of higher order, such as 64- and 256-QAM. Moreover, increasing the convergence speed of these algorithms is an open challenge for their application in high-mobility scenarios.

In contrast to the widely standardized CP-OFDM, no CP is necessary in an FBMC/OQAM transceiver. This feature can translate to significant gains in spectral efficiency under conditions of highly dispersive transmission. However, channel equalization is not as straightforward anymore. Equalization methods that reach an acceptable trade-off between performance, computational/implementation complexity, and spectral efficiency in a large range of scenarios and use cases are required if an equally widespread acceptance of the FBMC/OQAM waveform is desired.

# REFERENCES

[1] T. Karp and N. Fliege, "Modified DFT filter banks with perfect reconstruction," *IEEE Trans. Circuits Syst. II*, vol. 46, no. 11, pp. 1404–1414, Nov. 1999.

[2] P. Siohan, C. Siclet, and N. Lacaille, "Analysis design of OFDMA/OQAM systems based on filterbank theory," *IEEE Trans. Signal Process.*, vol. 50, no. 5, pp. 1170–1183, May 2002.

[3] B. Hirosaki, "An analysis of automatic equalizers for orthogonally multiplexed QAM systems," *IEEE Trans. Commun.*, vol. COM-28, no. 1, pp. 73–83, Jan. 1980.

[4] T. Wiegand and N. J. Fliege, "Equalizers for transmultiplexers in orthogonal multiple carrier data transmission," in *Proc. EUSIPCO-1996*, Trieste, Italy, Sep. 1996.

[5] S. Nedic, "An unified approach to equalization and echo cancellation in OQAM-based multi-carrier data transmission," in *Proc. GLOBECOM-1997*, Phoenix, AZ, Nov. 1997.

[6] T. Ihalainen, T. H. Stitz, M. Rinne, and M. Renfors, "Channel equalization in filter bank based multicarrier modulation for wireless communications," *EURASIP J. Adv. Signal Process.*, 2007. http://dx.doi.org/10.1155/2007/49389.

[7] N. Holte, "MMSE equalization of OFDM/OQAM systems for channels with time and frequency dispersion," in *Proc. WCSP-2009*, Nanjing, China, Nov. 2009.

[8] M. Al-Attraqchi, S. Boussakta, and S. L. Goff, "An enhanced OFDM/OQAM system exploiting Walsh–Hadamard transform," in *Proc. VTC-2011 (Spring)*, Budapest, Hungary, May 2011.

[9] G. Ndo, H. Lin, and P. Siohan, "FBMC/OQAM equalization: Exploiting the imaginary interference," in *Proc. PIMRC-2012*, Sydney, Australia, Sep. 2012.

[10] D. Katselis, E. Kofidis, and S. Theodoridis, "On OFDM/OQAM receivers," in *Proc. EUSIPCO-2012*, Bucharest, Romania, Aug. 2012.

[11] S. Jo and J. S. Seo, "Efficient LLR calculation for FBMC," *IEEE Commun. Lett.*, vol. 19, no. 10, pp. 1834–1837, Oct. 2015.

[12] D. Mattera and M. Tanda, "Optimum single-tap per-subcarrier equalization for OFDM/OQAM systems," *Digit. Signal Process.*, vol. 49, pp. 148–161, Nov. 2016. http://www.sciencedirect.com/science/article/pii/S1051200415003425 [Online].

[13] S. M. J. A. Tabatabaee and H. Zamiri-Jafarian, "Per-subchannel joint equalizer and receiver filter design in OFDM/OQAM systems," *IEEE Trans. Signal Process.*, vol. 64, no. 19, pp. 5094–5105, Oct. 2016.

[14] R. Nissel and M. Rupp, "OFDM and FBMC-OQAM in doubly-selective channels: Calculating the bit error probability," *IEEE Commun. Lett.*, Mar. 2017. http://dx.doi.org/10.1109/LCOMM.2017.2677941.

[15] C.-W. Chen and F. Maehara, "An enhanced MMSE subchannel decision feedback equalizer with ICI suppression for FBMC/OQAM systems," in *Proc. ICNC-2017*, Santa Clara, CA, Jan. 2017.

[16] Q. Duong and H. H. Nguyen, "Walsh–Hadamard precoded circular filterbank multicarrier communications," in *Proc. SigTelCom-2017*, Da Nang, Vietnam, Jan. 2017.

[17] J. Louveaux, *et al.*, "Equalization and demodulation in the receiver (single antenna)." Project PHYDYAS ICT-211887, Deliverable D3.1, Jul. 2008. [Online]. Available: http://www.ict-phydyas.org/.

[18] J. Louveaux, *et al.*, "Optimization of transmitter and receiver." Project PHYDYAS ICT-211887, Deliverable D3.2, Jul. 2009. [Online]. Available: http://www.ict-phydyas.org/.

[19] E. Kofidis, *et al.*, "Training design and algorithms for channel estimation." Project EMPhAtiC ICT-318362, Deliverable D3.1, Dec. 2013. [Online]. Available: http://www.ict-emphatic.eu/.

[20] S. Nedic, *et al.*, "Adaptive equalization and successive self-interference cancellation (SIC) methods." Project EMPhAtiC ICT-318362, Deliverable D3.2, Sep. 2014. [Online]. Available: http://www.ict-emphatic.eu/.

[21] J. C. Tu, "Optimum MMSE equalization for staggered modulation," in *Proc. ACSSC-1993*, Pacific Grove, CA, Nov. 1993.

[22] M. Bellanger, "FS-FBMC: A flexible robust scheme for efficient multicarrier broadband wireless access," in *Proc. GLOBECOM-2012*, Anaheim, CA, Dec. 2012.

[23] D. S. Waldhauser, L. G. Baltar, and J. A. Nossek, "MMSE subcarrier equalization for filter bank based multicarrier systems," in *Proc. SPAWC-2008*, Recife, Brazil, Jul. 2008.

[24] L. G. Baltar, A. Mezghani, and J. A. Nossek, "MLSE and MMSE subchannel equalization for filter bank based multicarrier systems: Coded and uncoded results," in *Proc. EUSIPCO-2010*, Aalborg, Denmark, Aug. 2010.

[25] L. G. Baltar, D. S. Waldhauser, and J. A. Nossek, "MMSE subchannel decision feedback equalization for filter bank based multicarrier systems," in *Proc. ISCAS-2009*, Taipei, Taiwan, May 2009.

[26] D. S. Waldhauser, L. G. Baltar, and J. A. Nossek, "Adaptive equalization for filter bank based multicarrier systems," in *Proc. ISCAS-2008*, Seattle, WA, May 2008.

[27] D. S. Waldhauser, L. G. Baltar, and J. A. Nossek, "Adaptive decision feedback equalization for filter bank based multicarrier systems," in *Proc. ISCAS-2009*, Taipei, Taiwan, May 2009.

[28] H. Jedda, L. G. Baltar, O. D. Candido, A. Mezghani, and J. A. Nossek, "DFE/THP duality for FBMC with highly frequency selective channels," in *Proc. EUSIPCO-2015*, Nice, France, Aug. 2015.

[29] L. Marijanovic, S. Schwarz, and M. Rupp, "MMSE equalization for FBMC transmission over doubly-selective channels," in *Proc. ISWCS-2016*, Poznan, Poland, Sep. 2016.

[30] B. Picinbono and P. Chevalier, "Widely linear estimation with complex data," *IEEE Trans. Signal Process.*, vol. 43, no. 8, pp. 2030–2033, Aug. 1995.

[31] B. Picinbono, "On circularity," *IEEE Trans. Signal Process.*, vol. 42, no. 12, pp. 3473–3482, Dec. 1994.

[32] P. Chevalier, "Optimal array processing for non-stationary signals," in *Proc. ICASSP-1996*, vol. 5, Atlanta, GA, May 1996.

[33] W. Brown and R. Crane, "Conjugate linear filtering," *IEEE Trans. Inf. Theory*, vol. 15, no. 4, pp. 462–465, Jul. 1969.

[34] W. A. Gardner, "Cyclic Wiener filtering: Theory and method," *IEEE Trans. Commun.*, vol. 41, no. 1, pp. 151–163, Jan. 1993.

[35] Z. Ding and G. Li, "Single-channel blind equalization for GSM cellular systems," *IEEE J. Sel. Areas Commun.*, vol. 16, no. 8, pp. 1493–1505, Oct. 1998.

[36] D. Mandic and V. Goh, *Complex Valued Nonlinear Adaptive Filters, Noncircularity, Widely Linear and Neural Models*. Wiley, 2009.

[37] P. Schreier and L. Scharf, *Statistical Signal Processing of Complex-Valued Data – The Theory of Improper and Noncircular Signals*. Cambridge University Press, 2010.

[38] T. Adali, P. J. Schreier, and L. L. Scharf, "Complex-valued signal processing: The proper way to deal with impropriety," *IEEE Trans. Signal Process.*, vol. 59, no. 11, pp. 5101–5125, Nov. 2011.

[39] P. Chevalier and F. Pipon, "New insights into optimal widely linear array receivers for the demodulation of BPSK, MSK, and GMSK signals corrupted by noncircular interferences-application to SAIC," *IEEE Trans. Signal Process.*, vol. 54, no. 3, pp. 870–883, Mar. 2006.

[40] P. Chevalier and F. Dupuy, "Widely linear Alamouti receiver for the reception of real-valued constellations corrupted by interferences – the Alamouti-SAIC/MAIC concept," *IEEE Trans. Signal Process.*, vol. 59, no. 7, pp. 3339–3354, Jul. 2011.

[41] H. Gerstacker, R. Schober, and A. Lampe, "Receivers with widely linear processing for frequency-selective channels," *IEEE Trans. Commun.*, vol. 51, no. 9, pp. 1512–1523, Sep. 2003.

[42] G. Dietl and W. Utschick, "MMSE turbo equalisation for real-valued symbols," *Eur. Trans. Telecommun.*, vol. 17, no. 3, pp. 351–359, May 2006.

[43] D. Mattera, L. Paura, and F. Sterle, "MMSE WL equalizer in presence of receiver IQ imbalance," *IEEE Trans. Signal Process.*, vol. 56, no. 4, pp. 1735–1740, Apr. 2008.

[44] P. Xiao, R. A. Carrasco, and I. J. Wassell, "Generalized equalization algorithm utilizing improper ISI," *IEEE Trans. Veh. Technol.*, vol. 58, no. 2, pp. 788–799, Feb. 2009.

[45] D. Raphaeli, "A reduced complexity equalizer for OQPSK," *IEEE Trans. Commun.*, vol. 58, no. 1, pp. 46–51, Jan. 2010.

[46] H. Trigui and D. Slock, "Performance bounds for cochannel interference cancellation within the current GSM," *Signal Process.*, vol. 80, pp. 1335–1346, Jul. 2000.

[47] R. Meyer, W. H. Gerstacker, R. Schober, and J. B. Huber, "A single antenna interference cancellation algorithm for increased GSM capacity," *IEEE Trans. Wireless Commun.*, vol. 5, no. 7, pp. 1616–1621, Jul. 2006.

[48] J. C. Olivier and W. Kleynhans, "Single antenna interference cancellation for synchronised GSM networks using a widely linear receiver," *IET Commun.*, vol. 1, no. 1, pp. 131–136, Feb. 2007.

[49] D. Darsena, G. Gelli, L. Paura, and F. Verde, "Widely linear equalization and blind channel identification for interference-contaminated multicarrier systems," *IEEE Trans. Signal Process.*, vol. 53, no. 3, pp. 1163–1177, Mar. 2005.

[50] D. Darsena, G. Gelli, and F. Verde, "Universal linear precoding for NBI-proof widely linear equalization in MC systems," *EURASIP J. Wirel. Commun. Netw.*, 2008. http://dx.doi.org/10.1155/2008/321450.

[51] M. Konrad and W. Gerstacker, "Interference robust transmission for the downlink of an OFDM-based mobile communications system," *EURASIP J. Wirel. Commun. Netw.*, 2008. http://dx.doi.org/10.1155/2008/549371.

[52] D. S. Waldhauser, "Multicarrier systems based on filter banks." Ph.D. thesis (Technische Universität München). Aachen: Shaker Verlag, 2009.

[53] S. Josilo, M. Narandzic, S. Tomic, and S. Nedic, "Widely linear filtering based kindred co-channel interference suppression in FBMC waveforms," in *Proc. ISWCS-2014*, Barcelona, Spain, Aug. 2014.

[54] A. Ishaque and G. Ascheid, "Widely linear receivers for SMT systems with TX/RX frequency-selective I/Q imbalance," in *Proc. PIMRC-2014*, Washington, DC, Sep. 2014.

[55] Y. Cheng and M. Haardt, "Widely linear processing in MIMO FBMC/OQAM systems," in *Proc. ISWCS-2013*, Ilmenau, Germany, Aug. 2013.

[56] M. Caus and A. I. Pérez-Neira, "Multi-stream transmission for highly frequency selective channels in MIMO-FBMC/OQAM systems," *IEEE Trans. Signal Process.*, vol. 62, no. 4, pp. 786–796, Feb. 2014.

[57] R. Chauvat, P. Chevalier, and J. P. Delmas, "How to make quasi-rectilinear signals (MSK, GMSK, OQAM) almost equivalent to rectilinear ones (BPSK, ASK) for widely linear filtering in the presence of CCI," in *VDE ITG WSA 2015*, Ilmenau, Germany, Mar. 2015.

[58] P. Chevalier, R. Chauvat, and J. P. Delmas, "Quasi-rectilinear (MSK, GMSK, OQAM) co-channel interference mitigation by three inputs widely linear FRESH filtering," in *Proc. ICASSP-2015*, Brisbane, Australia, Apr. 2015.

[59] R. Chauvat, P. Chevalier, and J. P. Delmas, "Widely linear FRESH receiver for SAIC/MAIC with frequency offsets," in *Proc. ISWCS-2015*, Brussels, Belgium, Aug. 2015.

[60] P. Chevalier, J. P. Delmas, and R. Chauvat, "Reception filter impact on widely linear FRESH receiver performance for SAIC/MAIC with frequency offsets," in *Proc. SAM-2016*, Rio de Janeiro, Brazil, Jul. 2016.

[61] A. Ikhlef and J. Louveaux, "An enhanced MMSE per subchannel equalizer for highly frequency selective channels for FBMC/OQAM systems," in *Proc. SPAWC-2009*, Perugia, Italy, Jun. 2009.

[62] X. Mestre, M. Majoral, and S. Pfletschinger, "An asymptotic approach to parallel equalization of filter bank based multicarrier signals," *IEEE Trans. Signal Process.*, vol. 61, no. 14, pp. 3592–3606, Jul. 2013.

[63] X. Mestre and D. Gregoratti, "Parallelized structures for MIMO FBMC under strong channel frequency selectivity," *IEEE Trans. Signal Process.*, vol. 64, no. 5, pp. 1200–1215, Mar. 2016.

[64] M. Bellanger, "Specification and design of a prototype filter for filter bank based multicarrier transmission," in *Proc. ICASSP-2001*, Salt Lake City, UT, May 2001.

[65] M. 3GPP TS 36.101, "User equipment (UE) radio transmission and reception." 3rd Generation Partnership Project; Technical Specification Group Radio Access Network; Evolved Universal Terrestrial Radio Access (E-UTRA), http://www.3gpp.org, Tech. Rep., 2013.

[66] M. Renfors and J. Yli-Kaakinen, "Channel equalization in fast-convolution filter bank based receivers for professional mobile radio," in *Proc. EW-2014*, Barcelona, Spain, May 2014.

[67] T. Ihalainen, A. Ikhlef, J. Louveaux, and M. Renfors, "Channel equalization for multi-antenna FBMC/OQAM receivers," *IEEE Trans. Veh. Technol.*, vol. 60, no. 5, pp. 2070–2085, Apr. 2011.

[68] T. H. Stitz, T. Ihalainen, A. Viholainen, and M. Renfors, "Pilot-based synchronization and equalization in filter bank multicarrier communications," *EURASIP J. Adv. Signal Process.*, 2010. http://dx.doi.org/10.1155/2010/741429.

[69] J. R. Barry, E. A. Lee, and D. G. Messerschmitt, *Digital Communication*, 3rd ed. Boston: Kluwer Academic Publishers, 2004.

[70] Y. Yang, T. Ihalainen, M. Rinne, and M. Renfors, "Frequency-domain equalization in single-carrier transmission: Filter bank approach," *EURASIP J. Adv. Signal Process.*, 2007. http://dx.doi.org/10.1155/2007/10438.

[71] K. Shao, J. Alhava, J. Yli-Kaakinen, and M. Renfors, "Fast-convolution implementation of filter bank multicarrier waveform processing," in *Proc. ISCAS-2015*, Lisbon, Portugal, May 2015.

[72] J. Yli-Kaakinen and M. Renfors, "Flexible fast-convolution implementation of single-carrier waveform processing," in *Proc. ICC-2015 Workshops*, London, UK, Jun. 2015.

[73] A. Viholainen, M. Bellanger, and M. Huchard, "Prototype filter and structure optimization." Project PHYDYAS ICT-211887, Deliverable D5.1, Jan. 2009. [Online]. Available: http://www.ict-phydyas.org/.

[74] D. N. Godard, "Self-recovering equalization and carrier tracking in two-dimensional data communication systems," *IEEE Trans. Commun.*, vol. COM-28, no. 11, pp. 1867–1875, Nov. 1980.

[75] J. Treichler and B. Agee, "A new approach to multipath correction of constant modulus signals," *IEEE Trans. Acoust., Speech, Signal Process.*, vol. 31, no. 2, pp. 459–472, Apr. 1983.

[76] R. Johnson Jr., P. Schniter, T. Endres, J. Behm, D. Brown, and R. Casas, "Blind equalization using the constant modulus criterion: A review," *Proc. IEEE*, vol. 86, no. 10, pp. 1927–1950, Oct. 1998.

[77] J. Yang, J.-J. Werner, and G. A. Dumont, "The multimodulus blind equalization and its generalized algorithms," *IEEE J. Sel. Areas Commun.*, vol. 20, no. 5, pp. 997–1015, Jun. 2002.

[78] J. Mendes Filho, M. D. Miranda, and M. T. M. Silva, "A regional multimodulus algorithm for blind equalization of QAM signals: Introduction and steady-state analysis," *Signal Process.*, vol. 92, no. 11, pp. 2643–2656, Nov. 2012.

[79] H. Bölcskei, P. Duhamel, and R. Hleiss, "A subspace-based approach to blind channel identification in pulse shaping OFDM/OQAM systems," *IEEE Trans. Signal Process.*, vol. 49, no. 7, pp. 1594–1598, Jul. 2001.

[80] B. Farhang-Boroujeny, "Multicarrier modulation with blind detection capability using cosine modulated filter banks," *IEEE Trans. Commun.*, vol. 51, no. 12, pp. 2057–2070, Dec. 2003.

[81] L. Lin and B. Farhang-Boroujeny, "Convergence analysis of blind equalizer in a cosine modulated filter bank-based multicarrier communication system," in *Proc. SPAWC-2003*, Roma, Italy, Jun. 2003.

[82] K. N. Oh and Y. O. Chin, "Modified constant modulus algorithm: Blind equalization and carrier phase recovery algorithm," in *Proc. ICC-1995*, Seattle, WA, Jun. 1995.

[83] S. Barbarossa, "Blind equalization using cost function matched to the signal constellation," in *Proc. ACSSC-1997*, Pacific Grove, CA, Nov. 1997.

[84] A. Goupil and J. Palicot, "New algorithms for blind equalization: The constant norm algorithm family," *IEEE Trans. Signal Process.*, vol. 55, no. 4, pp. 1436–1444, Apr. 2007.

[85] V. Savaux, F. Bader, and J. Palicot, "OFDM/OQAM blind equalization using CNA approach," *IEEE Trans. Signal Process.*, vol. 9, no. 64, pp. 2324–2333, May 2016.

[86] S. Abrar and A. K. Nandi, "Blind equalization of square-QAM signals: A multimodulus approach," *IEEE Trans. Commun.*, vol. 58, no. 6, pp. 1674–1685, Jun. 2010.

[87] P. Bouboulis, "Wirtinger's calculus in general Hilbert spaces," May 2010. arXiv:1005.5170v1.

[88] A. M. Nassar and W. El Nahal, "New blind equalization technique for constant modulus algorithm (CMA)," in *Proc. CQR-2010*, Vancouver, BC, Jun. 2010.

[89] Y. Sato, "A method of self-recovering equalization for multilevel amplitude-modulation systems," *IEEE Trans. Commun.*, vol. 23, no. 6, pp. 679–682, Jun. 1975.

# Multi-Antenna and Multi-User Techniques

# MIMO-FBMC Transceivers

**Màrius Caus\*, Xavier Mestre\*, David Gregoratti\*, Ana I. Pérez-Neira\*,†,**
**Martin Haardt‡, Yao Cheng‡, Leonardo Gomes Baltar§**

*Centre Tecnològic de Telecomunicacions de Catalunya (CTTC/CERCA), Barcelona, Spain\**
*Universitat Politècnica de Catalunya, Barcelona, Spain†*
*Ilmenau University of Technology, Ilmenau, Germany‡*
*Intel Deutschland GmbH, Neubiberg, Germany§*

## CONTENTS

The objective of this chapter is to provide a concise overview and a valuable insight into the recent signal processing developments that allow the combination of Filter-Bank MultiCarrier with Offset-QAM subcarrier modulation (FBMC/OQAM) with Multiple-Input Multiple-Output (MIMO) technology, which is referred to as MIMO FBMC/OQAM. The adoption of FBMC/OQAM is endorsed by a good spectral confinement exhibited by each subcarrier, which is crucial to achieve a fine granularity of the spectrum and wireless access when tight synchronization may not be attained. The transmission of FBMC/OQAM relies on delaying half the symbol period the imaginary (real) part of data symbols on the even-indexed (odd-indexed) subcarriers. This results in a real-domain orthogonality characterized by the presence of a

Orthogonal Waveforms and Filter Banks for Future Communication Systems. DOI: 10.1016/B978-0-12-810384-5.00013-X self-interference term at the received signal, which is also known as intrinsic interference. Under ideal propagation conditions, the intrinsic interference is in quadrature with the desired data symbol thanks to the Perfect Reconstruction (PR) property. In this scenario, the unwanted signal can be easily removed. When the signal is propagated in a dispersive media, the orthogonality is usually destroyed, highlighting the necessity of perfectly equalizing the channel to exploit the PR property.

If the receiver and the transmitter are equipped with multiple antennas, the additional degrees of freedom provided by the spatial diversity are usually exploited to multiplex several streams on each subcarrier. However, each stream represents a potential source of distortion, and, thus, the detrimental effects induced by the intrinsic interference are intensified in MIMO architectures. This issue is circumvented if the input–output relation can be modeled like a set of parallel flat fading channels. Then, similarly to the Single-Input Single-Output (SISO) case, the receiver can straightforwardly leverage on the PR property to get rid of the interference. Unfortunately, the channel does not exhibit this structure. The aim of this chapter is to provide insight into the design of MIMO precoding and decoding matrices, so that the combined response after performing the transmit and the receive processing is as similar as possible to the target impulse response. To this end, we cannot rely on a mere generalization of those techniques applied in Orthogonal Frequency-Division Multiplexing (OFDM), and the characteristics of FBMC/OQAM have to be taken into account. Indeed, how to unleash the MIMO potential in the FBMC/OQAM context is an open research problem. In the following sections, the state-of-the art techniques and the characterization of the distortion caused by the channel are presented. First, this chapter addresses the solutions that are devised for low- and mild-frequency selective scenarios, where the channel frequency response can be assumed flat at each subcarrier. Next, the most general case is considered, and no assumptions are made about the channel frequency selectivity. It is worth mentioning that, similarly to [1], only the solutions that offer the best performance have been selected for each scenario. However, this chapter is a bit more specific than [1] and only focuses on the situations where the channel state information is available at both transmitter and receiver, with special emphasis on strong frequency selective channels. For this type of scenario, this chapter goes a bit beyond the technical contents in [1], and includes the description of additional designs and their associated performance analyses. In this sense, it provides more insight into the transceiver design problem when the channel cannot be assumed flat fading at the subcarrier level. Before going into details, the next section provides the basics of the system model when FBMC/OQAM is implemented in MIMO architectures.

## 13.1 SYSTEM MODEL

This section is devoted to providing a compact notation for the baseband model of a MIMO communication system, the air interface of which is based on the FBMC/OQAM modulation scheme. The MIMO setup is composed of $N_T$ transmit

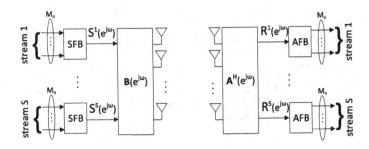

**FIGURE 13.1**

General architecture of a multicarrier MIMO transceiver.

antennas and $N_R$ receive antennas. Fig. 13.1 illustrates the general architecture when the spatial diversity is exploited to simultaneously transmit $S$ multicarrier signals. Let $S^i(e^{j\omega})$ denote the Discrete-Time Fourier Transform (DTFT) of the signal generated by the $i$th Synthesis FilterBank (SFB), namely,

$$S^i\left(e^{j\omega}\right) = \sum_{m \in \mathcal{M}_u} G_m\left(e^{j\omega}\right) \check{D}^i_m\left(e^{j\omega L}\right), \quad 1 \le i \le S,$$

where $\mathcal{M}_u$ is the set of used subcarriers. On the $m$th subcarrier, the DTFT of the synthesis filter and the data symbol streams are denoted by $G_m(e^{j\omega})$ and $\check{D}^i_m(e^{j\omega})$, respectively. Note that the spectrum of the symbols is shrunk by a factor $L$ to fit into the bandwidth of $G_m(e^{j\omega})$, which enables parallel transmission in the frequency domain. The factor $L$ depends on the sampling frequency and the symbol period. In the FBMC/OQAM context, the rate conversion is performed by upsampling the symbols by a factor $L = \frac{M}{2}$, where $M$ is the number of subcarriers. The frequency representation of the signal at the input of the Analysis FilterBank (AFB) is given by the receive vector $\mathbf{R}(e^{j\omega}) = \left[R^1(e^{j\omega}) \cdots R^S(e^{j\omega})\right]^T$. The input–output relation is formulated as follows:

$$\mathbf{R}(e^{j\omega}) = \mathbf{A}^H(e^{j\omega})\mathbf{H}(e^{j\omega})\mathbf{B}(e^{j\omega})\mathbf{S}(e^{j\omega}) + \mathbf{A}^H(e^{j\omega})\mathbf{W}(e^{j\omega}), \quad (13.1)$$

where the transmit vector is denoted by $\mathbf{S}(e^{j\omega}) = \left[S^1(e^{j\omega}) \cdots S^S(e^{j\omega})\right]^T$, and the noise vector that contaminates the reception is $\mathbf{W}(e^{j\omega}) = \left[W^1(e^{j\omega}) \cdots W^{N_R}(e^{j\omega})\right]^T$. The precoder, the equalizer and the channel are respectively denoted by $\mathbf{B}(e^{j\omega}) \in \mathbb{C}^{N_T \times S}$, $\mathbf{A}(e^{j\omega}) \in \mathbb{C}^{N_R \times S}$ and $\mathbf{H}(e^{j\omega}) \in \mathbb{C}^{N_R \times N_T}$. It becomes evident that we can leverage on the Singular Value Decomposition (SVD) of the channel to force the combined response $\mathbf{A}^H(e^{j\omega})\mathbf{H}(e^{j\omega})\mathbf{B}(e^{j\omega})$ to exhibit a diagonal structure. This approach paves the way to independently processing each multicarrier signal to recover the information conveyed in the frequency bins. It must be mentioned that in a time-dispersive media, the SVD beamforming design involves a lot of arithmetic operations and yields excessively long impulse responses. As a consequence, the

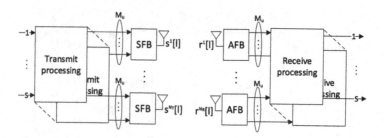

**FIGURE 13.2**

General architecture of a MIMO FBMC/OQAM transceiver with precoders and equalizers working on a per subcarrier basis.

complexity may not be affordable, rendering the solution impractical. To overcome the complexity issues, one typically performs transmit and receive processing on a per-subcarrier basis. This is tantamount to moving the precoder to the input of the SFB and the equalizer to the output of the AFB, as Fig. 13.2 shows. The rationale behind this choice stems form the fact that the channel frequency selectivity is less severe on the bandwidth of a single subcarrier than on the whole transmission bandwidth. Hence, the number of taps required to implement precoders and equalizers will definitely be reduced. It is important to emphasize that the simplest architecture relies on single-tap precoders and equalizers.

In the discrete-time domain, the signal generated by the SFB of the $p$th antenna is expressed as

$$s^p[l] = \sum_{m \in \mathcal{M}_u} \sum_{n=-\infty}^{\infty} v_{m,n}^p g_m \left[ l - n\frac{M}{2} \right], \quad 1 \le p \le N_T,$$

where $v_{m,n}^p$ denotes the signal transmitted by the $p$th antenna on the frequency-time position $(m, n)$, and $g_m[l]$ is the subband pulse that shapes the symbols transmitted on the $m$th subcarrier. To use a sufficiently general notation that is able to accommodate any solution, the vector of precoded symbols $\mathbf{v}_{m,n} \in \mathbb{C}^{N_T \times 1}$ is obtained by mapping $S$ data symbols

$$\check{\mathbf{d}}_{m,n} = e^{j\varphi_{m,n}} \left[ d_{m,n}^1 \cdots d_{m,n}^S \right]^{\mathrm{T}} = e^{j\varphi_{m,n}} \mathbf{d}_{m,n}$$

onto $N_T$ transmit antennas by using the function $Q_{\mathrm{tx},m}(.)$, i.e.,

$$\mathbf{v}_{m,n} = \left[ v_{m,n}^1 \cdots v_{m,n}^{N_T} \right]^{\mathrm{T}} = Q_{\mathrm{tx},m} (\check{\mathbf{d}}_{m,n}). \quad (13.2)$$

The phase term $e^{j\varphi_{m,n}}$ is designed to guarantee that adjacent symbols in the frequency-time grid have a difference of phase equal to $\frac{\pi}{2}$.

In the presence of multipath fading, the signal received by the $q$th antenna is given by

$$r^q[l] = \sum_{p=1}^{N_T} s^p[l] \star h^{q,p}[l] + w^q[l], \quad 1 \le q \le N_R.$$

Concerning the sources of distortion, $w^q[l]$ corresponds to the noise at the input of the $q$th receive antenna, and $h^{q,p}[l]$ accounts for the channel impulse response between the $p$th transmit antenna and the $q$th receive antenna. To recover the information conveyed on each subcarrier, $r^q[l]$ is fed into the AFB, which is composed of a bank of filters that are matched to the synthesis filters. Assuming that SFB and AFB use the same prototype pulse, the $m$th output of the AFB can be expressed as follows:

$$\begin{aligned}
y^q_{m,n} &= \sum_l r^q[l] e^{-j\varphi_{m,n}} g^*_m\left[l - n\frac{M}{2}\right] \\
&= \sum_{p=1}^{N_T} \sum_{m'=m-1}^{m+1} e^{-j\varphi_{m,n}} \left(\Gamma^{q,p}_{m,m';n} \star v^p_{m',n}\right) + \eta^q_{m,n}
\end{aligned}$$

with

$$\eta^q_{m,n} = \sum_l w^q[l] e^{-j\varphi_{m,n}} g^*_m\left[l - n\frac{M}{2}\right].$$

It can be verified that $y^q_{m,n}$ can be compactly expressed as a function of the impulse response between the $m$th subcarrier and the $m'$th subcarrier, which is given by $\Gamma^{q,p}_{m,m';n}$. Since the energy of the prototype pulse is usually concentrated in this frequency range $\left[-\frac{2\pi}{M}, \frac{2\pi}{M}\right]$, only adjacent subcarriers lead to Inter-Carrier Interference (ICI). Depending on the propagation conditions, the equivalent response that includes the channel and the subband pulses can be characterized by the following two models:

$$\begin{aligned}
&1) \quad \Gamma^{q,p}_{m,m';n} = \left(g^*_m[-l] \star h^{q,p}[l] \star g_{m'}[l]\right)_{l=n\frac{M}{2}}, \\
&2) \quad \Gamma^{q,p}_{m,m';n} = H^{q,p}_{m'}\left(g^*_m[-l] \star g_{m'}[l]\right)_{l=n\frac{M}{2}} = H^{q,p}_{m'}\alpha_{m,m';n},
\end{aligned} \tag{13.3}$$

where

$$\alpha_{m,m';n} = \sum_l g^*_m\left[l - n\frac{M}{2}\right] g_{m'}[l]$$

denotes the transmultiplexer response, and the variable $H^{q,p}_{m'}$ corresponds to the DTFT of $h^{q,p}[l]$ evaluated at the radial frequency $\frac{2\pi m'}{M}$. Model 1) is valid regardless of the degree of the channel frequency selectivity. Conversely, model 2) is only accurate when the channel can be assumed flat on each subcarrier. Note that, in model 2), the channel seen by the signal transmitted on the $m'$th subcarrier depends on $H_{m'}^{q,p}$ and the transmultiplexer response $\alpha_{m,m';n}$. It will be shown that model 2) is preferable to model 1) because it offers a better analytical tractability. It is worth emphasizing that the set of functions $\{e^{j\varphi_{m,n}} g_m[l - n\frac{M}{2}]\}$ constitutes an orthogonal basis in the real domain if the PR property is satisfied. In this case, it can be verified that the following property is satisfied:

$$\text{Re}\{e^{-j\varphi_{m,n}} e^{j\varphi_{m,n-\tau}} \alpha_{m,m';\tau}\} = \delta_{m,m'}\delta_{\tau,0}. \tag{13.4}$$

To exploit the potential of MIMO, it is customary to stack columnwise the samples received by each antenna. Resorting to the matrix notation, the interplay from the input of the SFB to the output of the AFB can be compactly formulated as

$$\mathbf{y}_{m,n} = \left[y_{m,n}^1 \cdots y_{m,n}^{N_R}\right]^T = \sum_{m'=m-1}^{m+1} e^{-j\varphi_{m,n}} \left(\mathbf{\Gamma}_{m,m';n} \star \mathbf{v}_{m',n}\right) + \boldsymbol{\eta}_{m,n}$$

$$= \sum_{m'=m-1}^{m+1} \sum_{\tau} e^{-j\varphi_{m,n}} \left(\mathbf{\Gamma}_{m,m';\tau} \mathbf{v}_{m',n-\tau}\right) + \boldsymbol{\eta}_{m,n}$$

with

$$\mathbf{\Gamma}_{m,m';n} = \begin{bmatrix} \Gamma_{m,m';n}^{1,1} & \cdots & \Gamma_{m,m';n}^{1,N_T} \\ \vdots & \ddots & \vdots \\ \Gamma_{m,m';n}^{N_R,1} & \cdots & \Gamma_{m,m';n}^{N_R,N_T} \end{bmatrix}, \quad -\left\lfloor\frac{L_g-1}{M/2}\right\rfloor \leq n \leq \left\lfloor\frac{L_g-1+L_{ch}}{M/2}\right\rfloor. \tag{13.5}$$

Note that the memory of $\mathbf{\Gamma}_{m,m';n}$ depends on the number of subcarriers, the duration of the pulse, and the maximum channel excess delay, which is denoted by $L_{ch}$. When model 2) in (13.3) holds, (13.5) becomes

$$\mathbf{\Gamma}_{m,m';n} = \alpha_{m,m';n}\mathbf{H}_{m'} = \alpha_{m,m';n}\begin{bmatrix} H_{m'}^{1,1} & \cdots & H_{m'}^{1,N_T} \\ \vdots & \ddots & \vdots \\ H_{m'}^{N_R,1} & \cdots & H_{m'}^{N_R,N_T} \end{bmatrix}. \tag{13.6}$$

The noise vector is defined analogously to $\mathbf{y}_{m,n}$ as follows:

$$\boldsymbol{\eta}_{m,n} = \left[\eta_{m,n}^1 \cdots \eta_{m,n}^{N_R}\right]^T.$$

When the pulses satisfy the PR property and the noise samples at the input of the receiver are spatially uncorrelated and modeled like zero-mean symmetric complex Gaussian random variables, i.e., $w^q[l] \sim \mathcal{CN}(0, \sigma_w^2)$, it follows that $\boldsymbol{\eta}_{m,n} \sim$

$\mathcal{CN}\left(\mathbf{0}, \sigma_w^2 \mathbf{I}_{N_R}\right)$. In the last step, the function $Q_{m,\mathrm{rx}}(\cdot)$ provides the symbol estimates, i.e.,

$$\hat{\mathbf{d}}_{m,n} = \left[\hat{d}_{m,n}^1 \cdots \hat{d}_{m,n}^S\right]^{\mathsf{T}} = Q_{m,\mathrm{rx}}(\mathbf{y}_{m,n}). \qquad (13.7)$$

Next sections address the design of $Q_{m,\mathrm{rx}}(\cdot)$ and $Q_{m,\mathrm{tx}}(\cdot)$ given different optimization criteria and propagation conditions. It will be shown that using the channel state information at both ends of the link, the rich scattering of the environment can be exploited to achieve multiplexing and diversity gains.

## 13.2 LOW- AND MILD-FREQUENCY SELECTIVITY SCENARIOS

In this section, we assume that the channel frequency response is flat at the subcarrier level and, thus, the simplified model 2) presented in (13.3) and (13.6) is used to characterize the equivalent channel. The MIMO techniques that are studied in this section perform linear processing. Without loss of generality, the MIMO precoding and decoding matrices that are employed on the $m$th subcarrier are denoted by $\mathbf{B}_m \in \mathbb{C}^{N_T \times S}$ and $\mathbf{A}_m \in \mathbb{C}^{N_R \times S}$, respectively. Borrowing the notation from (13.2), the linearly precoded symbols can be written as

$$\mathbf{v}_{m,n} = \mathbf{B}_m \check{\mathbf{d}}_{m,n}.$$

Since useful information is only conveyed in the real domain, the estimated symbols can be obtained by extracting the real part of the equalized vector. Therefore, (13.7) becomes

$$\hat{\mathbf{d}}_{m,n} = \mathbb{Re}\left\{\mathbf{A}_m^{\mathsf{H}} \mathbf{y}_{m,n}\right\}. \qquad (13.8)$$

According to the mild-frequency selective channel model, when linear processing is performed, the following equation can be used to describe the system model at the $(m,n)$th frequency-time position:

$$\hat{\mathbf{d}}_{m,n} = \sum_{m'=m-1}^{m+1} \sum_{\tau} \mathbb{Re}\left\{e^{-j\varphi_{m,n}} e^{j\varphi_{m',n-\tau}} \alpha_{m,m';\tau} \mathbf{A}_m^{\mathsf{H}} \mathbf{H}_{m'} \mathbf{B}_{m'}\right\} \mathbf{d}_{m',n-\tau} \qquad (13.9)$$
$$+ \mathbb{Re}\left\{\mathbf{A}_m^{\mathsf{H}} \boldsymbol{\eta}_{m,n}\right\}.$$

It becomes evident that all interfering symbols can be removed if the combined response $\mathbf{A}_m^{\mathsf{H}} \mathbf{H}_{m'} \mathbf{B}_{m'}$ is real valued thanks to the PR property (13.4). To achieve the desired response, the pair of matrices $\{\mathbf{A}_m, \mathbf{B}_m\}$ should take into account $\mathbf{H}_m$ and the channel experienced on the adjacent subcarriers, i.e., $\mathbf{H}_{m+1}$ and $\mathbf{H}_{m-1}$. This observation suggests that subcarriers cannot be independently processed. Indeed, the optimal solution consists in jointly processing all subcarriers. Nevertheless, when the number of subcarriers is large, the complexity burden that is required is too high. To overcome the complexity issues and effectively mitigate the interference, we cannot rely on solutions applied in narrowband MIMO communication systems. It is of paramount importance to take into account all sources of interference that appear in (13.9). With the aim of making progress toward the application of MIMO to FBMC/OQAM systems, next sections detail how to deal with self-interference and provide diversity gains.

## 13.2.1 LOW-FREQUENCY SELECTIVITY SCENARIOS

This section focuses on the design of MIMO techniques in low-frequency selective scenarios. It is worth mentioning that the term low-frequency selective is used to identify those systems, where the subcarrier spacing is such that the channel frequency response is flat at least in three consecutive subcarriers, i.e., $\mathbf{H}_{m+1} = \mathbf{H}_m = \mathbf{H}_{m-1}$. Therefore, the channel seen by all the interfering signals in a given frequency-time position is the same. This means that the channel is flat in the neighborhood around the frequency-time position of interest. As a consequence, it can be inferred that the precoders applied on a given subcarrier and the adjacent ones are virtually the same. Therefore, (13.9) is modified as

$$\hat{\mathbf{d}}_{m,n} = \sum_{m'=m-1}^{m+1} \sum_{\tau} \mathrm{Re}\left\{e^{-j\varphi_{m,n}} e^{j\varphi_{m',n-\tau}} \alpha_{m,m';\tau} \mathbf{A}_m^H \mathbf{H}_m \mathbf{B}_m \right\} \mathbf{d}_{m',n-\tau}$$
$$+ \mathrm{Re}\left\{\mathbf{A}_m^H \boldsymbol{\eta}_{m,n}\right\}.$$

Essentially, the difference with respect to (13.9) is that the streams go through the MIMO channel $\mathbf{A}_m^H \mathbf{H}_m \mathbf{B}_m$ rather than $\mathbf{A}_m^H \mathbf{H}_{m'} \mathbf{B}_{m'}$. Under these conditions, the MIMO techniques originally devised for OFDM can be directly applied to FBMC/OQAM without any performance degradation [2,3]. To see it, it is important to recall that the optimal solution in the Mean-Squared Error (MSE) sense for OFDM systems is achieved by the filters

$$\mathbf{B}_m = \begin{bmatrix} \mathbf{b}_{m,1} \cdots \mathbf{b}_{m,S} \end{bmatrix} = \begin{bmatrix} \sqrt{p_{m,1}} \mathbf{u}_{m,1} \cdots \sqrt{p_{m,S}} \mathbf{u}_{m,S} \end{bmatrix},$$
$$\mathbf{A}_m = \begin{bmatrix} \mathbf{a}_{m,1} \cdots \mathbf{a}_{m,S} \end{bmatrix} = \left(\mathbf{H}_m \mathbf{B}_m \mathbf{B}_m^H \mathbf{H}_m^H + \mathbf{R}_{\eta_m}\right)^{-1} \mathbf{H}_m \mathbf{B}_m.$$

The precoder has as columns the eigenvectors of $\mathbf{H}_m^H \mathbf{R}_{\eta_m} \mathbf{H}_m \in \mathbb{C}^{N_T \times N_T}$ [4], where $\mathbf{R}_{\eta_m} = \mathbb{E}\left\{\boldsymbol{\eta}_{m,n} \boldsymbol{\eta}_{m,n}^H\right\} = \sigma_w^2 \mathbf{I}_{N_R}$ is the noise covariance matrix. In other words, $\mathbf{u}_{m,i} \in \mathbb{C}^{N_T \times 1}$ is associated with the $i$th dominant eigenvalue $\lambda_{m,i}$. This closed-form solution is derived under the assumption that symbols are uncorrelated and that the mean energy symbol is one, i.e., $\mathbb{E}\left\{\mathbf{d}_{m,n} \mathbf{d}_{m',n'}^H\right\} = \delta_{m,m'} \delta_{n,n'} \mathbf{I}_S$. This beamforming design allows transmitting up to $S \leq \min\{N_T, N_R\}$ streams per subcarrier with the possibility of detecting them separately. The solution is governed by $\mathbf{H}_m$, the noise covariance matrix $\mathbf{R}_{\eta_m}$, and the coefficients $\{p_{m,i}\}$, which determine the power allocated to each stream. Therefore, beamformers are computed on a per-subcarrier basis. Assuming that $\mathbf{H}_{m+1} = \mathbf{H}_m = \mathbf{H}_{m-1}$, the channel seen by the stream $\mathbf{d}_{m',n-\tau}$ is real valued, for $m' = m - 1, m, m + 1$, and the end-to-end communication system reads

$$\hat{d}_{m,n}^i = \sum_{m'=m-1}^{m+1} \sum_{\tau} \mathrm{Re}\left\{e^{-j\varphi_{m,n}} e^{j\varphi_{m',n-\tau}} \alpha_{m,m';\tau}\right\} \frac{p_{m,i}\lambda_{m,i}}{1 + p_{m,i}\lambda_{m,i}} d_{m',n-\tau}^i$$
$$+ \mathrm{Re}\left\{\mathbf{a}_{m,i}^{\mathrm{H}} \boldsymbol{\eta}_{m,n}\right\}, \quad 1 \le i \le S.$$

Knowing from Section 13.1 that $\boldsymbol{\eta}_{m,n} \sim \mathcal{CN}\left(\mathbf{0}, \sigma_w^2 \mathbf{I}_{N_R}\right)$, it can be readily verified that the variance of the equalized noise is given by

$$\mathbb{E}\left\{\left|\mathrm{Re}\left\{\mathbf{a}_{m,i}^{\mathrm{H}} \boldsymbol{\eta}_{m,n}\right\}\right|^2\right\} = \frac{\sigma_w^2}{2} \left\|\mathbf{a}_{m,i}^{\mathrm{H}}\right\|^2.$$

Based on the statistical information of the noise and the design of the beamformers, the Signal-to-Noise Ratio (SNR) is formulated as

$$\mathrm{SNR}_{m,i} = 2 p_{m,i} \lambda_{m,i}.$$

In low-frequency selective channel scenarios, channels associated with adjacent subcarriers share the same SVD. Hence, symbols can be detected in the absence of Inter-Symbol Interference (ISI) and ICI if (13.4) is satisfied. When $\mathbf{H}_{m+1} \ne \mathbf{H}_m \ne \mathbf{H}_{m-1}$, beamformers differ from subcarrier to subcarrier, so that $\mathbf{A}_m^{\mathrm{H}} \mathbf{H}_m \mathbf{B}_{m'}$ is not a real-valued matrix for $m \ne m'$ anymore. As a consequence, the received symbols are contaminated by a residual interference term that leads to an error floor. The analysis reveals that channel frequency selectivity is the main inhibitor to a direct application of OFDM solutions [4] to FBMC/OQAM systems.

### 13.2.2 MILD-FREQUENCY SELECTIVITY SCENARIOS

The objective of this section is to present the techniques that exhibit a higher degree of robustness against the channel frequency selectivity compared to the solution described in the previous section. To this end, it is not assumed that the channel keeps constant on adjacent subcarriers, albeit flatness is exhibited on the pass band region of each subcarrier. Hence, the system model that characterizes the input–output relation is defined by (13.9). As it is pointed out in [5], the fulfillment of (13.4) enables differentiating desired and unwanted signals by recasting (13.9) as

$$\hat{\mathbf{d}}_{m,n} = \mathrm{Re}\left\{\mathbf{A}_m^{\mathrm{H}} \mathbf{H}_m \mathbf{B}_m\right\} \mathbf{d}_{m,n}$$
$$- \sum_{m'=m-1}^{m+1} \mathrm{Im}\left\{\mathbf{A}_m^{\mathrm{H}} \mathbf{H}_{m'} \mathbf{B}_{m'}\right\} \mathbf{i}_{m,m';n} + \mathrm{Re}\left\{\mathbf{A}_m^{\mathrm{H}} \boldsymbol{\eta}_{m,n}\right\}.$$

To use easy-to-handle expressions, the following vectors are defined:

$$\mathbf{i}_{m,m;n} = \sum_{\substack{\tau \\ \tau \neq 0}} \mathbb{Im}\left\{ e^{-j\varphi_{m,n}} e^{j\varphi_{m,n-\tau}} \alpha_{m,m;\tau} \right\} \mathbf{d}_{m,n-\tau},$$

$$\mathbf{i}_{m,m';n} = \sum_{\tau} \mathbb{Im}\left\{ e^{-j\varphi_{m,n}} e^{j\varphi_{m',n-\tau}} \alpha_{m,m';\tau} \right\} \mathbf{d}_{m',n-\tau}, \quad m \neq m'. \tag{13.10}$$

Separating real and imaginary parts of MIMO precoding and decoding matrices as

$$\bar{\mathbf{A}}_m = \left[ \mathbb{Re}\left\{\mathbf{A}_m^{\mathrm{T}}\right\} \ \mathbb{Im}\left\{\mathbf{A}_m^{\mathrm{T}}\right\} \right]^{\mathrm{T}}, \quad \bar{\mathbf{B}}_m = \left[ \mathbb{Re}\left\{\mathbf{B}_m^{\mathrm{T}}\right\} \ \mathbb{Im}\left\{\mathbf{B}_m^{\mathrm{T}}\right\} \right]^{\mathrm{T}} \tag{13.11}$$

and resorting to the augmented channel matrices

$$\mathbf{H}_m^{\mathrm{d}} = \begin{bmatrix} \mathbb{Re}\left\{\mathbf{H}_m\right\} & -\mathbb{Im}\left\{\mathbf{H}_m\right\} \\ \mathbb{Im}\left\{\mathbf{H}_m\right\} & \mathbb{Re}\left\{\mathbf{H}_m\right\} \end{bmatrix}, \quad \mathbf{H}_m^{\mathrm{i}} = \begin{bmatrix} \mathbb{Im}\left\{\mathbf{H}_m\right\} & \mathbb{Re}\left\{\mathbf{H}_m\right\} \\ -\mathbb{Re}\left\{\mathbf{H}_m\right\} & \mathbb{Im}\left\{\mathbf{H}_m\right\} \end{bmatrix}, \tag{13.12}$$

the real-valued representation can be compactly expressed as

$$\begin{aligned}
\hat{\mathbf{d}}_{m,n} = {}& \bar{\mathbf{A}}_m^{\mathrm{T}} \mathbf{H}_m^{\mathrm{d}} \bar{\mathbf{B}}_m \mathbf{d}_{m,n} \\
& - \sum_{m'=m-1}^{m+1} \bar{\mathbf{A}}_m^{\mathrm{T}} \mathbf{H}_{m'}^{\mathrm{i}} \bar{\mathbf{B}}_{m'} \mathbf{i}_{m,m';n} + \mathbb{Re}\left\{\mathbf{A}_m^{\mathrm{H}} \eta_{m,n}\right\}.
\end{aligned} \tag{13.13}$$

By getting rid of real and imaginary operators, matrix algebra tools can be exploited to design transmit and receive beamformers. Another important aspect that deserves consideration is the fact that the estimated symbols are obtained by linearly combining $\mathbb{Re}\left\{\mathbf{y}_{m,n}\right\}$ and $\mathbb{Im}\left\{\mathbf{y}_{m,n}\right\}$. This approach can be considered as an especial case of widely linear processing when data symbols only convey useful information in the real dimension [6]. It can be concluded that mild- and strong-frequency selective channels call for a procedure to estimate the data different from that proposed in low-frequency selective scenarios, where symbols are assumed to be complex valued. Building upon the system model described in (13.13) and taking into account the peculiarities of FBMC/OQAM, the Zero Forcing (ZF) and the coordinated beamforming methods are derived to combat the channel frequency selectivity.

### 13.2.2.1 *Zero Forcing*

The rationale behind the adoption of the ZF criterion is to achieve a multistream communication system free of interference. To this end, the interference and the desired signal should be projected onto orthogonal subspaces. Therefore, the design of the MIMO precoding and decoding matrices is governed by the zero-interference constraints, i.e., $\bar{\mathbf{A}}_m^{\mathrm{T}} \mathbf{H}_{m'}^{\mathrm{i}} \bar{\mathbf{B}}_{m'} = \mathbf{0}$ for $m' = m - 1, m, m + 1$. Unfortunately, both matrices $\mathbf{H}_{m'}^{\mathrm{d}} \in \mathbb{R}^{2N_{\mathrm{R}} \times 2N_{\mathrm{T}}}$ and $\mathbf{H}_{m'}^{\mathrm{i}} \in \mathbb{R}^{2N_{\mathrm{R}} \times 2N_{\mathrm{T}}}$ span the same space. Hence, it is not possible to remove the interference without eliminating the desired signal as well. To circumvent this problem, the authors in [7] propose to constrain the equalizers to be real valued. Then, (13.11) and (13.12) are recast as

$$\bar{\mathbf{A}}_m = \mathrm{Re}\left\{\mathbf{A}_m\right\}, \quad \bar{\mathbf{B}}_m = \left[\mathrm{Re}\left\{\mathbf{B}_m^{\mathsf{T}}\right\} \quad \mathrm{Im}\left\{\mathbf{B}_m^{\mathsf{T}}\right\}\right]^{\mathsf{T}}, \tag{13.14}$$

$$\mathbf{H}_m^{\mathrm{d}} = \left[\mathrm{Re}\left\{\mathbf{H}_m\right\} \quad -\mathrm{Im}\left\{\mathbf{H}_m\right\}\right], \quad \mathbf{H}_m^{\mathrm{i}} = \left[\mathrm{Im}\left\{\mathbf{H}_m\right\} \quad \mathrm{Re}\left\{\mathbf{H}_m\right\}\right]. \tag{13.15}$$

Now, by performing the SVD of $\mathbf{H}_m^{\mathrm{i}}$ it is possible to find a matrix $\mathbf{U}_m^0 \in \mathbb{R}^{2N_{\mathrm{T}} \times 2N_{\mathrm{T}}-N_{\mathrm{R}}}$ that fulfills $\bar{\mathbf{A}}_m^{\mathsf{T}}\mathbf{H}_m^{\mathrm{i}}\mathbf{U}_m^0 = \mathbf{0}$, $\bar{\mathbf{A}}_m^{\mathsf{T}}\mathbf{H}_m^{\mathrm{d}}\mathbf{U}_m^0 \neq \mathbf{0}$ and $\left(\mathbf{U}_m^0\right)^{\mathsf{T}}\mathbf{U}_m^0 = \mathbf{I}_{2N_{\mathrm{T}}-N_{\mathrm{R}}}$. The fulfillment of these conditions relies on the fact that $\mathbf{H}_m^{\mathrm{d}}$ and $\mathbf{H}_m^{\mathrm{i}}$ span different row subspaces. Notice that the dimensionality constraint imposed by the ZF on the number of antennas is $2N_{\mathrm{T}} > N_{\mathrm{R}}$. Without loss of generality, the precoder can be split into an inner precoder $\mathbf{B}_m^{\mathrm{i}} \in \mathbb{R}^{2N_{\mathrm{T}} \times 2N_{\mathrm{T}}-N_{\mathrm{R}}}$ and an outer precoder $\mathbf{B}_m^{\mathrm{o}} \in \mathbb{R}^{2N_{\mathrm{T}}-N_{\mathrm{R}} \times S}$ that results in this factorization $\mathbf{B}_m = \mathbf{B}_m^{\mathrm{i}}\mathbf{B}_m^{\mathrm{o}}$. If $\mathbf{B}_m^{\mathrm{i}} = \mathbf{U}_m^0$, then it follows that the input–output relation can be modeled like an $N_{\mathrm{R}} \times 2N_{\mathrm{T}} - N_{\mathrm{R}}$ MIMO communications system, namely,

$$\hat{\mathbf{d}}_{m,n} = \bar{\mathbf{A}}_m^{\mathsf{T}}\mathbf{H}_m^{\mathrm{d}}\mathbf{U}_m^0\bar{\mathbf{B}}_m^{\mathrm{o}}\mathbf{d}_{m,n} + \bar{\mathbf{A}}_m^{\mathsf{T}}\mathrm{Re}\{\boldsymbol{\eta}_{m,n}\}.$$

After eliminating ICI and ISI terms, the rest of matrices can be designed using the optimization framework developed in [4]. Following the same philosophy as Section 13.2.1, which is based on the Minimum Mean-Squared Error (MMSE), the matrix pair $\left\{\bar{\mathbf{A}}, \bar{\mathbf{B}}^{\mathrm{o}}\right\}$ is given by

$$\bar{\mathbf{B}}_m^{\mathrm{o}} = \left[\bar{\mathbf{b}}_{m,1}^{\mathrm{o}} \cdots \bar{\mathbf{b}}_{m,S}^{\mathrm{o}}\right] = \left[\sqrt{p_{m,1}}\mathbf{u}_{m,1} \cdots \sqrt{p_{m,S}}\mathbf{u}_{m,S}\right],$$

$$\bar{\mathbf{A}}_m = \left[\bar{\mathbf{a}}_{m,1} \cdots \bar{\mathbf{a}}_{m,S}\right] = \left(\mathbf{H}_m^{\mathrm{d}}\bar{\mathbf{B}}_m \left(\mathbf{H}_m^{\mathrm{d}}\bar{\mathbf{B}}_m\right)^{\mathsf{T}} + \frac{\sigma_w^2}{2}\mathbf{I}_{N_{\mathrm{R}}}\right)^{-1}\mathbf{H}_m^{\mathrm{d}}\bar{\mathbf{B}}_m.$$

The column vector $\mathbf{u}_{m,i}$ represents the eigenvector of the $2N_{\mathrm{T}} - N_{\mathrm{R}} \times 2N_{\mathrm{T}} - N_{\mathrm{R}}$ square matrix $\frac{2}{\sigma_w^2}\left(\mathbf{H}_m^{\mathrm{d}}\mathbf{U}_m^0\bar{\mathbf{B}}_m^{\mathrm{o}}\right)^{\mathsf{T}}\mathbf{H}_m^{\mathrm{d}}\mathbf{U}_m^0\bar{\mathbf{B}}_m^{\mathrm{o}}$ that is associated with the $i$th dominant eigenvalue, denoted by $\beta_{m,i}$. To find the solution, it is important to realize that the noise covariance matrix is formulated as $\frac{\sigma_w^2}{2}\mathbf{I}_{N_{\mathrm{R}}} = \mathbb{E}\left\{\mathrm{Re}\{\boldsymbol{\eta}_{m,n}\}\mathrm{Re}\{\boldsymbol{\eta}_{m,n}^{\mathsf{T}}\}\right\}$. The maximum number of streams that can be distinguished at the receiver depends on the dimensions of the equivalent channel, which includes the inner precoder. As a consequence, the number of streams and the number of antennas are related as follows: $S \leq \min\{2N_{\mathrm{T}} - N_{\mathrm{R}}, N_{\mathrm{R}}\}$. When the number of antennas is such that $N_{\mathrm{T}} > N_{\mathrm{R}}$, both the ZF method and the solution described in Section 13.2.1 are able to spatially multiplex the same number of streams. The difference lies in the SNR achieved by each solution. The quality of the received streams when the ZF is implemented can be measured as

$$\mathrm{SNR}_{m,i} = 2p_{m,i}\beta_{m,i}.$$

The theoretical analysis conducted in [7] reveals that the spatial channel gains of the ZF and the technique described in Section 13.2.1 are different. In MIMO systems where $N_R \leq N_T$, it has been shown that the eigenvalues are less spread out when the ZF approach is applied. Therefore, the following inequalities are satisfied:

$$\lambda_{m,1} \geq \beta_{m,1}, \quad \lambda_{m,N_R} \leq \beta_{m,N_R}.$$

From the inequalities it can be inferred that if $S = 1$ and $N_R \leq N_T$, then the technique described in this section gives poor performance. By contrast, if $S = N_R \leq N_T$, then we cannot predict which design will give the best performance because at least in one spatial channel the ZF achieves the highest gain. Therefore, it can be concluded that satisfactory performance is not guaranteed unless all the modes are active, i.e., $S = N_R \leq N_T$. Especial attention must be paid to the case where $N_R = 1$, because $\lambda_{m,1} = \beta_{m,1}$. In the light of this result, it can be resolved that in Multiple-Input Single-Output (MISO) communication systems the ZF does not exhibit any performance degradation.

### 13.2.2.2 Coordinated Beamforming

A coordinated beamforming scheme has been proposed in [8] to alleviate the dimensionality constraint on the ZF method (as it was just explained, its performance is only satisfactory when $S = N_R \leq N_T$). The enhancement lies on establishing a dependency between the precoders and the decoders. The precoding matrix and the decoding matrix are updated jointly in an iterative manner considering the real-valued notation written in (13.14) and (13.15). Based on such a coordinated beamforming concept, the decoding matrices are first initialized as $\bar{\mathbf{A}}_m^{(0)} \in \mathbb{R}^{N_R \times S}$. In the $k$th iteration, for $k > 1$, the ZF method is applied on the equivalent channel matrices $\mathbf{H}_{e_m}^{i(k)} = \bar{\mathbf{A}}_m^{(k-1)^T} \mathbf{H}_m^i \in \mathbb{R}^{S \times 2N_T}$ and $\mathbf{H}_{e_m}^{d(k)} = \bar{\mathbf{A}}_m^{(k-1)^T} \mathbf{H}_m^d \in \mathbb{R}^{S \times 2N_T}$. Consequently, the precoding matrix in the $k$th iteration is factorized as $\bar{\mathbf{B}}_m^{(k)} = \bar{\mathbf{B}}_m^{i(k)} \bar{\mathbf{B}}_m^{o(k)} \in \mathbb{R}^{2N_T \times S}$. The inner precoder is obtained such that $\mathbf{H}_{e_m}^{i(k)} \bar{\mathbf{B}}_m^{i(k)} = \mathbf{0}$ to achieve the mitigation of the intrinsic interference. Fixing $\bar{\mathbf{B}}_m^{i(k)} \in \mathbb{R}^{2N_T \times 2N_T - S}$, the outer precoder has the $S$ right singular vectors of $\mathbf{H}_{e_m}^{d(k)} \bar{\mathbf{B}}_m^{i(k)} \in \mathbb{R}^{S \times 2N_T - S}$ as columns, which are associated to the dominant singular values. Then, considering a certain criterion for the receive processing at the receiver, e.g., MMSE, or ZF, $\bar{\mathbf{A}}_m^{(k)} \in \mathbb{R}^{N_R \times S}$ is computed based on the equivalent channel that takes the form $\mathbf{H}_m^d \bar{\mathbf{B}}_m^{(k)} \in \mathbb{R}^{N_R \times S}$. To determine the termination of this iterative procedure, the change of the precoding matrix is tracked via the term

$$\Delta(\bar{\mathbf{B}}_m) = \left\| \bar{\mathbf{B}}_m^{(k+1)} - \bar{\mathbf{B}}_m^{(k)} \right\|_F^2.$$

If $\Delta(\bar{\mathbf{B}}_m) < \epsilon$, where $\epsilon$ is a predefined threshold, then the convergence is assumed to be achieved. Otherwise, the precoding matrix and the decoding matrix are further updated.

It is worth noting that the authors of [8] have pointed out a particular case where $N_T = N_R = S + 1$ and have proposed a choice of the initialization $\bar{\mathbf{A}}_m^{(0)}$. Then it is observed that the coordinated beamforming approach only needs two iterations to converge, leading to a two-step method instead of an iterative algorithm [8].

In general, due to the fact that the channel exhibits frequency selectivity, it cannot be assumed that $\bar{\mathbf{A}}_{m-1} = \bar{\mathbf{A}}_m = \bar{\mathbf{A}}_{m+1}$. Thus, $\bar{\mathbf{A}}_{m'}^{\mathrm{T}} \mathbf{H}_m^{\mathrm{i}} \bar{\mathbf{B}}_m \neq \mathbf{0}$ for $m' = \{m-1, m+1\}$. As a consequence, the aforementioned coordinated beamforming scheme may not totally get rid of the ICI, leading to an error floor. To achieve an effective interference cancellation without substantially worsening the spatial channel seen by the desired symbols, i.e., the diagonal elements of $\bar{\mathbf{A}}_m^{\mathrm{T}} \mathbf{H}_m^{\mathrm{d}} \bar{\mathbf{B}}_m$, a Signal-to-Leakage-plus-Noise Ratio (SLNR)-based metric is tailored for the precoder design [9]. The motivation behind this choice is based on the results provided in [5], which confirm that the best strategy is not to cancel out the interference (e.g., in a ZF manner). It is enough to attenuate the unwanted signals 20 dB below the desired signal. Therefore, the SLNR-based precoder becomes an attractive solution since unlike the coordinated beamforming approach, both the ISI and ICI can be mitigated. The metric to be optimized is formulated as

$$\text{SLNR}_m = \frac{\left\| \bar{\mathbf{A}}_m^{\mathrm{T}} \mathbf{H}_m^{\mathrm{d}} \bar{\mathbf{B}}_m \right\|_{\mathrm{F}}^2}{\displaystyle\sum_{m'=m-1}^{m+1} \sigma_{m'm}^2 \left\| \bar{\mathbf{A}}_{m'}^{\mathrm{T}} \mathbf{H}_m^{\mathrm{i}} \bar{\mathbf{B}}_m \right\|_{\mathrm{F}}^2 + \frac{\sigma_w^2}{2} \left\| \bar{\mathbf{A}}_m^{\mathrm{T}} \right\|_{\mathrm{F}}^2}.$$

Under the assumption that $\mathbb{E}\{\mathbf{d}_{m,n} \mathbf{d}_{m',n'}^{\mathrm{H}}\} = \delta_{m,m'} \delta_{n,n'} \mathbf{I}_S$, the second-order moments of the interfering signals, which are defined in (13.10), can be expressed in the form $\mathbb{E}\{\mathbf{i}_{m',m;n} \mathbf{i}_{m',m;n}^{\mathrm{H}}\} = \sigma_{m'm}^2 \mathbf{I}_S$. In the SLNR definition, the relation $\left\| \bar{\mathbf{A}}_m \right\|_{\mathrm{F}}^2 = \mathrm{tr}\left( \bar{\mathbf{A}}_m^{\mathrm{T}} \bar{\mathbf{A}}_m \right)$ is used with $\mathrm{tr}(\cdot)$ denoting the trace operator. Note that to compute the precoding matrix, $\bar{\mathbf{A}}_m \in \mathbb{R}^{N_{\mathrm{R}} \times S}$ is fixed and given beforehand. To simplify the design, the power is uniformly distributed among the streams, yielding $\left\| \bar{\mathbf{B}}_m \right\|_{\mathrm{F}}^2 = S$. Therefore, the SLNR can be reformulated as

$$\text{SLNR}_m = \frac{\mathrm{tr}\left( \bar{\mathbf{B}}_m^{\mathrm{T}} \mathbf{D}_m \bar{\mathbf{B}}_m \right)}{\mathrm{tr}\left( \bar{\mathbf{B}}_m^{\mathrm{T}} \mathbf{R}_m \bar{\mathbf{B}}_m \right)},$$

$$\mathbf{D}_m = \mathbf{H}_m^{\mathrm{d}^{\mathrm{T}}} \bar{\mathbf{A}}_m \bar{\mathbf{A}}_m^{\mathrm{T}} \mathbf{H}_m^{\mathrm{d}},$$

$$\mathbf{R}_m = \sum_{m'=m-1}^{m+1} \sigma_{m'm}^2 \mathbf{H}_m^{\mathrm{i}^{\mathrm{T}}} \bar{\mathbf{A}}_{m'} \bar{\mathbf{A}}_{m'}^{\mathrm{T}} \mathbf{H}_m^{\mathrm{i}} + \frac{1}{S} \frac{\sigma_w^2}{2} \left\| \bar{\mathbf{A}}_m \right\|_{\mathrm{F}}^2 \mathbf{I}_{2N_{\mathrm{T}}}.$$

Now the following optimization problem is constructed to obtain the precoding matrix for each subcarrier:

$$\underset{\bar{\mathbf{B}}_m}{\text{argmax}} \quad \text{SLNR}_m$$

$$\text{s.t.} \quad \left\| \bar{\mathbf{B}}_m \right\|_F^2 = S$$

$$\bar{\mathbf{B}}_m^T \mathbf{D}_m \bar{\mathbf{B}}_m \rightarrow \text{diagonal},$$

which can be solved by computing the generalized eigenvalue decomposition of the matrix pair $\{\mathbf{D}_m, \mathbf{R}_m\}$ [10]. Then the $S$ columns of $\bar{\mathbf{B}}_m \in \mathbb{R}^{2N_T \times S}$ are obtained as the $S$ generalized eigenvectors associated to the $S$ strongest generalized eigenvalues. To fulfill the power constraint, the precoders are properly scaled so that its columns have unit norm [9]. It is worth mentioning that $\bar{\mathbf{B}}_m^T \mathbf{D}_m \bar{\mathbf{B}}_m$ exhibits a diagonal structure because $\mathbf{D}_m$ is Hermitian and $\mathbf{R}_m$ is Hermitian and positive definite [10].

The SLNR-based coordinated beamforming algorithm [9] is summarized as follows:

- **Step 1**: Initialize the real-valued decoding matrices $\mathbf{A}_m^{(0)} = \bar{\mathbf{A}}_m^{(0)} \in \mathbb{R}^{N_R \times S}$ for all the subcarriers randomly, set the iteration index $k$ to zero, and set a threshold $\epsilon$ for the stopping criterion.

- **Step 2**: Set $k \leftarrow k + 1$ and compute the precoding matrices $\bar{\mathbf{B}}_m^{(k)} \in \mathbb{R}^{2N_T \times S}$ for all the subcarriers based on the SLNR-based beamforming design. The columns of $\bar{\mathbf{B}}_m^{(k)} \in \mathbb{R}^{2N_T \times S}$ correspond to the generalized eigenvectors associated with the $S$ largest generalized eigenvalues of the pair

$$\left\{ \mathbf{D}_m^{(k)}, \mathbf{R}_m^{(k)} \right\},$$

where

$$\mathbf{D}_m^{(k)} = \mathbf{H}_m^{\text{d}^T} \bar{\mathbf{A}}_m^{(k-1)} \bar{\mathbf{A}}_m^{(k-1)^T} \mathbf{H}_m^{\text{d}},$$

$$\mathbf{R}_m^{(k)} = \sum_{m'=m-1}^{m+1} \sigma_{m'm}^2 \mathbf{H}_m^{\text{i}^T} \bar{\mathbf{A}}_{m'}^{(k-1)} \bar{\mathbf{A}}_{m'}^{(k-1)^T} \mathbf{H}_m^{\text{i}} + \frac{1}{S} \frac{\sigma_w^2}{2} \left\| \bar{\mathbf{A}}_m^{(k-1)} \right\|_F^2 \mathbf{I}_{2N_T}.$$

The precoding matrices are normalized so that the transmit power constraint is fulfilled.

- **Step 3**: Update the decoding matrices $\bar{\mathbf{A}}_m^{(k)} \in \mathbb{R}^{N_R \times S}$ for all the subcarriers. Assume that the ISI and ICI are effectively suppressed via precoding, leading to the fact that the noise is the dominant source of interference. This allows us to adopt the following simplified form of the decoding matrices that minimize the MSE as introduced in [9]:

$$\bar{\mathbf{A}}_m^{(k)} = \left( \mathbf{H}_m^{\text{d}} \bar{\mathbf{B}}_m^{(k)} \bar{\mathbf{B}}_m^{(k)^T} \mathbf{H}_m^{\text{d}^T} + \mathbf{R}_{\eta_m} \right)^{-1} \mathbf{H}_m^{\text{d}} \bar{\mathbf{B}}_m^{(k)}.$$

- **Step 4**: Track the variation of the residual interference to determine the termination of the algorithm. The following term that provides a measure of the inter-stream interference, the ISI, and the ICI at the $k$th iteration is defined:

$$\xi^{(k)} = \frac{1}{M_{\mathrm{u}}} \sum_{m \in \mathcal{M}_{\mathrm{u}}} \left\| \mathrm{off}\left( \bar{\mathbf{A}}_m^{(k)\mathrm{T}} \mathbf{H}_m^{\mathrm{d}} \bar{\mathbf{B}}_m^{(k)} \right) \right\|_{\mathrm{F}}^2$$

$$+ \frac{1}{M_{\mathrm{u}}} \sum_{m \in \mathcal{M}_{\mathrm{u}}} \sum_{m'=m-1}^{m+1} \sigma_{m'm}^2 \left\| \bar{\mathbf{A}}_m^{(k)\mathrm{T}} \mathbf{H}_{m'}^{\mathrm{i}} \bar{\mathbf{B}}_{m'}^{(k)} \right\|_{\mathrm{F}}^2,$$

where $\mathrm{off}(\cdot)$ denotes an operation of replacing all elements on the diagonal of the input matrix by zeros. Then, the slope of the variation of the residual interference is approximated via [5]

$$\xi^{(k)'} = \frac{|\xi^{(k)} - \xi^{(k-1)}|}{\xi^{(k-1)}}.$$

If $\xi^{(k)'} < \epsilon$, then terminate the iterative procedure. Otherwise, go back to **Step 2** and further update the precoding matrices and decoding matrices. Alternatively, a stopping criterion based on the change of the precoding matrices can also be employed.

It is worth noting that a joint-subcarrier processing is involved, i.e., the precoding matrices and decoding matrices for all the subcarriers are updated jointly. By contrast, the ZF-based coordinated beamforming algorithm is on a per-subcarrier basis. The lack of the joint-subcarrier processing leads to its failure of effectively mitigating the ICI. It has been observed that rendering the decoding matrices as complex-valued in this SLNR-based coordinated beamforming algorithm results in a much worse performance compared to the real-valued design described above [9]. On the one hand, adopting complex-valued decoding matrices keeps the full degrees of freedom. On the other hand, it becomes even more challenging to suppress the interference in both real and imaginary domains. As the impact of the latter consequence outweighs that of the former, the performance is degraded compared to the real-valued design of the decoding matrices. In addition, similarly to the ZF-based coordinated beamforming scheme, for MIMO settings where $N_{\mathrm{T}} = N_{\mathrm{R}} = S + 1$, the SLNR-based coordinated beamforming technique can also be regarded as a two-step approach with a judiciously initialized decoding matrix $\bar{\mathbf{A}}_m^{(0)}$ [9]. No iterative procedures are required in such cases, leading to a moderate computational complexity.

### 13.2.3 NUMERICAL RESULTS

To evaluate the performance of the precoding techniques introduced in this section, let us consider two MIMO systems where $N_{\mathrm{T}} = 4$, $N_{\mathrm{R}} = 2$, $S = 2$ and $N_{\mathrm{T}} = N_{\mathrm{R}} = 6$, $S = 5$, respectively. The total number of subcarriers is 1024, and the subcarrier spacing is 15 kHz. The Extended Vehicular-A (EVA) channel is employed.

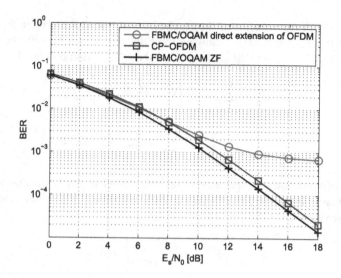

**FIGURE 13.3**

Comparison of the BER performances of different schemes in a MIMO system where $N_T = 4$, $N_R = 2$, and $S = 2$.

We present in Figs. 13.3 and 13.4 the Bit Error Rate (BER) performances of three precoding schemes for FBMC/OQAM systems. OFDM systems where the Cyclic Prefix (CP) has the length of $\frac{1}{8}$ of the symbol period are also simulated for comparison. The BER is represented as a function of the energy symbol-to-noise ratio, which is defined as $E_S/N_0 = 2/\sigma_w^2$ when in-phase and quadrature components of Offset Quadrature Amplitude Modulation (OQAM) and Quadrature Amplitude Modulation (QAM) symbols have energy equal to one. The transmit and receive beamformers in [4] are employed in the CP-OFDM-based systems. Equal power allocation is assumed on all data streams and subcarriers.

The transmission scheme detailed in Section 13.2.1 is a straightforward extension of the CP-OFDM case and relies on the assumption that the channel frequency responses remain the same across adjacent subcarriers. As the EVA channel exhibits frequency selectivity and such an assumption is therefore violated, the performance of this scheme degrades severely especially in the low-noise regime. By comparison, the ZF approach is able to achieve an effective cancellation of the intrinsic interference as shown in Fig. 13.3. However, in MIMO settings where $S = N_R \leq N_T$ is not satisfied, e.g., the example considered in Fig. 13.4, the ZF scheme fails to offer a good performance as anticipated in Section 13.2.2.2. In such multiantenna configurations, the coordinated beamforming schemes lead to a significant performance improvement. When employing the ZF-based precoder, the performance of the coordinated beamforming scheme is slightly worse compared to the case of CP-OFDM-based systems in the low-noise regime due to the residual intrinsic interference. On the other

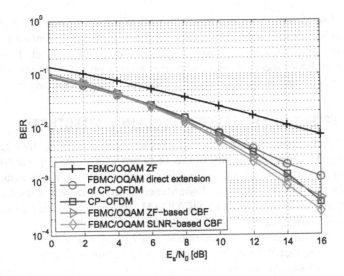

**FIGURE 13.4**

Comparison of the BER performances of different schemes in a MIMO system where $N_T = N_R = 6$, and $S = 5$; CBF-coordinated beamforming.

hand, the SLNR-based coordinated beamforming scheme motivated by the aforementioned observation achieves a satisfactory performance thanks to its stronger capability in intrinsic interference mitigation.

## 13.3 STRONG-FREQUENCY SELECTIVITY SCENARIOS

The numerical results provided in Section 13.2 reveal that (13.6) yields a negligible modeling error as long as the channel frequency response is sufficiently smooth. In this case, all the interfering signals that come from the same subcarrier go through the same channel, except for a constant, thanks to the PR property. Then, it is possible to completely remove the intrinsic interference by resorting to single-tap precoders and single-tap equalizers. As the subcarrier spacing widens and/or the channel frequency selectivity becomes stronger, the adoption of model 2) in (13.3) induces a mismatch error that degrades the system performance. Therefore, either (13.1) or model 1) in (13.3) have to be used to accurately characterize the system. These two models highlight that narrowband beamforming techniques, which rely on single-tap precoding and decoding matrices, cannot totally eliminate the interference in strong frequency selectivity scenarios. To determine to what extent the beamforming techniques specifically designed for the architecture depicted in Fig. 13.2 are able to cope with frequency selective channels, this section first provides an analytical characterization of the distortion. Based on this analysis, the MIMO techniques presented in this section do not make any assumption about the flatness of the channel, and, thus, the solutions that are proposed do not rely on the premise that interfering signals coming from the same subcarrier experience the same channel. To restore the orthogonality under these circumstances, this section proposes several joint Tx/Rx beamforming designs that are governed by different optimization criteria.

## 13.3.1 EFFECT OF THE SEVERE-FREQUENCY SELECTIVITY

As it has been explained in previous chapters, the design of transceivers for FBMC/OQAM under the simplifications of (13.3) lead to an important distortion effects when the channel presents strong-frequency selectivity. This was analyzed in more detail in Chapter 9, where a Taylor-based approximation was used to characterize this effect. The analysis of FBMC/OQAM intrinsic interference under strongly frequency-selective channels can be easily extended from the SISO case considered in Chapter 9 to the MIMO configuration studied here. Even without full details, we promptly realize that ICI and ISI are much more detrimental in multiantenna systems due to the superposition effect of all parallel streams, as also observed in [2, 11,12]. FilterBank MultiCarrier (FBMC) literature proposes complex strategies to try to mitigate this effect, such as sophisticated equalization architectures [13–15] or successive interference cancellation receivers [15–17]. Other works try to optimize specific transceiver structures focusing on particular objectives such as the diversity order while minimizing the residual distortion at the output of the receiver [18–20].

A more general approach is presented in the following subsection. Capitalizing on the distortion analysis introduced in Chapter 9 and following the idea of the multistage equalization architecture of Section 12.4, we propose a model for a generic linear precoder that performs asymptotically close to the ideal one when the number of subcarriers grows large and is the natural MIMO extension of the multistage linear receiver presented in Chapter 12.

## 13.3.2 MULTISTAGE APPROXIMATIONS AND DISTORTION ANALYSIS

Ideally, MIMO linear precoders would consist of $N_T S$ filters (one for each matrix entry) that process the entire signal bandwidth and do not take advantage of its multicarrier nature, as depicted in Fig. 13.1. Unfortunately, these filters are characterized by very long, or even infinite, impulse responses, and practical implementations are often infeasible.

As explained in the previous sections of this chapter, a common solution replaces the frequency dependent precoder matrix $\mathbf{B}(e^{j\omega})$ by a set of constant per-subcarrier precoding matrices $\mathbf{B}_m, m = 0, 1, \ldots, M-1$, placed before the SFBs (see Fig. 13.5). This choice typically comes with the assumption that the channel frequency response is flat at each subcarrier, and we can thus set $\mathbf{B}_m = \mathbf{B}(e^{j\omega_m})$, where $\omega_m$ is the central frequency of subcarrier $m$. Following the guidelines of Chapter 9, we will now show that this approach can be considered as the first-order truncation of a more general multistage approximation based on the Taylor expansion of the precoder matrix $\mathbf{B}(e^{j\omega})$.

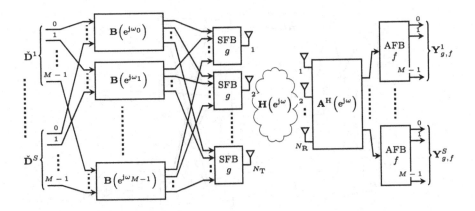

**FIGURE 13.5**

Conventional linear MIMO precoder based on a single matrix per subcarrier.

Consider a MIMO FBMC system where all the SFBs have the same transmit prototype pulse $g$. Next, let us assume that each entry of the precoding matrix $\mathbf{B}(z)$ is a transfer function that is analytic in an open set containing the boundary of the unit circle. Then, for subcarrier $m$, the concatenation of the prototype pulse with the precoder can be written as

$$\mathbf{B}\left(e^{j\omega}\right)G\left(e^{j(\omega-\omega_m)}\right) \approx \sum_{r=0}^{R_T} \frac{j^r}{r!}\mathbf{B}^{(r)}\left(e^{j\omega_m}\right)(\omega-\omega_m)^r G\left(e^{j(\omega-\omega_m)}\right)$$

$$\approx \sum_{r=0}^{R_T} \frac{1}{r!M^r}\mathbf{B}^{(r)}\left(e^{j\omega_m}\right)G^{(r)}\left(e^{j(\omega-\omega_m)}\right), \qquad (13.16)$$

where $G(e^{j\omega})$ is the discrete-time Fourier transform of $g$, and $G^{(r)}(e^{j\omega})$ is the discrete-time Fourier transform of $g^{(r)}$, the $r$th derivative of $g$. The expression in (13.16) follows from

$$G^{(r)}\left(e^{j\omega}\right) \approx (jM\omega)^r G\left(e^{j\omega}\right),$$

which is a fairly good approximation when the prototype pulse satisfies assumption (**As1**) of Chapter 9 and $M \to \infty$.

Eq. (13.16) implies that the ideal transmitter with a single broadband precoder of Fig. 13.1 can be approximated by the superposition of multiple stages, as depicted in Fig. 13.6. At stage $r$, for each subcarrier $m = 0, 1, \ldots, M-1$, the contributions of the $S$ streams are combined together by constant $N_T \times S$ matrix $(r!M^r)^{-1}\mathbf{B}^{(r)}(e^{j\omega_m})$. The result is a set of $N_T$ multicarrier symbols, which are fed into the corresponding antennas after being processed by $N_T$ SFBs, all with prototype pulse $g^{(r)}$. As explained in more detail below, the accuracy of the approximation increases with the number of

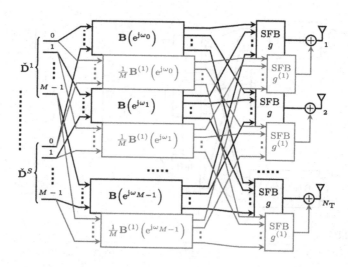

**FIGURE 13.6**

Parallel multistage MIMO precoder based on two stages ($R_T = 1$). In the second stage (red colored in the web version), the SFBs are built using the first time derivative of the prototype pulse (i.e., $g^{(1)}$), whereas the precoder matrices are given by the first derivative of the precoding transfer function matrix evaluated at $\omega = \omega_m$, $m = 0, 1, \ldots, M - 1$.

stages (i.e., $R_T$), $R_T = 0$ being the trivial single-tap per-subcarrier precoding scheme depicted in Fig. 13.5.

To understand the impact of the above decomposition on the quality of the received signal, we focus now on a specific stream, namely the $n_S$th one. From the input real-valued symbols, namely, $\{d_{m,n}^{ns}, m = 0, 1, \ldots, M - 1, n = 0, 1, \ldots, 2N_s - 1\}$ (with $N_s$ the number of QAM symbols to be transmitted), let us build a staggered matrix $\check{\mathbf{D}}^{ns}$ such that $[\check{\mathbf{D}}^{ns}]_{m,n} = d_{m,n}^{ns}$ if $n$ is even and $[\check{\mathbf{D}}^{ns}]_{m,n} = j d_{m,n}^{ns}$ if $n$ is odd. Assuming that the channel is ideal and there is no interference from other streams, denote by $\mathbf{X}_{g,f}(\check{\mathbf{D}}^{ns})$ the $M \times (2N_s + 4K)$ matrix of symbols produced by the $n_S$th AFB when the input of the $n_S$th SFB is $\check{\mathbf{D}}^{ns}$. Recall that $K$ is the overlapping factor of both the transmit and receive prototype pulses. Subindices $g$ and $f$ are introduced as a reminder of the prototype pulses used at the transmit and receive side, respectively. Recall that, for any $f$ and $g$ satisfying the PR constraints [see, e.g., (9.9) of Chapter 9], the $n_S$th data stream $\{d_{m,n}^{ns}\}$ can be reconstructed free of errors taking

$$\hat{d}_{m,n}^{ns} = \begin{cases} \mathbb{Re}\left\{\left[\mathbf{X}_{g,f}^{ns}\right]_{m,n+K-1}\right\}, & n \text{ even,} \\ \mathbb{Im}\left\{\left[\mathbf{X}_{g,f}^{ns}\right]_{m,n+K-1}\right\}, & n \text{ odd.} \end{cases}$$

More generally, assume that each of the $S$ multiplexed streams goes through a linear time-invariant system and let $\mathcal{T}(e^{j\omega})$ denote the $S \times S$ matrix containing the transfer functions between the $S$ transmit SFBs and the $S$ receive AFBs. On the other hand, let $\bar{\mathbf{Y}}_{g,f}^{ns}(\check{\mathbf{D}}^i; [\mathcal{T}]_{ns,i})$ be the contribution to the $ns$th output stream from the $i$th input stream through $[\mathcal{T}(e^{j\omega})]_{ns,i}$. Note that this is a point-to-point transfer function from the $i$th SFB at the transmit side to the $ns$th AFB at the receive side, which incorporates precoder and equalizer besides the channel. As a consequence of the superposition principle, we can write the compound output of the $ns$th AFB as

$$\mathbf{Y}_{g,f}^{ns}(\mathcal{T}) = \sum_{i=1}^{S} \bar{\mathbf{Y}}_{g,f}^{ns}\left(\check{\mathbf{D}}^i; [\mathcal{T}]_{ns,i}\right). \tag{13.17}$$

In the ideal case where $\mathcal{T}(e^{j\omega}) = \mathbf{I}_S$, it is straightforward to see that

$$\mathbf{Y}_{g,f}^{ns}(\mathbf{I}_S) = \bar{\mathbf{Y}}_{g,f}^{ns}\left(\check{\mathbf{D}}^{ns}; 1\right) = \mathbf{X}_{g,f}\left(\check{\mathbf{D}}^{ns}\right). \tag{13.18}$$

Since, in general, $\mathcal{T}(e^{j\omega}) = \mathbf{A}^H(e^{j\omega})\mathbf{H}(e^{j\omega})\mathbf{B}(e^{j\omega})$, (13.18) implies that the $ns$th stream (as well as all other streams) can be perfectly recovered by designing precoder matrix $\mathbf{B}(e^{j\omega})$ and equalizer $\mathbf{A}(e^{j\omega})$ such that $\mathbf{A}^H(e^{j\omega})\mathbf{H}(e^{j\omega})\mathbf{B}(e^{j\omega}) = \mathbf{I}_S$. To show how the multistage precoding scheme introduced above affects the performance of the system, we can use Proposition 9.1 of Chapter 9 to write

$$\bar{\mathbf{Y}}_{g,f}^{ns}\left(\check{\mathbf{D}}^i; \mathbf{A}^H\mathbf{H}\mathbf{B}\right)$$

$$= \sum_{n_T=1}^{N_T} \sum_{r=0}^{R_T} \frac{1}{r! M^r} \bar{\mathbf{Y}}_{g^{(r)},f}^{ns}\left(\mathbf{\Delta}\left(\mathbf{B}_{n_T,i}^{(r)}\right)\check{\mathbf{D}}^i; [\mathbf{A}^H\mathbf{H}]_{ns,n_T}\right) + o\left(M^{-R_T}\right), \tag{13.19}$$

where we introduced the diagonal matrix

$$\mathbf{\Delta}\left(\mathbf{B}_{n_T,i}^{(r)}\right) = \mathrm{diag}\left\{\left[\mathbf{B}^{(r)}\left(e^{j\omega_m}\right)\right]_{n_T,i}\right\}_{m=0,1,\dots,M-1},$$

and where $o\left(M^{-R_T}\right)$ denotes a matrix of appropriate size with entries that decay faster than $M^{-R_T}$ as $M \to \infty$. After careful inspection, we straightforwardly observe that the right-hand side of (13.19) corresponds to the scheme in Fig. 13.6 (for $R_T = 1$ and up to some quantity that is negligible when $M$ is large).

It is worth remarking that the decomposition into parallel stages can be carried out further at the receive side: it will involve the derivatives of the equalization matrix, i.e., $\mathbf{A}^{(r)}(e^{j\omega})$, and the derivatives of the receive prototype pulse, i.e., $f^{(r)}$. The multistage receiver structure is introduced in Section 14.3.4, whereas a detailed overview of this approach, together with a rigorous proof of (13.19), can be found in [21]. We report here the complete expression for the received signal when the parallel multistage configuration includes $R_T$ stages at the transmit side and $R_R$ stages at the receive side, namely,

$$\mathbf{Y}^{ns}\left(\mathbf{A}^H\mathbf{H}\mathbf{B}\right) = \mathbf{Z}^{ns}\left(\mathbf{A}^H\mathbf{H}\mathbf{B}\right) + o\left(M^{-R}\right),$$

where $R = \min(R_R, R_T)$

$$Z^{ns}\left(A^H HB\right) = \sum_{i=1}^{S}\sum_{n_T=1}^{N_T}\sum_{n_R=1}^{N_R}\sum_{r=0}^{R_T}\sum_{u=0}^{R_R}\frac{1}{r!u!M^{(r+u)}}\Delta^*\left(A_{n_R,n_S}^{(u)}\right)$$
$$\times \bar{Y}_{g^{(r)},f^{(u)}}^{ns}\left(\Delta\left(B_{n_T,i}^{(r)}\right)\check{D}^i;[H]_{n_R,n_T}\right)$$

with

$$\Delta\left(A_{n_R,n_S}^{(u)}\right) = \text{diag}\left\{\left[A^{(r)}\left(e^{j\omega_m}\right)\right]_{n_R,n_S}\right\}_{m=0,1,\ldots,M-1}.$$

Next, we compute the distortion that is generated when symbols $Z^{ns}\left(A^H HB\right)$ are used in place of $Y^{ns}\left(A^H HB\right)$ to recover the data stream.

Assume that all transmitted symbols $\{d_{m,n}^i, m = 0, 1, \ldots, M - 1, n = 0, 1, \ldots, 2N_s - 1, i = 1, \ldots, S\}$ are generated as independent and identically distributed (i.i.d.) random variables with zero mean and variance $P_s/2$. Moreover, let

$$\hat{d}_{m,n}^{ns} = \begin{cases} \text{Re}\left\{\left[Z_{g,f}^{ns}\left(A^H HB\right)\right]_{m,n+K-1}\right\}, & n \text{ even}, \\ \text{Im}\left\{\left[Z_{g,f}^{ns}\left(A^H HB\right)\right]_{m,n+K-1}\right\}, & n \text{ odd}, \end{cases}$$

be the estimated symbols of the $n_S$th stream at the receive side. Then, for both odd and even values of $n$, [21] shows that, as $M \to \infty$,

$$\mathbb{E}\left[\left|\hat{d}_{m,n}^{ns} - d_{m,n}^{ns}\right|^2\right] = P_e(m, n_S) + o\left(M^{-2R}\right), \qquad (13.20)$$

where $R = \min\{R_T, R_R\}$, and

$$P_e(m, n_S) = \frac{P_s}{2M^{2R}}\sum_{i=1}^{S}\text{Re}^T\left\{\xi_{ns,i}(m, R)\right\}\Psi_R^{(+,-)}\text{Re}\left\{\xi_{ns,i}(m, R)\right\}$$
$$+ \frac{P_s}{2M^{2R}}\sum_{i=1}^{S}\text{Im}^T\left\{\xi_{ns,i}(m, R)\right\}\Psi_R^{(-,+)}\text{Im}\left\{\xi_{ns,i}(m, R)\right\}.$$

In the last definition, we used

$$\Psi_R^{(+,-)} = \begin{bmatrix} v^{(+,-)}(g^{(R)}, f, g^{(R)}, f) & v^{(+,-)}(g^{(R)}, f, g, f^{(R)}) \\ v^{(+,-)}(g^{(R)}, f, g, f^{(R)}) & v^{(+,-)}(g, f^{(R)}, g, f^{(R)}) \end{bmatrix}$$

with $v^{(+,-)}(g^{(r_0)}, f^{(r_1)}, g^{(r_2)}, f^{(r_3)})$ as defined in (9.19). Similarly, the matrix $\Psi_R^{(-,+)}$ has an equivalent definition with all instances of $(+, -)$ changed by $(-, +)$. Finally, the components of the vector $\xi_{ns,i}(m, R) = \begin{bmatrix} \alpha_{ns,i}(m, R) & \beta_{ns,i}(m, R) \end{bmatrix}^T$ are given by

$$\alpha_{n_S,i}(m, R) = \frac{\sqrt{2}}{R!}\left[\mathbf{A}^H\!\left(e^{j\omega_m}\right)\mathbf{H}\!\left(e^{j\omega_m}\right)\mathbf{B}^{(R)}\!\left(e^{j\omega_m}\right)\right]_{n_S,i},$$

$$\beta_{n_S,i}(m, R) = \frac{\sqrt{2}}{R!}\left[\mathbf{A}^{(R)H}\!\left(e^{j\omega_m}\right)\mathbf{H}\!\left(e^{j\omega_m}\right)\mathbf{B}\!\left(e^{j\omega_m}\right)\right]_{n_S,i},$$

where $\mathbf{A}^{(R)}(z)$ and $\mathbf{B}^{(R)}(z)$ denote the $R$th derivatives of the transfer matrices $\mathbf{A}(z)$ and $\mathbf{B}(z)$, respectively. Note that this is a simplified distortion expression that holds only under the assumption that the original broadband precoder and equalizer functions achieve a perfect equivalent channel (i.e., $\mathbf{A}^H(e^{j\omega})\mathbf{H}(e^{j\omega})\mathbf{B}(e^{j\omega}) = \mathbf{I}_S$) and that prototype pulses $f$ and $g$ meet the PR conditions. A more general case is detailed in [21].

As an interesting remark on (13.20), note that the amount of distortion is dictated by $R = \min\{R_T, R_R\}$, the minimum number of stages at either the transmitter or the receiver. In other words, increasing the number of parallel processing stages at one end is only of little help since it will not improve significantly the performances of the MIMO link. This is corroborated by the simulation results of Fig. 13.7, which reports the Cumulative Distribution Function (CDF) of the Signal-to-Noise-plus-Distortion Ratio (SNDR) for a $4 \times 4$ MIMO link with four streams and different numbers of parallel stages at the transmit/receive side. The design of the precoder/receiver pair follows the classic SVD/matrix inversion paradigm. The curves are obtained averaging over 100 multicarrier symbols and 100 Extended Typical Urban (ETU) channel realizations, with a nominal SNR of 50 dB. We readily observe that the $(R_T = 1, R_R = 1)$, $(R_T = 1, R_R = 2)$, and $(R_T = 2, R_R = 1)$ configurations all have similar performances. Conversely, the curves quickly approach the ideal one as stages are added on both sides: at probability 0.5, the loss with respect to the optimal case is around 5 dB for the $(R_T = 2, R_R = 2)$ scheme and less than 3 dB for the $(R_T = 3, R_R = 3)$ scheme.

### 13.3.3 MSE-BASED MULTITAP JOINT TX/RX BEAMFORMING DESIGNS

The design of MIMO precoding and decoding matrices presented in this section is governed by the minimization of the MSE. The proposed technique jointly designs precoders and equalizers using an iterative algorithm for the Multi-User (MU)-MIMO DownLink (DL) scenario. In this scenario, one precoder needs to be designed per user, per antenna, and per subcarrier. We make use of the MSE-duality [22,23] between UpLink (UL) and DL such that we only need to design MMSE-based MIMO equalizers, which can be easily computed with well-known closed-form expressions, and then transform them into precoders where necessary. For the ease of exposition, we assume here a single stream transmission per user. However, the extension to multistreaming is straightforward.

Due to the overlap of adjacent subcarriers in FBMC/OQAM systems, a direct design of the precoders would need to take into account the interference generated into

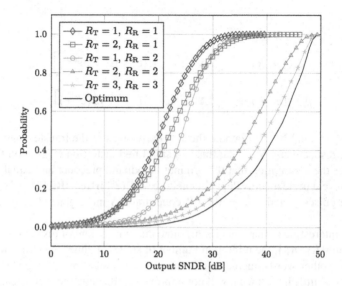

**FIGURE 13.7**

SNDR cumulative distribution function for input SNR of 50 dB and ETU channel model.

the adjacent subcarriers. To mitigate all the interfering signals, a joint optimization of all subcarriers has to be performed, but with practical restrictions in the solution due to the size of the problem to be solved. One way to avoid this is to look at the dual problem of designing the equalizers on a per-subcarrier basis at the UL. We can then transform the problem into the dual DL and obtain the corresponding precoders. The typical scenario where this can be applied is a Time-Division Duplex (TDD), where reciprocity can be employed to obtain transmitter Channel State Information (CSI).

In the following, we will use the notation $(\check{\bullet})$ to indicate DL precoders/equalizers and $(\hat{\bullet})$ for the UL ones. Consider Algorithm 1 as given in the separate description further note that each step only depends on variables from the same iteration (i), and thus to simplify notation, we will exclude the iteration index in the derivations that follow. The precoders/equalizers are implemented as complex-valued Finite Impulse Response (FIR) filters, but the derivations below consider a vector stacking of the real and imaginary parts of the coefficients.

The algorithm starts by initializing multitap equalizer filters in the DL $\check{\mathbf{a}}_m^v = [\check{\mathbf{a}}_{m,1}^{v,\mathrm{T}}, ..., \check{\mathbf{a}}_{m,N_{\mathrm{R}}}^{v,\mathrm{T}}]^{\mathrm{T}} \in \mathbb{R}^{2L_{\mathrm{eq}}N_{\mathrm{R}}}$ as simple delays, i.e., the unit vector $\mathbf{e}_{\lceil L_{\mathrm{eq}}/2 \rceil} \in \{0,1\}^{2L_{\mathrm{eq}}}$ with a 1 at the position $\lceil L_{\mathrm{eq}}/2 \rceil$, where $L_{\mathrm{eq}}$ is the number of taps of each Mobile Station (MS) equalizer/precoder per user $v$ and per subcarrier $m$. $N_{\mathrm{T}}$ and $N_{\mathrm{R}}$ are the numbers of antennas at the Base Station (BS) and at each MS. The initial DL to UL MSE-duality transformation in Step 2 sets the scaling factor $\check{\beta}$ in the zeroth iteration equal to 1, which means that the dual UL precoders $\hat{\mathbf{b}}_{m,(1)}^v \in \mathbb{R}^{2L_{\mathrm{eq}}N_{\mathrm{R}}}$ are also delays, and then the iterations start. The positive real-valued scaling factor $\check{\beta}$ is

---

**Algorithm 1** Joint precoder and equalizer design using the MSE-duality

---

1: **Initialization:**
2: $\hat{\mathbf{b}}^v_{m,(1)} = \check{\beta}^v_{(0)}\check{\mathbf{a}}^v_{m,(0)} = \mathbf{1}_{N_R} \otimes \mathbf{e}_{\lceil L_{eq}/2 \rceil}$ with $\check{\beta}^v_{(0)} = 1$   $\forall v, m$
3: $i = 1$
4: **repeat**

5:     $\hat{\mathbf{a}}^v_{m,(i)} = \arg\min\limits_{\hat{\mathbf{a}}^v_{m,(i)}} \mathrm{E}\left[\left|\hat{\tilde{d}}^v_{m,(i)}[n] - \hat{d}^v_{m,(i)}[n - \nu_{pre}]\right|^2_2\right]$

6:     $\hat{\beta}_{(i)}$                                   ▷ UL/DL MSE-duality transformation

7:     $\check{\mathbf{b}}^v_{m,(i)} = \hat{\beta}_{(i)}\hat{\mathbf{a}}^v_{m,(i)}$

8:     $\check{\mathbf{a}}^v_{m,(i)} = \arg\min\limits_{\check{\mathbf{a}}^v_{m,(i)}} \mathrm{E}\left[\left|\check{\tilde{d}}^v_{m,(i)}[n] - \check{d}^v_{m,(i)}[n - \nu_{eq}]\right|^2_2\right]$

9:     $\check{\beta}_{(i)}$                                   ▷ DL/UL MSE-duality transformation

10:    $\hat{\mathbf{b}}^v_{m,(i+1)} = \check{\beta}^v_{(i)}\check{\mathbf{a}}^v_{m,(i)}$
11:    $i = i + 1$
12: **until** $i = N_{iter}$

---

necessary to limit the transmit power at the transmitter side in the DL because for the MSE-duality, the same MSE and norm are set for equalizers/precoders design. In Step 5, we design MMSE UL equalizers $\hat{\mathbf{a}}^v_m \in \mathbb{R}^{2L_{pre}N_T}$, where $L_{pre}$ is the number of taps of the BS equalizers/precoders, and $\hat{d}^v_{m,(i)}$ and $\hat{\tilde{d}}^v_{m,(i)}$ are the UL real-valued transmitted and estimated received symbols as in (13.7) and (13.8). $\nu_{pre}$ is the precoder delay, which can also be subject to optimization, but we heuristically choose the value $\nu_{pre} = 2K + \lceil L_{pre}/2 \rceil$, where $K$ denotes the FBMC/OQAM overlapping factor. Through the same reasoning, the equalizer delay is set to $\nu_{eq} = 2K + \lceil L_{eq}/2 \rceil$. In Steps 6 and 7, we use the UL/DL MSE-duality transformation to calculate DL precoders $\check{\mathbf{b}}^v_{m,(i)} \in \mathbb{R}^{2L_{pre}N_T}$, what we call BS perspective. In Step 8, we design MMSE DL equalizers, where $\check{d}^v_{m,(i)}$ and $\check{\tilde{d}}^v_{m,(i)}$ are the DL real-valued transmitted and estimated received symbols, and in Step 9, we transform the DL equalizers into the UL precoders as start values for the next iteration, what we call MS perspective. Finally, our algorithm ends after a predefined number of iterations $N_{iter}$. Although not analytically proven here, the algorithm always converges at least to a local minimum. This can be confirmed by link level simulations as shown at the end of this chapter.

The real-valued $\beta$s are obtained by setting the MSE expressions for DL and UL equal and solving the linear system of equations [24]. The objective function may represent the total system-wide sum-MSE, where the MSE in all subcarriers and for all users are summed up, the partial sums across users or subcarriers, or the individual MSE per user and per subcarrier. Details on how the linear systems of equations for each variant look like are given in [24].

From the BS perspective the MMSE UL equalizer from Step 5 is given by

$$\hat{\mathbf{a}}_m^v = \left( \sum_{s=1}^{U} \sum_{l=m-1}^{m+1} \sigma_{\mathrm{M}}^2 \hat{\mathbf{\Gamma}}_l^s \hat{\mathbf{B}}_l^{s,\mathrm{T}} \hat{\mathbf{B}}_l^s \hat{\mathbf{\Gamma}}_l^{s,\mathrm{T}} + \hat{\mathbf{R}}_{\eta_m} \right)^{-1} \sigma_{\mathrm{M}}^2 \hat{\mathbf{\Gamma}}_m^v \hat{\mathbf{B}}_l^{s,\mathrm{T}} \mathbf{e}_{v_{\mathrm{pre}}},$$

where $U$ is the number of users, $\sigma_{\mathrm{M}}^2 = \sigma_d^2 /(2U)$ with $\sigma_d^2$ being the variance of the input symbols $d_m^v$, $\hat{\mathbf{\Gamma}}_l^s$ is a convolution matrix composed by the transmit, receive filter, and propagation channel as in 1) from (13.3), $\hat{\mathbf{B}}_l^s$ is the convolution matrix composed by the UL precoder impulse response, and $\hat{\mathbf{R}}_{\eta_m}$ is the noise covariance matrix.

It is important to remark that Algorithm 1 can be applied to all possible MSE-duality transformations. This comes from the fact that different MSE values can be defined according to the number subcarriers and users. The simplest possible definition is a single system-wise MSE value that is given by the sum of the MSEs for all subcarriers and users, i.e., a double summation. Other two possible MSE definitions take either only the sum over all subcarriers for each user, i.e., $U$ MSE values, or over all users for each subcarrier, i.e., $M_\mathrm{u}$ MSE values. Finally, a last MSE definition takes the MSE per subcarrier and user, i.e., $M_\mathrm{u} \times U$ MSE values. These four MSE definitions can be employed in the MSE-duality transformation by setting them equal for precoder and equalizer, either for DL or UL, and solving the system of equations for the corresponding number of scaling factors $\check{\beta}$ ($\hat{\beta}$) for the DL (UL). In summary, we can identify the following four UL/DL MSE-duality transformations:

$A_1$ – UL/DL system-wise sum-MSE:
Single scaling factor $\hat{\beta} \in \mathbb{R}_+$ with $\check{\mathbf{b}}_l^s = \hat{\beta}\hat{\mathbf{a}}_l^s$ and $\check{\mathbf{a}}_m^v = \hat{\beta}^{-1}\hat{\mathbf{b}}_m^v$.

$B_1$ – UL/DL user-wise sum-MSE:
$U$ scaling factors $\hat{\beta}^1, \ldots, \hat{\beta}^U \in \mathbb{R}_+$ with $\check{\mathbf{b}}_l^s = \hat{\beta}^s\hat{\mathbf{a}}_l^s$ and $\check{\mathbf{a}}_m^v = (\hat{\beta}^v)^{-1}\hat{\mathbf{b}}_m^v$.

$C_1$ – UL/DL subcarrier-wise sum-MSE:
$M_\mathrm{u}$ scaling factors $\hat{\beta}_1, \ldots, \hat{\beta}_{M_u} \in \mathbb{R}_+$ with $\check{\mathbf{b}}_l^s = \hat{\beta}_l\hat{\mathbf{a}}_l^s$ and $\check{\mathbf{a}}_m^v = \hat{\beta}_m^{-1}\hat{\mathbf{b}}_m^v$.

$D_1$ – UL/DL user and subcarrier-wise MSE:
$U \times M_\mathrm{u}$ scaling factors $\hat{\beta}_1^1, \ldots, \hat{\beta}_{M_u}^U \in \mathbb{R}_+$ with $\check{\mathbf{b}}_l^s = \hat{\beta}_l^s\hat{\mathbf{a}}_l^s$ and $\check{\mathbf{a}}_m^v = (\hat{\beta}_m^v)^{-1}\hat{\mathbf{b}}_m^v$.

For transformation $A_1$, there is a global power constraint, whereas for $B_1$, $C_1$, and $D_1$, power constraints are respectively defined for each user, subcarrier, and user and subcarrier.

From the MS perspective the MMSE DL equalizer from Step 8 is given by

$$\check{\mathbf{a}}_m^v = \left( \sum_{s=1}^{U} \sum_{l=m-1}^{m+1} \sigma_{\mathrm{M}}^2 \check{\mathbf{\Gamma}}_l^v \check{\mathbf{B}}_l^{s,\mathrm{T}} \check{\mathbf{B}}_l^s \check{\mathbf{\Gamma}}_l^{v,\mathrm{T}} + \check{\mathbf{R}}_\eta \right)^{-1} \sigma_{\mathrm{M}}^2 \check{\mathbf{\Gamma}}_m^v \check{\mathbf{B}}_l^{s,\mathrm{T}} \mathbf{e}_{v_{\mathrm{eq}}}.$$

Matrices are analogously defined to the BS perspective, although the dimensions may not be the same. Then, we can identify the following two MSE-duality transformations:

$B_2$ – DL/UL user-wise sum-MSE:

$U$ scaling factors $\check{\beta}^1, \ldots, \check{\beta}^U \in \mathbb{R}_+$ with $\hat{\mathbf{b}}_l^s = \check{\beta}^s \check{\mathbf{a}}_l^s$ and $\hat{\mathbf{a}}_m^v = (\check{\beta}^v)^{-1} \check{\mathbf{b}}_m^v$.

$D_2$ – DL/UL user and sub-carrier-wise MSE:

$U \times M_u$ scaling factors $\check{\beta}_1^1, \ldots, \check{\beta}_{M_u}^U \in \mathbb{R}_+$ with $\hat{\mathbf{b}}_l^s = \check{\beta}_l^s \check{\mathbf{a}}_l^s$ and $\hat{\mathbf{a}}_m^v = (\check{\beta}_m^v)^{-1} \check{\mathbf{b}}_m^v$.

From the MS perspective, it only makes sense to design the transmit filters under user power constraints and not subcarrier power constraints for obvious reasons.

It is worth mentioning that the algorithm can be simplified if we assume that either equalizers in the MS or precoders in the BS are already known a priori, for example, if they are precalculated using another optimization criteria. In this case, no iterations and only one of the MSE-duality transformations are necessary. Another special case of the algorithm was presented in [25] for MU-MISO, where no iterations are necessary and only the $\beta$s are necessary at the MS.

The algorithm can be easily extended to consider multiple streams per user, depending on the rank of the MIMO channel of course, and also take into account channel estimation errors for a robust design. Moreover, it can be also extended to nonlinear precoding (Tomlinson–Harashima Precoding (THP)) and/or equalizers (Decision Feedback Equalizer (DFE)) for even more challenging channels and/or overloaded scenarios, i.e., when the number of users is close to the number of transmit antennas.

### 13.3.4 SLR- AND SINR-BASED JOINT TX/RX BEAMFORMING DESIGNS

Now we present an iterative algorithm, where the precoders maximize the Signal-to-Leakage Ratio (SLR) at the BS and equalizers the SINR at the MS. In Algorithm 2, we again initialize the equalizers as simple delay vectors with a one at position $\lceil L_{eq}/2 \rceil$. The SLR-maximizing precoders are then designed taking the equalizer from the previous iteration into account, and the SINR-maximizing equalizer is designed taking the precoders from the current iteration into account. Analogously to the MSE-duality, we have omitted the iteration index (i) in the following derivations to simplify the notation.

---

**Algorithm 2** Joint SLR-based precoders and SINR-based equalizers design

---

1: **Initialization:**

2: $\check{\mathbf{a}}_{m,(0)}^v = \mathbf{e}_{\lceil L_{eq}/2 \rceil} \quad \forall v, m$

3: $i = 1$

4: **repeat**

5: $\quad \check{\mathbf{b}}_{m,(i)}^v = \underset{\check{\mathbf{b}}_{m,(i)}^v}{\arg \max}\ \mathrm{SLR}_{m,(i)}^v$

6: $\quad \check{\mathbf{a}}_{m,(i)}^v = \underset{\check{\mathbf{a}}_{m,(i)}^v}{\arg \max}\ \mathrm{SINR}_{m,(i)}^v$

7: $\quad i = i + 1$

8: **until** $i = N_{\text{iter}}$

---

The closed-form SLR expression of user $v$ in subcarrier $m$ is given by

$$
\text{SLR}_m^v = \frac{\breve{\mathbf{b}}_m^{v,\text{T}} \overbrace{\left( \breve{\boldsymbol{\Gamma}}_m^v \breve{\mathbf{A}}_m^{v,\text{T}} \mathbf{e}_{v_{\text{pre}}} \mathbf{e}_{v_{\text{pre}}}^{\text{T}} \breve{\mathbf{A}}_m^v \breve{\boldsymbol{\Gamma}}_m^{v,\text{T}} \right)}^{\mathcal{A}} \breve{\mathbf{b}}_m^v}{\underbrace{\breve{\mathbf{b}}_m^{v,\text{T}} \left( \displaystyle\sum_{s=1}^{U} \sum_{l=m-1}^{m+1} \breve{\boldsymbol{\Gamma}}_l^s \breve{\mathbf{A}}_l^{s,\text{T}} \breve{\mathbf{A}}_l^s \breve{\boldsymbol{\Gamma}}_l^{s,\text{T}} - \breve{\boldsymbol{\Gamma}}_m^v \breve{\mathbf{A}}_m^{v,\text{T}} \breve{\mathbf{A}}_m^v \breve{\boldsymbol{\Gamma}}_m^{v,\text{T}} \right. }_{\mathcal{C}} },
$$

$$
\left. + \breve{\boldsymbol{\Gamma}}_m^v \breve{\mathbf{A}}_m^{v,\text{T}} (\mathbf{I} - \mathbf{e}_{v_{\text{pre}}} \mathbf{e}_{v_{\text{pre}}}^{\text{T}})^2 \breve{\mathbf{A}}_m^v \breve{\boldsymbol{\Gamma}}_m^{v,\text{T}} \right) \breve{\mathbf{b}}_m^v
$$

where $\breve{\mathbf{A}}_m^v$ is a convolution matrix composed by the DL equalizer impulse response. The nominator contains the effective channel of user $v$, and the denominator contains ICI, ISI, and MultiUser Interference (MUI) terms. The solution is calculated as the generalized eigenvector corresponding to the maximum generalized eigenvalue of the matrix pencil $(\mathcal{A}, \mathcal{C})$.

The closed-form SINR expression of user $v$ in subcarrier $m$ is given by

$$
\text{SINR}_m^v = \frac{\breve{\mathbf{a}}_m^{v,\text{T}} \overbrace{\left( \breve{\boldsymbol{\Gamma}}_m^v \breve{\mathbf{B}}_m^{v,\text{T}} \mathbf{e}_{v_{\text{eq}}} \mathbf{e}_{v_{\text{eq}}}^{\text{T}} \breve{\mathbf{B}}_m^v \breve{\boldsymbol{\Gamma}}_m^{v,\text{T}} \right)}^{\tilde{\mathcal{A}}} \breve{\mathbf{a}}_m^v}{\underbrace{\breve{\mathbf{a}}_m^{v,\text{T}} \left( \displaystyle\sum_{s=1}^{U} \sum_{l=m-1}^{m+1} \breve{\boldsymbol{\Gamma}}_l^v \breve{\mathbf{B}}_l^{s,\text{T}} \breve{\mathbf{B}}_l^s \breve{\boldsymbol{\Gamma}}_l^{v,\text{T}} - \breve{\boldsymbol{\Gamma}}_m^v \breve{\mathbf{B}}_m^{v,\text{T}} \breve{\mathbf{B}}_m^v \breve{\boldsymbol{\Gamma}}_m^{v,\text{T}} \right. }_{\tilde{\mathcal{C}}} },
$$

$$
\left. + \breve{\boldsymbol{\Gamma}}_m^v \breve{\mathbf{B}}_m^{v,\text{T}} (\mathbf{I} - \mathbf{e}_{v_{\text{eq}}} \mathbf{e}_{v_{\text{eq}}}^{\text{T}})^2 \breve{\mathbf{B}}_m^v \breve{\boldsymbol{\Gamma}}_m^{v,\text{T}} + \sigma_{\text{M}}^{-2} \breve{\mathbf{R}}_\eta \right) \breve{\mathbf{a}}_m^v
$$

where the nominator contains the effective channel of user $v$, and the denominator contains ICI, ISI, MUI, and noise terms. The solution is calculated as the generalized eigenvector corresponding to the maximum generalized eigenvalue of the matrix pencil $(\tilde{\mathcal{A}}, \tilde{\mathcal{C}})$.

### 13.3.5 SIMULATION RESULTS

In this section, we discuss the simulation results of the MSE-duality-based design and the SLR/SINR-based design for the DL MU-MIMO scenario. The channel realizations are from the *Wireless World Initiative New Radio* (WINNER II) project [26]. We transmit data across $M_u = 210$ of the available $M = 256$ subcarriers per user and per transmitter antenna with a sampling rate of $f_s = 15.36$ MHz, giving a subcarrier spacing of 60 kHz. We randomly generate 16-QAM symbols and take a block length of 1000 symbols per subcarrier. The channel impulse response is $L_{\text{ch}} = 169$ taps. With these system configurations, especially due to the high-frequency selective channel, an OFDM system would have required a CP with a minimum length

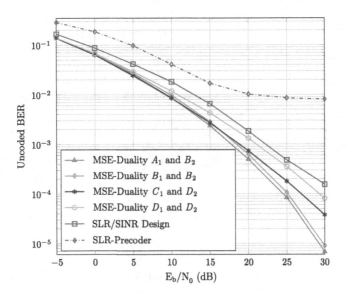

**FIGURE 13.8**

BER of the two iterative designs for $N_T = 4$, $N_R = 2$, and $U = 2$.

of 168 taps [27,28]. This represents an overhead higher than 50%; therefore, we do not include a direct comparison with OFDM in the simulation results. We take the quantity of $E_b/N_0$ for the MU-MIMO simulations. We take the uncoded BER, MSE, and SINR as an average over all users, and we average over 500 randomly generated channel realizations.

We have a precoder length of $L_{pre} = 5$ taps and an equalizer length of $L_{eq} = 3$ taps. Throughout our simulations, we have a system with $N_T = 4$ BS transmitter antennas and $U = 2$ users, each with two receiver antennas. We stopped both of our iterative algorithms after $N_{iter} = 4$ iterations.

In Fig. 13.8, we see the uncoded BER versus $E_b/N_0$ for the two iterative precoder and equalizer design algorithms introduced in this paper. We have compared our two iterative designs with an SLR-based precoder design with a real-valued single-tap equalizer from [29], referred as SLR-precoder in the figure. We observe that the MSE-duality-based designs show a better performance over the whole $E_b/N_0$ regime. Furthermore, in the high $E_b/N_0$ regime, the MSE-duality transformations with the *system-wide sum-MSE* and *user-wise sum-MSE* UL/DL transformation show performance gains of more than 5 dB compared with the other designs. This is attributed to the fact that these methods allow the total transmit power to be spread across all subcarriers depending on the channel conditions, somewhat like an inverse waterfilling power allocation scheme.

Fig. 13.9 shows the MSE versus $E_b/N_0$ of the two different iterative designs. We observe that all four MSE-duality-based designs outperform the SLR/SINR-based

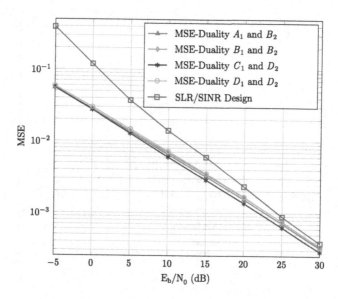

**FIGURE 13.9**

MSE of the two iterative designs for $N_T = 4$, $N_R = 2$, and $U = 2$.

design in the low $E_b/N_0$ regime due to the fact that the SLR-precoder design does not take the noise variance into account, leading to worse MSE values.

In Figs. 13.10 and 13.11, we see the MSE and SINR convergence, in dB, of the two different iterative designs. The curves in Fig. 13.10 show the MSE convergence of the MSE-duality-based designs, and the curves in Fig. 13.11 show the SINR convergence of the SLR/SINR design. After four iterations, the values do not significantly improve anymore, which is why we stopped our algorithms after $N_{iter} = 4$.

## 13.4 CONCLUDING REMARKS

This chapter reviews state-of-the art MIMO techniques for FBMC/OQAM systems. It has been shown that CP-OFDM-based solutions can be directly applied to FBMC/OQAM systems, as long as the channel frequency response is sufficiently smooth in the passband region of each subcarrier. Otherwise, the orthogonality is destroyed, inducing ISI and ICI and, thus, severely degrading the system performance. In mild-frequency selective scenarios, the orthogonality can be preserved, and the degrees of freedom provided by the spatial dimension can be exploited by implementing the ZF technique. However, the number of streams ($S$), transmit antennas ($N_T$), and receive antennas ($N_R$) should be related as $S = N_R \leq N_T$ to achieve a similar performance to OFDM. To overcome this dimensionality constraint, the coordinated beamforming scheme has been proposed. By jointly updating precoding

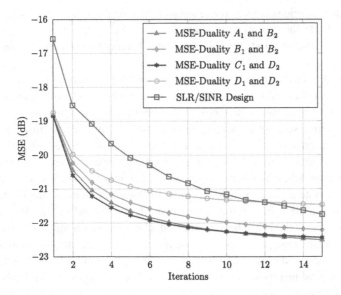

**FIGURE 13.10**

MSE convergence at $E_b/N_0 = 5$ dB of the iterative designs for $N_T = 4$, $N_R = 2$, and $U = 2$.

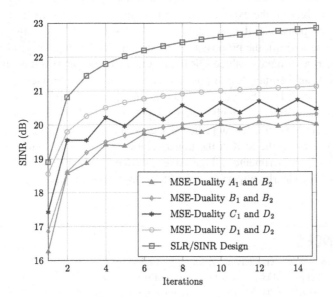

**FIGURE 13.11**

SINR convergence at $E_b/N_0 = 5$ dB of the iterative designs for $N_T = 4$, $N_R = 2$, and $U = 2$.

and decoding matrices in an iterative manner the intrinsic interference is mitigated to a high extent. In the special case where $S + 1 = N_R \leq N_T$, only two iterations are necessary to achieve satisfactory performance, leading to a moderate computational complexity.

One of the most attractive features of ZF and the coordinated beamforming solutions is that they allow restoring the orthogonality in mild-frequency selective scenarios by employing single-tap precoders and equalizers. However, in strong frequency selective scenarios, these techniques fail to eliminate intrinsic interference and exhibit an error floor. This is because the techniques conceived for low- and mild-frequency scenarios rely on the premise that the channel frequency response is flat at the subcarrier level. To offer competitive results, it is necessary to accurately characterize the system model without making any assumption about the flatness of the channel. To gain insight into the detrimental effects induced by the channel, a thorough distortion analysis has been conducted. Capitalizing on the distortion analysis, the idea of multistage precoding and equalization has been introduced to asymptotically approximate the ideal MIMO transceiver as the number of subcarriers grows large. Numerical results show that with two and three stages, which leads to an affordable complexity increase, the loss with respect to the optimal case is around 5 dB and 3 dB, respectively.

Shifting away from the typical assumption that the channel frequency response is flat at each subcarrier, an alternative to the multistage processing has been devised. The idea is to rely on the exact equivalent time response, which includes the channel impulse response and the subband pulses, to design the transmit and the receive processing. As a result, the proposed technique is based on implementing precoders and equalizers as complex-valued FIR filters. Owing to the fact that the joint transmit and receive beamforming design pose a complex nonconvex problem, two simpler iterative algorithms have been devised. Leveraging the MSE-duality between UL and DL, the first algorithm separately computes precoders and equalizers with the aim of reducing the MSE at each step. In the second iterative algorithm, precoders and equalizers maximize the SLR at the transmit side and the SINR at the receive side, respectively. The experimental validation reveals that the algorithm that benefits from the MSE-duality clearly outperforms the SLR/SINR-based design.

# REFERENCES

[1] A. I. Perez-Neira, M. Caus, R. Zakaria, D. Le Ruyet, E. Kofidis, M. Haardt, X. Mestre, and Y. Cheng, "MIMO signal processing in offset-QAM based filter bank multicarrier systems," *IEEE Trans. Signal Process.*, vol. 64, pp. 5733–5762, Nov. 2016.

[2] I. Estella, A. Pascual-Iserte, and M. Payaro, "OFDM and FBMC performance comparison for multistream MIMO systems," in *Proc. FutureNetw 2010*, Jun. 2010.

[3] M. Caus and A. I. Perez-Neira, "Comparison of linear and widely linear processing in MIMO-FBMC systems," in *Proc. ISWCS 2013*, Aug. 2013.

[4] D. P. Palomar, J. M. Cioffi, and M. A. Lagunas, "Joint Tx-Rx beamforming design for multicarrier MIMO channels: A unified framework for convex optimization," *IEEE Trans. Signal Process.*, vol. 51, pp. 2381–2401, Sep. 2003.

[5] M. Caus and A. I. Perez-Neira, "Experimental performance bounds of MIMO-FBMC/OQAM systems," in *Proc. EW 2014*, May 2014, pp. 1–6.

[6] B. Picinbono and P. Chevalier, "Widely linear estimation with complex data," *IEEE Trans. Signal Process.*, vol. 43, pp. 2030–2033, Aug. 1995.

[7] M. Caus and A. I. Perez-Neira, "Multi-stream transmission in MIMO-FBMC systems," in *Proc. IEEE ICASSP 2013*, May 2013.

[8] Y. Cheng, P. Li, and M. Haardt, "Coordinated beamforming in MIMO FBMC/OQAM systems," in *Proc. IEEE ICASSP 2014*, Florence, Italy, May 2014.

[9] M. Caus, A. I. Perez-Neira, Y. Cheng, and M. Haardt, "Towards a non-error floor multi-stream beamforming design for FBMC/OQAM," in *Proc. IEEE ICC 2015*, London, UK, Jun. 2015.

[10] M. Sadek, A. Tarighat, and A. Sayed, "A leakage-based precoding scheme for downlink multi-user MIMO channels," *IEEE Trans. Wireless Commun.*, vol. 6, pp. 1711–1721, May 2007.

[11] M. Nájar, M. Payaró, E. Kofidis, M. Tanda, J. Louveaux, M. Renfors, T. Hidalgo, D. Le Ruyet, C. Lélé, R. Zacaria, and M. G. Bellanger, "MIMO techniques and beamforming." ICT PHYDYAS (PHYsical layer for DYnamic AccesS and cognitive radio) project, Deliverable D4.2, Feb. 2010.

[12] M. Payaró, A. Pascual-Iserte, and M. Nájar, "Performance comparison between FBMC and OFDM in MIMO systems under channel uncertainty," in *Proc. EW 2010*, Lucca, Italy, Apr. 2010.

[13] E. Kofidis and A. A. Rontogiannis, "Adaptive BLAST decision-feedback equalizer for MIMO-FBMC/OQAM systems," in *Proc. IEEE PIRMC 2010*, Istanbul, Turkey, Sep. 2010.

[14] T. Ihalainen, A. Ikhlef, J. Louveaux, and M. Renfors, "Channel equalization for multi-antenna FBMC/OQAM receivers," *IEEE Trans. Veh. Technol.*, vol. 60, pp. 2070–2085, Jun. 2011.

[15] M. Nájar, C. Bader, F. Rubio, E. Kofidis, M. Tanda, J. Louveaux, M. Renfors, and D. Le Ruyet, "MIMO channel matrix estimation and tracking." ICT PHYDYAS (PHYsical layer for DYnamic AccesS and cognitive radio) project, Deliverable D4.1, Jan. 2009.

[16] A. Ikhlef and J. Louveaux, "Per-subchannel equalization for MIMO FBMC/OQAM systems," in *Proc. IEEE PACRIM 2009*, Victoria, BC, Canada, Aug. 2009.

[17] M. El Tabach, J.-P. Javaudin, and M. Hèlard, "Spatial data multiplexing over OFDM/OQAM modulations," in *Proc. IEEE ICC 2007*, Glasgow, Scotland, Jun. 2007.

[18] M. Caus and A. I. Perez-Neira, "Transmitter–receiver designs for highly frequency selective channels in MIMO FBMC systems," *IEEE Trans. Signal Process.*, vol. 60, pp. 6519–6532, Dec. 2012.

[19] N. Moret, A. Tonello, and S. Weiss, "MIMO precoding for filter bank modulation systems based on PSVD," in *Proc. IEEE VTC Spring 2011*, Yokohama, Japan, May 2011.

[20] S. Weiss, N. Moret, A. P. Millar, A. Tonello, and R. W. Stewart, "Initial results on an MMSE precoding and equalisation approach to MIMO PLC channels," in *Proc. IEEE ISPLC 2011*, Udine, Italy, Apr. 2011.

[21] X. Mestre and D. Gregoratti, "Parallelized structures for MIMO FBMC under strong channel frequency selectivity," *IEEE Trans. Signal Process.*, vol. 64, pp. 1200–1215, Mar. 2016.

[22] A. Mezghani, M. Joham, R. Hunger, and W. Utschick, "Transceiver design for multi-user MIMO systems," in *Proc. WSA 2006*, Ulm, Germany, 2006.

[23] R. Hunger, M. Joham, and W. Utschick, "On the MSE-duality of the broadcast channel and the multiple access channel," in *Proc. WSA 2009, vol. 57, no. 2*, 2009.

[24] O. De Candido, S.-A. Cheema, L. G. Baltar, M. Haardt, and J. A. Nossek, "Downlink precoder and equalizer designs for multi-user MIMO FBMC/OQAM," in *Proc. WSA 2016*, Mar. 2016.

[25] O. De Candido, L. G. Baltar, A. Mezghani, and J. A. Nossek, "SIMO/MISO MSE-duality for multi-user FBMC with highly frequency selective channels," in *Proc. WSA 2015*, Mar. 2015.

[26] L. Hentilä, P. Kyösti, M. Käske, M. Narandzic, and M. Alatossava, "MATLAB implementation of the WINNER Phase II Channel Model ver1.1," Dec. 2007.

[27] L. G. Baltar, D. Waldhauser, and J. A. Nossek, "Out-of-band radiation in multicarrier systems: A comparison," in *Proc. MC-SS 2007*, Herrsching, Germany, May 2007.

[28] L. G. Baltar and J. A. Nossek, "Multicarrier systems: A comparison between filter bank based and cyclic prefix based OFDM," in *Proc. InOWo'12*, 2012.

[29] Y. Cheng, L. Baltar, M. Haardt, and J. Nossek, "Precoder and equalizer design for multi-user MIMO FBMC/OQAM with highly frequency selective channels," in *Proc. IEEE ICASSP 2015*, Apr. 2015.

# MIMO-FBMC Receivers

# 14

**Eleftherios Kofidis**[*,†], **Markku Renfors**[‡], **Jérôme Louveaux**[§], **Xavier Mestre**[¶],
**David Gregoratti**[¶], **Didier Le Ruyet**[‖], **Rostom Zakaria**[‖]

*University of Piraeus, Piraeus, Greece*[*]
*Computer Technology Institute & Press "Diophantus" (CTI), Patras, Greece*[†]
*Tampere University of Technology, Tampere, Finland*[‡]
*Université catholique de Louvain, Louvain-la-Neuve, Belgium*[§]
*Centre Tecnològic de Telecomunicacions de Catalunya (CTTC/CERCA), Barcelona, Spain*[¶]
*CEDRIC Laboratory, CNAM, Paris, France*[‖]

## CONTENTS

Orthogonal Waveforms and Filter Banks for Future Communication Systems. DOI: 10.1016/B978-0-12-810384-5.00014-1

## 14.1 INTRODUCTION

Channel estimation and equalization in systems employing FilterBank MultiCarrier with Offset-QAM subcarrier modulation (FBMC/OQAM) can be quite challenging tasks, significantly more difficult than in Orthogonal Frequency-Division Multiplexing (OFDM), mainly due to the intrinsic interference effect, which translates into both Inter-Symbol Interference (ISI) and Inter-Carrier Interference (ICI). Nonflat subchannel responses, resulting from the Cyclic Prefix (CP)-free transmission over channels of nonnegligible (in relation to the MultiCarrier (MC) symbol duration) delay spread, are an additional source of difficulty. The challenge is intensified in Multiple-Input Multiple-Output (MIMO) systems, where the multiantenna interference has also to be taken into account [1]. It is therefore of no surprise that most of the research in FBMC/OQAM receiver design has relied on the assumption of channels that are slowly varying in frequency and time with respect to the subcarrier spacing and MC symbol duration, respectively. The motivation behind this simplification is to view the problems in a way similar to OFDM. However, especially in propagation conditions envisaged in future networks, this assumption may be quite inaccurate in communication environments involving, e.g., high data rate and/or mobility. In such cases, relying on the above assumption results in severe error floors at medium to high Signal-to-Noise Ratio (SNR) values, which cancel the advantage of the FBMC/OQAM modulation over OFDM [2].

Training schemes and associated channel estimation methods can be categorized as preamble-based and scattered pilots-based ones. A review of the Single-Input Single-Output (SISO) case was recently given in [3] (and in Chapter 11 of this book). The MIMO case was reviewed in [2,4], where the focus was on preamble-based methods, mainly intended for slowly varying channels. Scattered pilots-based techniques are of special interest in fast fading environments, where they allow tracking the channel variations throughout the frame [5]. Blind methods (i.e., not relying on training signals), mostly applicable in low-mobility environments, are also relevant in multiantenna FilterBank MultiCarrier (FBMC), especially in the light of the increased interest in massive MIMO systems and the corresponding challenges raised by training-based techniques in that context [6].

Designing an FBMC/OQAM equalizer is based on the principle that the purely real/imaginary nature of the Offset Quadrature Amplitude Modulation (OQAM) data must be explicitly taken into account, as stated in [7,8] for staggered modulation systems. Compared with OFDM [9], equalization in FBMC/OQAM systems faces the additional challenge of having to cope with the intrinsic interference effect, which leads to both ISI and ICI and to the generally frequency selective nature of the subchannels. An account of both linear and nonlinear (decision feedback) schemes for SISO FBMC/OQAM equalization, designed on the basis of Zero Forcing (ZF) and Minimum Mean-Squared Error (MMSE) criteria, is given in Chapter 12 of this book. Widely Linear (WL) MMSE equalization and its possible gains over strictly linear processing are also discussed therein.

This chapter overviews existing work in MIMO-FBMC/OQAM channel estimation and equalization schemes with emphasis on the more recent developments. Preamble-based and scattered pilot-based channel estimation is discussed in Sections 14.2.1 and 14.2.2, respectively. Section 14.3 is devoted to channel equalization in such systems. Maximum Likelihood Sequence Estimation (MLSE) and its intrinsic challenges in this context are presented in Section 14.3.1. Linear per-subchannel ZF and MMSE equalization using the idea of computing the equalizer on the basis of samples of the Channel Frequency Response (CFR) (i.e., frequency sampling) within the subband is discussed for Single-Input Multiple-Output (SIMO) and MIMO systems in Section 14.3.2. Section 14.3.3 focuses on Successive Interference Cancellation (SIC) structures coupled with MMSE MIMO-FBMC/OQAM per-subchannel equalization to improve the performance of such equalizers in high frequency selective channels. It also includes the so-called Ordered Successive Interference Cancellation (OSIC) schemes for recovering the input streams in the order of maximum to minimum Signal-to-Interference-plus-Noise Ratio (SINR). In Section 14.3.4, the parallel multistage equalization structure of Section 12.4 is extended to a MIMO setup. Its ability to equalize high frequency selective channels with good performance and low latency is demonstrated. Doubly dispersive channels, calling for adaptive equalization solutions, are considered in Section 14.3.5. Both linear/nonlinear and channel estimate-based/direct equalizers are considered. Recent results on efficient adaptive decision feedback equalization following the Bell Labs Layered Space Time (BLAST) idea (similar to OSIC) [10,11] are summarized. Notably, these works can be seen to effectively perform WL processing [12]. WL equalization for MIMO-FBMC/OQAM systems was also elaborated in [13] through the design of a number of MIMO decision feedback equalization schemes. Conclusions and future research directions are given in Section 14.4.

Notations and definitions closely follow the SISO-FBMC/OQAM case, and hence it would be beneficial for the reader to first have a look at Chapters 11 and 12.

## 14.2 CHANNEL ESTIMATION
### 14.2.1 PREAMBLE-BASED METHODS

Consider an FBMC/OQAM system with $N_T$ transmit and $N_R$ receive antennas, involving $N_T N_R$ channels, as shown in Fig. 14.1. The aim is to estimate their responses with the aid of $N_T$ training preambles, one per each Transmitter (TX) antenna. These preambles are constructed so as to consist of a number of pilot FBMC/OQAM symbols, which are preceded and/or followed by one (or more[1]) symbol(s) of all zeros, as in the SISO FBMC/OQAM system studied in Chapter 11. Similarly to the SISO

---

[1] Nevertheless, one guard FBMC/OQAM symbol is not always sufficient, and nonnegligible interference may still exist in practice, as demonstrated, for example, in [14]. A sufficiently large interframe gap is assumed here, which renders the guard FBMC symbol(s) preceding the pilots unnecessary.

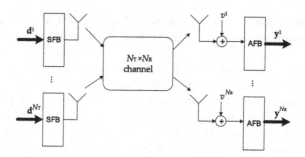

**FIGURE 14.1**

Block diagram of an $N_T \times N_R$ MIMO-FBMC system.

case, two different pilot configurations may be considered, sparse (comb-type) and full (block-type). For simplicity and analytical tractability, we will assume that there are no inactive (*virtual* [15]) subcarriers at the edges of the signal spectrum. As it is common in preamble-based estimation, the channel is taken to be time invariant over the training sequence duration.[2] Moreover, perfect synchronization is assumed.

### 14.2.1.1 *Low Frequency Selective Channels*

With channels of (relatively to the subband width) low frequency selectivity, under the common assumption that interference contributions to a given Frequency-Time (F-T) point $(m, n)$ only come from its first-order neighborhood [16,17], namely, $\bar{\Omega}_{m,n} = \{(m \pm 1, n \pm 1), (m, n \pm 1), (m \pm 1, n)\}$, the Analysis FilterBank (AFB) output at the F-T point $(m, n)$ (multiplied by $e^{-j\varphi_{m,n}}$ with $\varphi_{m,n} \bmod \pi = (m + n)\frac{\pi}{2}$) can be expressed as

$$\mathbf{y}_{m,n} = \mathbf{H}_{m,n}\mathbf{d}_{m,n} + \sum_{(m',n') \in \bar{\Omega}_{m,n}} \mathbf{H}_{m',n'}\Gamma_{m,m'}^{n,n'}\mathbf{d}_{m',n'} + \boldsymbol{\eta}_{m,n}, \qquad (14.1)$$

where $\mathbf{y}_{m,n} = \begin{bmatrix} y_{m,n}^1 & y_{m,n}^2 & \cdots & y_{m,n}^{N_R} \end{bmatrix}^T$ collects the corresponding samples from all Receiver (RX) antennas, $\boldsymbol{\eta}_{m,n}$ is similarly built from the noise samples, $\mathbf{d}_{m,n} = \begin{bmatrix} d_{m,n}^1 & d_{m,n}^2 & \cdots & d_{m,n}^{N_T} \end{bmatrix}^T$ is the vector of corresponding transmitted symbols, $\mathbf{H}_{m,n}$ is the $N_R \times N_T$ CFR matrix at $(m, n)$, and $\Gamma_{m,m'}^{n,n'}$ denotes the TransMUltipleXer (TMUX) transfer function as in Chapter 11. Assuming, moreover, that the CFR is (almost) invariant over the useful Time-Frequency (T-F) neighborhood, the above

---

[2]Thus, to be able to cope with fast fading environments, the latter should be as small as possible.

$$
\begin{array}{cccc}
1 & 0 & 1 & 0 \\
-j & 0 & -j & 0 \\
-1 & 0 & -1 & 0 \\
j & 0 & j & 0 \\
1 & 0 & 1 & 0 \\
-j & 0 & -j & 0 \\
-1 & 0 & -1 & 0 \\
j & 0 & j & 0 \\
\end{array}
\qquad
\begin{array}{cccc}
1 & 0 & -1 & 0 \\
-j & 0 & j & 0 \\
-1 & 0 & 1 & 0 \\
j & 0 & -j & 0 \\
1 & 0 & -1 & 0 \\
-j & 0 & j & 0 \\
-1 & 0 & 1 & 0 \\
j & 0 & -j & 0 \\
\end{array}
$$

(A)                    (B)

**FIGURE 14.2**

IAM-C preamble for a $2 \times X$ system with $M = 8$ subcarriers, with (A) and (B) corresponding to the two transmit antennas. OQPSK modulation is assumed.

input–output equation can take the simpler form

$$
\mathbf{y}_{m,n} = \mathbf{H}_m \underbrace{\left( \mathbf{d}_{m,n} + \sum_{(m',n') \in \bar{\Omega}_{m,n}} \Gamma_{m,m'}^{n,n'} \mathbf{d}_{m',n'} \right)}_{\mathbf{x}_{m,n}} + \boldsymbol{\eta}_{m,n}
$$

$$
= \mathbf{H}_m \mathbf{x}_{m,n} + \boldsymbol{\eta}_{m,n}, \tag{14.2}
$$

where the common assumption of time invariance of the channel over the preamble duration has been used to drop the time index from the CFR matrix. As in the SISO case, the usefulness of the latter model lies in its simplicity and similarity with what holds in Cyclic Prefix Orthogonal Frequency-Division Multiplexing (CP-OFDM) and has thus been extensively used in the development of channel estimation and equalization techniques for FBMC/OQAM systems admitting such a simplification.

In view of the above assumptions, if the immediate neighbors of $(m, n)$ carry training (hence known) symbols, then we can compute an approximation of the interference term in (14.2) and hence construct (in fact, only approximate) the pseudo-pilot $\mathbf{x}_{m,n}$. This in turn leads to the Interference Approximation Method (IAM) [17, 18] known from the SISO case. MIMO versions of IAM were first studied in [19], where it was proposed to construct the preambles for the $N_T$ antennas in the following way: for each TX antenna, repeat the SISO preamble $N_T$ times by also inserting sign changes that ensure orthogonality among the different antennas (in a way similar to MIMO-OFDM orthogonal training [20]). Fig. 14.2 provides an example for the case $N_T = 2$ using the IAM-C preamble (known to be optimal in the minimum channel estimation Mean-Squared Error (MSE) sense in SISO-FBMC/OQAM [2]). Taking the structure of this preamble into account, an estimate of the CFR matrix at subcarrier $m$ can be computed as [2,19]

$$\hat{\mathbf{H}}_m = \begin{bmatrix} \mathbf{y}_{m,0} & \mathbf{y}_{m,2} \end{bmatrix} \frac{1}{x_m} \mathbf{A}_2^{-1} = \mathbf{H}_m + \frac{1}{2x_m} \begin{bmatrix} \boldsymbol{\eta}_{m,0} & \boldsymbol{\eta}_{m,2} \end{bmatrix} \mathbf{A}_2, \qquad (14.3)$$

where $\mathbf{A}_2 = \begin{bmatrix} 1 & 1 \\ 1 & -1 \end{bmatrix}$, and $x_m$ is the corresponding scalar (SISO) pseudo-pilot. This can be easily generalized to any $N_T$ that is a power of two [2]. Observe that the orthogonality of the matrix $\mathbf{A}_2$ ensures that the estimation noise power is again (as in the SISO case) exclusively controlled by the magnitude of the pseudo-pilot $x_m$. However, in practice, there may exist a nonnegligible interference between time instants 0 and 2, and hence the channels from the $N_T$ TX antennas are not all estimated with the same accuracy [19]. Of course, this can be easily overcome, at an extra cost in the training overhead, if more than one guard symbols are placed among the repetitions of the IAM preamble.

Shorter preambles, consisting of only one pilot FBMC symbol and a guard one per antenna (independently of $N_T$), were reported in [21], and they are of the sparse type. One such example, known as *Frequency-Division Multiplexing (FDM)*[3] [22], results from sharing the pilot subcarriers among the antennas and, for a given receive antenna, allows one to separately estimate the channels from each of the transmit antennas. The frequency response values at the inactive frequencies are then found via interpolation in the frequency direction. This idea reappeared recently in [23,24], where it was enriched with auxiliary and side pilots (similarly with [26]) and shown to result in a significantly lower Peak-to-Average Power Ratio (PAPR) compared to the full preambles of [19]. Another sparse scheme, which is based on earlier work on orthogonal training design for MIMO-OFDM systems so as to be optimal in the MSE sense, was developed and presented in [2,21] and was later adopted in [27].

The literature on (semi)blind methods for FBMC/OQAM channel estimation is almost exclusively devoted to single-antenna systems (see Section 11.5 for a review) with the MIMO case first treated in [28]. The idea therein is to employ linear precoding that is so chosen as to allow the estimation of the channel directly from the estimated covariance matrix of the AFB output signal. More recently, the application of FBMC in MIMO systems employing a large number of antennas (massive MIMO) was also considered [6,29]. It was intuitively argued and experimentally verified in [29] that the large number of antennas has a *self-equalization* effect, which results in effectively flat subchannels. As demonstrated in [30] (in the context of equalizing such a system with the aid of a Frequency Spreading FBMC (FS-FBMC) equalizer), this has positive implications in a number of aspects of the system that include PAPR, latency, complexity, bandwidth efficiency, and robustness to Doppler

---

[3]The preambles proposed in [19] are instead of the *Code-Division Multiplexing (CDM)* type [22]. CDM preambles of duration independent of $N_T$, which consist of two pilot FBMC symbols per antenna (plus the null guards against interference), were devised in [25].

spread and frequency offsets. A theoretical analysis of this effect was recently presented in [31]. On this basis, a blind algorithm for channel estimation (and possibly tracking) was proposed and tested in [6] as a rather straightforward means of dealing with the pilot contamination problem in such systems. A (semi)blind estimation scheme that relies on the data symbol estimates to iteratively improve the estimate of the CFR was recently presented in [32,33]. It is based on a tensor (multiway array) modeling of the MIMO-FBMC system and relies on well-known uniqueness properties and decomposition procedures of tensor models to achieve significant performance improvements over the preamble-based approach in both MIMO-FBMC and MIMO-OFDM systems.

### 14.2.1.2 *Highly Frequency Selective Channels*

For channels of significant (with respect to the MC symbol duration) time dispersion, the previous estimators do not perform well, particularly, at medium to high SNR values. The results of Chapter 11 for high frequency selective channels are extended here to the MIMO setup, resulting in an estimation procedure that does not rely on any simplifying assumption and is thus applicable in more demanding transmission scenarios, such as those envisaged in future high-rate networks. Let $\mathbf{h}^{q,p}$ be the (finite length) Channel Impulse Response (CIR) of the link from the transmit antenna $p$ to the receive antenna $q$ and assume (without loss of generality) that all $N_T N_R$ channels have the same length, $L_h$. It is not difficult to extend the corresponding input/output equation from Chapter 11 (Eq. (11.18)) to the present context. Indeed, we can write

$$\mathbf{y}^q = \sum_{p=1}^{N_T} \mathbf{\Gamma}(\mathbf{d}^p)\mathbf{h}^{q,p} + \boldsymbol{\eta}^q, \quad q = 1, 2, \ldots, N_R, \tag{14.4}$$

where $\mathbf{\Gamma}(\mathbf{d}^p)$ is the $M \times L_h$ matrix defined as in (11.19). Recall that it is a priori known since it only depends on the FilterBank (FB) parameters and the FBMC pilot symbol $\mathbf{d}^p$ for the transmit antenna $p$. The above can be written more compactly as

$$\mathbf{y}^q = \check{\mathbf{\Gamma}}\mathbf{h}^{q,\cdot} + \boldsymbol{\eta}^q, \tag{14.5}$$

where $\check{\mathbf{\Gamma}} = \begin{bmatrix} \mathbf{\Gamma}(\mathbf{d}^1) & \mathbf{\Gamma}(\mathbf{d}^2) & \cdots & \mathbf{\Gamma}(\mathbf{d}^{N_T}) \end{bmatrix}$, and $\mathbf{h}^{q,\cdot} \in \mathbb{C}^{N_T L_h \times 1}$ results from concatenating the responses of the channels from all TX antennas to the RX antenna $q$. Assuming, as usual, that the noise signals at different receive antennas are uncorrelated with each other (i.e., spatially white noise), it is sufficient to address the estimation problem separately for each receive antenna, as suggested by (14.5). Clearly, the covariance matrix of the noise signal at the $q$th receive antenna is given by

$$\mathbf{R}_{\eta^q} = \sigma_\eta^2 \mathbf{B}$$

with $\mathbf{B}$ being a circulant matrix depending on the amount of correlation with adjacent subcarriers and defined in (11.21). Hence the Gauss–Markov estimate of $\mathbf{h}^{q,\cdot}$ results from (14.5) as

$$\hat{\mathbf{h}}^{q,\cdot} = (\tilde{\check{\mathbf{\Gamma}}}^H \tilde{\check{\mathbf{\Gamma}}})^{-1} \tilde{\check{\mathbf{\Gamma}}}^H \tilde{\mathbf{y}}^q \tag{14.6}$$

with the notation $\tilde{\phantom{a}}$ meaning a prewhitening transformation, that is, multiplication by a square root of $\mathbf{B}$. The above presumes that the matrix $\check{\mathbf{\Gamma}}$ is full column rank and hence tall, that is,

$$M \geq N_T L_h. \tag{14.7}$$

Furthermore, the optimality results from the SISO case (see Section 11.3.2) suggest the following choice for the pilot FBMC symbol (cf. (11.36)):

$$\mathbf{d}^p = \pm\sqrt{\frac{\mathcal{E}}{N_T \lambda_0[m_p]}}\mathbf{f}_{m_p} \tag{14.8}$$

with $m_p$ determined as in (11.37), where $\mathbf{f}_{m_p}$ and $\lambda_0[m_p]$ are the corresponding column of the $M \times M$ normalized Discrete Fourier Transform (DFT) matrix $\mathcal{F}$ and the eigenvalue of $\mathbf{B}$, respectively, and $\mathcal{E}$ denotes the total transmit energy budget (at the Synthesis FilterBank (SFB) output, which in this context can be written as $\sum_{p=1}^{N_T}(\mathbf{d}^p)^H\mathbf{B}\mathbf{d}^p \leq \mathcal{E}$) and is assumed to be equally shared among the $N_T$ TX antennas. In addition to (14.8), we must take care not to choose the same $m$ for two antennas since this would result in a rank-deficient matrix $\check{\mathbf{\Gamma}}$. Once $m_p$ has been determined, the possible values for $m_s$, $s \neq p$, exclude the set $\{m_p, ((m_p + 1))_M, ((m_p + 2))_M, \ldots, ((m_p + L_h - 1))_M\}$, where $((\cdot))_M$ denotes modulo $M$. This implies that, as also suggested by (14.7), the more transmit antennas and CIR taps there are, the fewer degrees of freedom are available in designing the corresponding preambles. This limitation can be mitigated if longer (than one pilot FBMC symbol) preambles are considered [34]. It is of interest to specialize the above in the particular but highly realistic case of two transmit antennas and real-valued pilot symbols:

$$\mathbf{d}^0 = \sqrt{\frac{\mathcal{E}}{2}}\mathbf{f}_0 = \sqrt{\frac{\mathcal{E}}{2M}}\mathbf{1}_M, \tag{14.9}$$

$$\mathbf{d}^1 = \sqrt{\frac{\mathcal{E}}{2}}\mathbf{f}_{\frac{M}{2}} = \sqrt{\frac{\mathcal{E}}{2M}}[(-1)^i]_{i=0}^{M-1}, \tag{14.10}$$

where the fact that $\lambda_0[0] = \lambda_0[\frac{M}{2}] = 1$ was also used.[4] It should be also noted that the channel estimation procedure is greatly simplified when optimal preambles are employed. For details on such simplifications in the SISO case, the reader is referred to Section 11.3.2. An example of the performance of the above method, with pilots as in (14.9), (14.10), as compared to that of the MSE-optimal MIMO-OFDM

---

[4]Note that this choice respects the constraint on the column indices stated above in the sense that $\frac{M}{2}$ is permissible in view of assumption (14.7).

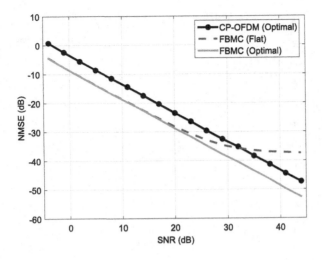

**FIGURE 14.3**

Estimation performance of preamble-based methods for 2 × 2 ITU Veh-A channels [36]. The PHYDYAS TMUX [37] was employed with $M = 64$ and $K = 3$. The subcarrier spacing is 15 kHz.

method of [35], is given in Fig. 14.3, where the CIR Normalized Mean-Squared Error (NMSE) is plotted vs. transmit SNR. The result of assuming flat subchannels in the preamble and estimator design is also shown. As it is typical in methods relying on the flat subchannel model, a severe error floor is observed at medium to high SNR values, due to the fact that the residual intrinsic interference becomes more apparent in such SNR regimes.

## 14.2.2 SCATTERED PILOTS-BASED METHODS

In fast fading environments, of special interest for future mobile networks, we also need to track the channel as it dynamically varies throughout the frame. To this end, pilots are scattered over the payload in both the frequency and time directions. This helps sampling the CFR and, with the aid of some interpolation technique, providing estimates of it at the data positions as well [5]. If it were not for the intrinsic interference effect, then pilot-assisted channel estimation for FBMC/OQAM could be very similar with that for CP-OFDM [5]. However, as explained in Chapter 11 (Section 11.4), to ensure an accurate channel estimate, a so-called Help Pilot (HP) (or auxiliary pilot [38]) is also needed to mitigate as much as possible the intrinsic interference effect at the pilot position. In a MIMO context, care must be also taken of the multiantenna interference when placing and setting pilots and HPs. Such an estimation scheme was proposed in [39] and relies on the assumptions underlying (14.1). This is the case for all of the MIMO-FBMC/OQAM scattered pilots-based channel estimation techniques reported in the literature. Obviously, (14.1) is an inaccurate

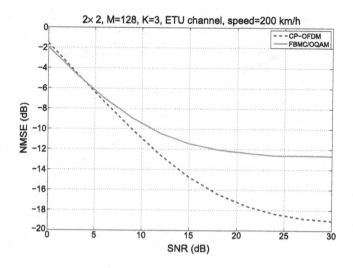

**FIGURE 14.4**

Estimation performance of a T-F interpolation scheme for a 2 × 2 ETU channel at a mobile speed of 200 km/h. Carrier frequency was set to 2.5 GHz. A PHYDYAS FB [37] with $M = 128$ and $K = 3$ was employed, and the subcarrier spacing was 15 kHz.

model for channels with considerable frequency and/or time selectivity. In [39], the case of $N_T = 2$ was studied. It was proposed to use pairs of adjacent pilots, which are chosen to be orthogonal among antennas since they are placed at the same F-T positions. An HP for each of the two pilots in a pair was used to reduce the associated interference. The performance of this channel estimation scheme was also evaluated in [40] when used along with the iterative receivers developed in [41]. The same idea reappeared later in [42]. A more straightforward pilot configuration was proposed in [42,43], consisting of pilot-HP pairs placed at different positions for each of the antennas and thus allowing the pilot values to be independently chosen. In the simulation results reported in [42], this scheme was the only one to outperform CP-OFDM. Similar pilot formats were employed in [44] for equalizing fast fading MIMO-FBMC/OQAM systems. The pilot format adopted in [11] for the purpose of adaptive equalization can also be used for channel estimation purposes. In fact, it was successfully employed in [11] for tracking the channel and accordingly reinitializing the equalizer filters. The pilot pattern follows the Long-Term Evolution (LTE) Cell-specific Reference Signal (CRS) frame format [12]. An example of the result of employing this pilot arrangement for estimating a highly frequency and time selective 3rd Generation Partnership Project (3GPP) Extended Typical Urban (ETU) channel [45] is given in Fig. 14.4. The performance of the CP-OFDM system is also shown. A simple interpolation scheme was employed, consisting of DFT and linear interpolation in the frequency and time directions, respectively. There was no pilot boosting. Observe the error floor exhibited by the FBMC system at higher SNR values, typical consequence of the inadequacy of (14.2) to accurately model such highly selective scenarios.

A less bandwidth efficient scheme, which can be roughly described as locally applying the IAM idea in the pilot neighborhood and involves four so-called *aided pilot carriers* per pilot position, was also tested in [42,43], without promising performance results. Among the techniques evaluated in [42], there was also the Pair Of Pilots (POP) method [2] (applied here for isolated pairs of pilots as in [38]), but, as expected from its application in SISO systems (Section 11.3.1), it exhibited a quite poor performance. Pilot placement schemes that exploit the MIMO channel's spatial correlation to reduce pilot overhead and/or achieve performance improvements were investigated in [46].

## 14.3 CHANNEL EQUALIZATION

### 14.3.1 MAXIMUM LIKELIHOOD SEQUENCE ESTIMATION

The intrinsic interference in (14.1) can be viewed as two-dimensional (2D) ISI in the T-F plane since it depends on the transmitted data symbols in the neighborhood around the considered position $(m, n)$. The received FBMC signal can be seen as a 2D convolution of the transmitted data frame with the TMUX impulse response weighted by the channel matrix. Efficient detection in 2D-ISI channels has been largely studied (e.g., [47]), but in the context of MIMO-FBMC, the problem of MLSE is much more difficult due to the intrinsic interference effect. Consequently, $\mathbf{y}_{m,n}$ has to be further processed before being fed into the detector. In this regard, most of the existing MLSE strategies involve equalization and interference cancellation.

#### 14.3.1.1 *Single-Tap Equalization*

Linear equalization based on ZF or MMSE criteria can be straightforwardly applied in (14.2), as described in [41]. The equalization matrix $\mathbf{W}_m \in \mathbb{C}^{N_R \times N_T}$ targets the virtually transmitted vector $\mathbf{x}_{m,n} = \mathbf{d}_{m,n} + j\mathbf{u}_{m,n}$, and since $\mathbb{E}\left\{\mathbf{x}_{m,n}\mathbf{x}_{m,n}^H\right\} \approx 2E_S\mathbf{I}_{N_T}$ with $E_S$ denoting the average symbol power, the MMSE equalizer becomes

$$\mathbf{W}_m = \left(\mathbf{H}_m\mathbf{H}_m^H + \frac{\sigma_\eta^2}{2E_S}\mathbf{I}_{N_R}\right)^{-1}\mathbf{H}_m^H. \tag{14.11}$$

The real part retrieval of the equalized symbol $\tilde{\mathbf{x}}_{m,n} = \mathbf{W}_m^H\mathbf{y}_{m,n}$ yields the equalized data vector

$$\tilde{\mathbf{d}}_{m,n} = \mathbb{Re}(\tilde{\mathbf{x}}_{m,n}), \tag{14.12}$$

and the transmitted Pulse Amplitude Modulation (PAM) symbols are then independently detected.

### 14.3.1.2 *Full Interference Cancellation*

The first attempt to outperform single-tap equalization is based on first estimating the intrinsic interference and then performing Interference Cancellation (IC) to either apply Maximum Likelihood (ML) detection [48] or exploit WL processing [49]. Two different approaches have been proposed to estimating the interference. In the first approach, since the MMSE equalizer provides an estimate of the virtual transmitted symbol vector $\mathbf{x}_{m,n}$, we can directly evaluate the interference term by taking only the imaginary part of $\tilde{\mathbf{x}}_{m,n}$, i.e.,

$$j\tilde{\mathbf{u}}_{m,n} = j\,\mathbb{Im}(\tilde{\mathbf{x}}_{m,n}). \tag{14.13}$$

In the second approach, the intrinsic interference is evaluated as follows:

$$j\tilde{\mathbf{u}}_{m,n} = \sum_{(m',n')\in\bar{\Omega}_{m,n}} \Gamma_{m,m'}^{n,n'}\mathbf{d}_{m',n'}', \tag{14.14}$$

where the tentative symbols $\left\{\mathbf{d}_{m',n'}'\right\}$ are the estimates of (14.12). The first option is called MMSE-ML, whereas the second option using tentative symbols is named IC-ML. Once the interference has been reconstructed, it can be canceled, yielding

$$\begin{aligned}
\mathbf{z}_{m,n} &= \mathbf{y}_{m,n} - j\mathbf{H}_m\tilde{\mathbf{u}}_{m,n} \\
&= \mathbf{H}_m\left(\mathbf{d}_{m,n} + j\boldsymbol{\epsilon}_{m,n}\right) + \boldsymbol{\eta}_{m,n},
\end{aligned} \tag{14.15}$$

where $\boldsymbol{\epsilon}_{m,n} = \mathbf{u}_{m,n} - \tilde{\mathbf{u}}_{m,n}$ is the interference estimation error. Assuming perfect estimation, i.e., $\boldsymbol{\epsilon}_{m,n} = \mathbf{0}$, the vector at the output of the IC stage is given by

$$\mathbf{z}_{m,n} = \mathbf{H}_m\mathbf{d}_{m,n} + \boldsymbol{\eta}_{m,n}, \tag{14.16}$$

and conventional ML detection can be applied. The latter has a high complexity of order $\mathcal{O}\left(Q^{N_T}\right)$, where $Q$ is the size of the symbol alphabet. Low-complexity, near ML alternatives can be employed instead, such as the K-best algorithm proposed in [50].

### 14.3.1.3 *Partial Interference Cancellation*

Since the intrinsic interference and the desired symbol have almost the same power in the FBMC/OQAM context, the error $\boldsymbol{\epsilon}$ will be nonzero in moderate- and strong-noise regimes. To deal with this, Partial Interference Cancellation (PaIC) was proposed in [51]. The receiver is composed of a tentative detector that serves to partially canceling the interference, followed by a Viterbi detector. In this scheme, the set $\bar{\Omega}_{m,n}$ is split into subsets $\bar{\Omega}_{m,n}'$ and $\bar{\Omega}_{m,n}''$. Then, (14.1) becomes

$$\begin{aligned}
\mathbf{y}_{m,n} &= \mathbf{H}_m\left(\mathbf{d}_{m,n} + \sum_{(m',n')\in\bar{\Omega}_{m,n}'} \mathbf{d}_{m',n'}\Gamma_{m,m'}^{n,n'} + \sum_{(m',n')\in\bar{\Omega}_{m,n}''} \mathbf{d}_{m',n'}\Gamma_{m,m'}^{n,n'}\right) + \boldsymbol{\eta}_{m,n} \\
&= \mathbf{H}_m\left(\mathbf{d}_{m,n} + j\mathbf{u}_{m,n}' + j\mathbf{u}_{m,n}''\right) + \boldsymbol{\eta}_{m,n}.
\end{aligned} \tag{14.17}$$

In a way analogous to Section 14.3.1.2, the decided estimates of (14.12), which are given by $\{\mathbf{d}'_{m,n}\}$, are utilized to reconstruct the interference associated with the subset $\bar{\Omega}''_{m,n}$ as follows:

$$j\tilde{\mathbf{u}}''_{m,n} = \sum_{(m',n')\in\bar{\Omega}''_{m,n}} \Gamma^{n,n'}_{m,m'}\mathbf{d}'_{m',n'}. \tag{14.18}$$

Assuming that the intrinsic interference resulting from the set $\bar{\Omega}''_{m,n}$ has been completely removed, a Viterbi detector is then applied to match the noncanceled interference that comes from $\bar{\Omega}'_{m,n}$. In [51], the authors have found that a satisfactory trade-off between the high complexity of the Viterbi detector and the effectiveness of IC is achieved by the set $\bar{\Omega}'_{m,n} = \{(m, n-1), (m, n+1)\}$. After removing the interference and assuming perfect interference estimation, the vector $\mathbf{z}_{m,n} = \mathbf{y}_{m,n} - j\mathbf{H}_m\mathbf{u}''_{m,n}$ is expressed as

$$\mathbf{z}_{m,n} = \mathbf{H}_m \left(\mathbf{d}_{m,n-1}\Gamma^{n,n-1}_{m,m} + \mathbf{d}_{m,n} + \mathbf{d}_{m,n+1}\Gamma^{n,n+1}_{m,m}\right) + \boldsymbol{\eta}_{m,n}. \tag{14.19}$$

Finally, $\mathbf{z}_{m,n}$ is fed into the Viterbi detector.

### 14.3.1.4 *Simulation Results*

The detection performance of the different receivers was evaluated in a $2 \times 2$ system using the 3GPP Extended Vehicular-A (EVA) channel model [45] and PHYDYAS FBs. The number of subcarriers is $M = 1024$, the subcarrier spacing is 15 kHz, and the data symbols are OQPSK modulated. Perfect Channel Information (PCI) is assumed to be available at the receiver side. The following receivers were tested: linear MMSE, MMSE-ML, IC-ML using an MMSE tentative detector, and PaIC. The OFDM-ML receiver is also included as a reference. The uncoded Bit Error Rate (BER) curves are shown in Fig. 14.5. Although the MMSE-ML receiver already performs better than the MMSE one, an additional 1 dB is gained when using IC-ML. However, the performance of OFDM-ML is still not reached. It should be noted that the FBMC-MMSE and OFDM-MMSE receivers perform similarly. The PaIC/Viterbi method, however, outperforms the other schemes and attains almost the same performance as the OFDM-ML receiver, except at very high SNR values.

To reduce the interference, it was proposed in [52] to modify the FBMC structure and use Quadrature Amplitude Modulation (QAM) instead of OQAM. However, in this way, the orthogonality is lost, and the receiver has to perform iterative IC even in SISO case. An alternative approach, called Fast Fourier Transform (FFT)-FBMC, was proposed in [53], where the data are precoded in a subcarrier-wise manner using extra Inverse Fast Fourier Transform (IFFT) and employing a CP. The interference within a subcarrier is addressed via simple equalization, whereas for the interference coming from the adjacent subcarriers, it is shown that it can be avoided with the aid of a special data transmission strategy. Hence, thanks to the extra IFFTs in each subcarrier, the channel equalization coefficients are simply obtained from the CFR with higher frequency resolution [54]. Consequently, MIMO techniques can be straight-

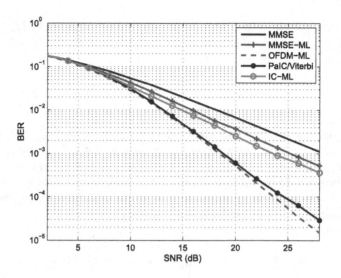

**FIGURE 14.5**

BER performance comparison of different IC receivers in a 2 × 2 system employing
PHYDYAS FBs with $M = 1024$ and $K = 4$. EVA channels are considered. The subcarrier
spacing is 15 kHz.

forwardly applied. Instead of using an IFFT, it was also proposed in [55,56] to spread
the symbols with a truncated Hadamard matrix.

## 14.3.2 FREQUENCY SAMPLING-BASED LINEAR MULTIANTENNA DETECTION SCHEMES

Frequency Sampling (FS) based subcarrier equalizer design was explained in Section 12.5.2 for the SISO case, but it has also been extended to the SIMO and Spatial Multiplexing (SM) MIMO-FBMC/OQAM systems in [57]. In this approach, the equalizer adaptation is based on a frequency sampled model, and subcarrier equalizers are implemented as multitap complex Finite Impulse Response (FIR) filters for each RX antenna per data stream. Frequency-sampled SISO equalization has also been proposed in the context of the FS-FBMC approach [58].

Actually, the frequency-sampled equalization scheme is very flexible, and the same design principle and the same subcarrier equalizer structure can be used in all Single-Carrier (SC) systems with frequency-domain equalization and FBMC systems in which significant frequency selectivity appears at the subcarrier/subband level. The key idea is to find the linear MMSE solution, in the same way as in MIMO-OFDM, in a number of frequency points within each subband. Then the multitap subcarrier equalizers are designed to reach the target frequency response at those frequencies [57]. At each frequency point used in the design, the target frequency response is obtained using the ZF or the MMSE criterion, as explained in Section 14.3.1.1.

For example, when using three-tap subcarrier equalizers, the frequency points correspond to subcarrier center frequencies and the subband cross-over frequencies. In this case, the number of equally spaced frequency points is about twice the number of used subcarriers, $M_u$. Generally, when using subcarrier equalizers with $L_{eq} > 1$ taps, the number of equally spaced frequency points used in the equalizer adaptation is $M_u(L_{eq} - 1) + 1$.

### 14.3.2.1 Embedded Multiantenna Subcarrier Equalization With Fast-Convolution FilterBank (FC-FB)

Embedded equalization with FC-FB implementation of SISO FBMC systems was explained in Section 12.5.1. This idea can be also applied in linear detection for multiantenna receivers, i.e., SIMO and MIMO systems [59]. Now the equalizer adaptation is based on the frequency sampling approach explained above, and the resulting subcarrier equalizers are implemented by modifying the weight coefficients in Fast-Convolution (FC) processing. With a fast fading channel, the weight coefficients are recalculated for each FC processing block. With low mobility, the rate of adaptation should be sufficient to match the channel variations.

Fig. 14.6 shows the FC-FB structure with embedded subcarrier equalizers in the $2 \times 2$ MIMO case with two transmitted data streams. Each FC-FB data path from a receiver antenna to an output data stream implements subchannel filtering using the multirate FC processing principle: long ($N$-point with $N = \frac{ML}{2}$) FFT, selection of the $L$ FFT bins corresponding to the subchannel frequency band, weighting of the FFT bins, and short ($L$-point) IFFT to obtain a filtered and decimated block of samples. Overlap-save processing by the factor of $1 - N_S/N = 1 - L_S/L$ is used in the process to implement linear convolution through a cyclic convolution process. Here $N_S$ and $L_S$ are the nonoverlapped (useful) data block lengths on the high-rate and low-rate sides, respectively.

In the embedded equalizer, the FS-based equalizer coefficients are directly combined with the weights of the basic FC-FB design. The weights are implemented separately for each signal path from one of the RX antennas to an output stream, and IFFT processing is implemented after combining the antenna signals. This embedded equalizer structure is independent of the transmitted waveform, subcarrier bandwidth, and center frequency. The frequency domain weights are computed from the CFRs and the predesigned basic weight mask of each subcarrier. The weight for each frequency bin is obtained as the product of the weight coming from the basic weight mask and the linear equalizer weight computed from the CFRs:

$$w_{q,p,m,i} = w_{m,i}^{(FB)} w_{q,p,m,i}^{(EQ)}. \tag{14.20}$$

Here $q$ is the index of the RX antenna, $p$ is the index of the data stream, $m$ is the subcarrier/subband index, and $i$ is the IFFT bin index. We can see that the FS-based subcarrier equalizer can be embedded in a natural way in the FC AFB. With the ZF criterion, the equalizer weights would be obtained for each frequency bin from the

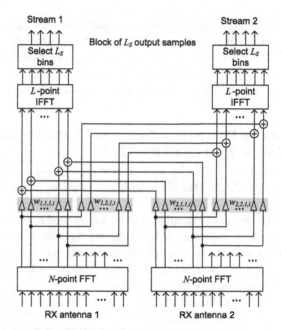

**FIGURE 14.6**

2 × 2 spatial multiplexing MIMO receiver utilizing the FC-FB structure with embedded equalization.

pseudo-inverse of the channel matrix $\mathbf{H}_{q,p,m,i}$ at the corresponding bin:

$$\mathbf{W}_{q,p,m,i}^{(ZF)} = \left(\mathbf{H}_{q,p,m,i}^{H}\mathbf{H}_{q,p,m,i}\right)^{-1}\mathbf{H}_{q,p,m,i}^{H}. \tag{14.21}$$

With the MMSE criterion, the corresponding expression is

$$\mathbf{W}_{q,p,m,i}^{(MSE)} = \left(\mathbf{H}_{q,p,m,i}^{H}\mathbf{H}_{q,p,m,i} + \frac{\sigma_{\eta}^{2}}{2E_S}\mathbf{I}_{N_R}\right)^{-1}\mathbf{H}_{q,p,m,i}^{H}. \tag{14.22}$$

It can be noted that the ZF solution would completely cancel the interference from other data streams and also from adjacent subcarriers in the FBMC/OQAM case. However, the MMSE solution results in lower MSE in the equalized signal by balancing the effects of imperfect equalization and noise enhancement.

Here clarification concerning the linear receiver structures used in different antenna configurations is appropriate. In SISO and SIMO configurations, it is possible to utilize the frequency-domain equalizer to closely approximate the classical optimal receiver structure where the fractionally spaced equalizer implements the matched fil-

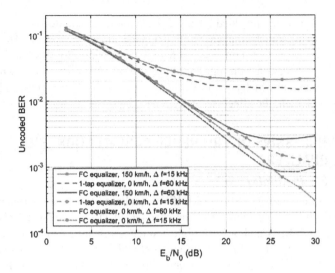

**FIGURE 14.7**

MMSE detection performance for 2 × 2 SM MIMO with QPSK modulation, ETU channel, and PCI at the receiver.

ter, matched both to the transmitted pulse shape and the channel response, along with linear channel equalizer [60,61]. However, in the SM MIMO configuration, each antenna chain receives multiple data streams affected by different channel responses, and it is not possible to match the receiver processing for multiple channel responses at the same time. Therefore, the receiver structure consists of the pulse shape matched filter, implemented by the basic weight mask of the FC-FB structure, and the MMSE criterion-based weights obtained from (14.22).

### 14.3.2.2 *Simulation-Based Performance Evaluation*

The 5 MHz LTE-like scenario with FBMC/OQAM waveform, 512 subcarriers out of which 300 are active, and subcarrier spacing of 15 kHz [62] is considered. Also, the cases with extended subcarrier spacing of 60 kHz and 72 active subcarriers are included. Results are shown for the 2 × 2 SM MIMO configuration using Quadrature Phase-Shift Keying (QPSK) modulation, ETU channel model with 0 km/h and 150 km/h mobilities, and the carrier frequency of 2.5 GHz. The FC-FB design uses 16 bins per subcarrier (i.e., $L = 16$) and 10 nonoverlapping samples (5 OQAM symbols) per block. In the simulations, independent instances of the corresponding channel model are used for the four propagation paths of the 2 × 2 MIMO configuration. In addition to the embedded equalizer, the single-tap subcarrier equalization model (using the same equalizer weights in the FC-FB implementation for all bins of a subcarrier) is included in the comparisons. Fig. 14.7 shows the results assuming PCI at the receiver. Although the single-tap equalizer model works quite well with 15 kHz subcarrier spacing, its performance with 60 kHz subcarrier spacing is significantly degraded. The embedded equalizer is also able to handle very well the high frequency selective subchannel case with extended subcarrier spacing. However, the equalizer weights are constant over each FFT processing block, which introduces performance degradation with high mobility. The effect is quite severe with 15 kHz subcarrier spacing, in which case the FC block length is 0.333 ms, but with four times reduced block length, the effect is greatly reduced.

### 14.3.3 SUCCESSIVE INTERFERENCE CANCELLATION (SIC) EQUALIZERS

#### 14.3.3.1 SIC in MIMO-FBMC

As seen previously, linear (ZF and MMSE) equalizers exhibit a good performance in SM MIMO-FBMC systems when the channel selectivity and/or the SNR is low. However, at high SNR and/or with high frequency selective channels, the residual interference becomes the dominant impairment and limits the performance of these equalizers. One way to cope with this and improve the equalization performance is to use SIC techniques, as described in [63].

We consider an $N_T \times N_R$ MIMO system where each transmit antenna sends an independent stream of data. The idea behind the SIC technique is to recover the symbol streams coming from different antennas in a sequential way. Per-subcarrier processing is still considered here. The first step consists in detecting the first symbol stream using an MMSE equalizer or any other equalizer. Then the contribution of this first stream can be subtracted from the received signals to remove any interference it could create on the detection of other streams.[5] The second stream can then be detected from the $N_R$ received samples without having to cope with the interference of the first stream, which increases the degrees of freedom of the equalizer and allows us to attain better performance. This continues sequentially until all streams have been detected. In MIMO-FBMC, the interference from adjacent subcarriers has to be taken into account in addition to the multiantenna and the intersymbol interference. Therefore, the contributions from already recovered streams are not only subtracted from the $N_R$ received samples (one per antenna) corresponding to the current subcarrier $m$, but their contributions are also subtracted from the $N_R$ received samples of the two adjacent channels, $m - 1$ and $m + 1$. Provided that error propagation can be kept small and that the channel can be properly estimated, the SIC technique provides an easy way to improve the performance of any equalizer. In this subsection, the application of this technique in the MMSE equalizer for MIMO-FBMC/OQAM is discussed. It can be summarized in the following steps.

**Step 1:** Recover the first symbol stream. This can be done, for instance, using a single-tap MMSE equalizer such as (14.11) described in Section 14.3.1.1. A multitap MMSE equalizer can also be applied by extending the results of

---

[5]This obviously requires some channel estimation and is susceptible to error propagation.

Section 12.3.1 to the MIMO case. The multitap MMSE equalizer is considered in this section. Denote by $\hat{d}_{m,n}^1$ the decisions made on the transmitted symbols $d_{m,n}^1$ for the first stream.

**Step 2:** Assuming that the channel is known and that the tentative decisions are correct, it is possible to remodulate them and compute the convolution with the cascade of synthesis filter, MIMO channel, and analysis filter to obtain the contributions of each of the symbols on the received sample vector $\mathbf{y}_m^q = \begin{bmatrix} y_{m,0}^q & y_{m,1}^q & \cdots \end{bmatrix}^{\mathrm{T}}$ at subcarrier $m$ and RX antenna $q$ for $m = 0, 1, \ldots, M - 1$ and $q = 1, 2, \ldots, N_{\mathrm{R}}$. The contributions from symbols in the current subcarrier $m$ and adjacent subcarriers $m - 1$ and $m + 1$ can be completely removed from the received signals as follows:

$$\mathbf{y}_m^{q-} = \mathbf{y}_m^q - \sum_{m'=m-1}^{m+1} \mathbf{G}_{m,m'}^{\prime q,1} \hat{\mathbf{d}}_{m'}^1, \qquad (14.23)$$

where $\mathbf{y}_m^{q-} = \begin{bmatrix} y_{m,0}^{q-} & y_{m,1}^{q-} & \cdots \end{bmatrix}^{\mathrm{T}}$ is the received sample vector at subcarrier $m$ and RX antenna $q$ after removing the contributions of $\hat{\mathbf{d}}_{m'}^1 = \begin{bmatrix} \hat{d}_{m',0}^1 & \hat{d}_{m',1}^1 & \cdots \end{bmatrix}^{\mathrm{T}}$ for $m' = m - 1, m, m + 1$, and $\mathbf{G}_{m,m'}^{\prime q,1}$ denotes the (transposed) convolution matrix from stream no. 1 at subcarrier $m'$ to receive antenna $q$ at subcarrier $m$ (defined in a manner analogous to that in Section 12.3.1). This step obviously requires the estimation of the $\mathbf{G}'$ matrices. These are also necessary for the design of most equalizers, and hence SIC does not have any special needs for estimation.[6]

**Step 3:** Go back to Step 1 and recover the second stream. It is important to note that the MMSE equalizer used in this case no longer needs to take into account the interantenna interference coming from the first stream since its contribution has been completely removed. The expression of the MMSE equalizer is thus slightly different, and all terms related to the previous stream can be removed. For this reason, the MMSE equalizer for the second stream is also expected to perform slightly better than for the first stream.

This is repeated stream by stream until all of them have been recovered. After recovering the $i$th stream, $i = 1, 2, \ldots$, and assuming that all the decisions are correct (i.e., no error propagation occurs), it is clear from the remarks above that the system becomes equivalent to one with $N_{\mathrm{T}} - i$ TX antennas and $N_{\mathrm{R}}$ RX antennas, thus offering an increased diversity. In general, however, error propagation may occur, and the performance may not benefit from full diversity.

---

[6]There are several simple equalization schemes, however, that do not need the estimation of the full matrix, particularly, for mildly selective channels. In such cases, the SIC technique might not be appropriate as simple equalization already performs well.

### 14.3.3.2 *OSIC for MIMO-FBMC*

To improve the performance and minimize the error propagation problem, the ordered SIC technique [64] can be considered. Its principle is to choose the detection order of the different streams based on their respective SINR (higher SINR first), instead of fixing the order in advance. This allows us to detect and remove the contribution of the most reliable streams first and therefore minimize the risk of error propagation. The OSIC algorithm is almost the same as the SIC algorithm, and only few changes are required. However, in contrast to the SIC receiver, the order of detection may be different from one subcarrier to another depending on the SINR of each subcarrier. Hence, the subtraction of the contribution of the adjacent subcarriers often becomes impossible. Suppose, without loss of generality, that for subcarrier $m$, the first stream has the highest SINR. Then the second step (14.23) of the SIC procedure becomes

$$\mathbf{y}_m^{q-} = \mathbf{y}_m^q - \mathbf{G'}_{m,m}^{q,1} \hat{\mathbf{d}}_m^1.$$
(14.24)

Only the contributions of the same subcarrier are subtracted. Then $\mathbf{y}_m^{q-}$ is used to compute the equalizer for the next stream to be detected, selecting the equalizer that provides the highest SINR. After detecting the stream at the output of this equalizer, its contribution is again subtracted by applying (14.24). This process is repeated until all streams are recovered. Note that to determine the stream with the best SINR, it is theoretically necessary to compute the equalizer for each stream and then determine which one has the best performance. Doing so substantially increases the complexity of OSIC with respect to the initial SIC technique as new equalizers have to be computed for all streams at each step. In practice, it is better to use approximate evaluations of the SINR on each stream.

### 14.3.3.3 *Two-Stage Ordered Successive Interference Cancellation (TS-OSIC) for MIMO-FBMC*

To further improve the performance of the OSIC receiver, it is possible to combine it with the two-stage SIC presented in Section 12.3.3 for SISO equalizers. The main idea behind the TS-OSIC [63] is that, after applying the OSIC algorithm once, subtraction of the contribution of the adjacent subcarriers from the subcarrier of interest becomes possible since a first estimate of all streams on all subcarriers is already available. The received samples after removal of the contributions of the adjacent subcarriers are given by

$$\mathbf{y}_m^{q+} = \mathbf{y}_m^q - \sum_{i=1}^{N_T} \left( \mathbf{G'}_{m,m-1}^{q,i} \hat{\mathbf{d}}_{m-1}^i + \mathbf{G'}_{m,m+1}^{q,i} \hat{\mathbf{d}}_{m+1}^i \right).$$
(14.25)

The decisions used here are coming from the first application of the OSIC algorithm. Once the contribution of the adjacent subcarriers has been removed, the OSIC algorithm can be applied a second time, to enhance the performance, on the adjusted received samples (14.25). Note that, in (14.25), only the contributions of the adjacent

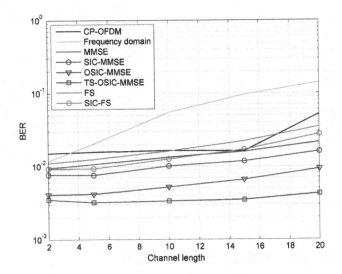

**FIGURE 14.8**

Effect of the channel length on the performance of the equalizers: BER vs. channel length. $E_b/N_0 = 20$ dB, MMSE equalizers of length 3.

subcarriers are removed. The contributions of the current subcarrier are removed during the second stage of the OSIC algorithm using the most recent estimates of the symbol streams. This means that the interference coming from adjacent subcarriers is always canceled using the results of the first stage, and the interference coming from the subcarrier of interest is canceled using the OSIC technique in the second stage. It is worth noting that, similarly to the TS-IC in the SISO case, the equalizer computation in the second stage is simplified since the interference coming from the adjacent subcarriers can be ignored. All in all, the complexity of the TS-OSIC still remains very high and is only justified if very high performance is required.

To assess the performance of these algorithms and compare them to CP-OFDM, the BER is evaluated for a $3 \times 3$ MIMO system using PHYDYAS FBs with $M = 64$ subcarriers, overlapping factor $K = 4$, and 16-QAM modulation. The CP length for the CP-OFDM system is $\frac{M}{4}$. The MIMO FIR channels are chosen randomly. Fig. 14.8 shows the effect of the channel length on the performance of the algorithms at $E_b/N_0 = 20$ dB. Various algorithms are also compared to the simple single-tap frequency domain equalizer, as well as the FS versions (see Section 14.3.2) of these equalizers. It can be seen that, for a short channel (length 2 or 3), the MMSE and FS equalizers almost have the same performance. The SIC-based technique in itself brings some small improvement. The OSIC and TS-OSIC further slightly improve the performance. For more selective channels, the MMSE equalizer seems to perform better, and all SIC-based algorithms are increasingly better at handling the interference. It is worth noting that SIC-based techniques are interesting even for short channels, especially when a large number of antennas is considered.

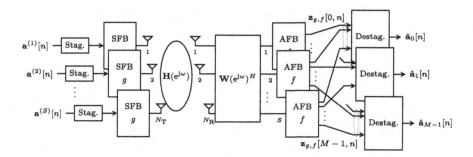

**FIGURE 14.9**

Ideal implementation of a MIMO equalizer prior to FBMC/OQAM demodulation. The equalizer processes the whole signal bandwidth.

## 14.3.4 PARALLEL MULTISTAGE EQUALIZATION

The parallel multistage equalization architecture presented in Section 12.4 of Chapter 12 can be trivially generalized in the MIMO context [65]. Call $\mathbf{H}\left(e^{j\omega}\right) \in \mathbb{C}^{N_R \times N_T}$ the frequency response of the MIMO channel and denote by $\mathbf{W}\left(e^{j\omega}\right)$ the $N_R \times N_T$ equalizer matrix that would be ideally implemented. For example, the ZF equalizer can be formulated as

$$\mathbf{W}\left(e^{j\omega}\right) = \mathbf{H}\left(e^{j\omega}\right)\left[\mathbf{H}\left(e^{j\omega}\right)^H \mathbf{H}\left(e^{j\omega}\right)\right]^{-1} \tag{14.26}$$

assuming that $N_T \leq N_R$. Fig. 14.9 illustrates a direct implementation of this equalizer. The equalization matrix $\mathbf{W}\left(e^{j\omega}\right)$ processes the whole signal bandwidth without exploiting the multicarrier structure of the FBMC/OQAM signal. It consists of a set of $N_T N_R$ filters (one for each entry of the matrix) and provides a set of $N_T$ signals that are demodulated using conventional FBMC/OQAM receivers.

In practice, the equalization in Fig. 14.9 is costly from the computational point of view because it processes the whole signal bandwidth as in SC modulations. To exploit the multicarrier structure of the signal, we might assume that the equalizer is sufficiently frequency flat around the subcarrier frequencies, so that

$$\mathbf{W}\left(e^{j\omega}\right) \approx \mathbf{W}_m = \mathbf{W}\left(e^{j\omega_m}\right) \tag{14.27}$$

when $\omega \approx \omega_m$. This allows us to approximate the original equalizer using a single matrix multiplication per subcarrier, as further illustrated in Fig. 14.10. The main idea behind the parallel multistage equalization architecture consists in refining the approximation in (14.27) and consider a more elaborate dependence on the frequency $\omega$. To further illustrate this point, consider the discrete-time Fourier transform of the receive prototype pulse $f$, denoted by $F\left(e^{j\omega}\right)$. The frequency response associated with the $m$th subcarrier of the AFB prior to downsampling is therefore proportional to $F\left(e^{j(\omega-\omega_m)}\right)$. Using the Taylor development of $\mathbf{W}\left(e^{j\omega}\right)$ up to order $R$ around $\omega_m$, we

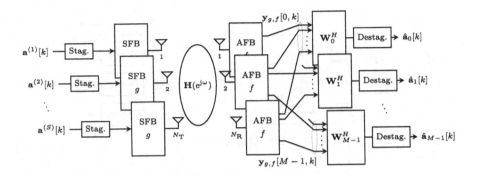

**FIGURE 14.10**

Conventional linear MIMO equalizer based on a single matrix per subcarrier.

may therefore approximate the cascade of the AFB and the ideal equalizer as

$$\mathbf{W}\left(e^{j\omega}\right)^{H} F\left(e^{j(\omega-\omega_m)}\right) \approx \sum_{r=0}^{R} \frac{j^r}{r!} \mathbf{W}^{(r)}\left(e^{j\omega_m}\right)^{H} (\omega-\omega_m)^r F\left(e^{j(\omega-\omega_m)}\right),$$

where $\mathbf{W}^{(r)}(z)$ is the $r$th derivative of the equalizer transfer function $\mathbf{W}(z)$. The above equation can be seen as a linear combination of $R+1$ transfer functions, each corresponding to the multiplication by a constant matrix $\mathbf{W}_m^{(r)} = \mathbf{W}^{(r)}\left(e^{j\omega_m}\right)$ and a *virtual* AFB with frequency response $\omega^r F\left(e^{j\omega}\right)$.

At this point, we can apply a trick that will bring out the physical meaning of the frequency response $\omega^r F\left(e^{j\omega}\right)$. This trick was already introduced in Chapter 9 when discussing the effect of channel frequency selectivity on the FBMC/OQAM modulation. Assume that the original prototype $f$ is obtained by discretizing a certain analog smooth waveform and denote by $f^{(r)}$ the discretized version of the $r$th derivative of this waveform. It turns out that, when the number of subcarriers unboundedly increases, i.e., $M \to \infty$, $\omega^r F\left(e^{j\omega}\right)$ can be well approximated (up to a constant) with the Fourier transform of $f^{(r)}$. Therefore, we can refine the approximative implementation in Fig. 14.10 by considering the more evolved architecture in Fig. 14.11. A set of $R+1$ parallel AFBs is considered for each receive antenna. Each AFB is constructed using a sequential time-domain derivative of the original prototype (namely $f, f^{(1)}, f^{(2)}, \ldots$). Furthermore, the output of these AFBs is then linearly combined using the sequential frequency derivatives of the original equalizer transfer function $\mathbf{W}, \mathbf{W}^{(1)}, \ldots$. The result is then linearly combined prior to the destaggering operation. It must be emphasized that, contrary to multitap equalization architectures, the multistage equalization technique considered here does not introduce additional latency in the receiver chain.

Under some regularity conditions, we can reduce the approximation error of the multistage architecture by simply adding additional parallel stages. To illustrate this

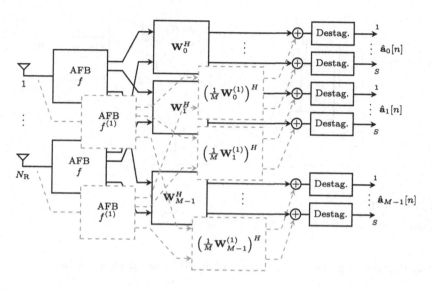

**FIGURE 14.11**

Parallel multistage MIMO equalizer based on two stages ($R = 1$). In the second stage (dotted line), the AFB are built using time derivatives of the original prototype pulses ($f^{(1)}$) and frequency derivatives of the linear receivers ($\mathbf{W}_m^{(1)}$).

point using formal mathematical arguments, let $\mathbf{y}_{g,f}[m, n] \in \mathbb{C}^{N_R \times 1}$ denote a column vector gathering the samples at the output of the $N_R$ receive AFBs corresponding to the $n$th real-valued symbol allocated to the $m$th subcarrier (see Fig. 14.10). Assume that the ideal equalizer $\mathbf{W}\left(e^{j\omega}\right)$ is implemented, so that the effective channel is given by $\mathbf{W}^H\left(e^{j\omega}\right)\mathbf{H}\left(e^{j\omega}\right)$, which is a square matrix of dimensions $N_T \times N_T$. In this situation, let $\mathbf{z}_{g,f}[m, l] \in \mathbb{C}^{N_T \times 1}$ consist of the samples at the output of the $N_T$ receive AFBs that demodulate the signals at the outputs of the ideal equalizer (see Fig. 14.9). We can now apply Proposition 9.1 in Chapter 9, replacing the role of the channel $H(z)$ by the entries of the equalizer matrix $\mathbf{W}(z)$, resulting in

$$\mathbf{z}_{g,f}[m, n] = \sum_{r=0}^{R} \frac{1}{r!M^r}\left(\mathbf{W}_m^{(r)}\right)^H \mathbf{y}_{g,f^{(r)}}[m, n] + o\left(M^{-R}\right)$$

as $M \to \infty$. This equation is formally stating that the response of the ideal equalizer ($\mathbf{z}_{g,f}[m, l]$) can be approximated by properly weighting the output of $R + 1$ FBs up to an error that decays to zero faster than $M^{-R}$ as $M$ increases to infinity. This proves that we can guarantee an approximation error as low as possible by simply increasing the number of parallel stages. Furthermore, we can study the behavior of the error term above to determine the order $R$ that guarantees a minimum performance across the spectrum (see [65] for more details).

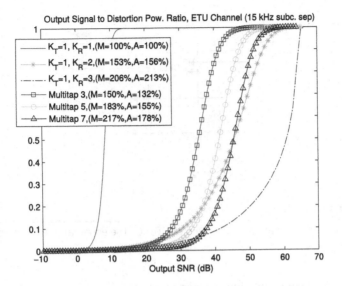

Output Signal to Distortion Pow. Ratio, ETU Channel (15 kHz subc. sep)

Legend:
- $K_T=1$, $K_R=1$, (M=100%, A=100%)
- $K_T=1$, $K_R=2$, (M=153%, A=156%)
- $K_T=1$, $K_R=3$, (M=206%, A=213%)
- Multitap 3, (M=150%, A=132%)
- Multitap 5, (M=183%, A=155%)
- Multitap 7, (M=217%, A=178%)

x-axis: Output SNR (dB)

**FIGURE 14.12**

Cumulative distribution function of the signal-to-distortion power ratio for different equalizers in a 2 × 2 MIMO system applying channel inversion at the receiver. In the legend, M and A respectively represent the relative increase in the number of real-valued multiplications and additions with respect to single matrix per-subcarrier multiplication.

To test the performance of the multistage equalizer, a MIMO-FBMC/OQAM system with $N_T = N_R = 2$ and $M = 256$ subcarriers was considered, where the two prototypes were PHYDYAS pulses [37] with overlapping factor $K = 4$. The intended equalizer was the ZF one, given in (14.26), and the channel was generated according to the ETU model with intercarrier separation equal to 15 kHz. Fig. 14.12 shows the cumulative distribution function of the Signal-to-Distortion power Ratio (SDR) at the output of the equalizer. Apart from the performance of the parallel multistage architecture for $R = 0, 1, 2$, the figure also depicts the performance of the multitap equalizer with 3, 5, and 7 matrix coefficients per subcarrier, computed according to the FS technique. The legend of the figure provides the relative increase in terms of real-valued multiplications (M) and additions (A) with respect to the single matrix per subcarrier equalizer in Fig. 14.10. Observe that the multistage implementation of the equalizer provides a very competitive performance/complexity trade-off with respect to multitap equalizer schemes.

### 14.3.5 ADAPTIVE EQUALIZATION OF DOUBLY DISPERSIVE CHANNELS

In highly mobile environments, where doubly dispersive channels have to be dealt with [66], the equalizer needs also to be adaptive and of low complexity. The literature on adaptive FBMC/OQAM equalization is quite limited for MIMO systems.

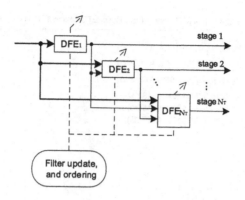

**FIGURE 14.13**

Block diagram of the adaptive BLAST DFE of [10,11]. The equalization structure for a given subcarrier is shown.

Most of the adaptive algorithms reported are concerned with the SISO case and rely on linear or decision-feedback structures employing the Least Mean Squares (LMS) algorithm (see [67] and references therein). Adaptive equalization for MIMO-FBMC/OQAM systems was first studied in [10,68]. An adaptive decision feedback equalization scheme was developed, whose structure is dictated by the BLAST idea and its equivalence with the generalized Decision Feedback Equalizer (DFE) [10]. It involves a MIMO DFE per subcarrier, which consists of $N_T$ serially connected stages, each equalizing one of the $N_T$ symbol streams according to BLAST ordering. Both the equalizer filters and the stream ordering are updated via an efficient Recursive Least Squares (RLS)-based algorithm. The latter is of the square-root type to ensure numerical stability in addition to fast convergence. This scheme, depicted schematically in Fig. 14.13, demonstrated a good performance in high frequency selective channels. However, long training preambles were required for the initial convergence of the equalizer filters, and no means of coping with fast varying channels was provided, thus restricting the applicability of this scheme to channels exhibiting slow fading. Significant improvements were attained in [11] by employing channel estimates to appropriately start up the equalizer and assist its continued adaptation in highly time/frequency selective environments. Considerable performance gains (also in comparison with the analogous MIMO-OFDM case) were shown to be achievable at the cost of (re)estimating the channel based on a quite short preamble and appropriately placed scattered (LTE-compliant) pilots. An example is depicted in Fig. 14.14 for a $2 \times 2$ system employing PHYDYAS FBs with $M = 128$ subcarriers and an overlapping factor of $K = 3$ at a subcarrier spacing of 15 kHz. ITU Veh-A channels [36] are considered, fading at a speed of 150 km/h, with QPSK input. The FBMC DFEs have forward and backward orders $K_f = K_b = 2$, and the equalization delay is taken as $\Delta = 1$. It is apparent how important is to rely on initial ("Init") and intermediate ("Reinit") channel estimates to keep a good detection performance.

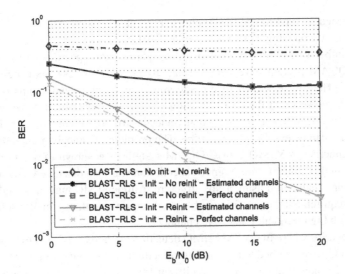

**FIGURE 14.14**

Uncoded BER of the BLAST RLS DFE algorithm for ITU Veh-A channels, with $K_f = K_b = 2$ and $\Delta = 1$, at a mobile speed of 150 km/h [12].

An adaptive linear equalization scheme for highly time/frequency selective channels in MIMO-FBMC/OQAM was reported in [44]. It relies, for the computation of the equalizer, on estimates of the channel obtained with the aid of pilots scattered throughout the frame (see also Section 14.2.2). The equalizer was integrated with the subcarrier processing of the FC-FB structure (see Chapter 8), leading to a highly efficient overall implementation. The performance of the proposed scheme was evaluated in [44] for both slow and fast fading channels in high frequency selective propagation scenarios.

## 14.4 CONCLUDING REMARKS

In a MIMO-FBMC/OQAM setup, the challenging task of FBMC/OQAM receiver design must also consider the need to cope with the multiantenna interference. An up-to-date overview of MIMO-FBMC/OQAM channel estimation and equalization approaches that aim at addressing these challenges was presented in this chapter. Emphasis was given to more recent developments, associated with transmission scenarios of high frequency and/or time selectivity.

Estimation of high frequency selective channels in a MIMO-FBMC/OQAM context has only been addressed so far through preamble-based techniques. Scattered pilots-based schemes that can be effective in such demanding scenarios are still to be developed. Extending the SISO scheme presented in Section 11.4.2 to the MIMO case seems a promising way to achieve this goal.

A number of ML detectors that are adapted to the specificity of the MIMO-FBMC/OQAM system have been discussed for more or less frequency selective channels. They were shown to provide significant performance gains over linear equalizers, however, at the cost of a computational complexity increase, which can become very significant. Variants that trade the ML optimality for a reduction in complexity can be employed in practice. The problem of canceling the interferences that are inherent in MIMO-FBMC/OQAM systems was shown to be of critical importance in this context. Interference cancellation was also discussed in combination with linear equalization using a SIC structure. Although SIC may have high computational requirements, it can bring substantial improvement, especially, at high SNR and/or with high frequency selective channels.

In scenarios where linear SM MIMO detection methods are found to be sufficient, FS- and FC-based embedded equalization methods were found to be promising approaches for handling cases with significant frequency selectivity at the subcarrier level. The main topic for future studies in this context is to develop effective methods of this type for MIMO channel estimation with frequency resolution exceeding the subcarrier spacing.

The parallel multistage equalization architecture presented in Section 14.3.4 is able to equalize the channel without introducing additional latency. This interesting feature may prove to be fundamental in facilitating the adoption of FBMC/OQAM in future wireless applications, such as tactile Internet or Industry 4.0. Furthermore, the architecture can be generalized to strongly selective channels both in time and frequency domains. This will prove to be useful in rapidly time varying channels subject to strong Doppler effects, such as in high speed trains or satellite links. In such applications, the adaptive equalization schemes discussed here can be directly applied, especially, once pilot schemes that can better cope with strong frequency- and time-selectivity have become available.

# REFERENCES

[1] B. Farhang-Boroujeny, "OFDM versus filter bank multicarrier," *IEEE Signal Process. Mag.*, vol. 28, no. 3, pp. 92–112, May 2011.

[2] E. Kofidis, D. Katselis, A. Rontogiannis, and S. Theodoridis, "Preamble-based channel estimation in OFDM/OQAM systems: A review," *Signal Process.*, vol. 93, no. 7, pp. 2038–2054, Jul. 2013.

[3] E. Kofidis, "Channel estimation in filter bank-based multicarrier systems: Challenges and solutions," in *Proc. ISCCSP-2014*, Athens, Greece, May 2014.

[4] A. I. Pérez-Neira, M. Caus, D. Le Ruyet, R. Zakaria, E. Kofidis, X. Mestre, M. Haardt, and Y. Cheng, "MIMO signal processing in offset-QAM based filter bank multicarrier systems," *IEEE Trans. Signal Process.*, vol. 64, no. 21, pp. 5733–5762, Nov. 2016.

[5] S. Coleri, M. Ergen, A. Puri, and A. Bahai, "Channel estimation techniques based on pilot arrangement in OFDM systems," *IEEE Trans. Broadcast.*, vol. 48, no. 3, pp. 223–229, Sep. 2002.

[6] A. Farhang, A. Aminjavaheri, N. Marchetti, L. E. Doyle, and B. Farhang-Boroujeny, "Pilot decontamination in CMT-based massive MIMO networks," in *Proc. ISWCS-2014*, Barcelona, Spain, Aug. 2014.

[7] J. C. Tu, "Optimum MMSE equalization for staggered modulation," in *Proc. 27th Asilomar Conf. Signals, Systems and Computers*, Pacific Grove, CA, Nov. 1993.

[8] S. Nedic and N. Popovic, "Per-bin DFE for advanced OQAM-based multicarrier wireless data transmission systems," in *Proc. Int. Zurich Seminar on Broadband Communications*, Zurich, Switzerland, Feb. 2002.

[9] V. van Zelst and T. C. W. Schenk, "Implementation of a MIMO OFDM-based wireless LAN system," *IEEE Trans. Signal Process.*, vol. 52, no. 2, pp. 483–494, Feb. 2004.

[10] E. Kofidis and A. A. Rontogiannis, "Adaptive BLAST decision-feedback equalizer for MIMO-FBMC/OQAM systems," in *Proc. PIMRC-2010*, Istanbul, Turkey, Sep. 2010.

[11] C. Mavrokefalidis, A. Rontogiannis, E. Kofidis, A. Beikos, and S. Theodoridis, "Efficient adaptive equalization of doubly dispersive channels in MIMO-FBMC/OQAM systems," in *Proc. ISWCS-2014*, Barcelona, Spain, Aug. 2014.

[12] E. Kofidis, *et al.*, "MIMO channel estimation and data detection." Project EM-PhAtiC ICT-318362, Deliverable D4.2, Apr. 2014. [Online]. Available: http://www.ict-emphatic.eu/images/deliverables/deliverable_d4.2_final.pdf.

[13] M. Lipardi, "Widely linear filtering in modern MIMO transceivers." Ph.D. dissertation. University of Naples Federico II, 2009.

[14] E. Kofidis, "Preamble-based estimation of highly frequency selective channels in MIMO-FBMC/OQAM systems," in *Proc. EW-2015*, Budapest, Hungary, May 2015.

[15] Q. Huang, M. Ghogho, and S. Freear, "Pilot design for MIMO OFDM systems with virtual carriers," *IEEE Trans. Signal Process.*, vol. 57, no. 5, pp. 2024–2029, May 2009.

[16] J.-P. Javaudin, D. Lacroix, and A. Rouxel, "Pilot-aided channel estimation for OFDM/OQAM," in *Proc. VTC-2003 (Spring)*, Jeju island, Korea, Apr. 2003.

[17] C. Lélé, J.-P. Javaudin, R. Legouable, A. Skrzypczak, and P. Siohan, "Channel estimation methods for preamble-based OFDM/OQAM modulations," in *Proc. EW-2007*, Paris, France, Apr. 2007.

[18] C. Lélé, P. Siohan, and R. Legouable, "The Alamouti scheme with CDMA-OFDM/OQAM," *EURASIP J. Adv. Signal Process.*, vol. 2010, pp. 2:1–2:11, Jan. 2010.

[19] E. Kofidis and D. Katselis, "Preamble-based channel estimation in MIMO-OFDM/OQAM systems," in *Proc. ICSIPA-2011*, Kuala Lumpur, Malaysia, Nov. 2011.

[20] G. Stüber, J. Barry, S. McLaughlin, Y. Li, M.-A. Ingram, and T. Pratt, "Broadband MIMO-OFDM wireless communications," *Proc. IEEE*, vol. 92, no. 2, pp. 271–294, Feb. 2004.

[21] M. Nájar, *et al.*, "MIMO channel matrix estimation and tracking." PHYDYAS D4.1 deliverable, 2009. [Online]. Available: http://www.ict-phydyas.org/delivrables/PHYDYAS-D4.1.pdf/view.

[22] H. Minn and N. Al-Dhahir, "Optimal training signals for MIMO OFDM channel estimation," *IEEE Trans. Wireless Commun.*, vol. 5, no. 5, pp. 1158–1168, May 2006.

[23] S. Taheri, M. Ghoraishi, and P. Xiao, "Overhead reduced preamble-based channel estimation for MIMO-FBMC systems," in *Proc. IWCMC-2015*, Dubrovnik, Croatia, Aug. 2015.

[24] S. Taheri, M. Ghoraishi, X. Pei, C. Aijun, and G. Yonghong, "Evaluation of preamble based channel estimation for MIMO-FBMC systems," *ZTE Commun.*, vol. 14, no. 4, pp. 3–10, Oct. 2016.

[25] S. Hu, Z. Liu, Y.-L. Guan, C. Jin, Y. Huang, and J.-M. Wu, "Training sequence design for efficient channel estimation in MIMO-FBMC systems," *IEEE Access*, vol. 5, pp. 4747–4758, Apr. 2017.

[26] E. Kofidis and D. Katselis, "Improved interference approximation method for preamble-based channel estimation in FBMC/OQAM," in *Proc. EUSIPCO-2011*, Barcelona, Spain, Aug.–Sep. 2011.

[27] M. Caus and A. I. Pérez-Neira, "Transmitter–receiver designs for highly frequency selective channels in MIMO FBMC systems," *IEEE Trans. Signal Process.*, vol. 60, no. 12, pp. 6519–6532, Dec. 2012.

[28] J. Du, P. Xiao, J. Wu, and Q. Chen, "Design of isotropic orthogonal transform algorithm-based multicarrier systems with blind channel estimation," *IET Commun.*, vol. 6, no. 16, pp. 2695–2704, Nov. 2012.

[29] A. Farhang, N. Marchetti, L. E. Doyle, and B. Farhang-Boroujeny, "Filter bank multicarrier for massive MIMO," Feb. 2014. arXiv:1402.5881v1 [cs.IT].

[30] A. Aminjavaheri, A. Farhang, N. Marchetti, L. E. Doyle, and B. Farhang-Boroujeny, "Frequency spreading equalization in multicarrier massive MIMO," in *Proc. ICC-2015*, London, UK, Jun. 2015.

[31] X. Mestre, F. Rottenberg, and M. Navarro, "Linear receivers for massive MIMO FBMC/OQAM under strong channel frequency selectivity," in *Proc. SSP-2016*, Palma de Mallorca, Spain, Jun. 2016.

[32] E. Kofidis, C. Chatzichristos, and A. L. F. de Almeida, "Tensor-based processing of filter bank-based multicarrier signals," in *Tensor Decompositions and Applications (TDA-2016)*, Leuven, Belgium, Jan. 2016.

[33] E. Kofidis, C. Chatzichristos, and A. L. F. de Almeida, "Joint channel estimation/data detection in MIMO-FBMC/OQAM systems – A tensor-based approach," Sep. 2016. arXiv:1609.09661v1 [cs.IT]. [Online]. Available: https://arxiv.org/abs/1609.09661.

[34] E. Kofidis, "On optimal multi-symbol preambles for highly frequency selective FBMC/OQAM channel estimation," in *Proc. ISWCS-2015*, Brussels, Belgium, Aug. 2015.

[35] I. Barhumi, G. Leus, and M. Moonen, "Optimal training designs for MIMO OFDM systems in mobile wireless channels," *IEEE Trans. Signal Process.*, vol. 51, no. 6, pp. 1615–1624, Jun. 2003.

[36] R. ITU-R M.1225, "Guidelines for evaluation of radio transmission technologies for IMT-2000," ITU, 1997. [Online]. Available: https://www.itu.int/dms_pubrec/itu-r/rec/m/R-REC-M.1225-0-199702-I!!PDF-E.pdf.

[37] M. G. Bellanger, "Specification and design of a prototype filter for filter bank based multicarrier transmission," in *Proc. ICASSP-2001*, Salt Lake City, UT, May 2011.

[38] J. Louveaux, *et al.*, "Equalization and demodulation in the receiver (single antenna)." PHYDYAS D3.1 deliverable, 2008. [Online]. Available: http://www.ict-phydyas.org/delivrables/PHYDYAS-D3.1.pdf/view.

[39] J.-P. Javaudin and Y. Jiang, "Channel estimation in MIMO OFDM/OQAM," in *Proc. SPAWC-2008*, Recife, Brazil, Jul. 2008.

[40] J.-P. Javaudin and Y. Jiang, "Channel estimation for iterative MIMO OFDM/OQAM transceivers," in *Proc. EW-2008*, Prague, Czech Republic, Jun. 2008.

[41] M. El Tabach, J.-P. Javaudin, and M. Hélard, "Spatial data multiplexing over OFDM/OQAM modulations," in *Proc. ICC-2007*, Glasgow, Scotland, Jun. 2007.

[42] F. Bader and M. Shaat, "Pilot pattern design for PUSC MIMO WiMAX-like filter banks multicarrier system," *Int. J. Adv. Telecommun.*, vol. 4, no. 1–2, pp. 156–165, 2011.

[43] F. Bader and M. Shaat, "Pilot pattern adaptation and channel estimation in MIMO WiMAX-like FBMC system," in *Proc. ICWMC-2010*, Valencia, Spain, Sep. 2010.

[44] M. Renfors and J. Yli-Kaakinen, "Fast-convolution implementation of linear equalization based multiantenna detection schemes," in *Proc. ISWCS-2014*, Barcelona, Spain, Aug. 2014.

[45] TS 36.104 V8.2, "3rd Generation Partnership Project; Technical Specification Group Radio Access Network; Evolved Universal Terrestrial Radio Access (E-UTRA); Base Station (BS) radio transmission and reception (Release 8)." 3GPP, 2008. [Online]. Available: http://www.arib.or.jp/IMT-2000/V700Sep08/5_Appendix/Rel8/36/36104-820.pdf.

[46] M. Kalil, M. M. Banat, and F. Bader, "Three dimensional pilot aided channel estimation for filter bank multicarrier MIMO systems with spatial channel correlation," in *Proc. 8th Jordanian Int. Electrical and Electronic Engineering Conf. (JIEEEC-2013)*, Amman, Jordan, Apr. 2013.

[47] B. M. Kurkoski, "Towards efficient detection of two-dimensional intersymbol interference channels," *IEICE Trans. Fundam. Electron. Commun. Comput. Sci.*, vol. E91-A, pp. 2696–2703, Oct. 2008.

[48] R. Zakaria, D. Le Ruyet, and M. Bellanger, "Maximum likelihood detection in spatial multiplexing with FBMC," in *Proc. EW-2010*, Lucca, Italy, Apr. 2010.

[49] Y. Cheng and M. Haardt, "Widely linear processing in MIMO FBMC/OQAM systems," in *Proc. ISWCS-2013*, Ilmenau, Germany, Aug. 2013.

[50] Y. Fang, J. Zhong, M. Zhao, and M. Lei, "Low complexity K-best algorithm based iterative detectors for MIMO-FBMC systems," in *Proc. WCSP-2015*, Nanjing, China, Oct. 2015.

[51] R. Zakaria and D. Le Ruyet, "Partial ISI cancellation with Viterbi detection in MIMO filter-bank multicarrier modulation," in *Proc. ISWCS-2011*, Aachen, Germany, Nov. 2011.

[52] R. Zakaria, D. Le Ruyet, and Y. Medjahdi, "On ISI cancellation in MIMO-ML detection using FBMC/QAM modulation," in *Proc. ISWCS-2012*, Paris, France, Aug. 2012.

[53] R. Zakaria and D. Le Ruyet, "A novel FBMC scheme for spatial multiplexing with maximum likelihood detection," in *Proc. ISWCS-2010*, York, UK, Sep. 2010.

[54] R. Zakaria and D. Le Ruyet, "Analysis of the FFT-FBMC equalization in selective channels," *IEEE Signal Processing Letters*, vol. 24, no. 6, pp. 897–901, Apr. 2017. http://doi.org/10.1109/LSP.2017.2697079.

[55] C. Lélé, P. Siohan, R. Legouable, and M. Bellanger, "CDMA transmission with complex OFDM/OQAM," *EURASIP J. Wirel. Commun. Netw.*, 2008. http://dx.doi.org/10.1155/2008/748063.

[56] R. Nissel and M. Rupp, "Enabling low-complexity MIMO in FBMC/OQAM." Tech. Rep., Vienna Technical University, 2016. [Online]. Available: https://publik.tuwien.ac.at/files/PubDat_251033.pdf.

[57] T. Ihalainen, A. Ikhlef, J. Louveaux, and M. Renfors, "Channel equalization for multi-antenna FBMC/OQAM receivers," *IEEE Trans. Veh. Technol.*, vol. 60, no. 5, pp. 2070–2085, Jun. 2011.

[58] M. Bellanger, "FS-FBMC: A flexible robust scheme for efficient multicarrier broadband wireless access," in *Proc. 2012 IEEE Globecom Workshops (GC Wkshps)*, Anaheim, CA, USA, Dec. 2012.

[59] M. Renfors and J. Yli-Kaakinen, "Fast-convolution implementation of linear equalization based multiantenna detection schemes," in *ISWCS-2014*, Barcelona, Spain, Aug. 2014.

[60] J. R. Barry, E. A. Lee, and D. G. Messerschmitt, *Digital Communication*, 3rd ed. Boston: Kluwer Academic Publishers, 2004.

[61] Y. Yang, T. Ihalainen, M. Rinne, and M. Renfors, "Frequency-domain equalization in single-carrier transmission: Filter bank approach," *EURASIP J. Adv. Signal Process.*, vol. 2007:010438, 2007.

[62] E. Dahlman, S. Parkvall, and J. Sköld, *4G LTE/LTE-Advanced for Mobile Broadband*. Academic Press, 2011.

[63] A. Ikhlef and J. Louveaux, "Per subchannel equalization for MIMO FBMC/OQAM systems," in *Proc. PACRIM-2009*, Victoria, Canada, Aug. 2009.

[64] A. Paulraj, R. Nabar, and D. Gore, *Introduction to Space-Time Wireless Communications*. Cambridge University Press, 2003.

[65] X. Mestre and D. Gregoratti, "Parallelized structures for MIMO FBMC under strong channel frequency selectivity," *IEEE Trans. Signal Process.*, vol. 64, no. 5, pp. 1200–1215, Mar. 2016.

[66] L. Zhang, P. Xiao, A. Zafar, A. Quddus, and R. Tafazolli, "FBMC system: An insight into doubly dispersive channel impact," *IEEE Trans. Veh. Technol.*, 2016. http://doi.org/10.1109/TVT.2016.2602096.

[67] D. S. Waldhauser, "Multicarrier systems based on filter banks." Ph.D. dissertation. Technical University of Munich, 2009.

[68] M. Nájar, *et al.*, "MIMO techniques and beamforming." PHYDYAS D4.2 deliverable, 2008. [Online]. Available: http://www.ict-phydyas.org/delivrables/PHYDYAS-D4-2.pdf/view.

# Space-Time Coding for FBMC

# 15

**Rostom Zakaria\*, Didier Le Ruyet\*, Markku Renfors**[†]

*CEDRIC Laboratory, CNAM, Paris, France\**
*Tampere University of Technology, Tampere, Finland*[†]

## CONTENTS

## 15.1 INTRODUCTION

The introduction of multiple antennas at the transmitter and/or at the receiver provides spatial diversity in the system. This spatial diversity can be exploited using Space Time Block Code (STBC) or Space-Time Trellis Coding (STTC).

The first research works on Space-Time Coding (STC) for FilterBank MultiCarrier with Offset-QAM subcarrier modulation (FBMC/OQAM) were carried out on STTC [1,2]. However, due to the difficult aspect of FBMC/OQAM interference management in STTC, only a single time-delay coding was considered. It was shown that the receiver obtains a sample sequence corresponding to a weighted sum of symbols in time domain. Thus, the data symbols are recovered from the received sequence through the maximum likelihood technique by means of the Viterbi algorithm.

Regarding STBC for FBMC/OQAM, most of the works have considered the well-known Alamouti code. The direct application of Alamouti coding to FBMC/OQAM makes an inherent interference to appear, which cannot be easily removed [3]. The

Orthogonal Waveforms and Filter Banks for Future Communication Systems. DOI: 10.1016/B978-0-12-810384-5.00015-3

difficulty in direct application of the Alamouti scheme to FBMC/OQAM can be conceptually explained by the fact that the Alamouti scheme relies on a complex orthogonality, whereas FBMC/OQAM technique has only a real orthogonality, which cannot lead to the same type of equations [3]. Many works have proposed potential solutions to this drawback, and we will review in this chapter the most significant contributions.

## 15.2 BASIC ALAMOUTI CODING PRINCIPLE

Alamouti STBC is a famous transmit diversity scheme for two transmit antennas [4]. We consider here mostly the basic scheme with a single receive antenna. Orthogonal sequences of two complex data symbols $a_1$ and $a_2$ are transmitted from both transmit antennas in two consecutive symbol intervals as follows:

- Antenna 1: $[a_1, -a_2^*]$;
- Antenna 2: $[a_2, a_1^*]$.

The corresponding received samples are:

$$y_1 = h_1 a_1 + h_2 a_2 + w_1, \tag{15.1}$$
$$y_2 = -h_1 a_2^* + h_2 a_1^* + w_2, \tag{15.2}$$

where $w_1$ and $w_2$ are samples from independent identically distributed circular complex white Gaussian noise processes, and $h_1$ and $h_2$ are the channel gains from the two transmit antennas to the receiver. The channels are assumed to be flat-fading. Assuming knowledge of the channel gains, the decision variables are obtained by combining the two observations as follows:

$$r_1 = h_1^* y_1 + h_2 y_2^* = \left(|h_1|^2 + |h_2|^2\right) a_1 + h_1^* w_1 + h_2 w_2^*, \tag{15.3}$$

$$r_2 = h_2^* y_1 - h_1 y_2^* = \left(|h_1|^2 + |h_2|^2\right) a_2 + h_2^* w_1 - h_1 w_2^*. \tag{15.4}$$

Then to recover the transmitted symbols, we just have to perform scaling and slicing. The Alamouti decoding was formulated here for the case of flat-fading channels, which is quite a valid model when coding is done within the subcarrier symbol sequences of a Cyclic Prefix Orthogonal Frequency-Division Multiplexing (CP-OFDM) system.

## 15.3 BLOCKWISE ALAMOUTI SCHEME FOR FBMC/OQAM

Renfors et al. in [5] have proposed a solution to combine the Alamouti scheme with FBMC/OQAM, where the Alamouti coding is performed in a blockwise manner while inserting gaps (zero-symbols and pilots) to isolate the blocks. This solution is feasible when the FBMC/OQAM transmultiplexer response $\mathbf{\Gamma}$ (see Section 11.2) is conjugate symmetric along the time axis. Let $\mathbf{A_1}$ and $\mathbf{A_2}$ be two data symbol blocks consisting of $N$ samples in $m$ subcarriers simultaneously transmitted from antenna 1 and antenna 2, respectively. After that, the first antenna transmits $-\overleftarrow{\mathbf{A_2}}^*$, whereas the second one transmits $\overleftarrow{\mathbf{A_1}}^*$. The left arrow on top of a variable denotes the time-reversal version of the corresponding sequence [5]. Assuming that the two channels are constant in time and frequency domains during the transmission of both symbol blocks, the first signal block collected at the receive antenna can be written as

$$\mathbf{Y_1} = h_1 \mathbf{\Gamma} * \mathbf{A_1} + h_2 \mathbf{\Gamma} * \mathbf{A_2} + \mathbf{W_1}, \qquad (15.5)$$

where $*$ operation stands for the 2D convolution, $h_1$ and $h_2$ denote the channel responses from the two transmit antennas to the receive one, and $\mathbf{W_1}$ contains the additive Gaussian noise terms. Likewise, the second received signal block is

$$\mathbf{Y_2} = h_2 \mathbf{\Gamma} * \overleftarrow{\mathbf{A_1}}^* - h_1 \mathbf{\Gamma} * \overleftarrow{\mathbf{A_2}}^* + \mathbf{W_2}. \qquad (15.6)$$

Naturally, the 2D convolution expands the data blocks, both in time and frequency directions depending on the range of significant Transmultiplexer (TMUX) response elements. We assume that $\mathbf{Y_1}$ and $\mathbf{Y_2}$ are truncated to the size of the transmitted data blocks. We also assume that there is sufficient guard-space in both directions around the transmitted data blocks to prevent leakage from adjacent transmission blocks. The channel is assumed to be constant over the whole transmission block. This is more critical in the time direction because the jointly coded elements appear symmetrically around the center gap. Different subcarriers are treated independently, so the effect of frequency selectivity is assumed to be similar to the case of Single-Input Single-Output (SISO) transmission using single-tap subcarrier equalizers.

Then, we can write

$$\overleftarrow{\mathbf{Y_2}}^* = h_2^* \overleftarrow{\mathbf{\Gamma}}^* * \mathbf{A_1} - h_1^* \overleftarrow{\mathbf{\Gamma}}^* * \mathbf{A_2} + \overleftarrow{\mathbf{W_2}}^*$$
$$= h_2^* \mathbf{\Gamma} * \mathbf{A_1} - h_1^* \mathbf{\Gamma} * \mathbf{A_2} + \overleftarrow{\mathbf{W_2}}^*. \qquad (15.7)$$

The last equality stands thanks to the fact that $\mathbf{\Gamma}$ is conjugate symmetric along the time axis ($\overleftarrow{\mathbf{\Gamma}}^* = \mathbf{\Gamma}$). Therefore, applying the Alamouti decoding [4], we have

$$\mathbf{R_1} = \frac{h_1^* \mathbf{Y_1} + h_2 \overleftarrow{\mathbf{Y_2}}^*}{|h_1|^2 + |h_2|^2} = \mathbf{\Gamma} * \mathbf{A_1} + \frac{h_1^* \mathbf{W_1} + h_2 \overleftarrow{\mathbf{W_2}}^*}{|h_1|^2 + |h_2|^2} \qquad (15.8)$$

and

$$\mathbf{R_2} = \frac{h_2^* \mathbf{Y_1} - h_1 \overleftarrow{\mathbf{Y_2}}^*}{|h_1|^2 + |h_2|^2} = \mathbf{\Gamma} * \mathbf{A_2} + \frac{h_2^* \mathbf{W_1} - h_1 \overleftarrow{\mathbf{W_2}}^*}{|h_1|^2 + |h_2|^2}. \qquad (15.9)$$

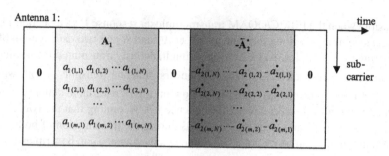

**FIGURE 15.1**

Alamouti-coded frame at SFB input.

The term $\mathbf{\Gamma} * \mathbf{A_i}$ contains the transmitted data symbols and also the intrinsic interference. We can write it as in [5]:

$$\mathbf{\Gamma} * \mathbf{A_i} = \mathbf{D_i} + \mathbf{U_i}, \tag{15.10}$$

where $\mathbf{U_i}$ is the block of the interference terms. When a data symbol in $\mathbf{A_i}$ is purely real, its corresponding interference term in $\mathbf{U_i}$ is purely imaginary. Then, the transmitted data symbols are easily estimated by a simple real part retrieval.

### 15.3.1 FRAME STRUCTURE

The frame structure for blockwise Alamouti transmission is illustrated in Fig. 15.1. The frame consists of a number of adjacent subcarriers and $2N$ Offset Quadrature Amplitude Modulation (OQAM) subsymbols of data in each subcarrier within each frame.

We consider FBMC/OQAM systems that are well localized in frequency. Typically, the transition bands of a subchannel are overlapping only with the immediately adjacent subchannels. Assuming relatively mild-frequency selectivity in the subcarrier bandwidth, the near-orthogonality of subcarriers is maintained by the Alamouti detection scheme, which basically implements one-tap subcarrier equalization through the maximum ratio combining process. In synchronized transmission, different users' code blocks can be placed right next to each other in the frequency direction. In asynchronous transmission, like multiuser uplink, the transmission blocks can be well isolated from each other in the frequency direction by leaving an empty subcarrier as a guardband between blocks.

In the basic scheme, zero-samples are inserted in all subcarriers between the Alamouti-coded frames and also between the two parts of each Alamouti coded frame (these are referred to as the edge gaps and center gap, respectively). Assuming that the filter bank overlapping factor is $K = 4$, a gap of four subcarrier samples (two OQAM symbols) is needed to isolate the blocks completely. With three zero samples, the crosstalk between the blocks is still at a very low level. Also shorter distances may be considered, depending on the targeted Signal-to-Noise Ratio (SNR) operation range. Since the intrinsic interference within each part of the coded frame is canceled by the block Alamouti scheme, it is not able to cope with the intrinsic interference across the two parts. Therefore, a sufficient guard space is needed between them. There is no such hard restriction for the edge gap length. Actually, nonzero repetitive sample sequences can be inserted in the edge gaps without distorting the symmetry required by the Alamouti-coded block pairs. A natural way to make use of this opportunity is to embed pilots (training sequences) in the edge blocks. Feasible block sequences for the two transmit antennas can be represented as follows:

$$
\begin{cases}
\mathbf{T_1} = \left[\mathbf{P_1}, \mathbf{A_1}, 0, -\overleftarrow{\mathbf{A_2}}^*, -\overleftarrow{\mathbf{P_2}}^*, \mathbf{A_3}, 0, -\overleftarrow{\mathbf{A_4}}^*, -\mathbf{P_1}, \cdots\right], \\
\mathbf{T_2} = \left[\mathbf{P_2}, \mathbf{A_2}, 0, \overleftarrow{\mathbf{A_1}}^*, \overleftarrow{\mathbf{P_1}}^*, \mathbf{A_4}, 0, \overleftarrow{\mathbf{A_3}}^*, -\mathbf{P_2}, \cdots\right].
\end{cases}
\tag{15.11}
$$

Here $\mathbf{P_1}$ and $\mathbf{P_2}$ are arbitrarily chosen pilot symbol blocks.

Pilot blocks following this structure introduce no more interference to the data symbols than zero samples in these places, assuming stationary nonfrequency-selective subchannels. Naturally, there is some interference between different Alamouti code frames if the length of the pilot block is lower than $K$ samples. The received secondary parts of the pilot symbols are not under control in this scheme, but this is not critical for the channel estimation scheme introduced in [5].

### 15.3.2 PERFORMANCE EVALUATION

The blockwise Alamouti scheme was tested using the International Telecommunication Union Radiocommunication (ITU-R) Vehicular-A (Veh-A) channel model and Worldwide Interoperability for Microwave Access (WiMAX)-like system parameters with $M = 1024$ subcarriers and subcarrier spacing of $\Delta f = 10.94$ kHz. The PHY-DYAS filterbank with overlapping factor $K = 4$ is applied in this study. Block-fading channel model is assumed, i.e., the channel remains constant over each Alamouti code frame, and the results show average performance over 1000 independent channel instances. With these parameters, the subchannels are essentially flat-fading, and single-tap subchannel equalizers provide sufficient performance. Quadrature Phase-Shift Keying (QPSK) subcarrier modulation is applied for data, and random QPSK sequences are used as pilots. We apply a specific channel estimation approach proposed in [5], which utilizes the pilot structure of Eqs. (15.11).

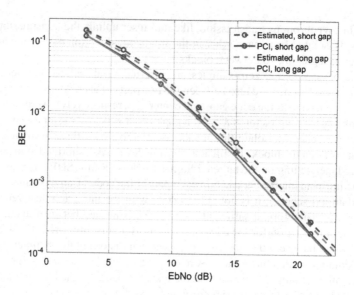

**FIGURE 15.2**

BER performance of blockwise Alamouti scheme with FBMC/OQAM, QPSK modulation, and Veh-A channel.

It was found that three subcarrier samples is the best choice for the pilot sequence length. We use the two pilot blocks that are closest to each Alamouti code frame for estimating the channel for that frame. We consider using pilots also from the immediately adjacent subchannels. More specifically, for an Alamouti code frame on subcarrier $m$ starting at time index $n$ and ending at $n + n_0$, we use twelve pilots samples with sample indices $n - 2, n - 1, n + n_0 + 1$, and $n + n_0 + 2$ at subcarriers $m - 1, m$, and $m + 1$ for estimating the channel gain. Regarding the center gap length, both 1 and 3 samples are considered, and as mentioned, the samples in the center gap are all zero. Aiming at pilot overhead of one pilot symbol per six data symbols, we choose Alamouti code block length as 9 OQAM symbols (18 subcarrier samples) or 6 OQAM symbols (12 samples) for the long gap and short gap, respectively. The distance between consecutive pilot blocks is about 20 or 13 OQAM symbols, respectively. The average pilot symbol energy is boosted by 2.5 dB above average data symbol energy.

Fig. 15.2 shows the average Bit Error Rate (BER) performance as a function of $E_b/N_0$ for both pilot schemes and QPSK data modulation. We can see that the performance is quite acceptable also with the shorter center gap. Fig. 15.3 shows the average BER performance as a function of mobile velocity, with 2.5 GHz carrier frequency for both pilots schemes. We can see that the shorter frame length, enabled by shorter gap, is clearly more robust with significant mobility.

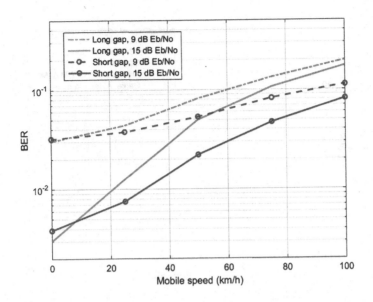

**FIGURE 15.3**

BER performance for blockwise Alamouti scheme with FBMC/OQAM as a function of mobility. QPSK modulation, $E_b/N_0$ of 9 or 15 dB, Veh-A channel, pilot block length of 3 samples, 0-block length of 1 or 3 samples.

## 15.4 INTERFERENCE CANCELLATION IN ALAMOUTI CODING SCHEME FOR FBMC/OQAM

To remove the interference term, some works based on iterative interference estimation and cancellation have also been carried out. Unfortunately, detection schemes with interference estimation and cancellation are subject to error propagation and are not always effective [6]. Therefore, the challenge in this kind of schemes is to mitigate the error propagation through iterations [7]. To counteract the error propagation and make the cancellation scheme effective, the authors in [7] have shown that a necessary condition to avoid the error propagation is to hold the interference power under a certain threshold, i.e., the interference cancellation technique can only be effective when the Inter-Symbol Interference (ISI) is small enough compared to the minimal distance between two different symbols.

To reduce the power of the intrinsic interference, it was proposed in [8] to modify the conventional FBMC/OQAM system by using Quadrature Amplitude Modulation (QAM) symbols instead of OQAM ones and transmitting the data symbols on each period $T$ instead of transmitting them on each half period $T/2$. That is, the authors utilize the conventional lattice structure of OFDM with complex-valued symbols and localized filters. Thus, the interference coefficients corresponding to the odd multiples of $T/2$ are avoided, and the overall interference power is reduced. However, as a consequence of this modification, the real orthogonality condition of

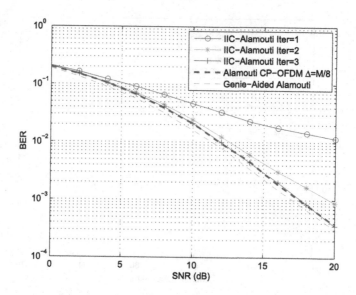

**FIGURE 15.4**

Performance of IIC-Alamouti receiver with FBMC-QAM using PHYDYAS filter and 4-QAM constellation.

FBMC/OQAM is no longer satisfied. Therefore, even in SISO system, interference cancellation methods must be applied at the receiver. As for the STBC decoding, a first Alamouti decoder is used as tentative detector providing tentative estimations of the data symbols. Based on these tentative estimates, the interference canceler calculates an estimation of the interference, and then its contribution is removed from the received vector. After that, the Alamouti decoding is again applied, and the operation is repeated several times. The authors in [8] show by simulations that Alamouti coding scheme with Iterative Interference Cancellation (IIC) applied in FBMC-QAM provides satisfactory performance. Fig. 15.4 depicts the BER performance obtained of the IIC-Alamouti in FBMC-QAM with three iterations and compare them to the OFDM and genie-aided Alamouti performance. This latter is obtained by assuming perfect intrinsic interference cancellation. We observe an SNR loss of only 0.5 dB compared to the genie-aided performance. Nevertheless, the IIC-Alamouti in the FBMC-QAM system exhibits almost the same performance as that obtained with OFDM; this is due to the $E_b/N_0$ loss caused by the Cyclic Prefix (CP) in OFDM.

Another technique aiming to reduce the intrinsic interference power is introduced in [9]. The authors propose some arrangements in the STBC and Space Frequency Block Code (SFBC) schemes to reduce the FBMC/OQAM intrinsic interference and improve the Signal-to-Interference Ratio (SIR). The idea behind these arrangements is to automatically remove an important part of the interference only by performing the adequate Alamouti decoding. Then, the remaining interference is canceled iteratively by interference estimation and cancellation procedure. The selected ar-

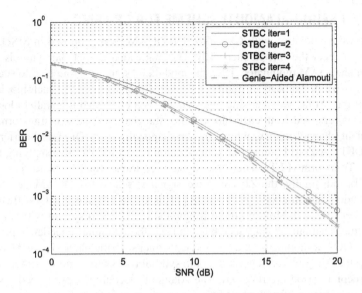

**FIGURE 15.5**

Performance of the proposed STBC scheme in FBMC/OQAM using OQPSK modulation.

rangement that minimizes the remaining interference operates on two interleaved Alamouti blocks (in time or frequency domain). Furthermore, there is an alternating rule in position of the minus sign between both blocks. The proposed Alamouti schemes applied in FBMC/OQAM were tested in [9], and it was shown that the BER performance curves converge to the interference-free performance. More specifically, the BER performance with Offset Quadrature Phase-Shift Keying (OQPSK) mapping is shown in Fig. 15.5. The BER performance of the first iteration corresponds to the tentative Alamouti decoding, and the BER floor at $7 \times 10^{-3}$ is due to the remaining interference. However, we can observe that the performance curves reach the genie-aided performance after three iterations.

It is worth noticing that the weakness of these IIC receivers is the fact that the performance is promptly decayed when higher mapping orders are considered [8,9].

## 15.5 OTHER SOLUTIONS

In the previous sections, two proposed solutions for FBMC/OQAM-Alamouti were presented. Both solutions deal with the intrinsic interference of FBMC/OQAM without any change in the modulator structure. On the other hand, some other solutions have been also proposed by slightly changing the FBMC/OQAM modulator/demodulator architecture. These proposals will be more specifically reviewed in this section.

## 15.5.1 PSEUDO-ALAMOUTI SCHEME FOR CP-FBMC

An OQAM modulation with CP, called CP-OQAM, was proposed in SISO case to perfectly cancel the interference [10]. The main idea is that the symbols first go through an OQAM modulator. Then the modulated signal is grouped into successive blocks of size $M$ (subcarrier number). Then a CP is appended to each block before transmitting. At the receiver side, the signal is first converted to parallel blocks, and then CP is removed from each block. Thus, the channel matrix is transformed into a circulant channel matrix, which can be diagonalized by Discrete Fourier Transform (DFT). A Zero Forcing (ZF) equalization is carried out after applying the Fast Fourier Transform (FFT). Then, the signals are transformed back to time domain by IFFT, and, finally, the ZF equalized signal is fed to the OQAM demodulator. The pseudo-Alamouti coding scheme in [11] is an extension of the CP-FBMC SISO transceiver to $2 \times 1$ Multiple-Input Single-Output (MISO) system with an adaptation of the Alamouti coding/decoding. Indeed, the Alamouti encoder is performed after the OQAM modulator in a blockwise manner with the block size of $M$ symbols. To keep the Alamouti orthogonality, both symbol blocks transmitted simultaneously at the second period are reversed. The Alamouti decoding is performed instead of the ZF equalization after the first FFT. The authors in [11] show by simulations that due to a specific and efficient channel estimation method [12], the proposed pseudo-Alamouti transceiver can outperform the Alamouti OFDM scheme. Nevertheless, it is worth noting that the simulation results in [11] are obtained by considering a prototype filter with an overlapping factor of $K = 1$. This avoids any interference between useful data in the time domain.

## 15.5.2 ALAMOUTI CODING SCHEME FOR CDMA-FBMC

It was shown in [3] that Alamouti coding can be employed in combination with Code-Division Multiple Access (CDMA). Indeed, it is shown in [13] that it is possible to have a complex orthogonality with FBMC thanks to Walsh–Hadamard CDMA codes. This is because the CDMA despreading operation allows us to recover a complex orthogonality property in FBMC/OQAM. Therefore, the principle in this proposal is to take advantage of the orthogonality property resulting from the CDMA-OFDM/OQAM combination to get a new Multiple-Input Multiple-Output (MIMO) Alamouti scheme with FBMC.

Two different approaches with Alamouti coding are proposed by considering a CDMA spreading either in the frequency or in the time domain. If the CDMA spreading is carried out in the frequency domain, the Alamouti decoding scheme can only be applied if the channel is assumed to be spectrally flat. For the Alamouti scheme with time spreading CDMA-OFDM/OQAM, two strategies are elaborated for implementing the MIMO space-time coding scheme. Strategy 1 implements the Alamouti over pairs of adjacent frequency domain samples, whereas Strategy 2 processes the Alamouti coding scheme over pairs of spreading codes from two successive time instants. The authors show that Strategy 2 appears to be more appropriate since it requires less restrictive assumptions on the channel variations across the frequencies. Therefore, under some channel hypotheses, the combination of Alamouti with complex CDMA-OFDM/OQAM is possible without increasing the complexity of the Alamouti decoding process. Furthermore, in the case of a frequency selective channel, FBMC/OQAM keeps its intrinsic advantage with an SNR gain in direct relation with the CP length.

### 15.5.3 ALAMOUTI SPACE-TIME CODE IN FFT-FBMC

To mitigate the problem of inherent interference in FBMC, a modified version of the system was developed in [14,15]. The basic idea was to view the interference term as the result of three linear Finite Impulse Response (FIR) filters, one for each subcarrier (the subcarrier under question plus the preceding and following ones). Then, OFDM is mimicked for each of the three terms to come up with a system that is formulated as OFDM and hence facilitates detection at the receiver. Hence, the authors introduce, before the (usual) IFFT block of size $M$ in the FBMC scheme, another IFFT precoding block of size $N$ for each subcarrier, optionally followed by a CP insertion. The CP prevents ISI in time domain, whereas it is shown that a specific arrangement of the block data symbols setting half of them to zero and the other ones to complex-values (instead of real-valued for the conventional FBMC/OQAM system) can prevent from Inter-Carrier Interference (ICI). Moreover, this specific arrangement makes the FFT-FBMC spectrum more confined than the one of FBMC [16]. The authors show that the equivalent system can be formulated as OFDM, and any MIMO technique can be applied in a straightforward manner. Indeed, it is shown that the equivalent system model can be given by [15]

$$y_q[n] \cong H_q F_{0,n}^{(q)} d_q[n] + w_q[n], \tag{15.12}$$

where $w_q[n]$ is the noise term at the output of the demodulator, $H_q$ is the channel coefficient, and $F_{0,n}^{(q)}$ is a real coefficient greater than 1, which depends on the prototype filter. Hence, according to the above expression of the system model, Alamouti coding could be performed straightforwardly. Indeed, the performance of the Alamouti coding with FFT-FBMC scheme is evaluated in [15]. Since the proposed FFT-FBMC includes CP at each subcarrier, an effort is also devoted to assess the performance loss when reducing the CP length. Simulation results showed that we can almost obtain the same performance as OFDM in some configurations. Fig. 15.6 depicts the BER performance of the Alamouti coding with FFT-FBMC using the PHYDYAS filter for different configurations according to extra FFT block length $N$ and the CP length $L$. The performance curves are also compared to the curve of OFDM-Alamouti performance.

We can observe that the FFT-FBMC obtains almost no degradation compared to OFDM, except for the case where $N = 16$ and $L = 0$, where we have about a 2.75 dB SNR loss with respect to OFDM at $BER = 10^{-4}$ and less than 1 dB at $BER = 10^{-2}$. Furthermore, the avoidance of the CP does not result in considerable degradations. We can notice, at worst, about 1 dB of SNR loss at $BER = 10^{-4}$ (except for $N = 16$).

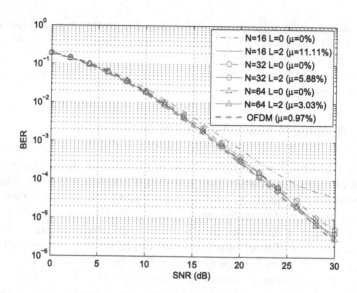

**FIGURE 15.6**

Performance of FFT-FBMC using PHYDYAS prototype filter with (2×1) Alamouti coding scheme and QPSK modulation in Ped-A channel.

## 15.6 CONCLUDING REMARKS

Because of its intrinsic interference, the full potential performance of FBMC/OQAM cannot be straightforwardly reached when combined with the Alamouti coding. Therefore, specific signal processing techniques had to be developed and used. In this chapter, an overview of known solutions for FBMC/OQAM with Alamouti coding was presented. We have mainly detailed two schemes dealing with the intrinsic interference without changing the structure of the FBMC/OQAM modulator/demodulator. Other solutions have been also presented, which require some modifications in the original FBMC/OQAM structure such as appending a CP. All the different overviewed solutions for FBMC/OQAM-Alamouti offer good performance at the expense of either complexity increase or spectral efficiency reduction.

## REFERENCES

[1] M. Bellanger, "Transmit diversity in multicarrier transmission using OQAM modulation," in *3rd International Symposium on Wireless Pervasive Computing, ISWPC 2008*, May 2008, pp. 727–730.

[2] C. Lele, D. L. Ruyet, and R. Zakaria, "On the decoding of single delay STTC using filter bank based multicarrier modulation," in *2009 6th International Symposium on Wireless Communication Systems*, Sep. 2009, pp. 86–90.

[3] C. Lélé, P. Siohan, and R. Legouable, "The Alamouti scheme with CDMA-OFDM/OQAM," *EURASIP Journal on Advances in Signal Processing*, vol. 2010, 2010. [Online]. Available: http://dblp.uni-trier.de/db/journals/ejasp/ejasp2010.html#LeleSL10.

[4] S. Alamouti, "A simple transmit diversity technique for wireless communications," *IEEE Journal on Selected Areas in Communications*, vol. 16, no. 8, pp. 1451–1458, Oct. 1998.

[5] M. Renfors, T. Ihalainen, and T. Stitz, "A block-Alamouti scheme for filter bank based multicarrier transmission," in *Wireless Conference (EW), 2010 European*, Apr. 2010, pp. 1031–1037.

[6] R. Zakaria and D. Le Ruyet, "Partial ISI cancellation with Viterbi detection in MIMO filter-bank multicarrier modulation," in *8th International Symposium on Wireless Communication Systems (ISWCS)*, Nov. 2011, pp. 322–326.

[7] O. Agazzi and N. Seshadri, "On the use of tentative decisions to cancel intersymbol interference and nonlinear distortion (with application to magnetic recording channels)," *IEEE Transactions on Information Theory*, vol. 43, pp. 394–408, Mar. 1997.

[8] R. Zakaria and D. L. Ruyet, "Intrinsic interference reduction in a filter bank-based multicarrier using QAM modulation," *Physical Communication*, vol. 11, pp. 15–24, 2014.

[9] R. Zakaria and D. Ruyet, "On interference cancellation in Alamouti coding scheme for filter bank based multicarrier systems," in *Proceedings of the Tenth International Symposium on Wireless Communication Systems (ISWCS 2013)*, Aug. 2013, pp. 1–5.

[10] H. Lin and P. Siohan, "A new transceiver system for the OFDM/OQAM modulation with Cyclic Prefix," in *IEEE 19th International Symposium on Personal, Indoor and Mobile Radio Communications, PIMRC 2008*, Sep. 2008, pp. 1–5.

[11] H. Lin, C. Lélé, and P. Siohan, "A pseudo Alamouti transceiver design for OFDM/OQAM modulation with cyclic prefix," in *IEEE 10th Workshop on Signal Processing Advances in Wireless Communications, SPAWC '09*, Jun. 2009, pp. 300–304.

[12] C. Lélé, J.-P. Javaudin, R. Legouable, A. Skrzypczak, and P. Siohan, "Channel estimation methods for preamble-based OFDM/OQAM modulations," in *European Wireless*, Paris, France, Apr. 2007.

[13] C. Lélé, P. Siohan, R. Legouable, and M. Bellanger, "CDMA transmission with complex OFDM/OQAM," *EURASIP Journal on Wireless Communications and Networking*, vol. 2008, 2008.

[14] R. Zakaria and D. Le Ruyet, "A novel FBMC scheme for spatial multiplexing with maximum likelihood detection," in *7th International Symposium on Wireless Communication Systems (ISWCS)*, Sep. 2010, pp. 461–465.

[15] R. Zakaria and D. Le Ruyet, "A novel filter-bank multicarrier scheme to mitigate the intrinsic interference: Application to MIMO systems," *IEEE Transactions on Wireless Communications*, vol. 11, pp. 1112–1123, Mar. 2012.

[16] R. Zakaria and D. Le Ruyet, "Theoretical analysis of the power spectral density for FFT-FBMC signals," *IEEE Communications Letters*, vol. 20, pp. 1748–1751, Sep. 2016.

# FBMC Distributed and Cooperative Systems

# 16

**Martin Haardt\*, Yao Cheng\*, François Rottenberg[†], Jérôme Louveaux[†], Eleftherios Kofidis[‡,§]**

*Ilmenau University of Technology, Ilmenau, Germany\**
*Université catholique de Louvain, Louvain-la-Neuve, Belgium[†]*
*University of Piraeus, Piraeus, Greece[‡]*
*Computer Technology Institute & Press "Diophantus" (CTI), Patras, Greece[§]*

## CONTENTS

## 16.1 INTRODUCTION

Distributed and cooperative systems exploit the benefits of Multiple-Input Multiple-Output (MIMO) in a flexible manner and are regarded as key enabling technologies for efficient spectrum utilization [1]. Perfect synchronization, though stringently required, proves to be difficult to guarantee in such sophisticated communication scenarios. One intriguing advantage of FilterBank MultiCarrier (FBMC) is its strong resilience against synchronization errors. This makes it a promising multicarrier modulation scheme for distributed and cooperative systems, which is the

Orthogonal Waveforms and Filter Banks for Future Communication Systems. DOI: 10.1016/B978-0-12-810384-5.00016-5

focus of this chapter. FilterBank MultiCarrier with Offset-QAM subcarrier modulation (FBMC/OQAM) is considered here. First, we present FBMC-based Coordinated Multi-Point (CoMP) schemes. Partial coordination of adjacent cells is enabled to assist cell edge users to combat the intracell interference, the intercell interference, and the greater path loss compared to cell interior users. This is achieved via an Intrinsic Interference Mitigating Coordinated Beamforming (IIM-CBF) scheme, which jointly and iteratively computes the precoding and the decoding matrices for each subcarrier. Second, channel state information acquisition in FBMC-based distributed systems is addressed, which is a requirement in order to implement the schemes presented in the first part. Preamble-based techniques are analyzed, and a study is performed on the influence of the assignment of pilot subcarriers to different base stations. Then, a method of tracking channel estimates for the distributed MIMO downlink is also investigated that does not require any pilot and relies on the feedback of Signal-to-Noise Ratio (SNR) measurements.

## 16.2 COOPERATIVE MULTI-POINT TECHNIQUES FOR FBMC

### 16.2.1 FBMC-BASED COORDINATED MULTI-POINT DOWNLINK WITH PARTIAL COOPERATION OF ADJACENT CELLS

CoMP is known as one of the advanced communication techniques that are able to provide benefits of reduced intercell interference and enhanced cell edge throughput [2–6]. When full cooperation between the base stations of adjacent cells is assumed, the Channel State Information (CSI) and signals for all users are shared by the base stations. In this case, a virtual multiuser MIMO downlink setting is formed, where the transmit antennas are geographically separated. Thereby, the transmission strategies that have been developed for the single-cell multiuser MIMO downlink can be employed. Nevertheless, such a full cooperation scheme is not practical due to a number of reasons. For example, it requires excessive information exchange resulting in a large signaling overhead, and the CSI of all users is very hard to acquire [4]. As a more realistic solution, partial cooperation schemes have been proposed in [4,5,7], where the users are classified into two categories, namely cell (or, in [4,5], cluster that consists of multiple cells) interior users and cell edge users. The base stations of adjacent cells transmit the same signals to the cell edge users, and coordinated beamforming techniques that rely on the limited cooperation between the cells (e.g., the exchange of the beamforming matrices for cell edge users) are employed to suppress the intracell and intercell interference. For these downlink CoMP scenarios, it is more likely that the total number of receive antennas of the users served by one base station is larger than the number of transmit antennas. Thus, transmission strategies that are able to tackle such a scenario are required. Note that in the aforementioned publications on Cyclic Prefix Orthogonal Frequency-Division Multiplexing (CP-OFDM)-based downlink CoMP, perfect synchronization is assumed. However, the interference in the downlink CoMP setting is of an asynchronous na-

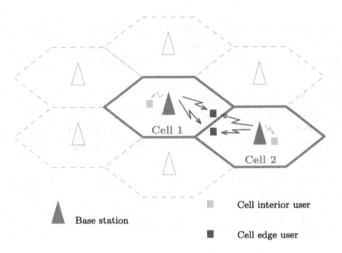

**FIGURE 16.1**

An example of a two-cell CoMP downlink scenario where a cell interior user and two cell edge users are served in each cell.

ture, as suggested in [8]. It was shown in [8] that the lack of perfect synchronization results in a significant performance degradation. Such a fact greatly motivates the use of FBMC as a replacement of CP-OFDM since the former is more robust against synchronization errors compared to the latter.

We focus on a CoMP downlink setting based on [4] and present a scheme that belongs to the category of joint transmission [2]. Note that in [4] a clustered cellular scenario is considered, where each cluster contains multiple cells. Since full cooperation is assumed among the base stations of every cluster, the downlink transmissions in each cluster resemble those of the single-cell multiuser MIMO downlink. Thereby, we simplify the scenario description of the CoMP downlink and only consider joint transmission in adjacent cells. The cell interior users only receive signals from their own base station and suffer only from the intra-cell interference, i.e., the multiuser interference as in the single-cell multiuser MIMO downlink scenarios. On the other hand, both intracell and intercell interferences have an impact on the cell edge users. To assist those users to combat the interference and also to deal with the greater path loss compared to the cell interior users, the base stations of the adjacent cells transmit the same signals to each cell edge user. An example of a two-cell FBMC-based CoMP downlink scenario is illustrated in Fig. 16.1. Here Cell 1 and Cell 2 are equipped with $N_T^{(BS1)}$ and $N_T^{(BS2)}$ transmit antennas, respectively. The number of users served by each cell is denoted by $U$ for simplicity of notation.[1] Assuming that the channel on each subcarrier can be treated as flat fading [9–11], the received

---

[1] A general case of the CoMP setting is considered in Section 16.2.2, where the intrinsic interference mitigating coordinated beamforming scheme is presented.

**Table 16.1** Coefficients $\Gamma_{m,m'}^{n,n'}$ representing the system impulse response determined by the synthesis and analysis filters [12] for the case of even subcarrier index $m$ (the PHYDYAS prototype filter [13] is used with overlapping factor of $K = 4$)

|       | $n-3$ | $n-2$ | $n-1$ | $n$ | $n+1$ | $n+2$ | $n+3$ |
|-------|-------|-------|-------|-----|-------|-------|-------|
| $m-1$ | $-j0.0429$ | $j0.125$ | $-j0.2058$ | $j0.2393$ | $-j0.2058$ | $j0.125$ | $-j0.0429$ |
| $m$   | $j0.0668$ | $-0.0002$ | $j0.5644$ | $1$ | $-j0.5644$ | $-0.0002$ | $-j0.0668$ |
| $m+1$ | $-j0.0429$ | $-j0.125$ | $-j0.2058$ | $-j0.2393$ | $-j0.2058$ | $-j0.125$ | $-j0.0429$ |

signal on the $m$th subcarrier and at the $n$th time instant of the $i$th user in Cell 1 as a cell interior user is expressed as

$$\mathbf{y}_{i,m}[n] = \mathbf{H}_{i,m,(1)}[n]\mathbf{F}_{m,(1)}[n]\mathbf{d}_{m,(1)}[n]$$

$$+ \sum_{n'=n-3}^{n+3} \sum_{m'=m-1}^{m+1} \mathbf{H}_{i,m',(1)}[n']\mathbf{F}_{m',(1)}[n']\Gamma_{m,m'}^{n,n'}\mathbf{d}_{m',(1)}[n']$$

$$+ \boldsymbol{\eta}_{i,m}[n], \quad (m', n') \neq (m, n), \tag{16.1}$$

where $\mathbf{H}_{i,m,(1)}[n] \in \mathbb{C}^{N_{R_i} \times N_T^{(BS1)}}$ represents the channel matrix between the base station of Cell 1 and the $i$th user with $N_{R_i}$ receive antennas, and $\mathbf{d}_{m,(1)}[n] \in \mathbb{R}^d$ contains the signals for all users served by Cell 1 on the $m$th subcarrier at the $n$th time instant, i.e.,

$$\mathbf{d}_{m,(1)}[n] = \left[\mathbf{d}_{1,m,(1)}^T[n] \quad \mathbf{d}_{2,m,(1)}^T[n] \quad \cdots \quad \mathbf{d}_{U,m,(1)}^T[n]\right]^T. \tag{16.2}$$

Here $\mathbf{d}_{i,m,(1)}[n] \in \mathbb{R}^{d_i}$ is the signal for the $i$th user with $d_i$ denoting the number of spatial streams sent to the $i$th user and $d = \sum_{i=1}^{U} d_i$. Each entry in the data vector corresponds to either the in-phase component or the quadrature component of a Quadrature Amplitude Modulation (QAM) symbol that is assumed to have unit energy. The terms $\Gamma_{m,m'}^{n,n'}\mathbf{d}_{m',(1)}[n']$ in (16.1) contribute to the intrinsic interference and are purely imaginary if the prototype pulse satisfies the perfect reconstruction property [12], where $m' = m - 1, m, m + 1, n' = n - 3, \ldots, n + 3$, and $(m', n') \neq (m, n)$. The coefficients $\Gamma_{m,m'}^{n,n'}$, as defined in Chapter 11, represent the system impulse response determined by the synthesis and analysis filters and are presented in Table 16.1.[2] The PHYDYAS prototype filter [13] is used, and the overlapping factor is chosen to be $K = 4$. Moreover, $\mathbf{F}_{m,(1)}[n] \in \mathbb{C}^{N_T^{(BS1)} \times d}$ symbolizes the precoding matrix for Cell 1. It can be seen that for a cell interior user, assuming that the intercell interference is negligible, the transmission from the base station in its own cell is the same as in a

---

[2]Here only the main system impulse response is considered, i.e., the coefficients with a very small value are ignored in the data model. See also Table 11.1.

single-cell multiuser MIMO downlink system. For the $i$th user, which is a cell edge user and receives the same signals from both Cell 1 and Cell 2, its received signal on the $m$th subcarrier at the $n$th time instant has the following form:

$$
\begin{aligned}
\mathbf{y}_{i,m}[n] = \ &\mathbf{H}_{i,m,(1)}[n]\mathbf{F}_{m,(1)}[n]\mathbf{d}_{m,(1)}[n] \\
&+ \sum_{n'=n-3}^{n+3}\sum_{m'=m-1}^{m+1} \mathbf{H}_{i,m',(1)}[n']\mathbf{F}_{m',(1)}[n']\Gamma_{m,m'}^{n,n'}\mathbf{d}_{m',(1)}[n'] \\
&+ \mathbf{H}_{i,m,(2)}[n]\mathbf{F}_{m,(2)}[n]\mathbf{d}_{m,(2)}[n] \\
&+ \sum_{n'=n-3}^{n+3}\sum_{m'=m-1}^{m+1} \mathbf{H}_{i,m',(2)}[n']\mathbf{F}_{m',(2)}[n']\Gamma_{m,m'}^{n,n'}\mathbf{d}_{m',(2)}[n'] \\
&+ \boldsymbol{\eta}_{i,m}[n], \quad (m',n') \neq (m,n).
\end{aligned} \tag{16.3}
$$

Here the channel matrix, precoding matrix, and the data vector with respect to Cell 2 are denoted as in (16.1) only with "(2)" in the subscripts. Note that $\mathbf{d}_{i,m,(1)}[n] = \mathbf{d}_{i,m,(2)}[n]$, i.e., the signals from Cell 1 and Cell 2 transmitted to the $i$th user[3] are the same. At each user terminal, decoding is performed, and the real part of the resulting signal is taken to cancel the intrinsic interference. To enable such FBMC-based CoMP downlink transmissions, the mitigation of the intracell, intercell, and intrinsic interference has to be achieved via the design of the precoding at the base stations and the decoding at the user terminals. To this end, we present the following IIM-CBF scheme, which is an extension of the IIM-CBF approaches for single-cell multiuser MIMO downlink systems [14] and is also the outcome of adapting the Extended FlexCoBF algorithm for CP-OFDM-based systems in [4] to FBMC-based systems.

## 16.2.2 INTRINSIC INTERFERENCE MITIGATING COORDINATED BEAMFORMING

Consider $N_c$ cells and suppose that the $j$th cell serves $U_j$ users simultaneously, $j = 1, 2, \ldots, N_c$. It is assumed that, for the $j$th cell, the users $1, 2, \ldots, U_j^{(\text{int})}$ are cell interior users, whereas the remaining $U_j - U_j^{(\text{int})}$ users are cell edge users. Notice that the IIM-CBF scheme is on a per-subcarrier basis, i.e., the precoding and decoding matrices at a certain time instant and on a certain subcarrier are solely determined by the channel at the same time instant and on the same subcarrier. Hence, to avoid cumbersome notation, in this section, we ignore the time and frequency indices in the channel matrices, the precoding matrices, and the decoding matrices [14]. For instance, $\mathbf{H}_{i,j}$ denotes the channel between the base station in the $j$th cell and the $i$th user in the same cell at a certain time instant and on a certain subcarrier. The IIM-CBF scheme is summarized as follows:

---

[3]In this example, we assume that the indices of each cell edge user in Cell 1 and Cell 2 are the same to facilitate the description of the scenario.

- **Step 1**: Initialize the real-valued decoding matrices $\mathbf{D}_{i,j}^{(0)}$ ($i = 1, 2, \ldots, U_j$) of all the users in the $j$th cell, where $j = 1, 2, \ldots, N_c$. Set the iteration index $\ell$ to zero, and set a threshold $\epsilon$ for the stopping criterion. If the current subcarrier is the first one, then the decoding matrices are generated randomly; otherwise, set the decoding matrices equal to those calculated for the previous subcarrier [14].
- **Step 2**: Set $\ell \leftarrow \ell + 1$ and calculate the equivalent channel matrix $\mathbf{H}_{ej}^{(\ell)}$ for the $j$th cell, where $j = 1, 2, \ldots, N_c$, at the $\ell$th iteration as

$$\mathbf{H}_{ej}^{(\ell)} = \left[\mathbf{H}_{e(1,j)}^{(\ell)^{\mathrm{T}}} \quad \mathbf{H}_{e(2,j)}^{(\ell)^{\mathrm{T}}} \quad \cdots \quad \mathbf{H}_{e(U_j,j)}^{(\ell)^{\mathrm{T}}}\right]^{\mathrm{T}}, \tag{16.4}$$

where $\mathbf{H}_{e(i,j)}^{(\ell)} = \mathbf{D}_{i,j}^{(\ell-1)^{\mathrm{T}}}\mathbf{H}_{i,j}$ is the equivalent channel matrix for the $i$th user of the $j$th cell at the $\ell$th iteration.
- **Step 3**: Calculate the precoding matrices $\mathbf{F}_j^{(\ell)}$ ($j = 1, 2, \ldots, N_c$) at the $\ell$th iteration,

$$\mathbf{F}_j^{(\ell)} = \left[\mathbf{F}_{1,j}^{(\ell)}\mathbf{G}_{1,j}^{(\ell)} \quad \mathbf{F}_{2,j}^{(\ell)}\mathbf{G}_{2,j}^{(\ell)} \quad \cdots \quad \mathbf{F}_{U_j,j}^{(\ell)}\mathbf{G}_{U_j,j}^{(\ell)}\right], \tag{16.5}$$

where $\mathbf{F}_{i,j}^{(\ell)}$ ($i = 1, 2, \ldots, U_j$) serve to mitigate the multiuser interference, and $\mathbf{G}_{i,j}^{(\ell)}$ ($i = 1, 2, \ldots, U_j$) are computed to achieve the suppression of the intrinsic interference [14].
First, calculate $\mathbf{F}_{i,j}^{(\ell)}$ for the $i$th user ($i = 1, 2, \ldots, U_j$) based on the block diagonalization (BD) algorithm [15] such that it lies in the null space of the combined channel matrix of all the other users in the $j$th cell ($j = 1, 2, \ldots, N_c$).
To further compute $\mathbf{G}_{i,j}^{(\ell)}$, define a matrix $\check{\mathbf{H}}_{e(i,j)}^{(\ell)}$ for the $i$th user based on its equivalent channel matrix $\mathbf{H}_{e(i,j)}^{(\ell)}\mathbf{F}_{i,j}^{(\ell)}$ after the cancellation of the multiuser interference

$$\check{\mathbf{H}}_{e(i,j)}^{(\ell)} = \left[\mathrm{Im}\left\{\mathbf{H}_{e(i,j)}^{(\ell)}\mathbf{F}_{i,j}^{(\ell)}\right\} \quad \mathrm{Re}\left\{\mathbf{H}_{e(i,j)}^{(\ell)}\mathbf{F}_{i,j}^{(\ell)}\right\}\right]. \tag{16.6}$$

Decompose $\mathbf{G}_{i,j}^{(\ell)}$ as $\mathbf{G}_{i,j}^{(\ell)} = \mathbf{G}_{i,j,1}^{(\ell)}\mathbf{G}_{i,j,2}^{(\ell)}$, and $\mathbf{G}_{i,j,1}^{(\ell)}$ is obtained such that $\begin{bmatrix} \mathrm{Re}\left\{\mathbf{G}_{i,j,1}^{(\ell)}\right\} \\ \mathrm{Im}\left\{\mathbf{G}_{i,j,1}^{(\ell)}\right\} \end{bmatrix}$ lies in the null space of $\check{\mathbf{H}}_{e(i,j)}^{(\ell)}$. Consequently,

$$\mathrm{Im}\left\{\mathbf{H}_{e(i,j)}^{(\ell)}\mathbf{F}_{i,j}^{(\ell)}\mathbf{G}_{i,j,1}^{(\ell)}\right\} = \mathbf{0} \tag{16.7}$$

is fulfilled to mitigate the intrinsic interference [9], [14]. Then $\mathbf{G}_{i,j,2}^{(\ell)}$ is calculated for spatial mapping [14].
- **Step 4**: Update the decoding matrix for each user based on the real-valued equivalent channel matrix where the processing at the transmitter and the step of taking the real part of the receive signal are taken into account. For the $i$th user that is a cell interior user of the $j$th cell, its equivalent channel matrix $\mathbf{H}_{etx(i,j)}^{(\ell)}$ is calculated as

$$\mathbf{H}^{(\ell)}_{\mathrm{etx}_{(i,j)}} = \mathrm{Re}\left\{\mathbf{H}_{i,j}\mathbf{F}^{(\ell)}_{i,j}\mathbf{G}^{(\ell)}_{i,j}\right\}. \tag{16.8}$$

For the $i$th user that is a cell edge user, define a set $\mathcal{S}_{i,j}$ that contains the indices of the cells that simultaneously transmit the same signals to the $i$th user. Then its equivalent channel matrix is expressed as

$$\mathbf{H}^{(\ell)}_{\mathrm{etx}_{(i,j)}} = \sum_{r\in\mathcal{S}_{i,j}} \mathrm{Re}\left\{\mathbf{H}_{i_r,r}\mathbf{F}^{(\ell)}_{i_r,r}\mathbf{G}^{(\ell)}_{i_r,r}\right\}, \tag{16.9}$$

where $i_r$ represents the index of the $i$th user of the $j$th cell in the $r$th cell, and $i_j = i$ following this definition. The cooperation of the adjacent cells involves the knowledge of the signals for all cell edge users. It also requires the exchange of these real-valued equivalent channel matrices that are used to compute the decoding matrix for each cell edge user, which can be achieved by adopting the two exchange mechanisms proposed in [4].

Afterwards, when a single data stream is transmitted to each user, the decoding matrix for the $i$th user at the $\ell$th iteration can be obtained by employing the Maximum Ratio Combining (MRC) receiver or the Minimum Mean-Squared Error (MMSE) receiver of its equivalent channel matrix $\mathbf{H}^{(\ell)}_{\mathrm{etx}_{(i,j)}}$. On the other hand, when there exist users to which multiple data streams are transmitted, the Zero Forcing (ZF) receiver is used.

- **Step 5**: Calculate the term $\xi^{(\ell)}_j$ for the $j$th cell that measures the residual multiuser and the interstream interference for the $\ell$th iteration [16]. If $\xi^{(\ell)}_j < \epsilon$, then the convergence is achieved, and the iterative procedure terminates. Otherwise, go back to **Step 2**.

This coordinated beamforming scheme is designed based on the CoMP technique in [4]. Nevertheless, due to the fact that the intrinsic interference is inherent in FBMC systems, additional processing has been incorporated to suppress this interference. The reader is referred to [16] for different choices of the stopping criterion that are recommended for single-stream and multiple-stream transmissions.

### 16.2.3 NUMERICAL RESULTS AND DISCUSSIONS

In the sequel, the performance of the FBMC-based CoMP downlink is assessed. Transmissions in cooperative cells are assumed to be perfectly synchronized. The sum rate performance is used as the evaluation approach [16]. Let us start with a two-cell scenario consisting of five users in total. Each cell equipped with four transmit antennas serves three users, each with two receive antennas, which forms a $4 \times 6$ multiuser downlink setting. Among the three users served in each cell, two are cell interior users, and one is a cell edge user. A single data stream is transmitted to each cell interior user, and two data streams are transmitted to the cell edge user, i.e., for each cell, full spatial multiplexing is considered. The path loss of the transmission to the cell edge users is assumed to be 10 times greater than that for the cell interior users [4]. The ITU Ped-A channel is considered.

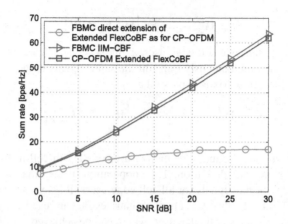

**FIGURE 16.2**

Comparison of the sum rate performances of different schemes in the CoMP downlink where the total number of users in two adjacent cells is 5, and the ITU Ped-A channel is considered.

To this end, we use Fig. 16.2 to illustrate the sum rate performances of two transmission schemes for the aforementioned FBMC-based CoMP downlink system. Here SNR $= P_T/\sigma_\eta^2$, where $P_T$ is the per-subcarrier transmit power of each cell, and $\sigma_\eta^2$ represents the noise variance. The CP-OFDM-based system, where Extended FlexCoBF[4] is employed, is also considered for comparison. It can be seen that the IIM-CBF scheme leads to a slightly better performance of the FBMC-based system in contrast to the CP-OFDM-based system. The reason is that the former has a higher spectral efficiency compared to the latter since no cyclic prefix is required. This observation also implies the effectiveness of the proposed transmission strategy in mitigating the intracell and intercell interference as well as the intrinsic interference. By comparison, when a transmission strategy originally designed for CP-OFDM, here Extended FlexCoBF, is straightforwardly extended to the FBMC system [14], and the channel is frequency selective, the performance is much worse than that of the proposed IIM-CBF scheme due to a much higher level of the residual interference.

Moreover, Fig. 16.3 depicts the Complementary Cumulative Distribution Function (CCDF) of the number of iterations required for the IIM-CBF technique to converge. For the evaluated schemes, $\epsilon$ used for the stopping criterion is set to $10^{-5}$, and the iterative procedure is manually terminated if the number of iterations exceeds 50.

---

[4]In the implementation of Extended FlexCoBF [4], we adopt the same mechanism of initializing the decoding matrices as in the LoCCoBF algorithm [14] such that the correlation of the channels of adjacent subcarriers is exploited, and consequently the number of iterations required for the convergence is reduced.

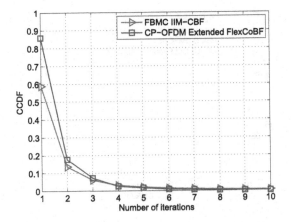

**FIGURE 16.3**

CCDF of the number of iterations required for different coordinated beamforming schemes in the CoMP downlink system where the total number of users in two adjacent cells is 5, and the ITU Ped-A channel is considered.

As an iterative algorithm, the complexity of the IIM-CBF scheme is acceptable. In addition, compared to the case of CP-OFDM, this FBMC-based CoMP technique requires a comparable number of iterations to reach the convergence.

Finally, a three-cell CoMP downlink scenario is considered. Each cell equipped with four transmit antennas serves three users, each with two receive antennas. This leads to a $4 \times 6$ multiuser downlink setting. Among the three users, two are cell interior users, and one is a cell edge user. A single data stream is transmitted to each cell interior user, and two data streams are transmitted to the cell edge user. The numerical results with respect to the sum rate performance are shown in Fig. 16.4. Similar observations are obtained as in Fig. 16.2.

## 16.3 DISTRIBUTED CHANNEL ESTIMATION
### 16.3.1 PREAMBLE-BASED MULTIUSER CHANNEL ESTIMATION

The precoding and beamforming methods presented in the previous section rely on CSI. It is therefore necessary to investigate how this channel information can be obtained, in particular, in the distributed scenario where full coordination and full synchronization between the base stations is not necessarily available.

This section addresses channel estimation relying on a preamble for the downlink of a distributed system having $B$ base stations (transmitters) and $U$ users (receivers), as shown in Fig. 16.5. The base stations and users use FBMC/OQAM, and the channel estimation performance will be compared to a corresponding CP-OFDM system. Each base station and user is assumed to be equipped with a single antenna. This can be straightforwardly generalized to multiple antennas. During the training phase,

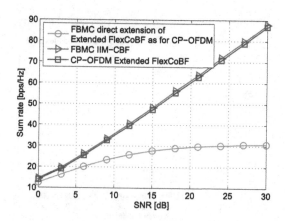

**FIGURE 16.4**

Comparison of the sum rate performances of different coordinated beamforming schemes in the CoMP downlink where the total number of users in three adjacent cells is 7, and the ITU Ped-A channel is considered.

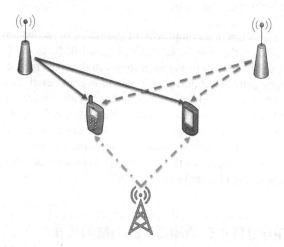

**FIGURE 16.5**

FBMC-based distributed system: $B$ base stations (transmitters) and $U$ users (receivers).

the base stations cannot precode their streams to avoid interfering with one another. Therefore, the pilot subcarriers are distributed among different base stations following a certain Subcarriers Assignment Scheme (SAS). After this estimation phase, each base station is assumed to use simultaneously the entire frequency range to transmit its data stream to the different users.

Two specific SASs are investigated, as depicted in Fig. 16.6, namely, the *equispaced SAS* and the *block SAS*:

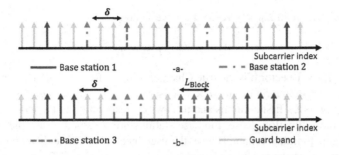

**FIGURE 16.6**

Subcarriers assignment schemes: -a- equispaced (guard band of $\delta = 2$ subcarriers), -b- block (block size $L_{\text{Block}} = 3$ subcarriers).

- Equispaced SAS: the subcarriers are interleaved among the base stations in such a way that the pilot subcarriers assigned to each base station are allocated uniformly over the entire frequency range. To avoid interference, some subcarriers are left inactive and serve as guard bands between neighboring pilot subcarriers. The guard band size in number of subcarriers is denoted by $\delta$ and will directly depend on the roll-off factor of the prototype filter. In this scheme, the pilot tones are well spread over the whole band inducing low correlation with each other. However, this results in a large number of subcarriers used as guard bands. No more than $L_p = \lfloor \frac{M}{(\delta+1)B} \rfloor$ pilot subcarriers can be assigned to each base station.
- Block SAS: several blocks of consecutive subcarriers in the preamble are allocated to each base station, and $\delta$ free subcarriers adjacent to each block of $L_{\text{Block}}$ subcarriers serve as guard bands between the different blocks. The ratio of pilot subcarriers, $L_p = \lfloor \frac{M}{(\delta+L_{\text{Block}})B} \rfloor L_{\text{Block}}$, per base station to guard band subcarriers is much more favorable in this case compared to the previous one. However, this scheme may degrade the estimation performance given that the subcarriers allocated to a base station are more correlated.

Since the transmissions in different base stations are protected with guard bands, they are not interfering. Hence, the estimation of the channel between each base station and each user may be performed independently. Assuming a relatively (to the filterbank size) low channel selectivity, the part of each base station-user Channel Frequency Response (CFR) corresponding to the pilot subcarriers can be estimated as

$$\hat{H}_m = \frac{y_m}{x_m} = H_m + \frac{\eta_m}{x_m}, \tag{16.10}$$

where $x_m$ is the corresponding pseudo-pilot, $\eta_m$ is the noise sample, and $m \in \mathcal{M}$ with $\mathcal{M}$ being the set of subcarrier indices for this particular base station. Stacking the CFR estimates $\hat{H}_m$, $m \in \mathcal{M}$, in an $L_p \times 1$ vector $\hat{\mathbf{H}}_{L_p}$ and the estimation error samples $\eta_m/c_m$ in an $L_p \times 1$ vector $\boldsymbol{\eta}_{L_p}$, we obtain

$$\hat{\mathbf{H}}_{L_p} = \mathcal{F}_{L_p \times L_h} \mathbf{h} + \boldsymbol{\eta}_{L_p},$$

where the $L_h \times 1$ vector $\mathbf{h}$ is the true Channel Impulse Response (CIR), and $\mathcal{F}_{L_p \times L_h}$ is the $L_p \times L_h$ submatrix of the $M \times M$ Discrete Fourier Transform (DFT) matrix $\mathcal{F}$ consisting of its $L_h$ first columns and its rows corresponding to the indices in $\mathcal{M}$.

After the estimation phase, the base stations are assumed to use the whole frequency range to transmit data. Hence, the users need to estimate their channels at all subcarriers. Since the receiver can sense only a part of the CFR for each transmitter, the matrix $\mathcal{F}_{L_p \times L_h}$ can get ill-conditioned due to, e.g., a lack of pilots or high correlation between neighboring subcarriers. A classical Least Squares (LS) estimator requiring the inversion of such an ill-conditioned matrix would amplify the noise and is therefore not an appropriate solution. To estimate the entire CFR while taking care of the possibly ill-conditioned matrix $\mathcal{F}_{L_p \times L_h}$, a Linear Minimum Mean Squared Error (LMMSE) estimator is considered here. It computes the CIR estimate as $\hat{\mathbf{h}} = \mathbf{G}\hat{\mathbf{H}}_{L_p}$, with

$$\begin{aligned}
\mathbf{G} &= \arg\min \mathbb{E}\left\{ \|\mathbf{h} - \mathbf{G}\hat{\mathbf{H}}_{L_p}\|^2 \right\} \qquad (16.11)\\
&= \mathbf{C}_h \mathcal{F}_{L_p \times L_h}^{\mathrm{H}} \left( \mathcal{F}_{L_p \times L_h} \mathbf{C}_h \mathcal{F}_{L_p \times L_h}^{\mathrm{H}} + \mathbf{C}_\eta \right)^{-1},
\end{aligned}$$

where $\mathbf{C}_h = \mathbb{E}\{\mathbf{h}\mathbf{h}^{\mathrm{H}}\}$ and $\mathbf{C}_\eta = \mathbb{E}\{\boldsymbol{\eta}_{L_p}\boldsymbol{\eta}_{L_p}^{\mathrm{H}}\}$ are the channel and noise correlation matrices, respectively. With nonadjacent pilot tones, $\mathbf{C}_\eta$ reduces to a diagonal matrix. However, when pilots are placed at adjacent subcarriers, the estimation error samples become correlated, and the diagonal elements of the estimation error covariance matrix may be different due to the varying amplitude of the pseudo-pilots. Finally, the CFR at all frequencies is found from the estimated CIR as $\hat{\mathbf{H}} = \mathcal{F}_{M \times L_h} \hat{\mathbf{h}}$ (i.e., through DFT interpolation). The interested reader can refer to [17,18] for more details on LMMSE channel estimation in FBMC/OQAM systems. Another related work is presented in [19], where the interpolation in frequency is performed by an iterative process using inverse DFT, windowing, and DFT successively until convergence.

In Figs. 16.7 and 16.8, an exponentially decaying power delay profile is assumed for the channel, namely, the last tap of the channel has a variance 20 dB lower relatively to the first tap. The channel is assumed to be sufficiently slowly varying so as to be constant over the preamble duration. The preamble is assumed protected from the data by a sufficient guard time [20]. The subcarrier spacing is set to $\Delta f = 15$ kHz with $M = 128$ subcarriers. The prototype filter used is the PHYDYAS filter [21] with roll-off factor of one, so that a one-subcarrier guard band is sufficient ($\delta = 1$) to prevent the base stations from interfering. The performance of a corresponding fully synchronized CP-OFDM system is considered. Note, however, that establishing this synchronization in an Orthogonal Frequency Division Multiple Access (OFDMA) context would require a large overhead in contrast to FBMC/OQAM [22].

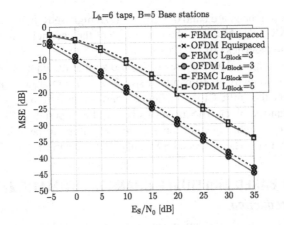

**FIGURE 16.7**

The equispaced SAS and the $L_{\mathrm{Block}} = 3$ SAS are optimal. However, the $L_{\mathrm{Block}} = 5$ SAS faces a lack of pilot blocks since $\lfloor \frac{M}{(\delta + L_{\mathrm{Block}})B} \rfloor < L_h$.

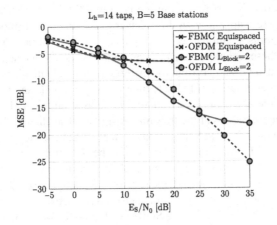

**FIGURE 16.8**

The equispaced SAS lacks pilots ($\lfloor \frac{M}{(\delta + 1)B} \rfloor < L_h$) due to the $\delta = 1$ subcarrier guard band. Then, the block SAS at high SNR performs better because its total number of pilots is larger than the channel length, $L_p = \lfloor \frac{M}{(\delta + L_{\mathrm{Block}})B} \rfloor L_{\mathrm{Block}} > L_h$.

Fig. 16.7 illustrates the fact that equispaced and block SAS might achieve the same performance if the size $L_{\mathrm{Block}}$ of the block is not too large. Unsynchronized FBMC/OQAM performs slightly better than fully synchronized CP-OFDM because it does not waste energy for transmitting the CP. As discussed before, FBMC/OQAM does not suffer from interference between base stations thanks to the well-localized prototype filter and the guard band of size $\delta = 1$. At high SNR, however, an error floor appears in FBMC/OQAM due to the inaccuracy of the channel frequency selectivity assumption that underlies (16.10) [20].

In Fig. 16.8, longer channels are considered so that the equispaced SAS is short of pilots due to the many guard bands of size $\delta = 1$. There are no sufficiently many pilots to estimate the entire CFR, and hence a performance floor occurs at high SNR. The result is that the block SAS performs better after a certain SNR value. Indeed, it still has a total number of pilot tones sufficiently large to be able to estimate the entire CFR, whereas the equispaced SAS only allows a part of it to be estimated.

## 16.3.2 SNR-BASED MULTIUSER CHANNEL ESTIMATION

### 16.3.2.1 Introduction

Preamble-based channel estimation has been investigated in Section 16.3.1. It is based on the assumption that the channels remain constant on the duration of the packet. Depending on the mobility in the system and the length of the packets, some additional adaptation of the channel estimation may be required in many scenarios. Most estimation methods for this purpose in multicarrier communications systems are based on pilots [20]. However, in the case of FBMC/OQAM, the complicated interference structure appearing at the receiver makes the use of pilots slightly more difficult and requires additional techniques such as Pair of Pilots (POP) or auxiliary pilots [23]. All these techniques are presented in more detail in Chapter 11.

In this section, a different estimation method is presented that is no longer making use of pilots. It is appropriate for slowly time varying channels and has the advantage of being applicable in a distributed scenario. It is based on the application of several well-designed perturbations of the transmitted signals. The changes brought by these perturbations on the observed SNRs at different receivers are measured and fed back to the base stations. The information for different perturbations is collected at the base station and then used to perform the channel estimation at regular intervals during data transmission. With this method, the receivers are simply estimating their SNR in a normal way and need not be aware of the created perturbation. It has the advantage of requiring a low overhead (only the SNRs need to be fed back). On the other hand, the perturbation has an impact on the transmission of normal data and has to be kept small enough to allow continuous transmission. Similar methods have already been presented in the context of Very high bit-rate Digital Subscriber Lines (VDSL) with Discrete Multitone (DMT) transmission [24] or for multicell OFDM systems [25]. In this section, it is shown how the method can be applied to the more complicated case of FBMC/OQAM with several adjustments.

### 16.3.2.2 System Model and Precoding

We consider a distributed MIMO environment with FBMC/OQAM modulation employed. The considered model is similar to the previous section and depicted in Fig. 16.5. For simplicity, however, the model is here limited to the case where the number of base stations is equal to the number of users. There are $N$ base stations (transmitters) in the system and $N$ users, each node being equipped with a single antenna (but the method can be easily generalized to multiple antennas). The base stations cooperatively send their signals (similarly to a single transmitter with multiple antennas) to transmit $N$ independent streams of information to the $N$ users. A linear precoder is considered. It is assumed that the channel is mildly selective and the number of subcarriers is sufficient so that the channel can be considered flat inside each subcarrier. The intersubcarrier interference is therefore negligible. All operations can be applied independently per subcarrier, so that *only one given subcarrier m is considered in this section*. It is also assumed that the channel is slowly time-varying, so that it can be considered constant on the duration of various measurements. This is a somewhat strong assumption, which restricts the use of this method to systems with low mobility.

The real information symbols to be transmitted to the different users $i = 0, 1, \ldots, N - 1$ on the subcarrier of interest $m$ at instant $n$ are denoted by $d_{m,n}^i$. They are grouped into a vector $\mathbf{d}_{m,n} = \begin{bmatrix} d_{m,n}^0 & \cdots & d_{m,n}^{N-1} \end{bmatrix}^{\mathrm{T}}$. The variance of the symbols is denoted by $\sigma_d^2$ and is assumed to be the same for all symbols. The information is precoded through different base stations and sent on the channel. The receivers only have access to the symbols received at their own antenna. It is assumed that each receiver (independently) applies a per-subcarrier zero-forcing equalizer. The channel matrix of interest $\mathbf{H}$ corresponds to the combination of the precoding, the channel $\mathbf{H}_m$ at the subcarrier of interest, and the set of independent per-user equalizers. This is a square $N \times N$ matrix. The method presented here aims at estimating this matrix. It is called the combined channel matrix in the remainder of this section. It is interesting to note that in a tracking scenario (i.e., when some preamble-based estimation has already been performed), the current precoding can be assumed to offer a reasonable interuser interference cancellation, and $\mathbf{H}$ is close to the unity matrix. Taking into account the particular interference structure resulting from the FBMC/OQAM modulation and assuming that the channel is flat inside each subcarrier, the overall transmission model can be written as [13,26]

$$\bar{\mathbf{d}}_{m,n} = \mathbf{H}\mathbf{d}_{m,n} + j\mathbf{H} \sum_{m'=m-1}^{m+1} \sum_{n'=n-3}^{n+3} t_{m,m'}(n - n')\mathbf{d}_{m',n'} + \bar{\mu}_{m,n}, \quad (16.12)$$

where $\bar{\mathbf{d}}_{m,n}$ stacks the equalizer outputs of the different users, where $\Gamma_{m,m'}^{n,n'} = jt_{m,m'}(n - n')$ denotes the transmultiplexer response of the filterbank, as in Chapter 11, responsible for the complex interference appearing at the receiver from the FBMC/OQAM modulation, and where $\bar{\mu}_{m,n}$ denotes the vector of noise samples. The variance of the noise samples for user $i$ is denoted by $\sigma_{\mu,i}^2$.

### 16.3.2.3 *Estimation Method Based on Small Perturbation*

For illustration purpose, consider the estimation of the combined channel matrix coefficients of index $(i, N)$, that is, estimation of $H_{iN}$ for a given user $i$ (the element

$(i, j)$ of matrix $\mathbf{H}$ is denoted as $H_{ij}$). The basic principle of this method is to add a small perturbation to the symbols of the victim user $i$ coming from the data stream of user $N$. The effect of this small perturbation on the SNR of user $i$ is observed, and from the corresponding changes in SNR the combined channel matrix coefficients of interest can be estimated.

First, the SNR is measured for normal transmission with all users active (including user $N$).[5] The SNR observed at receiver $i$ during normal transmission is given by

$$\frac{1}{\text{SNR}_{i,0}} = \sum_{l \neq i} H^2_{il,\mathcal{R}} + \frac{\sigma^2_{\mu,i}}{\sigma^2_d} + \sum_{m'} \sum_{n'} \sum_{l} t^2_{m,m'}(n - n') H^2_{il,\mathcal{I}}, \qquad (16.13)$$

where $H_{il,\mathcal{R}} = \Re\{H_{il}\}$ and $H_{il,\mathcal{I}} = \Im\{H_{il}\}$.

After this first SNR measurement, a second situation is considered, where, instead of transmitting the usual data symbol $d^i_{m,n}$ for user $i$, a perturbation is added proportionally to the symbols from user $N$. The transmitted symbols now are

$$d^i_{m,n}(1) = d^i_{m,n} + \epsilon_{iN} d^N_{m,n}, \qquad (16.14)$$

where the coefficients $\epsilon_{iN}$ are real coefficients of small amplitude to avoid disturbing the normal transmission too much. The new SNR for user $i$ can be approximated by

$$\frac{1}{\text{SNR}_{i,1}} = \left[ H_{iN,\mathcal{R}} + \epsilon_{iN} \right]^2 + \sum_{l \neq i; l \neq N} H^2_{il,\mathcal{R}} + \frac{\sigma^2_{\mu,i}}{\sigma^2_d}$$

$$+ \sum_{m'} \sum_{n'} \sum_{l} t^2_{m,m'}(n - n') H^2_{il,\mathcal{I}}. \qquad (16.15)$$

As can be observed, the perturbation has an impact on the SNR by modifying the apparent combined coefficient $H_{iN}$. From the receiver point of view and because of the perturbation, everything happens as if the coefficient had been changed to the new value $H_{iN} + \epsilon_{iN}$. All other coefficients remain unaffected.

A third measure of SNR is necessary to obtain the complete normalized matrix coefficient (real and imaginary parts) using a second perturbation. In this third situation, a slightly modified perturbation is added to the symbols. The new transmitted symbols are

$$d^i_{m,n}(2) = d^i_{m,n} + \epsilon_{iN} t_{m,m}(1) d^N_{m,n-1}. \qquad (16.16)$$

---

[5]To be more precise, we should call it the Signal-to-Interference-plus-Noise Ratio (SINR) since it includes both additive noise and interference. For simplicity of notation, however, we write it simply as SNR, emphasizing the fact that the receiver itself cannot make the difference between these two sources of degradation.

Note that $d_{m,n-1}^N$ is used here instead of $d_{m,n}^N$ as well as an additional coefficient $t_{m,m}(1)$, the purpose of which will be apparent later on. In this case, it can be easily observed that everything happens as if the combined channel coefficient $H_{iN}$ is replaced with $H_{iN} + j\epsilon_{iN}$. The SNR with the second perturbation satisfies

$$\frac{1}{\text{SNR}_{i,2}} = \left[H_{iN,\mathcal{I}} + \epsilon_{iN}\right]^2 t_{m,m}^2(1) + \sum_{l \neq i} H_{il,\mathcal{R}}^2 + \frac{\sigma_{\mu,i}^2}{\sigma_d^2}$$

$$+ \sum \sum_{(m',n',l) \neq (m,n-1,N)} \sum t_{m,m'}^2(n - n') H_{il,\mathcal{I}}^2. \tag{16.17}$$

Now based on these three measurements, it is possible to obtain the combined channel coefficients $H_{iN,\mathcal{R}}$ and $H_{iN,\mathcal{I}}$ (real and imaginary parts). After a few computations, the following estimators are obtained:

$$\hat{H}_{iN,\mathcal{R}} = \frac{1}{2\epsilon_{iN}} \left(\frac{1}{\text{SNR}_{i,1}} - \frac{1}{\text{SNR}_{i,0}}\right) - \epsilon_{iN}/2. \tag{16.18}$$

Similarly,

$$\hat{H}_{iN,\mathcal{I}} = \frac{1}{2\epsilon_{iN} t_{m,m}^2(1)} \left(\frac{1}{\text{SNR}_{i,2}} - \frac{1}{\text{SNR}_{i,0}}\right) - \epsilon_{iN}/2. \tag{16.19}$$

These equations provide an estimation method for the real and imaginary parts of the combined channel matrix coefficient. It is valid even when other users are present, as long as their power remains constant during all three SNR measurements. It only requires the three SNR measurements described above for a value of $\epsilon_{iN}$, which can be chosen arbitrarily. For the estimation to be as precise as possible, $\epsilon_{iN}$ has to be chosen as large as possible, so that a significant impact on the SNR can be measured. On the other hand, if $\epsilon_{iN}$ is too large, there is a risk to decrease the SNR excessively and prevent the normal transmission of the data symbols.

### 16.3.2.4 Simulation Results

To illustrate the performance of the method, in this section, we present some simulation results. A system with five users and five base stations is considered, and only one given subcarrier is investigated for simplicity. Several noise and interference power situations are investigated (corresponding to different accuracies or different stages of the channel tracking). For each situation, a large number of simulations (5000) is performed. The channel matrix is generated randomly for each simulation according to a log-normal model, fulfilling the noise and interference conditions. The perturbation coefficients $\epsilon_{ij}$ are chosen according to the rule developed in [27]. The variance of the channel coefficient estimates is computed by averaging over all simulations. Instead of representing the variance estimate itself, the results are presented with the more meaningful value of the resulting Signal-to-Interference Ratio (SIR). It provides an evaluation of how well the interuser interference is decreased

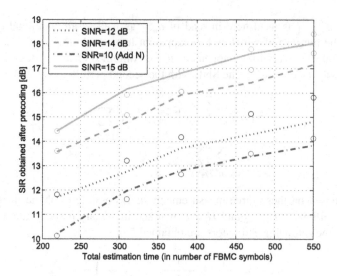

**FIGURE 16.9**

Results of the SNR method: resulting SIR as a function of the total time used for estimation for different noise and interference situations. In three out of four situations, the inter-user interference is dominant (the additive noise is approximately 15 dB below the signal). For the situation labeled "Add N", the additive noise is dominant, and the inter-user interference is low. The theoretical performance predictions are represented with circles.

by the obtained precoder with respect to the signal power. Fig. 16.9 shows the obtained SIR as a function of the number of blocks $K_e$ used for each measurement, expressed in terms of the total needed time (in the number of FBMC symbols). Three noise and interference situations are represented where the interuser interference is dominant. An additional situation is presented (labeled Add N) where the additive noise is dominant. Also shown in circles is the predicted performance, as obtained with results presented in [27]. As can be observed, the proposed method can improve the channel estimation and thereby reduce the remaining interuser interference by several dB. This, however, requires sufficient time and hence a slowly time-varying channel.

## 16.4 CONCLUDING REMARKS

This chapter presents advanced FBMC-based CoMP techniques together with channel estimation and tracking schemes for FBMC-based distributed networks. Based on a coordinated beamforming concept, the IIM-CBF algorithm enables partial coordination of adjacent cells. Simulation results have shown that the IIM-CBF technique is very effective in mitigating the intracell interference, the intercell interference, and the intrinsic interference inherent in FBMC/OQAM-based systems. By employing the IIM-CBF scheme the FBMC-based CoMP downlink system achieves a similar sum rate performance as its OFDM-based counterpart while exhibiting superiority in terms of a higher spectral efficiency, a greater robustness against synchronization errors, and a lower out-of-band radiation. In addition, the convergence of the IIM-CBF technique has been analyzed numerically, leading to the conclusion that its complexity is quite acceptable. Moreover, preamble-based channel estimation in an FBMC-based distributed MIMO downlink scenario has been investigated. A comparison of the equispaced SAS and the block SAS has been performed via extensive simulations. The numerical results have also corroborated the significant advantage of FBMC over OFDM in asynchronous scenarios. Finally, the tracking of distributed MIMO downlink channel estimates has been addressed. The presented method requires the feedback of SNR measurements and is based on small perturbations applied to the transmitted signal. It has been observed that a good estimate can be obtained in a reasonable amount of time with a limited impact on the data transmission. Theoretical performance evaluations have been given, which well match the numerical simulation results.

# REFERENCES

[1] M. Renfors, F. Bader, L. Baltar, D. Le Ruyet, D. Roviras, P. Mege, and M. Haardt, "On the use of filter bank based multicarrier modulation for professional mobile radio," in *Proc. VTC Spring 2013*, Dresden, Germany, Jun. 2013.

[2] D. Lee, H. Seo, B. Clerckx, E. Hardouin, D. Mazzarese, S. Nagata, and K. Sayana, "Coordinated multipoint transmission and reception in LTE-advanced deployment scenarios and operational challenges," *IEEE Commun. Mag.*, vol. 50, no. 2, pp. 148–155, Feb. 2012.

[3] L. Venturino, N. Prasad, and X. Wang, "Coordinated linear beamforming in downlink multi-cell wireless networks," *IEEE Trans. Wireless Commun.*, vol. 9, no. 4, pp. 1451–1461, Apr. 2010.

[4] B. Song, F. Roemer, and M. Haardt, "Flexible coordinated beamforming (FlexCoBF) for the downlink of multi-user MIMO systems in single and clustered multiple cells," *Signal Process.*, vol. 93, pp. 2462–2473, Sep. 2013.

[5] J. Zhang, R. Chen, and J. G. Andrews, "Networked MIMO with clustered linear precoding," *IEEE Trans. Wireless Commun.*, vol. 8, no. 4, pp. 1910–1921, Apr. 2009.

[6] K. Kim, J. Lee, C. Lee, N. Jeon, and S. Kim, "Coordinated beamforming with limited BS cooperation for multicell multiuser MIMO broadcast channel," in *Proc. VTC Spring 2009*, Barcelona, Spain, Apr. 2009.

[7] F. Boccardi and H. Huang, "Limited downlink network coordination in cellular network," in *Proc. PIMRC 2007*, Athens, Greece, Sep. 2007.

[8] H. Zhang, N. B. Mehta, A. F. Molisch, J. Zhang, and H. Dai, "Asynchronous interference mitigation in cooperative base station systems," *IEEE Trans. Wireless Commun.*, vol. 7, no. 1, pp. 155–164, Jan. 2008.

[9] M. Caus and A. I. Pérez-Neira, "Multi-stream transmission in MIMO-FBMC systems," in *Proc. ICASSP 2013*, Vancouver, Canada, May 2013.

[10] M. Caus and A. I. Pérez-Neira, "Comparison of linear and widely linear processing in MIMO-FBMC systems," in *Proc. ISWCS 2013*, Ilmenau, Germany, Aug. 2013.

[11] M. Caus, A. I. Pérez-Neira, and M. Moretti, "SDMA for FBMC with block diagonalization," in *Proc. SPAWC 2013*, Darmstadt, Germany, Jun. 2013.

[12] M. G. Bellanger, "FBMC physical layer: A primer," Jun. 2010. [Online]. Available: http://www.ict-phydyas.org/teamspace/internal-folder/FBMC-Primer_06-2010.pdf.

[13] FP7-ICT Project PHYDYAS – Physical Layer for Dynamic Spectrum Access and Cognitive Radio, http://www.ict-phydyas.org.

[14] Y. Cheng, S. Li, J. Zhang, F. Roemer, B. Song, M. Haardt, Y. Zhou, and M. Dong, "An efficient and flexible transmission strategy for the multicarrier multiuser MIMO downlink," *IEEE Trans. Veh. Technol.*, vol. 63, no. 2, pp. 628–642, Feb. 2014.

[15] Q. H. Spencer, A. L. Swindlehurst, and M. Haardt, "Zero-forcing methods for downlink spatial multiplexing in multi-user MIMO channels," *IEEE Trans. Signal Process.*, vol. 52, no. 2, pp. 461–471, Feb. 2004.

[16] Y. Cheng, P. Li, and M. Haardt, "Intrinsic interference mitigating coordinated beamforming for the FBMC/OQAM based downlink," *EURASIP J. Adv. Signal Process.*, vol. 2014:86, May 2014.

[17] F. Rottenberg, Y. Medjahdi, E. Kofidis, and J. Louveaux, "Preamble-based channel estimation in asynchronous FBMC-OQAM distributed MIMO systems," in *Proc. ISWCS 2015*, Brussels, Belgium, Aug. 2015.

[18] L. Caro, V. Savaux, D. Boiteau, M. Djoko-Kouam, and Y. Louët, "Preamble-based LMMSE channel estimation in OFDM/OQAM modulation," in *Proc. VTC Spring 2015*, Glasgow, UK, May 2015.

[19] M. Belotserkovsky, "An equalizer initialization algorithm for OFDM receivers," in *Proc. ICCE 2002*, Indianapolis, USA, Jun. 2002.

[20] E. Kofidis, D. Katselis, A. Rontogiannis, and S. Theodoridis, "Preamble-based channel estimation in OFDM/OQAM systems: A review," *Signal Process.*, vol. 93, no. 7, pp. 2038–2054, Jul. 2013. http://dx.doi.org/10.1016/j.sigpro.2013.01.013 [Online].

[21] M. Bellanger, "Specification and design of a prototype filter for filter bank based multicarrier transmission," in *Proc. ICASSP 2001*, Salt Lake City, USA, May 2001.

[22] J.-B. Doré, V. Berg, N. Cassiau, and D. Kténas, "FBMC receiver for multi-user asynchronous transmission on fragmented spectrum," *EURASIP J. Adv. Signal Process.*, vol. 2014, no. 1, pp. 1–20, 2014.

[23] C. Lélé, J.-P. Javaudin, R. Legouable, A. Skrzypczak, and P. Siohan, "Channel estimation methods for preamble-based OFDM/OQAM modulation," *Eur. Trans. Telecommun.*, vol. 19, pp. 741–750, Sep. 2008.

[24] J. Louveaux, A. Kalakech, M. Guenach, J. Maes, M. Peeters, and L. Vandendorpe, "An SNR-assisted crosstalk channel estimation technique," in *Proc. ICC 2009*, Dresden, Germany, Jun. 2009.

[25] L. Vandendorpe and J. Louveaux, "ML co-channel interference estimation from SINR measurements for multicell OFDM downlink: Bounds and performance analysis," in *Proc. ICASSP 2010*, Dallas, USA, Mar. 2010.

[26] P. Siohan, C. Siclet, and N. Lacaille, "Analysis and design of OFDM/OQAM systems based on filterbank theory," *IEEE Trans. Signal Process.*, vol. 50, no. 5, pp. 1170–1183, May 2002.

[27] J. Louveaux, A. Bourdoux, and F. Horlin, "Low feedback downlink MIMO channel estimation for distributed FBMC systems using SNR measurements," in *Proc. ISWCS 2013*, Ilmenau, Germany, Aug. 2013.

# Multiuser PHY-MAC Interaction for FBMC

# 17

Dimitris Tsolkas*, Nikos Passas*, Dmitry Petrov†, Mylène Pischella‡

*University of Athens, Athens, Greece**
*Magister Solutions Ltd., Jyväskylä, Finland†*
*CEDRIC Laboratory, CNAM, Paris, France‡*

## CONTENTS

## 17.1 INTRODUCTION

In this chapter, the FilterBank MultiCarrier (FBMC) performance is examined from the Radio Resource Management (RRM) perspective by considering multiuser environments. Fundamental processes for performing system-level simulations over FBMC are analyzed and a representative quantification of FBMC performance in a cellular network is provided. Additionally, two key asynchronous communication scenarios are studied. The first scenario refers to direct transmissions among user devices, dealing with the very appealing concept of Device-to-Device (D2D) communications. For this scenario, a cross-layer power and resource allocation scheme is proposed and evaluated over FBMC physical layer. The second scenario refers to multiple asynchronous transmissions toward a receiving node and provides useful insights on the radio resource overhead required to compensate for multiuser synchronization errors.

## 17.2 SYSTEM-LEVEL SIMULATIONS FOR FBMC CELLULAR NETWORK PERFORMANCE EVALUATION

In practice, there are many ways to evaluate and optimize the performance of wireless networks. Some methods such as drive tests or probes can be used on the already deployed Radio Access Network (RAN) infrastructure. New features and software can be checked in smaller areas or test-beds that include few User Equipments (UEs) and Base Stations (BSs). However, since the technology is on its early stage and real devices are not yet available, these approaches are hardly applicable. Nowadays, flexible Software Defined Radios (SDRs) and Field Programmable Gate Array (FPGA) boards make the research and development process easier [1,2]. Purely analytic considerations that establish the basement of the technology are highly important but cannot be used directly in full-scale scenarios because of their multifactor and dynamic nature. Thus, in many circumstances, simulation is the only method of evaluation and optimization of real networks and also in prototyping [3]. Moreover, computer modeling gives much more flexibility in the selection of equipment configuration, scenario typologies, user mobility, and other parameters. Many of the new features proposed and discussed in standardization organizations like 3rd Generation Partnership Project (3GPP) are based on results of simulations.

Depending on the target of the study and due to computational complexity reasons, modeling process in many cases is divided into two main parts, link- and system-level simulations.

*Link-level simulations:* The scope of link-level simulations often covers a BS and one or few UEs connected through the common wireless media. Symbol resolution in Time Domain (TD) and subcarrier resolution in Frequency Domain (FD) allow the detailed modeling of realistic channels, noise, interference, and multiantenna techniques. Special focus is on Physical layer (PHY) procedures, such as modulation, channel estimation, equalization, etc. Transmitted data is modeled with the bit resolution, and therefore, it is possible to study (along with others) the effects of coding, interleaving, and error correction. Thus, one of the most important outputs of link-level simulations are Bit Error Rate (BER) or Block Error Rate (BLER) curves that demonstrate the joint performance of transmitter and receiver depending on the channel conditions and the implemented PHY algorithms. Additionally, the simulation results can be Peak-to-Average Power Ratio (PAPR), spectral characteristics of the signal, and many other figures of merit depending on the focus of a study.

*System-level simulations:* The main target of system-level simulations is to study the performance of the larger-scale network that may include up to thousands of nodes, e.g. UEs, BSs, retransmitters, and repeaters. Such simulation tools are, in fact, the combination of different models that determine the behavior of the network, for example, traffic generation, user mobility, power consumption, signal propagation, and so on. Realistic protocol stack and RRM procedures such as handing-off, scheduling, interference coordination, etc. can be modeled in detail. The complexity of the simulator usually does not allow actual bit payload in the packets and very detailed PHY modeling. Thus, it is assumed that Resource Block (RB) resolution can provide sufficient level of accuracy. Nevertheless, it is necessary to keep sensitivity of the results to the channel conditions and demonstrate dependence on the low-level equipment configuration and algorithms. This is the main task of the link-level abstraction model, also called link-to-system interface.

### 17.2.1 LINK-TO-SYSTEM INTERFACE FOR FBMC

There are two main targets of link-level abstraction model in system-level simulator: firstly, it is calculation of Block Error Probability (BLEP), which measures the probability of whether the Protocol Data Unit (PDU) was received correctly or not; and secondly, it is Channel Quality Indicator (CQI) measurement and link adaptation based on this information [4]. The particularity of a broadband MultiCarrier (MC) system is that the channel can change considerably from one subcarrier to another due to Fast Fading (FF) and Doppler effects. Therefore, simple mathematical average of the Signal-to-Interference-plus-Noise Ratios (SINRs) over scheduled RBs cannot provide accurate information about channel and interference conditions. On the other hand, it is impractical to develop a unique link performance model specifically for a given FF channel model.

One approach to address this problem is analytical. For example, a modified version of the Shannon formula for the theoretical spectral efficiency [5] can be used to define channel capacity. Other group of methods is based on the results of link-level simulations in Additive White Gaussian Noise (AWGN) channel as the only reference of BLERs performance. The set of such results is already quite extensive because it depends on the transmission mode, i.e., at least on Modulation and Coding Scheme (MCS) and Transport Block Size (TBS).

*Effective SINR Mapping (ESM) and Mean Mutual Information per coded Bit (MMIB):* The main idea of ESM approach is to construct such value $\text{SINR}_{\text{eff}}$ that causes the same BLER in AWGN channel as the combination of $M$ individual SINRs of each RB in the fading channel. Hence, the generic expression of ESM is the following:

$$\text{BLER}_{\text{AWGN}} \{\text{SINR}_{\text{eff}}\} \approx \text{BLER} \left\{ \alpha_1 I^{-1} \left( \frac{1}{N} \sum_{m=1}^{M} I \left( \frac{\text{SINR}_m}{\alpha_2} \right) \right) \right\},$$

where $I(\cdot)$ is the information measure function to be selected, and $\alpha_{1,2}$ are fitting coefficients. In the literature, heuristic and theoretically validated approaches for the selection of $I(\cdot)$ can be found [6]. Between the most popular methods is Exponential ESM (EESM) with $I(\text{SINR}_m) = e^{\text{SINR}_m}$ [7]. The downside of this method is the necessity to use fitting coefficients for every MCS. This disadvantage was addressed in the methods based on the Mutual Information (MI), i.e., on the information shared

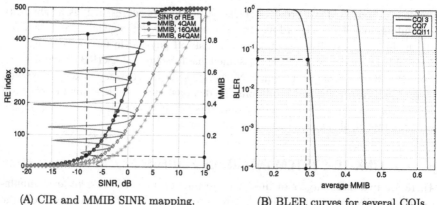

(A) CIR and MMIB SINR mapping.                    (B) BLER curves for several CQIs.

**FIGURE 17.1**

Main steps of MMIB link abstraction model.

by transmitter and receiver. Depending on the considered data type, symbol or bit information channels can be used. The latter approach is the one followed by the popular MMIB method [8].

The main steps of MMIB algorithm are presented in Fig. 17.1. Firstly, the values of MI are calculated from the SINRs in the RBs allocated to a UE. Then, the arithmetical average of the MI is found. Finally, BLER is estimated from the AWGN channel curves.

*The impact of FBMC distortion:* FBMC link-level performance in AWGN channel is the same as for Orthogonal Frequency-Division Multiplexing (OFDM) because there is no need in equalization. Therefore, it is not necessary to modify the set of AWGN BLER curves if they are already available. However, additional intrinsic interference appears when FBMC signal is transmitted over highly frequency selective channels. The main asymptotic component of the arising distortion power depends on the shape of the pulse and on the first derivative $H'_m$ of the Channel Frequency Response (CFR) $H_m$:

$$P_e[m] = P_s \left| \frac{H'_m}{H_m} \right|^2 G, \qquad (17.1)$$

where $P_s$ is the symbol power, and $G$ is a coefficient defined by the pulse autocorrelation properties [9] (see also Chapter 9). The contribution of the distortion component in terms of interference can be considerable. Therefore, a Signal-to-Interference-plus-Noise-and-Distortion Ratio (SINDR) should be used instead of the conventional SINR at the input of link abstraction model:

$$SINDR[m] = \frac{P_s}{\frac{N_0}{|H_m|^2} + P_e[m]}, \qquad (17.2)$$

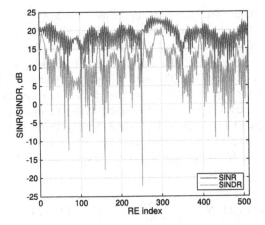

**FIGURE 17.2**

The contribution of distortion into interference at 3GPP Hilly Terrain (3GPP-HT) channel.

where $N_0$ is the noise power, and the distortion power $P_e[m]$ is defined in (17.1). As demonstrated in Fig. 17.2, the difference between SINR and SINDR when the signal goes through a 3GPP Hilly Terrain (3GPP-HT) channel can be more than 15 dB. If FBMC interference is not compensated by the equalizer, then it should be taken explicitly into account. Otherwise, the link-level abstraction algorithm is not capable of producing correct average MMIB values well described by AWGN results. With SINDR instead of SINR, the MMIB method can be used almost directly for FBMC [10].

## 17.2.2 SYSTEM-LEVEL EVALUATION OF CELLULAR FBMC NETWORK

The classical scenario for system-level evaluation of cellular networks is based on uniform hexagonal topology of BSs [11]. This choice is explained by the necessity to produce reference simulation results that can clearly demonstrate the impact of FBMC on system performance. Along with that, higher layers can be adopted directly from Long-Term Evolution (LTE) specification. Modifications in the simulation models are needed in two main aspects. Firstly, it is the utilization of FBMC technology instead of OFDM on the physical layer. This issue was already addressed in the previous section. Next, the frame structure of FBMC can be different from that used in LTE. In particular, additional TD symbol in each subframe is present in FilterBank MultiCarrier with Offset-QAM subcarrier modulation (FBMC/OQAM) due to the absence of the Cyclic Prefix (CP). These resources can be used to transmit more data or to increase the reliability by using more robust coding. Thus, TBS selection algorithm on the Medium Access Control (MAC) layer also requires changes and more flexibility.

*TBS selection:* One of the principle functions of the MAC layer is to decide upon the MCS and to select physical resources that can be allocated to the user. Then, the TBS can be defined explicitly for some given MCS index and number of UE's RBs. The corresponding DownLink (DL) and UpLink (UL) look-up tables are defined in the LTE PHY specification [12]. The idea behind this matching is to keep the same spectral efficiency for a given MCS regardless of the amount of allocated resources. The table in the standard is precalculated based on the assumption of RB consisting of 12 subcarriers by 14 TD symbols. Excluding three symbols for control channel and six resource elements for Control Symbols (CSs), we get 126 physical Resource Elements (REs) for data. If a different amount of TD symbols or subcarriers is available per RB, then TBS should be compensated to keep the same spectral efficiency:

$$\text{TB}_{\text{FBMC}}(I, N) = \left\lfloor \frac{\alpha_{\text{FBMC}}}{126} (\text{TB}_{\text{LTE}}(I, N) + 24) \right\rfloor - 24,$$

where $\text{TB}_{\text{FBMC}}(I, N)$ is the TBS in bits for FBMC with $\alpha_{\text{FBMC}}$ REs per RB available for data. It is calculated based on LTE TBS from look-up table for given Transport Block (TB) index $I$ and number of RBs $N$. The existence of 24-bit Cyclic Redundancy Check (CRC) information is also taken into account. Thus, $\text{TB}_{\text{FBMC}}$ is in average 9.62% larger than $\text{TB}_{\text{LTE}}$.

*Distortion calculation and performance results:* The modifications described above should be implemented in the LTE cellular network simulator to model the FBMC technology. For example, ns-3 [13] is an appropriate publicly available tool for this purpose that was used in this study. The full set of simulation parameters can be found in [14]. Usually, for the sake of computation time, sufficiently large but limited number of channel realizations are precalculated and stored before the main simulation loop. Since distortion calculations are dependent on channel variations, it should be computed together with FF. Then, the SINDR per RB in DL at the input of the link-to-system interface will be

$$\text{SINDR} = \frac{P_{\text{Rx}}}{I + P_e \cdot \text{SF} \cdot \text{PL} \cdot P_{\text{Tx}} + N_0},$$

where the received power $P_{\text{Rx}}$ from the serving BS is a product of slow fading (SF), path loss (PL), Power Spectral Density (PSD) of the channel $H$, and transmit power $P_{\text{Tx}}$. The distortion power $P_e$ is defined by (17.1), and the interference power $I$ is calculated as the sum of received powers of all other BSs. Note that the distortion is taken into account only for the serving BS.

The comparison of DL user throughputs of FBMC- and OFDM-based broadband networks are presented in Fig. 17.3. The combined effects of the considered modifications can be observed in this figure. Regardless of larger TBSs, throughput in the FBMC based system with simple one-tap Zero Forcing (ZF) equalizer is comparable but slightly lower than in LTE. However, the utilization of more advanced equalizers that effectively cancel FBMC distortions [9,15] improves the performance. It can be concluded that, at the system level, FBMC technology does not concede to OFDM

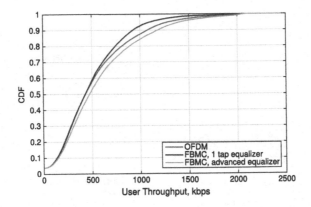

**FIGURE 17.3**

DL throughput in cellular LTE- and FBMC-based networks.

and can even surpass it when advanced equalization is used. It is also necessary to keep in mind that additional flexibility of FBMC was not taken into account. In particular, the effective bandwidth of the network can be extended because of a much better spectral containment of the FBMC signal.

## 17.3 RADIO RESOURCE MANAGEMENT OVER FBMC

The key drawbacks of CP-based multicarrier modulation schemes (like OFDM) from the Radio Resource Management perspective (RRM) are: (i) the fact that the CP reduces the spectral efficiency, and (ii) the rectangular pulse shape applied on each subcarrier that has a very large frequency response, with significant side-lobes. As a consequence, if the received interference signal is not synchronized with the received data signal, then orthogonality is lost. This generates high levels of interference, spanning on several adjacent subcarriers, as shown in [16]. Asynchronous transmissions to a receiver may take place in various scenarios, including D2D communications and multiple transmissions to a cluster head or BS. The following sections focus on these two cases.

### 17.3.1 PHY-MAC CROSS-LAYER OPTIMIZATION IN D2D TRANSMISSIONS WITH FBMC

Consider the case where D2D transmission are performed during the UL period of a multicarrier cellular system, assuming that D2D pairs are already established. Each D2D receiver is subject to several interferences coming from D2D transmitters with different clocks since each D2D transmitter is only synchronized with its own receiver (see Fig. 17.4). The time delay from each interference source cannot be

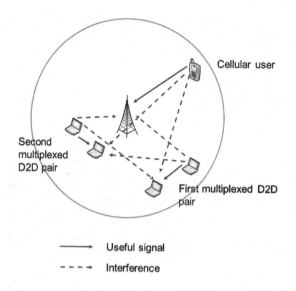

Cellular user

Second
multiplexed
D2D pair

First multiplexed D2D
pair

⟶ Useful signal

- - - → Interference

**FIGURE 17.4**

PHY-MAC cross-layer optimization in D2D transmissions with FBMC scenario.

foreseen and varies from one source to the other. It is thus necessary to consider an average interference that takes into account all possible delays. Interchannel interference due to asynchronicity is then modeled with power vectors, as they are defined in [16]. In this context, FBMC [17] is a good candidate for multicarrier transmissions since the performance of FBMC in unsynchronized environments is quite better than its opponents [18–20]. To emphasize the efficiency of FBMC in D2D communications, we now consider $K$ D2D couples occupying the same frequency band. Each D2D transmitter and its receiver are within a distance $d_{D2D}$ set to 100 m. There are $L$ subcarriers of 15 kHz each. Let $\Omega_k$ be the set of subcarriers allocated to D2D transmitter $k$. PHY-MAC optimization is performed through a cross-layer weighted sum rate maximization algorithm that takes into account the transmitter buffer sizes. Each D2D transmitter has a buffer of data to transmit to its receiver. Packets arrive in this buffer according to a Poisson law. The size of packets has a lognormal distribution. The weight $\alpha_k$ of each user $k$ is equal to its buffer size $Q_k$ relatively to all users' buffer sizes: $\alpha_k = \frac{Q_k}{\sum_{j=1}^{K} Q_j}$. In each Time Transmit Interval (TTI), once data rates have been computed, the buffer size is decreased of the data that have been transmitted. At each TTI, the objective is to maximize the weighted sum rate to avoid any buffer overflow. In the high SINR regime (with the approximation $\log_2(1 + \text{SINR}) \approx \log_2(\text{SINR})$), the weighted sum rate optimization problem is written as

$$\max_{\mathbf{P}} \sum_{k=1}^{K} \sum_{l \in \Omega_k} \alpha_k \log_2 \left( \frac{G_{k,k}^l P_k^l}{n_k^l + I_k^l} \right)$$

$$\text{subject to} \quad \sum_{l \in \Omega_k} P_k^l \leq P_{\max}, \ \forall k \in \{1, ..., K\}$$

$$\text{subject to} \quad P_k^l \geq 0, \forall k \in \{1, ..., K\}, \ \forall l \in \Omega_k, \tag{17.3}$$

where $G_{k,j}^l$ is the channel gain between D2D transmitter $j$ and D2D receiver $k$ (including path loss, shadowing, and flat fading) in subcarrier $l$, $I_k^l$ is the total interference at receiver $k$ in subcarrier $l$,

$$I_k^l = \sum_{\substack{j=1 \\ j \neq k}}^{K} \sum_{l' \in \Omega_j} G_{k,j}^{l'} V_{|l-l'|} P_j^{l'}, \tag{17.4}$$

and $n_k^l$ is the noise power. $\mathbf{P} = \left[ P_k^l \right]_{\substack{k \in \{1, ..., K\} \\ l \in \{1, ..., L\}}}$ is the matrix of transmit powers of D2D transmitters in all subcarriers. Resource allocation is separated into two steps, users clustering and power allocation. Users clustering aims at achieving low interference levels through graph coloring. Then D2D power is allocated under the high SINR assumption.

### 17.3.1.1 *D2D Clustering With Graph Coloring*

D2D couples are grouped within clusters where no D2D transmitter highly interferes the other D2D receivers. To determine these clusters, graph coloring is used. Let $G = (V, E)$ be a graph without any loop, defined by its vertices $V$ and its edges $E$. Graph coloring assigns colors to the vertices such that no two adjacent vertices share the same color. It has been widely used in the literature to mitigate interference in multicell scenarios [21,22] and in D2D scenarios [23,24]. In the studied case, vertices are all D2D transmitters. They are adjacent, which means that they are connected by a vertex if they are not allowed to transmit in the same subcarriers because they would be interfering each other's receivers too heavily. A D2D transmitter $n$ is supposed to highly interfere a D2D receiver $k$ if their distance is lower than a given threshold $D_{\text{int}}$. This implies that transmitter $n$ and the transmitter corresponding to receiver $k$ are forbidden to transmit in the same subcarriers. Once all edges have been identified, graph-coloring is performed, for instance, with greedy Degree SATURation (DSATUR) algorithm [25]. After graph coloring, each transmitter is assigned a color that corresponds to a cluster of subcarriers. Let $N_C$ be the number of colors. Then the bandwidth is split into $N_C$ clusters composed of $L_{\text{per cluster}} = \left\lfloor \frac{L}{N_C} \right\rfloor$ adjacent subcarriers. Each D2D transmitter $k$ is allocated all the subcarriers in its cluster $\Omega_k$.

### 17.3.1.2 *Weighted Sum Rate Maximization Algorithm*

The proposed power allocation algorithm is an adaptation of the Dual Asynchronous Distributed Pricing (DADP) algorithm for the multichannel case from Huang et al. [26] when interchannel interference is taken into account. It can be shown that the interference information per user with interchannel interference is then equal to

$$\Pi_k^l(\mathbf{P}^*) = \sum_{l' \in \Omega_k} \frac{\alpha_k V_{|l-l'|}}{n_k^{l'} + I_k^{l'}(\mathbf{P}^*)}. \tag{17.5}$$

Similarly to [26,27], the sum power constraint is relaxed by introducing a dual price per BS, $\mu_k$. In the dual space, the initial problem is separated into $L$ problems, one per subcarrier. The iterative algorithm performs power control at each subcarrier independently, taking into account the dual prices. The dual prices are then updated, depending on whether the sum power constraint per BS is fulfilled or not. $\kappa$ is the step for the dual price evaluation. The following iterative algorithm is used:

*Initialization*: at $T_d = 0$, set the initial power and price for all subcarriers $l$ and all users $k \in \Omega_l$, and the initial dual price per BS $\mu_k(0) \geq 0$ for all $k \in \{1, ..., K\}$.

*Iterative process*:

1. Dual price update: at each iteration $T_d$, each user $k \in \{1, ..., K\}$ updates its dual price according to

$$\mu_k(T_d) = \left[ \mu_k(T_d - 1) + \kappa \left( \sum_{l \in \Omega_k} P_k^l(T_d - 1) - P_{\max} \right) \right]^+, \tag{17.6}$$

where $[a]^+ = \max\{0, a\}$.

2. Iterative power and interference information update: for a given dual price setting, an iterative process is used independently on each subcarrier $l \in \{1, ..., L\}$.

*Initialization*: at iteration $T = 0$, set $P_k^l(0) = \frac{P_{\max}}{L_{\text{per cluster}}}$ for all $k \in \Omega_l$ and compute the corresponding interference information $\Pi_k^l(0)$ using (17.5).

*Iterative process*:

(a) Power update: Compute the power of each user $k \in \Omega_l$, depending on its channel state, weight, the interference information of the previous iteration, and on its dual price $\mu_k(T_d)$:

$$P_k^l(T + 1) = \frac{\alpha_k}{\left( \sum_{\substack{j=1 \\ j \neq k}}^{K} G_{n,k}^l \Pi_n^l(\mathbf{P}(T)) + \mu_k(T_d) \right)}. \tag{17.7}$$

(b) Interference information update: Compute the interference information of each user $k \in \Omega_l$, depending on its weight and received noise plus interference,

$$\Pi_k^l(\mathbf{P}(T + 1)) = \sum_{l' \in \Omega_k} \frac{\alpha_k V_{|l-l'|}}{n_k^{l'} + I_k^{l'}(\mathbf{P}(T + 1))}. \tag{17.8}$$

The initial problem belongs to the class of geometric programming and thus has a unique solution. It can be shown that the iterative algorithm does converge. Since any

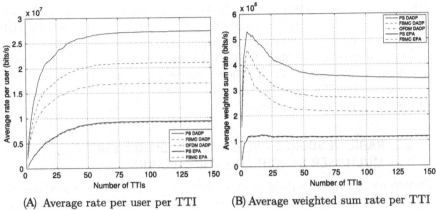

(A) Average rate per user per TTI      (B) Average weighted sum rate per TTI

**FIGURE 17.5**

Average rate per user per TTI and weighted sum rate per TTI.

obtained set of power values **P** then fulfill the Karush–Kuhn–Tucker conditions of the initial problem (17.3), it corresponds to its global optimum [28]. This algorithm can be further improved to include per Resource Block power control and additional constraints, such as an interference constraint at the Base Station in the context of underlay D2D communications [29].

### 17.3.1.3 *Simulation Results*

The proposed algorithm's performances are assessed with the following parameters: the maximum distance for graph coloring is $D_{int} = 250$ m, the path loss model is $PL = 140 + 36.8 \log_{10}(d)$, where $d$ is expressed in km, shadowing follows a log-normal distribution with standard deviation 4 dB, and the multipath channel adds Rayleigh fading. Only thermal noise with spectral power density $-174$ dBm/Hz is considered. The packet size's mean is 5 kbits, the packet size's standard deviation is 1 kbits/s, the average time packet arrival is 20 TTIs, and the total number of TTIs is 200. A transmitter is in buffer overflow if its buffer size exceeds 1024 kbits. In this case, any new incoming packet is discarded. A TTI lasts 1 ms, and there are $L = 8$ subcarriers with bandwidth 15 kHz. $K = 8$ D2D pairs are located in a cell with radius $R = 1$ km having an omnidirectional antenna. The D2D transmitters' angle with the BS is between 0 and $\pi/4$. Figs. 17.5A and 17.5B show the average rate per user and the weighted sum rate per TTI, respectively. The proposed method (referred to as DADP on the figures) is compared with an equal power allocation algorithm (referred to as EPA on the figures), where after graph coloring, the power per subcarrier is simply set to $P_k^l = \frac{P_{max}}{L_{per\ cluster}}$ in each subcarrier $l$ in the cluster of user $k$. Fig. 17.5A shows that users rates reach a stable state in approximately the 150th TTI, which corresponds to the moment when all users are in full buffer. In this case, the average rate degradation compared to the perfect synchronization case is 23% with

FBMC and 38% with OFDM when the DADP algorithm is used. Using equal power allocation also decreases the rate of 66%, which proves the effectiveness of DADP. In Fig. 17.5B, we can see that the weighted sum rate is maximized even with varying buffer sizes with DADP and that FBMC allows us to mitigate the rate loss brought by asynchronicity thanks to lower interchannel interference. As a consequence, FBMC is a good candidate for future physical layer of 5th Generation (5G) systems when considering D2D communications.

## 17.3.2 RADIO RESOURCE OVERHEAD FOR COMPENSATING MULTIUSER SYNCHRONIZATION ERRORS

Consider the case where asynchronous transmissions are performed from a set of devices to a receiving node. When no close-loop procedures or interference cancellation schemes for compensating multiuser synchronization errors are used, potential multiuser synchronization errors are compensated by applying guard subcarriers. The actual radio resource overhead introduced to compensate synchronization errors in this case is due to three factors: (i) the CP added (when the multicarrier waveform is CP-based), (ii) the adopted Carrier Assignment Scheme (CAS), i.e., the resource allocator, and (iii) the diversity of the synchronization error among the multiple transmitters in the system. The latter factor refers to the fact that the synchronization error is not the same for all the devices that transmit in neighboring subcarriers. Let the transmissions to be organized in subframes in the time domain. For each subframe, a central node is in charge of allocating RBs to devices (i.e., in charge of performing the CAS) and of inserting guard subcarriers between transmissions of different devices. For the aforementioned scenario, the radio resource overhead due to the use of CP (denoted here by $CP_{loss}$) is given by

$$CP_{loss} = \frac{CP_{length}}{Sym + CP_{length}}, \tag{17.9}$$

where $CP_{length}$ and $Sym$ are the duration of the CP and the data symbol, respectively. As it can be observed in (17.9), the radio resource overhead due to the CP is fixed and independent of the actual asynchronism level. However, in asynchronous communication scenarios, the proactive selection of the CP size is a challenging problem [30]. The calculation of the radio resource overhead due to the CAS, i.e., the way that the RBs are allocated to devices, is more complicated. To deal with this challenge, we model the subcarrier allocation as a set of simpler allocations, where in each one of them, a specific allocation unit is used. This is a convenient approach since after any allocation, the available bandwidth can be represented by a conjunction of $X$ independent subbands, where in subband $x$ ($x = 1, 2, 3, \ldots, X$), a specific allocation unit of $p(x)$ RBs has been used (recall that, according to the adopted system model, the minimum allocation unit is one RB, but more than one contiguous RBs can be allocated to a device). To give an example, in Fig. 17.6, a generic CAS has assigned the available spectrum to six devices with $X = 3$, $p(1) = 1$, $p(2) = 2$, $p(3) = 4$.

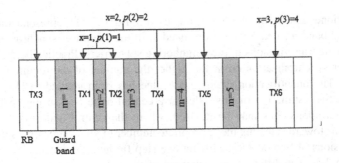

**FIGURE 17.6**

Example of a generic CAS and the representative subbands.

Consider now a fragmented spectrum of $BW_{RB}$ RBs. Let $N_{RB}$ be the total number of RBs used for data transmission in a subframe, and $N_{RB}(x)$ the number of RBs used for data transmission in subband $x$. Also, let each RB be composed by $RB_{sc}$ contiguous subcarriers. Clearly, $BW_{RB} \geq N_{RB}$, and $\sum_{x=1}^{X} N_{RB}(x) = N_{RB}$. Consequently, the number of guard bands needed for a subframe, denoted here by $N_{GB}$, is calculated as follows:

$$N_{GB} = \sum_{x=1}^{X} \frac{N_{RB}(x)}{p(x)} - 1. \tag{17.10}$$

In the general case, the number of guard subcarriers in each guard band is related to the specific devices that transmit to neighboring subcarriers since the distribution of the devices in space strongly affects the synchronization errors, i.e., the required guard size. Let $n_{Gsc}(m)$ denote the number of guard subcarriers needed in guard band $m$, $m = 1, 2, 3, \ldots, N_{GB}$. The total number of guard subcarriers needed to fill all the guard bands, denoted here by $N_{Gsc}$, is simply given by

$$N_{GB} = \sum_{m=1}^{N_{RB}} n_{Gsc}(m). \tag{17.11}$$

Consequently, the overhead due to the guard subcarriers, denoted here by $GS_{loss}$ is as follows:

$$GS_{loss} = \frac{N_{Gsc}}{N_{Gsc} + N_{RB}RB_{sc}} \tag{17.12}$$

$$\text{s.t.} \quad BW_{RB}RB_{sc} \geq N_{RB}RB_{sc} + N_{Gsc}. \tag{17.13}$$

Taking into account Eqs. (17.9) and (17.11), the overall radio resource overhead is given by

$$Overhead = GS_{loss} + CP_{loss}(1 - GS_{loss}). \tag{17.14}$$

As mentioned above, the $n_{\text{Gsc}}(m)$ value used in Eq. (17.11) represents the length of the guard band $m$, i.e., a value correlated with the multiuser synchronization error between the transmissions in the neighboring subcarriers. Practically, the higher the multiuser synchronization error, the higher the number of the required guard subcarriers. The quantification of $n_{\text{Gsc}}(m)$ value has been studied in [16]. However, the proactive estimation of this value at a central node is a strong challenge. Sophisticated approaches may be devised, where the central node may exploit past allocations toward selecting the appropriate number of guard subcarriers between the transmissions of two devices. Moving one step further, the dependence of $n_{\text{Gsc}}(m)$ on $m$ implies that different permutations of the CAS allocation may have different impact on the radio resource overhead. For instance, users with smaller synchronization deviations may be given neighboring subcarriers, and, consequently, shorter guard bands could be used. To find the permutation that minimizes the number of guard subcarriers (maximizes reallocation gain), an additional procedure should be added at the resource allocator. This procedure should go through all the available permutations and calculate the reallocation gain for each of them. To face the problem of generating all the permutations, the lexicographic order [31] can be used. Overall, the exploitation of the multiuser diversity is a vehicle for an OFDM-based system to reduce the number of guard subcarriers. However, it requires the consumption of extra processing and radio resources, making the compensation of the synchronization errors much more complicated. Additionally, if we consider the multiuser channel diversity, then a potential reallocation may degrade the system's throughput performance. Note that for other multicarrier waveforms such as the FBMC, the problem is much more straightforward. As explained in [32], in the FBMC case, one guard subcarrier is adequate, i.e., $n_{\text{Gsc}}(m) = 1$ for all $m$.

### 17.3.2.1 *Evaluation Results*

Assume that a fixed amount of data resources equal to $N_{\text{RB}} = 75$ RBs is requested from the UEs, whereas the available bandwidth is fragmented and equal to 100 RBs. We consider a resource allocation scheme where each RB is allocated to a different UE. Fig. 17.7 depicts the radio resource overhead for different CPs and guard band lengths, indicating the performance of the OFDM for LTE-compliant values (loss $= 0.075$ and loss $= 0.20$ for normal and extended CP, respectively). As it can be observed in Fig. 17.7, the radio resource overhead ranges from 0.2 (for CP length equal to 2% of the symbol length and guard subcarriers 25% of RB size) to 0.6 (for CP length 20% of the symbol length and guard subcarriers 100% of RB size). The result in Fig. 17.7 may be considered as a comprehensive quantification of the maximum radio resource overhead required by any CP-based scheme since guard subcarriers are allocated next to each RB that carries data. It is worth noting that, due to the fact that the channel gain is not the same for the whole bandwidth and for all the UEs, the radio resource overhead in the physical layer (depicted in Fig. 17.7) may not be reflected linearly to throughput losses in the MAC layer. To quantify the impact of the guard subcarriers to throughput performance, we consider OFDM with nor-

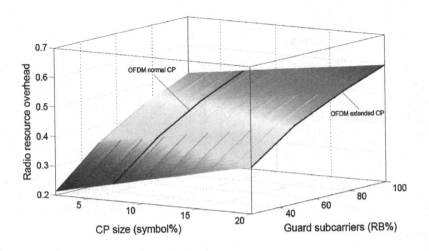

**FIGURE 17.7**

Radio resource overhead for various CPs and guard lengths.

mal CP in the simulation environment described in [33]. An urban multicell network with 20 uniformly distributed devices per cell was selected. We also considered two different CAS approaches, namely, the Round Robin (RR) and the Proportional Fair (PF), for the allocation of in total 50 RBs. The former approach is a channel-blind allocator that divides the available RB to equally sized portions and allocates them to UEs, whereas the latter is a channel-aware allocator where each RB is assigned to the UE with the best expected data rate moderated by a fairness factor. For both cases, two scenarios are defined, regarding the size of the guard band: scenario A, where one guard subcarrier is used, and scenario B, which uses six guard subcarriers (referring to reasonable choices for the FBMC and OFDM case, respectively, as explained in [16]).

Fig. 17.8 depicts the throughput performance for different Signal-to-Noise Ratio (SNR) values for the RR and PF allocators. As expected, for both allocators and scenarios, the lower the SNR, the lower the throughput losses are. As it can be observed in Fig. 17.8, the behavior of the RR allocator leads to an almost linear relation between throughput and SNR, whereas the higher throughput losses are up to 17%. The result in Fig. 17.8 shows that when a channel-aware allocator is used, the impact on the throughput performance is much more severe. Although the overall throughput is higher than in the case of the RR allocator, when the SNR is high, the PF allocator can lead to a per RB allocation and thus to the need of using guards for each transmitting RB. This explains the decreasing part of the curve that represents scenario B (dashed line) in Fig. 17.8.

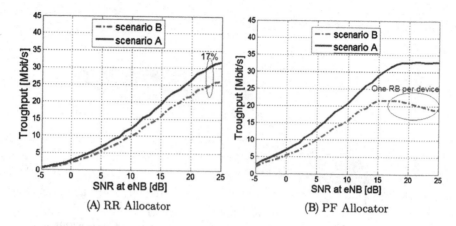

(A) RR Allocator  (B) PF Allocator

**FIGURE 17.8**

Throughput performance of the RR and the PF allocators.

## 17.4 CONCLUDING REMARKS

This chapter offers a comprehensive study on FBMC from the RRM perspective for multiuser environments. More specifically, insights on system-level evaluation of FBMC have been described, providing the reader with a useful guide on how to map link-level simulations to system-level ones. Additionally, the RRM procedure for asynchronous scenarios has been examined using system-level simulations. For the scenarios under study, FBMC is proven to be highly robust in terms of radio resource consumption, endorsing the argument that FBMC is a good candidate for 5G communications.

## REFERENCES

[1] A. Virdis, N. Iardella, G. Stea, and D. Sabella, "Performance analysis of OpenAirInterface system emulation," in *Proc. FiCloud 2015*, Rome, Italy, Aug. 2015.

[2] O. Font-Bach, N. Bartzoudis, X. Mestre, D. López, P. Mège, L. Martinod, V. Ringset, and T. A. Myrvoll, "When SDR meets a 5G candidate waveform: Providing efficient use of fragmented spectrum and interference protection for PMR networks," in *Proc. IEEE Wireless Communications*, vol. 22, no. 6, pp. 56–66, Jan. 2015.

[3] K. Wehrle, M. Günes, and J. Gross, *Modeling and Tools for Network Simulation*. Springer Science & Business Media, 2010.

[4] M. Mezzavilla, M. Miozzo, M. Rossi, N. Baldo, and M. Zorzi, "A lightweight and accurate link abstraction model for the simulation of LTE networks in ns-3," in *Proc. ACM MSWiM 2012*, Paphos, Cyprus Island, Aug. 2012.

[5] P. Mogensen, W. Na, I. Kovács, F. Frederiksen, A. Pokhariyal, K. I. Pedersen, T. Kolding, and K. H. M. Kuusela, "LTE capacity compared to the Shannon bound," in *Proc. VTC2007-Spring*, Dublin, Ireland, Apr. 2007.

[6] D. Petrov, P. Gonchukov, and T. H. Stitz, "Link to system mapping for FBMC based systems in SISO case," in *ISWCS 2013*, Ilmenau, Germany, Aug. 2015.

[7] "System-level evaluation of OFDM – further considerations," in *3GPP TSG-RAN WG1 R1-031303*, Nov. 2003.

[8] R. Srinivasan, J. Zhuang, L. Jalloul, R. Novak, and J. Park, "IEEE 802.16m evaluation methodology document," in *IEEE 802.16 Broadband Wireless Access Working Group*, 2008.

[9] X. Mestre, M. Majoral, and S. Pfletschinger, "An asymptotic approach to parallel equalization of filter bank based multicarrier signals," *IEEE Trans. Signal Process.*, vol. 61, no. 14, pp. 3592–3606, 2013.

[10] D. Petrov, A. Oborina, L. Giupponi, and T. H. Stitz, "Link performance model for filter bank based multicarrier systems," *EURASIP Journal on Advances in Signal Processing*, vol. 2014, 2014.

[11] "LTE; Evolved Universal Terrestrial Radio Access (E-UTRA); Radio Frequency (RF) system scenarios," in *3GPP TR 36.942 version 8.2.0 Release 8*, Jul. 2009.

[12] "LTE; E-UTRA; Physical layer procedures," in *3GPP TS 36.213 version v8.8.0*, Oct. 2009.

[13] "Network simulator version 3 (ns-3)." [Online]. Available: https://www.nsnam.org/.

[14] D. Petrov, B. Herman, T. Hämäläinen, and S. Melnik, "On the system level performance of cellular FBMC-based wideband PMR network," in *ISWCS 2015*, Brussels, Belgium, Aug. 2015.

[15] T. Ihalainen, T. H. Stitz, M. Rinne, and M. Renfors, "Channel equalization in filter bank based multicarrier modulation for wireless communications," *EURASIP Journal on Applied Signal Processing*, vol. 2007, no. 1, 2007.

[16] Y. Medjahdi, M. Terre, D. L. Ruyet, D. Roviras, J. A. Nossek, and L. Baltar, "Inter-cell interference analysis for OFDM/FBMC systems," in *Proc. 10th IEEE Signal Processing Workshop (SPAWC 2009)*, Perugia, Italy, Jun. 2009.

[17] M. G. Bellanger, "Specification and design of a prototype filter for filter bank based multicarrier transmission," in *Proc. ICASSP 2001*, Salt Lake City, UT, USA, May 2001.

[18] M. Shaat and F. Bader, "Computationally efficient power allocation algorithm in multicarrier-based cognitive radio networks: OFDM and FBMC systems," *EURASIP Journal on Advances in Signal Processing*, vol. 2010, 2010, article ID 528378.

[19] H. Zhang, D. L. Ruyet, D. Roviras, Y. Medjahdi, and H. Sun, "Spectral efficiency comparison of OFDM/FBMC for uplink cognitive radio networks," *EURASIP Journal on Advances in Signal Processing*, vol. 2010, 2010, article ID 621808.

[20] M. Pischella, D. L. Ruyet, and Y. Medjahdi, "Sum rate maximization in asynchronous ad hoc networks: Comparison of multi-carrier modulations," in *Proc. ISWCS 2013*, Ilmenau, Germany, Aug. 2013.

[21] M. Pischella and J.-C. Belfiore, "Graph-based weighted sum throughput maximization in OFDMA cellular networks," in *Proc. IWCLD 2009*, Palma de Mallorca, Spain, Jun. 2009.

[22] A. Lamiable and J. Tomasik, "Spatial frequency reuse in a novel generation of PMR networks," in *Proc. WCNC 2013*, Shanghai, China, Apr. 2013.

[23] D. Tsolkas, E. Liotou, N. Passas, and L. Merakos, "A graph-coloring secondary resource allocation for D2D communications in LTE networks," in *Proc. CAMAD 2012*, Barcelona, Spain, Sep. 2012.

[24] C. Lee, S.-M. Oh, and J.-S. Shin, "Resource allocation for device-to-device communications based on graph-coloring," in *Proc. ISPACS 2015*, Bali, Indonesia, Nov. 2015.

[25] D. Brélaz, "New methods to color the vertices of a graph," *Communications of the ACM*, vol. 22, no. 4, pp. 251–256, Apr. 1979.

[26] J. Huang, R. Berry, and M. Honig, "Distributed interference compensation for wireless networks," *IEEE J. Sel. Areas Commun.*, vol. 24, no. 5, pp. 1074–1084, May 2006.

[27] M. Pischella and J.-C. Belfiore, "Weighted sum throughput maximization in multicell OFDMA networks," *IEEE Trans. Veh. Technol.*, pp. 896–905, Feb. 2010.

[28] S. Boyd and L. Vanderbergue, *Convex Optimization*. Cambridge University Press, 2004.

[29] M. Pischella, R. Zakaria, and D. Le Ruyet, "Resource Block-level power allocation in asynchronous multi-carrier D2D communications," *IEEE Commun. Lett.*, pp. 813–816, Apr. 2017.

[30] G. Wunder, *et al.*, "5GNOW: Non-orthogonal, asynchronous waveforms for future mobile applications," *IEEE Commun. Mag.*, vol. 52, no. 2, pp. 97–105, Feb. 2014.

[31] R. Sedgewick, "5GNOW: Non-orthogonal, asynchronous waveforms for future mobile applications," *Computing Surveys*, vol. 9, no. 2, pp. 137–164, Jun. 1977.

[32] "Enhanced multicarrier techniques for professional Ad-hoc and Cell-based communications – Cooperative communications and synchronization." FP7-ICT 318362 EMPhAtiC project, Deliverable 6.1, Aug. 2013.

[33] "Vienna LTE simulator." LTE-A Uplink Link Level Simulator. [Online]. Available: www.nt.tuwien.ac.at/ltesimulator, 2015.

# Implementation Aspects

# Power Amplifier Effects and Peak-to-Average Power Mitigation

**Krishna Bulusu\*, Hmaied Shaiek\*, Daniel Roviras\*, Rafik Zayani\*,**
**Markku Renfors†, Lauri Anttila†, Mahmoud Abdelaziz†**

*CEDRIC Laboratory, CNAM, Paris, France\**
*Tampere University of Technology, Tampere, Finland†*

## CONTENTS

*Orthogonal Waveforms and Filter Banks for Future Communication Systems.* DOI: 10.1016/B978-0-12-810384-5.00018-9

## 18.1 INTRODUCTION

Multicarrier systems suffer from high Peak-to-Average Power Ratio (PAPR). If the High-Power Amplifier (HPA) is operated in its quasilinear region, the high PAPR of the multicarrier signal has no influence on the quality of the transmission. Nevertheless, this situation has a high cost in terms of energy efficiency, especially, for mobile applications with batteries. To increase power efficiency, the HPA should be operated as close as possible to its saturation point, but this would introduce broadening of the amplified signal spectrum and a distortion over the transmitted signal itself.

For FilterBank MultiCarrier (FBMC) signals, spectral broadening is a more critical drawback compared to Orthogonal Frequency-Division Multiplexing (OFDM). Indeed, the Power Spectral Density (PSD) of an FBMC signal is supposed to have very narrow transition bands around the active passband region. If NonLinear (NL) HPA is used, then the advantage of FBMC compared to OFDM, in terms of frequency localization, is highly diminished.

In this chapter, we first review some common HPA nonlinearity models. Then, we study the impairments related to the NL HPA in terms of In-Band (IB) NL distortion and Out-Of-Band (OOB) spectral regrowth. Finally, we present how to mitigate HPA NL effects by linearizing the HPA conversion characteristics through Digital PreDistortion (DPD) and/or by reducing the PAPR of the FBMC signal.

## 18.2 HPA MODELS AND CHARACTERISTICS

In this section, we introduce the transmission scheme and describe the main characteristics of the HPA in terms of amplitude distortion, phase distortion, and memory effects. The multicarrier transceiver considered in this chapter is shown in Fig. 18.1.

The complex envelope of the signal at the input of HPA can be written as

$$i_h(t) = \rho(t)\exp\big(j\varphi(t)\big), \qquad (18.1)$$

where $\rho(t)$ and $\varphi(t)$ are respectively the input signal modulus and phase.

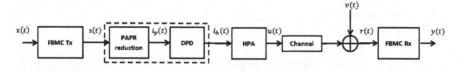

**FIGURE 18.1**

The transmission system model.

The HPA is commonly described by its input/output or conversion characteristics. Here we can distinguish two zones, linear and nonlinear one. In the linear zone, the signal is amplified with very good linearity, albeit with low energy efficiency. On the contrary, the NL region has high energy efficiency, but the amplified signal is distorted. The concept of energy efficiency can be understood by defining the power efficiency of an HPA $\eta$ as the part of Direct Current (DC) power ($P_{DC}$) that is converted to Radio Frequency (RF) power ($P_u$):

$$\eta = \frac{P_u}{P_{DC}}. \qquad (18.2)$$

The amplified signal $u(t)$ can be written as

$$u(t) = F_a\big(\rho(t)\big) \exp\big(jF_p\big(\rho(t)\big)\big) \exp\big(j\varphi(t)\big) = \widehat{S}\big(\rho(t)\big) \exp\big(j\varphi(t)\big), \qquad (18.3)$$

where $F_a(\rho(t))$ and $F_p(\rho(t))$ are the Amplitude Modulation/Amplitude Modulation (AM/AM) and Amplitude Modulation/Phase Modulation (AM/PM) characteristics of the HPA, respectively, and $\widehat{S}(\rho(t)) = F_a(\rho(t)) \exp(jF_p(\rho(t)))$ is the complex envelope of the amplified signal $u(t)$.

In practice, to avoid or at least to reduce the effects of nonlinearities, the HPA is operated at a given Input Back-Off (IBO) from its saturation point. In the log scale, the IBO is defined as

$$\text{IBO} = 10 \log_{10} \left( \frac{P_{\text{sat}}}{P_{i_h}} \right), \qquad (18.4)$$

where $P_{\text{sat}}$ is the input saturation power and $P_{i_h}$ is the mean input signal power.

## 18.2.1 SOME HPA MODELS

The HPA models can be broadly classified into two categories, memoryless HPA models and those with memory. In frequency domain, the zero-memory nonlinearity implies that the transfer characteristics are frequency independent. However, for wideband signals, the frequency dependency of the HPA characteristics becomes important, giving rise to so-called memory effects in the response of the amplifier. A detailed discussion of nonlinearities with and without memory can be found in [33]. In the following, we will introduce some HPA models for later use in this chapter. The focus is on baseband behavioral or black-box models, which lend themselves well to both analytic HPA modeling and predistortion studies [23]. For simplicity, we assume that the gain of the HPA (in the linear region) is 1.

### 18.2.1.1 Memoryless HPA Models

*Rapp model*: This model, commonly used for modeling Solid-State Power Amplifiers (SSPAs), was presented in [50] and exhibits only AM/AM conversion. It can be

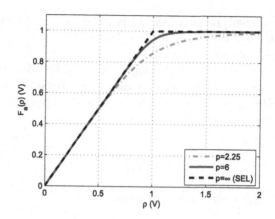

**FIGURE 18.2**

AM/AM characteristics of SEL and Rapp HPA models.

expressed as

$$F_a\big(\rho(t)\big) = \frac{\rho(t)}{\left(1 + \left(\frac{\rho(t)}{v_{\text{sat}}}\right)^{2p}\right)^{\frac{1}{2p}}}, \qquad (18.5)$$

where $p$ is a smoothness factor that controls the transition from the linear region to the saturation region ($p > 0$). As $p \to \infty$, the Rapp model converges toward a Soft Envelope Limiter (SEL) model. In Fig. 18.2, we plot the AM/AM characteristics of the SEL model and for the Rapp model with two values of the smoothness factor $p$.

*Saleh model*: Saleh's model is generally used for modeling Traveling Wave Tube Amplifiers (TWTA) [53]. According to this model, the AM/AM and AM/PM conversion characteristics can be expressed as follows:

$$F_a\big(\rho(t)\big) = v_{\text{sat}}^2 \frac{\rho(t)}{\rho(t)^2 + v_{\text{sat}}^2}, \qquad (18.6)$$

$$F_p\big(\rho(t)\big) = \varphi_0 \frac{\rho(t)^2}{\rho(t)^2 + v_{\text{sat}}^2}, \qquad (18.7)$$

where $\varphi_0$ controls the maximum phase distortion introduced by this HPA model.

In Figs. 18.3 and 18.4, we plot the AM/AM and AM/PM characteristics of the Saleh model for $v_{\text{sat}} = 1$ V and $\varphi_0 = \pi/3$.

*Polynomial model*: For the purpose of theoretical analysis of NL HPA effects, it is often suitable to approximate the complex soft envelope of the amplified signal $\widehat{S}(\rho(t))$ with a polynomial model. In this case, the signal $u(t)$ at the output of the NL device

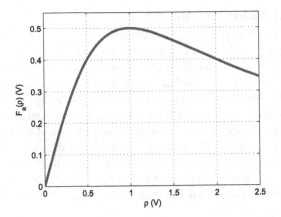

**FIGURE 18.3**

AM/AM characteristics of Saleh HPA model with $v_{sat} = 1$ V.

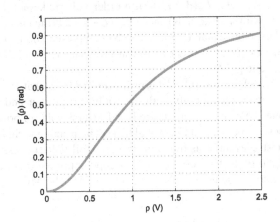

**FIGURE 18.4**

AM/PM characteristics of Saleh HPA model with $v_{sat} = 1$ V and $\varphi_0 = \pi/3$.

can be written as

$$u(t) = \sum_{n=1}^{L} a_n i_h(t) \left| i_h(t) \right|^{n-1}, \qquad (18.8)$$

where $L$ is the polynomial order, and $a_n$ are the complex coefficients of the polynomial approximation.

### 18.2.1.2 *HPA Models With Memory*

As the signal bandwidth becomes wider, RF power amplifiers begin exhibiting frequency-dependent behavior, often called *memory effects* [33]. Most baseband HPA memory models in the literature are parametric models, including those based on *basis function representations,* such as the Volterra series model, and *modular structures,* such as the Wiener and Hammerstein models [23,39,55].

The models based on basis function representations consider linear combinations of basis waveforms created from nonlinear transformations of the HPA input signal. The most simple example of this class of HPA models is the memoryless polynomial in (18.8). On the other end of the complexity spectrum, there is the Volterra series model, which can be used to represent a wide range of (mild) nonlinearities [55]. The truncated baseband Volterra model for bandpass nonlinearities is defined as [7,69]

$$u[l] = \sum_{p=0}^{P}\sum_{d_1=0}^{D}\cdots\sum_{d_{2p+1}=0}^{D} h_{2p+1}(\mathbf{d}_{2p+1})\prod_{i=1}^{p+1} i_h[l-d_i]\prod_{i=p+2}^{2p+1} i_h^*[l-d_i], \quad (18.9)$$

where $\mathbf{d}_n = [d_1, d_2, ..., d_n]$, $h_n(\mathbf{d}_n)$ is the $n$th-order Volterra kernel, $D$ is the memory depth (assumed equal for all kernels), and $P = (L-1)/2$ is the number of considered nonlinearity orders. Notice that the above baseband representation considers only odd nonlinearity orders since even-order nonlinear products are generally far away from the fundamental zone [33,69].

A major drawback of the full Volterra model (18.9) is that the number of basis waveforms grows exponentially with the nonlinearity order and memory depth. Therefore, simpler HPA models are desired, and several Volterra pruning techniques have been proposed to this end [17,20,35,41,71]. The pruning techniques aim to extract only the most relevant basis functions from the full Volterra basis set, e.g., based on physical insights of the HPA circuits [71]. A popular pruned version of the Volterra model is the Memory Polynomial (MP) model [20,35]. The MP model retains only the diagonal entries ($d_1 = d_2 = \cdots = d_{2P+1}$) and is defined as

$$u[l] = \sum_{p=0}^{P}\sum_{d=0}^{D} h_{2p+1}(d)i_h[l-d]^{p+1}i_h^*[l-d]^{p}$$

$$= \sum_{p=0}^{P}\sum_{d=0}^{D} h_{2p+1}(d)i_h[l-d]\big|i_h[l-d]\big|^{2p}. \tag{18.10}$$

The MP model has been found to provide a good trade-off between complexity and accuracy in both direct and inverse modeling of HPAs [3,20,29]. The greatest benefit of the basis function representation is that it provides linear-in-parameters models, and therefore the parameter identification can be performed with simple well-known algorithms, such as Least Squares (LS) fitting or Least Mean Squares (LMS) type of adaptive algorithms. The drawback is that a large number of parameters may be required for an accurate fit.

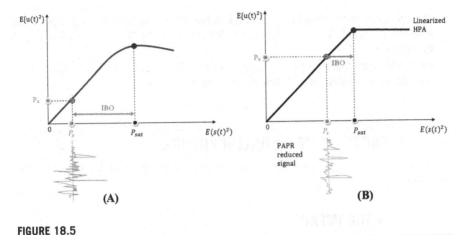

**FIGURE 18.5**

Linearized HPA and input signal with lowered PAPR.

Another popular class of HPA memory models are modular structures, which involve different combinations of linear time-invariant filters and (typically) static nonlinearities. The most elementary two-box models are the Wiener and Hammerstein models, which are formed by cascading a linear filter and a memoryless nonlinearity, or vice versa, respectively [23,55]. More elaborate two- or three-box models have been proposed, for example, in [24,40]. The benefit of such models is that the constituent blocks can be quite simple and therefore the overall number of model parameters stays reasonable. The downside is that the parameter identification procedure inevitably becomes more complex.

### 18.2.2 HPA EFFECTS AND NEED FOR NL EFFECTS MITIGATION

HPA linearity and energy efficiency are two vital parameters in the context of signals with strong fluctuations, as it is the case in MultiCarrier Modulation (MCM) techniques. To get rid of the amplified signal distortion, the HPA can be made to operate at high IBO, which leads to very poor energy efficiency. The fluctuation of the MCM signal envelope can be well understood by its PAPR. The PAPR of a continuous-time base-band signal $s(t)$ during a multicarrier symbol period $T$ is defined by

$$\text{PAPR}_{s(t)} = \frac{\max_{0 \leq t \leq T} |s(t)|^2}{\frac{1}{T} \int_0^T |s(t)|^2 dt}. \tag{18.11}$$

The presence of high peaks causes IB and OOB interferences when the MCM signal is passed through an HPA, which does not have enough linear range. Looking at Fig. 18.5, we can identify two approaches for mitigating the NL effects of the HPA at the transmitter side:

- we can decrease the distribution of high values of the instantaneous power of the transmitted signal. This is equivalent to lowering the PAPR of $s(t)$, as illustrated in Fig. 18.5;
- we can increase the HPA linear range by using HPA linearization techniques. This is equivalent to increasing $P_s$ toward $P_{sat}$ (see Fig. 18.5).

## 18.3 IMPACTS OF HPA NONLINEARITIES

In this section, we study the in-band and out-of-band impacts of NL HPA over FBMC waveforms.

### 18.3.1 IN-BAND IMPACT

A theoretical characterization of NL effects on OFDM systems was presented in [18], where the authors focused on the impact of the distortions induced by three HPA models: SEL, Rapp, and Saleh. Other contributions [46,47] used the results presented in [18] to study the effect of HPA on MIMO transmit diversity systems.

Let us consider the FBMC transceiver with memoryless NL HPA as shown in Fig. 18.1. Then, the signal $r(t)$ at the demodulator input can be written as

$$r(t) = h(t) \star u(t) + v(t), \tag{18.12}$$

where $h(t)$ is the channel impulse response, and $v(t)$ is a zero-mean white Gaussian noise.

### 18.3.2 NONLINEAR DISTORTION MODELING

When considering a large number of subcarriers $M$, thanks to the central limit theorem, the input signal $i_h(t)$ is assumed to be a zero-mean complex Gaussian random process. According to the Bussgang theorem [14], the NL HPA output $u(t)$ is related to the input $i_h(t)$ by the equation

$$u(t) = K_0 i_h(t) + d(t), \tag{18.13}$$

where $d(t)$ is a zero-mean noise, which is uncorrelated with $i_h(t)$, and $K_0$ is a complex gain.

For simplicity, in the following equations, we will discard the time variable $t$ from $\rho$ and $\varphi$. According to [18], $K_0$ can be computed analytically by

$$K_0 = \frac{1}{2}\mathbb{E}\left[\frac{\partial \widehat{S}(\rho)}{\partial \rho} + \frac{\widehat{S}(\rho)}{\rho}\right]. \tag{18.14}$$

The variance $\sigma_d^2$ of the NL distortion $d(t)$ is given by the equation

$$\sigma_d^2 = \mathbb{E}\left(\left|d(t)\right|^2\right) = \mathbb{E}\left(\left|\widehat{S}(\rho)\right|^2\right) - |K_0|^2\mathbb{E}\left(\rho^2\right). \tag{18.15}$$

As clearly shown by Eqs. (18.14) and (18.15), the analytical computation of the parameters $K_0$ and $\sigma_d^2$ depends on the complexity of the expression of $\widehat{S}(\rho)$. To simplify this computation, we can first approximate the HPA conversion characteristics by a polynomial model as per Eq. (18.8). With such HPA approximation, $\widehat{S}(\rho)$ is simplified to $\sum_{n=1}^{L} a_n \rho^n$. Then, the computation of $K_0$ and $\sigma_d^2$ is reduced to the computation of the expectation of Rayleigh random variables $\mathbb{E}[\rho^n]$, $n = 1, \ldots, 2L$. In [8], the analytical expressions of $K_0$ and $\sigma_d^2$ have been derived and are given by

$$
\begin{aligned}
K_0 = a_1 + \sqrt{\frac{\pi}{8}} \sum_{m=2,\, m \text{ even}}^{L} (m+1) a_m \sigma_{i_h}^{m-1} \prod_{i=0}^{\frac{m-2}{2}} (2i+1) \\
+ \frac{1}{2} \sum_{m=3,\, m \text{ odd}}^{L} (m+1) a_m (\sqrt{2}\sigma_{i_h})^{m-1} \left(\frac{m-1}{2}\right)!,
\end{aligned}
\tag{18.16}
$$

where ! stands for the factorial operator, and $\sigma_{i_h}^2$ of the variance of the signal at the input of the HPA.

The variance $\sigma_d^2$ of the NL noise $d(t)$ is given by the expression

$$
\begin{aligned}
\sigma_d^2 = \sum_{n=1}^{L} |a_n|^2 2^n \sigma_{i_h}^{2n} n! - 2 |K_0|^2 \sigma_{i_h}^2 \\
+ \sqrt{\frac{4\pi}{2}} \sum_{m,n=1, m \neq n, [m+n] \text{ odd}}^{L} \mathbb{Re}[a_m a_n^*] \sigma_{i_h}^{m+n} \prod_{i=0}^{\frac{m+n-1}{2}} (2i+1) \\
+ 2 \sum_{m,n=1, m \neq n, [m+n] \text{ even}}^{L} \mathbb{Re}[a_m a_n^*] (\sqrt{2}\sigma_{i_h})^{m+n} \left(\frac{m+n}{2}\right)!.
\end{aligned}
\tag{18.17}
$$

### 18.3.3 THEORETICAL BER ANALYSIS AND SIMULATION RESULTS

The received signal after HPA and channel filtering can be expressed, using (18.12) and (18.13), as

$$
r(t) = K_0 i_h(t) \star h(t) + d(t) \star h(t) + v(t).
\tag{18.18}
$$

Looking at Eq. (18.18), it is clear that the effect of the NL complex gain $K_0$ will be taken into account during channel equalization at the receiver side ($h(t)$ and $K_0$ will be estimated jointly).

For an $N$-level Quadrature Amplitude Modulation (QAM), the Bit Error Rate (BER) over Additive White Gaussian Noise (AWGN) channel, after compensation

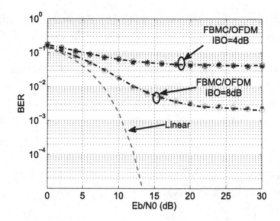

**FIGURE 18.6**

BER vs. $E_b/N_0$ for OFDM and FBMC systems. Saleh HPA model, 64 subcarriers, 16-QAM, $\varphi_0 = \pi/6$, and AWGN channel.

for the factor $K_0$, can be written as follows [8]:

$$\text{BER}_{\text{N-QAM}}^{\text{AWGN}} = \frac{2(\sqrt{N}-1)}{\sqrt{N}\log_2(N)}\text{erfc}\left(\sqrt{\left(\frac{3\log_2(N)|K_0|^2}{4(N-1)T(\sigma_v^2+\sigma_d^2)}\right)}\right). \qquad (18.19)$$

With the same QAM alphabet and over slowly varying flat fading Rayleigh channel, the BER, after compensation for the factor $K_0$, is given by [8]

$$\text{BER}_{\text{N-QAM}}^{\text{Rayleigh}} = \int\limits_0^{\frac{\gamma_c}{\sigma_d^2}} \frac{2(\sqrt{N}-1)}{\sqrt{N}\log_2(N)}$$

$$\times \text{erfc}\left(\sqrt{\left(\frac{3\log_2(N)\gamma}{(N-1)}\right)}\right)\frac{\sigma_v^2\gamma_c}{\Omega(\gamma_c-\sigma_d^2\gamma)^2}e^{-\frac{\sigma_v^2\gamma}{\Omega(\gamma_c-\sigma_d^2\gamma)}}d\gamma, \quad (18.20)$$

where $\Omega = \mathbb{E}[\alpha^2]$ is the average fading power ($\alpha$ being the Rayleigh fading amplitude), $\gamma_c = |K_0|^2 E_b$ ($E_b$ being the energy per bit), and $\gamma$ is the instantaneous Signal-to-Noise Ratio (SNR).

The impact of Saleh HPA model nonlinearity on the performance of FBMC with Offset-QAM subcarrier modulation (FBMC/OQAM) and OFDM systems under AWGN and Rayleigh fading channels is shown in Figs. 18.6 and 18.7. The BER is computed by averaging on $5 \times 10^7$ randomly generated FBMC and OFDM symbols with $M = 64$ subcarriers and 16-QAM symbols.

In Fig. 18.6, a Saleh HPA with $v_{\text{sat}} = 1$ V and $\varphi_0 = \pi/6$ is used. After correction for the gain factor $K_0$, both OFDM and FBMC/OQAM modulations show the same

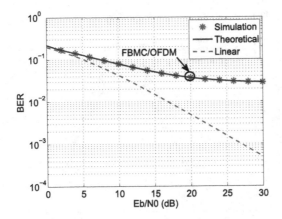

**FIGURE 18.7**

BER vs. $E_b/N_0$ for OFDM and FBMC systems. Saleh HPA model, 64 subcarriers, 16-QAM, $\varphi_0 = \pi/3$, IBO = 6 dB, and Rayleigh channel.

performance. In the case of Rayleigh channel, the BER performance of FBMC, taking the 16-QAM modulation scheme and Saleh's HPA model with $\varphi_0 = \pi/3$ an IBO of 6 dB, is shown in Fig. 18.7.

Based on Eq. (18.20) and simulation results shown in Fig. 18.7, we note that the BER, for relatively low $E_b/N_0$ (i.e., $E_b/\sigma_v^2 < 2$ dB), is very close to the BER performance of the Rayleigh channel with a linear HPA. At high values of $E_b/N_0$ (i.e., $E_b/\sigma_v^2 > 30$ dB), $\sigma_v$ is negligible, and the SNR tends to a constant ($\gamma \to \frac{E_b}{\sigma_d^2}$).

## 18.3.4 OUT-OF-BAND IMPACT

When an MCM signal is amplified by an NL HPA, OOB interference is introduced in the form of spectral leakage into the neighboring channels. This phenomenon is referred to as "spectral regrowth" or "out-of-band emission."

In general, when a modulated signal passes through an NL system, its bandwidth is broadened by odd-order nonlinearities. In any modern communication system, the spectral regrowth is a typical concern on the transmitter side. Spectral regrowth is mainly caused by the most nonlinear component of the transmit chain, which is almost always the HPA. Spectral regrowth can cause several problems in a communication system. Usually, the most important problem is creating inference to adjacent channels and thereby degrading system level performance. Radio spectrum is a very precious commodity, and the channels are tightly packed together to maximize the efficiency of the spectrum usage. Excessive spectral regrowth causes interference to the adjacent channels, and this is the primary reason why standards formulated by regulation bodies require conformance to a spectral mask.

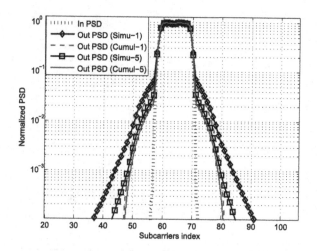

**FIGURE 18.8**

Prediction of spectral regrowth for IBO = {1, 5} dB with NL order 3.

There are many popular methods for predicting this spectral regrowth with the aid of a closed-form expression for the autocovariance function of the HPA output, such as based on the price theorem [34,54] and cumulants [10,44,68]. Because of the interesting properties of cumulants, they are used as an alternative to the moments of a distribution. In some cases, theoretical treatments of problems in terms of cumulants are simpler than those using moments. The Fourier transform of HPA output autocovariance function yields the PSD. For this, we need to construct a polynomial model of HPA from its input–output conversion characteristics, the coefficients of which are needed to derive the closed-form expression. To obtain this model, it is sufficient to consider only odd-order nonlinearities [6]. These coefficients are complex in general; however, in the absence of AM/PM conversion, they will be real. Once we know the conversion characteristics of the HPA, we can predict the spectral regrowth of the signal, provided that its statistical properties do not change.

The spectral regrowth prediction of an $M = 128$ subcarrier FBMC/OQAM signal with 4-QAM subcarrier modulation, amplified with a Saleh model of HPA can be seen in Figs. 18.8 and 18.9. We can notice from these figures that for a given IBO, a polynomial model with sufficient polynomial order can faithfully approximate the spectral regrowth. The order needs to be increased if the HPA is to be operated at a lower IBO. Here nonlinearity orders of 3 and 11 were used with two different IBO values, 1 and 5 dB. For more details about the analytical formulation used to obtain the results shown in Figs. 18.8 and 18.9, we refer to [10].

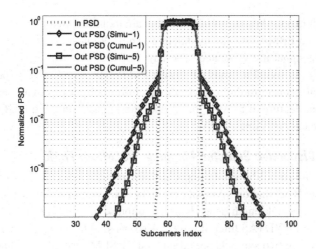

**FIGURE 18.9**

Prediction of spectral regrowth for IBO = {1, 5} dB with NL order 11.

## 18.4 PAPR REDUCTION TECHNIQUES
### 18.4.1 STATE OF THE ART

In MCM systems, the PAPR reduction remains to be one of the most crucial issues that need to be solved effectively with a reasonable complexity. Existing popular PAPR reduction schemes for OFDM signals include clipping [52], Tone Injection (TI) [59], Tone Reservation (TR) [58], Active Constellation Extension (ACE) [65], Partial Transmit Sequence (PTS) [42], Selected Mapping (SLM) [4], and block coding [64]. Comparison of most of these aforesaid schemes can be found in [26] and [31]. In general, the PAPR reduction techniques can be broadly classified into three varieties namely *clip effect transformations (clipping), block coding techniques, and probabilistic approaches*. All existing solutions involve some form of compromise. The most obvious one is the tradeoff between computational complexity and signal degradation (IB and OOB distortion).

The classical schemes, proposed for OFDM, cannot be directly applied to FBMC/OQAM. Indeed, FBMC/OQAM signals have overlapping symbol waveforms. Recently, various PAPR schemes have been suggested for FBMC/OQAM systems, such as ACE [62], iterative clipping [36,37], ACE combined with TR [28], TR [12, 38], SLM [11,13,16,56], and PTS [48].

### 18.4.2 SLM TECHNIQUES

SLM is a probabilistic technique that takes advantages of the fact that the PAPR of an MCM signal is heavily dependent on phase shifts in frequency-domain. Firstly, we generate $V$ complex phase rotation vectors $\boldsymbol{\phi}^{(v)}$ for $0 \leq v \leq V - 1$ of length $M$. The frequency-domain subcarrier symbol vectors are multiplied with these phase rotation vectors before FBMC/OQAM modulation. Thus, $V$ *independent* mappings of the time-domain signal are generated. The target of the optimization problem in SLM is to identify the signal that has the lowest PAPR. The index of its respective phase rotation vector is sent to the receiver as side information (SI) comprising of $\log_2 V$ bits. A salient feature of SLM is that it does not impact either the BER or the OOB emissions in case of a linear HPA.

### 18.4.2.1 *Dispersive Selected Mapping (DSLM) for FBMC*

DSLM uses a symbol-by-symbol approach for reducing the PAPR. FBMC/OQAM symbol extends beyond one symbol period, in sharp contrast with that of an OFDM symbol. The symbol length depends on the impulse response of the prototype filter (in the case of PHYDYAS filter, it is $4T$). Also, most of the energy of the symbol lies in the succeeding two symbols period intervals. DSLM for FBMC is quite similar to classical OFDM-SLM except that we take into account the past overlapping symbols for generating the time domain signal [13,56].

### 18.4.2.2 *Trellis-Based Selected Mapping (TSLM) for FBMC*

DSLM is a suboptimal approach due to the fact that when solving the optimization problem by considering one symbol only, whatever improvement has been achieved for that symbol, is likely to be hampered by its immediate next symbol. An optimal approach would be to test all possible phase rotation vectors in order to find the lowest PAPR in the overall signal composed of $P$ FBMC symbols. This would have huge complexity since there is a need to perform an exhaustive search over $V^P$ possible rotation vectors. A quasi-optimal solution is trellis-based dynamic programming. Thus, TSLM performs a trellis-based search to find the optimal phase rotations considering several successive symbols at a time. Stages of the trellis are related to different input symbol vectors to be transmitted. Within a given stage, different states represent all possible phase vector rotations. Path metric between two states is a function of the signal PAPR with corresponding phase rotations. An illustration of a trellis with $V = 3$ states and $P = 3$ stages is given in Fig. 18.10.

The performances of DSLM and TSLM for an FBMC/OQAM signal with 64 subcarriers and 4-QAM modulation are shown in Fig. 18.11 in terms of Complementary Cumulative Distribution Function (CCDF) plots. We can notice that FBMC/OQAM with TSLM is not only superior to FBMC/OQAM with DSLM scheme in PAPR reduction but also outperforms OFDM with classical SLM. This superior performance implies that when the overlapping nature of the FBMC/OQAM signals is well exploited, it can significantly impact the PAPR reduction, and such proper exploitation can be possible with trellis-based approach instead of symbol-by-symbol optimization. Also, the complexity of trellis-based approach is $\mathcal{O}(V^2(P - 1))$ compared to $\mathcal{O}(V^P)$ for exhaustive search.

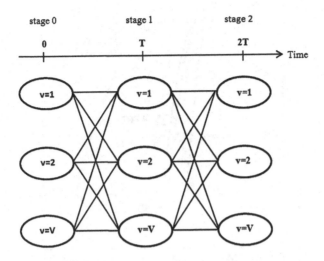

**FIGURE 18.10**

Trellis diagram illustration between three stages with three states.

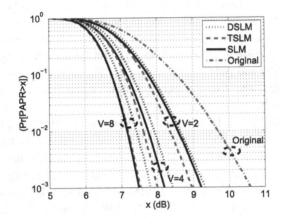

**FIGURE 18.11**

CCDF of PAPR for FBMC/OQAM symbols with DSLM and TSLM and OFDM with SLM.

### 18.4.3 TONE RESERVATION TECHNIQUES

The idea behind TR is to isolate energy used to cancel large peaks to a predefined set of $R$ tones. Such a Peak Reserved Tone (PRT) does not carry any useful information, i.e., PRTs are disjoint from the data tones (DTs). Stated mathematically, the resulting signal to be transmitted will be

$$i_p(t) = m(t) + c(t), \quad 0 \le t < \infty, \qquad (18.21)$$

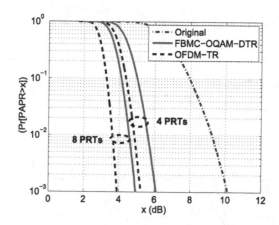

**FIGURE 18.12**

CCDF of PAPR for FBMC/OQAM symbols with DTR and OFDM with TR.

where $c(t)$ is the peak cancellation signal, and $m(t)$ is the data signal. The aim of TR scheme is to compute the optimal values of reserved tones in order to minimize the PAPR. This can be achieved by using convex programming algorithms such as QCQP [60], POCS [22], gradient search [60], etc. The QCQP has the computational complexity of $\mathcal{O}(RM^2)$ and yields the optimal result. The suboptimal approaches such as POCS and gradient search have the computational complexities of $\mathcal{O}(M \log M)$ and $\mathcal{O}(M)$, respectively.

For FBMC/OQAM, a Dispersive Tone Reservation (DTR) has been proposed, which takes into account the overlapped past symbols while calculating signal $c(t)$. To identify the optimal PRT locations, a kernel-based method has been proposed [58]. The PAPR reduction performance of DTR can be seen in Fig. 18.12. Also, in the same figure, it can be noticed that FBMC/OQAM with DTR is closely trailing OFDM with TR by roughly 1 dB (for CCDF of PAPR equal to $10^{-3}$), for $R = 4, 8$.

## 18.4.4 PEAK WINDOWING TECHNIQUE

The SEL-type nonlinearity introduces sharp clipping of the signal peaks at the saturation level, which introduces heavy spectral regrowth. This effect can be reduced by using smooth clipping instead. This is the basic idea of peak windowing, which is a simple and generic method for PAPR mitigation [45,49,51,57,61]. It can be seen as a multiplication of the original signal by a smoothly time-varying function, pushing the peaks below the threshold:

$$i_p[l] = \hat{p}[l] \cdot s[l], \qquad (18.22)$$

where $\hat{p}[l]$ is the envelope scaling function given by

$$\hat{p}[l] = 1 + \left(p[l] - 1\right) \star h[l].\tag{18.23}$$

Here $h[l]$ is the window function determining the smoothness of the peak control, and $p[l]$ is the discrete peak control sequence defined as

$$p[l] = \begin{cases} \frac{v_{\text{sat}}}{|s(\tau_k)|} & \text{for } l = \tau_k, \\ 1 & \text{otherwise.} \end{cases}\tag{18.24}$$

Each peak of the envelope of $s[l]$ exceeding the threshold is represented by a single sample $s[\tau_k]$ for the $k$th peak, with the envelope reaching a local maximum at discrete time index $\tau_k$. Sufficient oversampling (e.g., four times the active signal bandwidth) is required to observe the peak heights and locations sufficiently well. Since the peak control is achieved through multiplication in time domain, this corresponds to convolution in frequency domain:

$$I_p\left(e^{j\omega}\right) = \widehat{P}\left(e^{j\omega}\right) \star S\left(e^{j\omega}\right).\tag{18.25}$$

Here $\hat{P}(e^{j\omega})$ is the discrete-time Fourier transform of $\hat{p}[l]$, which is mostly determined by the choice of the window function since the spectrum of $p[l]$ can be assumed to be white. To minimize the spectrum broadening, $\hat{P}(e^{j\omega})$ should have narrow mainlobe (passband) together with sufficient stopband attenuation.

In contrast to methods like TR and SLM, peak windowing introduces interference in the passband of the transmitted signal, in addition to OOB emissions, and there is a tradeoff between these two effects depending on the used window. Generally, the in-band distortion is minimized with the SEL model without any windowing. With spectrum localized waveforms, the Dolph–Chebyshev window was found to give clearly better performance than the earlier results in [45] and [61]. This window function has two parameters, window length and sidelobe attenuation. The tradeoffs between spectrum broadening and in-band interference in terms of the window parameters were explored in [51].

In this method, occasionally, multiple windows will contribute to the peak reduction, and some of the peaks are pushed to an unnecessarily low level below the threshold. A recursive windowing method has been introduced in [61] to reduce this effect. However, this approach does not fully compensate the effect. Furthermore, since we are targeting at very low spectrum regrowth, the density of the peaks is relatively low, in which case this effect is not very significant.

## 18.5 HPA LINEARIZATION

HPA linearization aims to mitigate the in-band and out-of-band emissions below the limits determined by standardization and regulatory bodies. These emission limits are different for different technologies and standards and can be different for the uplink and downlink as well. For example, the Adjacent Channel Power Ratio (ACPR) limit in LTE uplink transmission is $-30$ dBc, which is usually obtainable without any HPA linearization, whereas for the downlink, it is generally $-45$ dBc, which usually cannot be met without some kind of HPA linearization.

HPA linearization techniques include feedback linearization, feedforward linearization, RF predistortion, and DPD [32,33]. Feedback techniques rely on a closed-loop feedback from the HPA output to its input in order to reduce the emissions and are limited in terms of bandwidth. Feedforward linearization is an inherently wideband technique but requires considerable additional RF hardware. The accuracy, flexibility, and practical bandwidth of feedforward techniques are all limited by the analog components. RF predistortion is a wideband linearization technique, but its fidelity is rather limited due to the analog implementation. Digital predistortion, on the other hand, is probably the most effective and cost efficient among all linearization techniques. It has the greatest flexibility, reconfigurability, and also performance in most cases. Its bandwidth is mostly limited by the digital hardware and will therefore benefit from the scaling of CMOS technology. The focus in the rest of this section will be on DPD; for more information on other linearization techniques, we refer the reader to [32,33] and the references therein.

### 18.5.1 PREDISTORTION TECHNIQUES

DPD aims to distort the HPA input signal by an additional nonlinear device called a *predistorter*, whose characteristics are the inverse of those of the amplifier. Predistortion approaches can be broadly categorized into the following types: look-up-table-based DPD, DPD using basis functions, cascaded or modular DPD structures, and Neural Network (NN) based DPD. The basis function and modular structures are defined in exactly the same manner as the behavioral HPA models in Section 18.2.1, only with the model input signal changed from $i_h[l]$ to $i_p[l]$ (refer to Fig. 18.1). These models are therefore not repeated here. A traditional predistortion approach is based on Look Up Table (LUT) [15,30]. The LUT is used to multiply the signal before feeding it to the HPA by a coefficient depending on the current signal amplitude and phase [43]. The LUT has a very low complexity compared to other techniques, but it is a memoryless system and thus cannot correct the memory effects in the power amplifier. Memory LUT techniques do exist (e.g., [25]), but their size grows exponentially with the memory depth, thus rendering LUT's undesirable for wideband applications.

For wideband applications, DPD techniques with memory need to be employed. These include DPDs based on basis functions such as Volterra and its pruned versions, modular DPD structures, and neural network-based DPD. In the rest of this chapter, the focus will be on basis function and neural network-based DPD techniques.

In addition to the DPD structure, an important choice is the *DPD learning architecture*. These can be broadly classified as open-loop learning architectures,

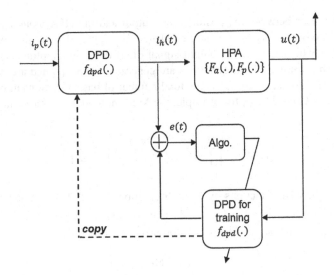

**FIGURE 18.13**

Indirect learning architecture for DPD.

including the so-called direct and indirect learning architectures, and closed-loop learning architectures [9]. For the simulations in this section, we utilize the indirect learning architecture (ILA) (see Fig. 18.13) [9,19,21] due to its simplicity of use and wide application in predistortion studies. The ILA model uses two identical nonlinear functions, one for the actual predistortion and the other one for training, the latter one connected as a post-distorter to the nonlinear HPA. The input to the predistorter training function is the amplifier output signal $u(t)$ divided by the linear voltage gain $G_0$ of the cascade of DPD and the HPA. The linear gain is taken into account in the input of the predistorter training function. In the case of an ideally linearized amplifier, $u(t) = G_0 \cdot i_p(t)$. The coefficients of the training function are estimated using the error signal and training algorithms. The coefficient estimation algorithms range from batch solutions such as least-squares [3,20] to adaptive solutions like LMS or recursive LS type of solutions [2,21,27,70].

## 18.5.2 DPD USING BASIS FUNCTIONS

The most widely used class of predistorters is the one based on nonlinear basis functions, where the output of the DPD is computed as a linear combination of a set of nonlinear transformations of the DPD input signal. The Volterra and MP models, introduced in Section 18.2.1, are well-known examples of this type.

The DPD structure with the indirect parameter learning architecture (Fig. 18.13) is illustrated next. In ILA, a postdistorter is found for the nonlinear device by minimizing the error between the postdistorter output and the HPA input signal and is then copied to be used as the predistorter. Denote by $y[l] = u[l]/G_0$ the postdistorter input signal, which is the HPA output signal divided by the amplitude gain $G_0$ of the HPA. The nonlinear basis functions are generated from $y[l]$ and are denoted by $\phi_k[l]$, $k = 1, \ldots, K_b$, where $K_b$ is the total number of basis functions in the considered basis set. Considering, for example, the Volterra model, the basis functions are given by

$$\phi_k[l] = \prod_{i=1}^{p+1} y[l - d_i] \prod_{i=p+2}^{2p+1} y^*[l - d_i]. \tag{18.26}$$

Then, write the basis functions over an observation period of $N$ samples as vectors $\boldsymbol{\phi}_k = \begin{bmatrix} \phi_k[0] & \phi_k[1] & \cdots & \phi_k[N-1] \end{bmatrix}^T$ and concatenate these into a single matrix $\boldsymbol{\Phi} = \begin{bmatrix} \boldsymbol{\phi}_1 & \boldsymbol{\phi}_2 & \cdots & \boldsymbol{\phi}_{K_b} \end{bmatrix}$. The postdistorter output is then written as

$$\mathbf{z} = \boldsymbol{\Phi}\mathbf{h}, \tag{18.27}$$

where $\mathbf{h} = \begin{bmatrix} h_1 & h_2 & \cdots & h_{K_b} \end{bmatrix}^T$ is the coefficient vector. The well-known least-squares solution to the parameter vector $\mathbf{h}$ minimizes the power of the error vector $\mathbf{e} = \mathbf{i}_h - \boldsymbol{\Phi}\mathbf{h}$ between the postdistorter output and the DPD output, yielding

$$\widehat{\mathbf{h}} = \arg\min_{\mathbf{h}} \|\mathbf{e}\|^2 = \arg\min_{\mathbf{h}} \|\mathbf{i}_h - \boldsymbol{\Phi}\mathbf{h}\|^2$$
$$= \left(\boldsymbol{\Phi}^H \boldsymbol{\Phi}\right)^{-1} \boldsymbol{\Phi}^H \mathbf{i}_h, \tag{18.28}$$

assuming full column rank in $\boldsymbol{\Phi}$.

To obtain the predistorted signal, the estimated parameters are then plugged into the DPD to compute the predistorter output as $\mathbf{i}_h = \boldsymbol{\Psi}\widehat{\mathbf{h}}$, where $\boldsymbol{\Psi}$ is now constructed from $i_p[l]$ in the same manner and order as $\boldsymbol{\Phi}$ was constructed from $y[l]$. This procedure is then typically iterated a few times for the parameters to fully convergence [3,20,21].

## 18.5.3 NEURAL NETWORK-BASED DPD

It has been discovered that NNs, which are nonlinear in their nature, can be good tools for compensating nonlinearities [66]. These predistortion modules are realized with a multilayer perceptron (MLP) neural networks to linearize stationary HPAs. Various adaptive algorithms can be applied, including the standard backpropagation (BP), conjugate gradient (CG), natural gradient (NG), and Levenberg–Marquardt (LM) [66].

Also, here we utilize the ILA described above since it is much more efficient than the direct architecture for DPD systems [66]. The coefficients of the training function are estimated using the error signal. The adopted training function is a neural network associated by the Levenberg–Marquardt algorithm, which has been demonstrated to provide the fastest convergence in terms of iteration number, lowest mean square error, and the lowest amount of computation than all other algorithms studied in the literature [66]. This advantage is mainly noticeable if a very accurate quality level is required. For the DPD architecture, we consider to compensate separately amplitude and phase distortions (Fig. 18.14) with two independent neural network predistortion functions [67].

Here the used NN structure is a multilayer perceptron neural network, which has two inputs, namely, the I and Q components of the input signal, two linear output neurons that are the predistorted signals (I and Q) and one hidden layer with nine nonlinear neurons. Activation functions used for the hidden layer are hyperbolic tangent, whereas the output layer is linear.

It is well known that each neuron in the network is composed of a linear combiner and an activation function, which gives the neuron output

$$
x_{\text{out},n,j} = f\left(\sum_{i=0}^{N_l-1} w_{n,j,i} x_{\text{in},n-1,i} + b_{n,j}\right),
\tag{18.29}
$$

where $w_{n,j,i}$ is the weight which connects the $i$th neuron in layer $n-1$ to the $j$th neuron in layer $n$, $b_{n,j}$ is the bias term, and $x_{\text{in},n-1,i}$ denotes the $i$th component of the input signal to the neuron. The weights of the Neural Network PreDistorter (NNPD) are adjusted using the LM algorithm, which was shown in [66] to exhibit a very good performance in terms of computational complexity and convergence speed, compared to other algorithms studied in the literature. The LM algorithm was designed to approach second-order training speed, and the weights are updated as follows:

$$
w_{n,j,i}(k+1) = w_{n,j,i}(k) - \left[\mathbf{J}^{\mathsf{T}}\mathbf{J} + \mu\mathbf{I}\right]\mathbf{J}^{\mathsf{T}}e,
\tag{18.30}
$$

where $\mathbf{J}$ is the Jacobian matrix that contains first derivatives of the network errors with respect to the weights and biases, $e$ is the error, and $\mu$ is the training rate.

### 18.5.4 SIMULATION RESULTS

In this section, we present simulation results illustrating the performance in terms of Symbol Error Rate (SER) of FBMC/OQAM and OFDM systems in the presence of HPA nonlinearities. The performance of basis-function-based predistortion and a neural network-based predistortion scheme in the compensation of the nonlinear distortions are also presented.

#### 18.5.4.1 *DPD Using Basis Functions*

Here, we consider FBMC/OQAM waveforms with the PHYDYAS prototype filter with overlap factor $K = 4$ [63]. The power amplifier model is a fifth-order Volterra series model extracted from a small-cell base-station power amplifier with output 1-dB compression point of 34.7 dBm. For the DPD models, the memoryless polynomial, MP, and Volterra models are adopted, all with nonlinearity order 7, and the MP

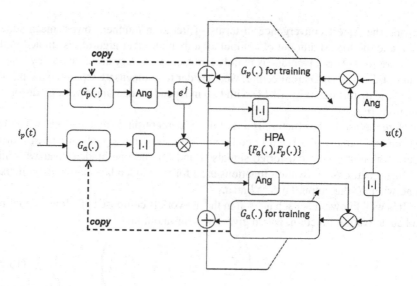

**FIGURE 18.14**

Second DPD scheme.

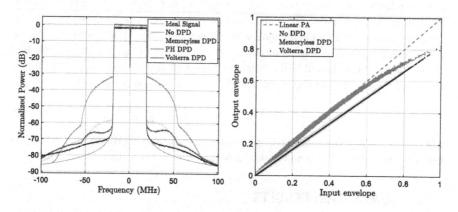

**FIGURE 18.15**

Performance results for different DPD structures (including memoryless, memory polynomial, and Volterra DPD's) under a HPA model extracted from a real small-cell base-station HPA. FBMC/OQAM with 64 subcarriers, 4-QAM modulation, output backoff of 8.5 dB from 1 dB compression point. Left: Comparison of power amplifier output PSD's. Right: Instantaneous AM-AM responses.

and Volterra models having a memory depth of 1. In the learning stage, the ILA is iterated three times, each time with $20k$ samples.

Fig. 18.15 shows the PSDs of the HPA output signal, both with and without the predistorters. Without DPD, the good spectral containment of the FBMC signal is completely lost. The memoryless DPD is already providing a good amount of improvement, whereas the MP and Volterra DPD models give additional 5 and 10 dB of adjacent channel leakage suppression, respectively, over the memoryless DPD. In Fig. 18.15, also the instantaneous AM-AM responses of the original and linearized HPA model are shown. The figure illustrates the improving linearization performance from memoryless DPD to Volterra. The small addition to the back-off, which is required in most DPD learning techniques to maintain stable operation, is also evident in the AM-AM plot as a smaller gain slope after linearization.

Fig. 18.16 shows the simulated PSDs of the HPA output when using the combination of peak windowing-based PAPR reduction and DPD [1]. The results are for 5 MHz LTE-like FBMC/OQAM case (DC + 300 active subcarriers) with 64-QAM modulation, and the Error Vector Magnitude (EVM) target is 8% (−22 dB). For DPD, the MP model with three ILA iterations is used and the backoff is 7.0 dB from the saturation level; otherwise, the DPD and HPA models are the same as above. A Dolph–Chebyshev window of length 257 (at 30.72 MHz sampling rate corresponding to 2048 FB subbands) and sidelobe attenuation of 55 dB is used. The resulting EVM value is −23.2 dB, clearly reaching the target.

It can be seen that using peak windowing for PAPR reduction along with DPD can significantly reduce the spectral regrowth. However, when peak windowing is used alone without DPD, almost no gain is achieved. On the other hand, the DPD algorithm fails when the input signal exceeds significantly the saturation range, and therefore some kind of PAPR reduction is necessary.

### 18.5.4.2 *DPD Based on Neural Network*

Herein, we consider the performance of the NN-based DPD scheme with the Saleh model for the nonlinear HPA as described in Section 18.2.1. A complex baseband FBMC/OQAM or OFDM with $M = 64$ subcarriers using $10^6$ randomly generated symbols was considered. An AWGN channel model was used to clearly observe the effect of nonlinearity and performance improvement by the considered DPDs. For the FBMC/OQAM system, we recall that we use the PHYDYAS prototype filter with an overlapping factor of 4 [5].

In Fig. 18.17, we investigate the performance of the NN DPD scheme over OFDM and FBMC/OQAM systems in presence of amplitude and phase distortions with an IBO of 6 dB. We can notice from these results that the FBMC/OQAM performance is more affected by the HPA nonlinearities than OFDM. This observation is expected, and it can be explained by the fact that the FBMC/OQAM system is more sensitive to the phase distortion than the OFDM system.

On the other hand, Fig. 18.17 shows that the proposed NN DPD scheme can reduce considerably the SER in the two considered systems, compared to the one without any predistortion. We clearly note an excellent match between the performance provided by the proposed predistortion scheme for both OFDM and FBMC/OQAM systems. Indeed, we can note from these results that this DPD

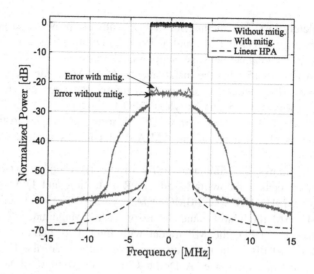

**FIGURE 18.16**

Example case of joint nonlinearity mitigation through peak windowing-based PAPR reduction and DPD. Baseband equivalent normalized PSDs of HPA output and error PSDs (indicating in-band distortion) are shown. FBMC/OQAM with 300 subcarriers, 64-QAM modulation, and IBO = 7 dB.

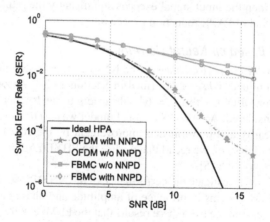

**FIGURE 18.17**

SER vs. SNR for OFDM/FBMC system with NN DPD scheme. IBO = 6 dB, 64 subcarriers, 4-QAM modulation, AWGN channel.

scheme is able to compensate perfectly the phase error due to the nonlinear power amplifier.

## 18.6 CONCLUDING REMARKS

A theoretical evaluation of the IB and OOB effects of nonlinear HPA on FBMC systems was carried out. The in-band distortion was modeled by a complex gain and uncorrelated additive noise, given by the Bussgang theorem. The parameters of the model are obtained from the HPA model through polynomial approximation. This approximation can also be used to characterize the spectral regrowth in the transmitted FBMC signals. With cumulants, we can compute the auto-correlation function of the amplified signal, the Fourier transform of which leads to the PSD. The OOB evaluation demonstrated advantage of FBMC formats over conventional OFDM.

Regarding PAPR mitigation for FBMC waveforms, an SLM-based scheme combined with trellis-based approach for joint block optimization was proposed. With this TSLM scheme, FBMC outperforms OFDM with classical SLM in terms of PAPR reduction.

Regarding power amplifier linearization, basis function, and neural network-based digital predistortion methods, both utilizing the indirect learning architecture, were introduced and tested for FBMC/OQAM signals using practical power amplifier models. A memory polynomial-based DPD method was tested together with a generic low-complexity PAPR reduction method, peak windowing. The results demonstrated the benefits of the joint application of PAPR control with HPA linearization for effective mitigation of HPA nonlinearities in case of spectrally well-localized waveforms, such as FBMC/OQAM.

## REFERENCES

[1] M. Abdelaziz, L. Anttila, M. Renfors, and M. Valkama, "Joint PAPR reduction and digital predistortion for non-contiguous waveforms with well-localized spectrum," in *Proc. ISWCS 2016*, Poznan, Poland, Sep. 2016.

[2] L. Anttila, P. Händel, O. Mylläri, and M. Valkama, "Recursive learning-based joint digital predistorter for power amplifier and I/Q modulator impairments," *Int. J. Microwave Wireless Technol.*, vol. 2, no. 2, pp. 173–182, Apr. 2010.

[3] L. Anttila, P. Händel, and M. Valkama, "Joint mitigation of power amplifier and I/Q modulator impairments in broadband direct-conversion transmitters," *IEEE Trans. Microw. Theory Tech.*, vol. 58, pp. 730–739, Apr. 2010.

[4] R. W. Bauml, R. F. H. Fischer, and J. B. Huber, "Reducing the peak-to-average power ratio of multicarrier modulation by selected mapping," *Electron. Lett.*, vol. 32, no. 22, pp. 2056–2057, Oct. 1996.

[5] M. Bellanger, "Specification and design of a prototype filter for filter bank based multicarrier transmission," in *Proc. ICASSP 2001*, Salt Lake City, UT, USA, May 2001.

[6] S. Benedetto and E. Biglieri, *Principles of Digital Transmission with Wireless Applications*. New York: Kluwer Academic/Plenum Publishers, 1999.

[7] S. Benedetto, E. Biglieri, and R. Daffara, "Modeling and performance evaluation of nonlinear satellite links a Volterra series approach," *IEEE Trans. Aerosp.*, vol. 15, no. 4, pp. 494–507, 1979.

[8] H. Bouhadda, H. Shaiek, D. Roviras, R. Zayani, Y. Medjahdi, and R. Bouallegue, "Theoretical analysis of BER performance of nonlinearly amplified FBMC/OQAM and OFDM signals," *EURASIP J. Adv. Signal Process.*, vol. 2014, pp. 1–16, 2014.

[9] R. Braithwaite, "General principles and design overview of digital predistortion," in *Digital Front-End in Wireless Communications and Broadcasting*. Cambridge: Cambridge University Press, 2011 (Chapter 6).

[10] K. C. Bulusu, H. Shaiek, and D. Roviras, "Prediction of spectral regrowth for FBMC-OQAM system using cumulants," in *Proc. WiMob 2014*, Larnaca, Cyprus, Oct. 2014.

[11] K. C. Bulusu, H. Shaiek, and D. Roviras, "Potency of trellis-based SLM over the symbol-by-symbol approach in reducing PAPR for FBMC-OQAM signals," in *Proc. ICC 2015*, London, UK, Jun. 2015.

[12] K. C. Bulusu, H. Shaiek, and D. Roviras, "Reduction of PAPR of FBMC-OQAM signals by dispersive tone reservation technique," in *Proc. ISWCS 2015*, Brussels, Belgium, Aug. 2015.

[13] K. C. Bulusu, H. Shaiek, D. Roviras, and R. Zayani, "PAPR reduction for FBMC-OQAM systems using dispersive SLM technique," in *Proc. ISWCS 2014*, Barcelona, Spain, Aug. 2014.

[14] J. Bussgang, *Crosscorrelation Functions of Amplitude-Distorted Gaussian Signals*. Cambridge: Research Laboratory of Electronics, Massachusetts Institute of Technology, 1952.

[15] J. Cavers, "Amplifier linearization using a digital predistorter with fast adaptation and low memory requirements," *IEEE Trans. Veh. Technol.*, vol. 39, pp. 374–382, Nov. 1990.

[16] G. Cheng, B. H. Li, and S. Li, "An overview: Peak-to-average power ratio of OFDM signals," *IEEE Trans. Broadcast.*, vol. 54, pp. 257–268, Jun. 2008.

[17] C. Crespo-Cadenas, J. Reina-Tosina, M. Madero-Ayora, and J. Munoz-Cruzado, "A new approach to pruning Volterra models for power amplifiers," *IEEE Trans. Signal Process.*, vol. 58, pp. 2113–2120, Apr. 2010.

[18] D. Dardari, V. Tralli, and A. Vaccari, "A theoretical characterization of nonlinear distortion effects in OFDM systems," *IEEE Trans. Commun.*, vol. 48, pp. 1755–1764, Oct. 2000.

[19] L. Ding, R. Raich, and G. Zhou, "A Hammerstein predistortion linearization design based on the indirect learning architecture," in *Proc. ICASSP 2002*, Orlando, FL, USA, May 2002.

[20] L. Ding, G. Zhou, D. Morgan, Z. Ma, J. Kenney, J. Kim, and C. Giardina, "A robust digital baseband predistorter constructed using memory polynomials," *IEEE Trans. Commun.*, vol. 52, pp. 159–165, Jan. 2004.

[21] C. Eun and E. Powers, "A new Volterra predistorter based on the indirect learning architecture," *IEEE Trans. Signal Process.*, vol. 45, pp. 223–227, Jan. 1997.

[22] A. Gatherer and M. Polley, "Controlling clipping probability in DMT transmission," in *Proc. 31st Asilomar Conf. on Signals, Systems and Computers*, vol. 1, Nov. 1997, pp. 578–584.

[23] F. Ghannouchi and O. Hammi, "Behavioral modeling and predistortion," *IEEE Microw. Mag.*, vol. 10, no. 7, pp. 52–64, Dec. 2009.

[24] O. Hammi and F. Ghannouchi, "Twin nonlinear two-box models for power amplifiers and transmitters exhibiting memory effects with application to digital predistortion," *IEEE Microw. Wireless Compon. Lett.*, vol. 19, no. 8, pp. 530–532, Aug. 2009.

[25] O. Hammi, F. Ghannouchi, S. Boumaiza, and B. Vassilakis, "A data-based nested LUT model for RF power amplifiers exhibiting memory effects," *IEEE Microw. Wireless Compon. Lett.*, vol. 17, no. 10, pp. 712–714, Oct. 2007.

[26] H. S. Hee and L. J. Hong, "An overview of peak-to-average power ratio reduction techniques for multicarrier transmission," *IEEE Trans. Wireless Commun.*, vol. 12, pp. 56–65, Apr. 2005.

[27] T. Hoh, G. Jian-Hua, G. Shu-Jian, and W. Gang, "A nonlinearity predistortion technique for HPA with memory effects in OFDM systems," *Nonlinear Anal., Real World Appl.*, vol. 8, no. 1, pp. 249–256, 2007.

[28] B. Horvath and P. Horvath, "Establishing lower bounds on the peak-to-average-power ratio in filter bank multicarrier systems," in *Proc. EW 2015*, Budapest, Hungary, May 2015.

[29] M. Isaksson, D. Wisell, and D. Rönnow, "A comparative analysis of behavioral models for RF power amplifiers," *IEEE Trans. Microw. Theory Tech.*, vol. 54, pp. 348–359, Jan. 2006.

[30] P. Jardin and G. Baudoin, "Filter look up table method for power amplifiers linearization," *IEEE Trans. Veh. Technol.*, vol. 56, no. 3, pp. 1067–1087, 2007.

[31] T. Jiang and Y. Wu, "An overview: Peak-to-average power ratio of OFDM signals," *IEEE Trans. Broadcast.*, vol. 54, pp. 257–268, Jun. 2008.

[32] A. Katz, J. Wood, and D. Chokola, "The evolution of PA linearization," *IEEE Microw. Mag.*, vol. 17, no. 2, pp. 32–40, Feb. 2016.

[33] P. Kenington, *High-Linearity RF Amplifier Design*. Boston: Artech House, 2000.

[34] M. Khodjet-Kesba, C. Saber, D. Roviras, and Y. Medjahdi, "Multicarrier interference evaluation with jointly non-linear amplification and timing errors," in *Proc. VTC 2011-Spring*, Budapest, Hungary, May 2011.

[35] J. Kim and K. Konstantinou, "Digital predistortion of wideband signals based on power amplifier model with memory," *Electron. Lett.*, vol. 37, no. 23, pp. 1417–1418, Nov. 2001.

[36] Z. Kollar and P. Horvath, "PAPR reduction of FBMC by clipping and its iterative compensation," *Int. J. Comput. Netw. Commun.*, vol. 2012, pp. 1–11, May 2012.

[37] Z. Kollar, L. Varga, B. Horvath, P. Bakki, and J. Bito, "Evaluation of clipping based iterative PAPR reduction techniques for FBMC systems," *Sci. World J.*, vol. 2014, pp. 1–12, Jan. 2014.

[38] S. Lu, D. Qu, and Y. He, "Sliding window tone reservation technique for the peak-to-average power ratio reduction of FBMC-OQAM signals," *IEEE Wireless Commun. Lett.*, vol. 4, no. 1, pp. 268–271, Aug. 2012.

[39] V. Mathews and G. Sicuranza, *Polynomial Signal Processing*. New York: Wiley, 2000.

[40] J. Moon and B. Kim, "Enhanced Hammerstein behavioral model for broadband wireless transmitters," *IEEE Trans. Microw. Theory Tech.*, vol. 59, pp. 924–933, Apr. 2011.

[41] D. Morgan, Z. Ma, J. Kim, M. Zierdt, and J. Pastalan, "A generalized memory polynomial model for digital predistortion of RF power amplifiers," *IEEE Trans. Signal Process.*, vol. 54, pp. 3852–3860, Oct. 2006.

[42] S. H. Muller and J. B. Huber, "OFDM with reduced peak-to-average power ratio by optimum combination of partial transmit sequences," *Electron. Lett.*, vol. 33, no. 5, pp. 368–369, Feb. 1997.

[43] M. Nizamuddin, P. Balister, W. Tranter, and J. Reed, "Nonlinear tapped delay line digital predistorter for power amplifiers with memory," in *Proc. WCNC 2003*, New Orleans, LA, USA, Mar. 2003.

[44] J. Nsenga, W. V. Thillo, A. Bourdoux, V. Ramon, F. Horlin, and R. Lauwereins, "Spectral regrowth analysis of band-limited Offset-QPSK," in *Proc. ICASSP 2008*, Apr. 2008.

[45] M. Pauli and H. Kuchenbecker, "Minimization of the intermodulation distortion of a nonlinearly amplified OFDM signal," *Wirel. Pers. Commun.*, vol. 4, pp. 93–101, Jan. 1997.

[46] J. Qi and S. Aissa, "On the effect of power amplifier nonlinearity on MIMO transmit diversity systems," in *Proc. ICC 2009*, Jun. 2009.

[47] J. Qi and S. Aissa, "Analysis and compensation of power amplifier nonlinearity in MIMO transmit diversity systems," *IEEE Trans. Veh. Technol.*, vol. 59, pp. 2921–2931, Jul. 2010.

[48] D. Qu, S. Lu, and T. Jiang, "Multi-block joint optimization for the peak-to-average power ratio reduction of FBMC-OQAM signals," *IEEE Trans. Signal Process.*, vol. 61, pp. 1605–1613, Apr. 2013.

[49] M. Rahim, T. Stitz, and M. Renfors, "Analysis of clipping-based PAPR-reduction in multicarrier systems," in *Proc. VTC 2009-Spring*, Apr. 2009.

[50] C. Rapp, "Effects of HPA nonlinearity on 4-DPSK-OFDM signal for digital sound broadcasting systems," in *Proc. Second European Conf. on Sat. Comm*, Oct. 1991.

[51] M. Renfors, J. Yli-Kaakinen, and M. Valkama, "Power amplifier effects on frequency localized 5G candidate waveforms," in *Proc. EW 2016*, May 2016.

[52] H. G. Ryu, B. I. Jin, and I. B. Kim, "PAPR reduction using soft clipping and ACI rejection in OFDM system," *IEEE Trans. Consum. Electron.*, vol. 48, pp. 17–22, Aug. 2002.

[53] A. Saleh, "Frequency-independent and frequency-dependent nonlinear models of TWT amplifiers," *IEEE Trans. Commun.*, vol. 29, pp. 1715–1720, Nov. 1981.

[54] S. Sall, H. Shaiek, D. Roviras, and Y. Medjahdi, "Analysis of the nonlinear spectral regrowth in FBMC systems for cognitive radio context," in *Proc. ISWCS 2013*, Aug. 2013.

[55] M. Schetzen, *The Volterra and Wiener Theories of Nonlinear Systems*. New York: Wiley, 1980.

[56] A. Skrzypczak, P. Siohan, and J. P. Javaudin, "Reduction of the peak-to-average power ratio for OFDM/OQAM modulation," in *Proc. VTC 2006-Spring*, vol. 4, May 2006, pp. 2018–2022, no. 841680.

[57] J. Song and H. Ochiai, "A low-complexity peak cancellation scheme and its FPGA implementation for peak-to-average power ratio reduction," *EURASIP J. Wirel. Commun. Netw.*, pp. 1–14, 2015.

[58] J. Tellado, "Peak to average ratio reduction for multi-carrier modulation." Ph.D. dissertation, Stanford, CA, USA: Stanford University, Sep. 1999.

[59] J. Tellado, *Multicarrier Modulation with Low PAR: Applications to DSL and Wireless*. Kluwer Academic Publishers, 2000.

[60] J. Tellado and J. Cioffi, "Peak power reduction for multicarrier transmission," in *CTMC, GLOBECOM*, Nov. 1998.

[61] O. Väänänen, J. Vankka, and K. Halonen, "Simple algorithm for peak windowing and its application in GSM, EDGE and WCDMA systems," *IEE Proc., Commun.*, vol. 152, pp. 357–362, Jun. 2005.

[62] N. van der Neut, B. Maharaj, F. de Lange, G. Gonzalez, F. Gregorio, and J. Cousseau, "PAPR reduction in FBMC using an ACE-based linear programming optimization," *EURASIP J. Adv. Signal Process.*, vol. 2014, no. 1, pp. 1–21, Dec. 2014.

[63] A. Viholainen, *et al.*, "Prototype filter and structure optimization." Project PHYDYAS ICT-211887, Deliverable D5.1, Jan. 2009.

[64] T. A. Wilkinson and A. E. Jones, "Minimisation of the peak-to-mean envelope power ratio of multicarrier transmission schemes by block coding," in *Proc. VTC 1995*, vol. 2, Feb. 1995, pp. 825–829.

[65] Z. X. Yang, H. D. Fang, and C. Y. Pan, "ACE with frame interleaving scheme to reduce peak-to-average power ratio in OFDM systems," *IEEE Trans. Broadcast.*, vol. 51, pp. 571–575, Mar. 2005.

[66] R. Zayani, R. Bouallegue, and D. Roviras, "Adaptive pre-distortions based on neural networks associated with Levenberg–Marquardt algorithm for satellite down links," *EURASIP J. Wirel. Commun. Netw.*, pp. 1–15, 2008.

[67] R. Zayaniand, Y. Medjahdi, H. Bouhadda, H. Shaiek, D. Roviras, and R. Bouallegue, "Adaptive predistortion techniques for non-linearly amplified FBMC-OQAM signals," in *Proc. VTC Spring-2014*, May 2014.

[68] G. T. Zhou, G. Tong, and R. Raich, "Spectral analysis of polynomial nonlinearity with applications to RF power amplifiers," *EURASIP J. Adv. Signal Process.*, vol. 2004, no. 12, pp. 1831–1840, 2004.

[69] G. Zhou, H. Gian, L. Ding, and R. Raich, "On the baseband representation of a bandpass nonlinearity," *IEEE Trans. Signal Process.*, vol. 53, pp. 2953–2957, Aug. 2005.

[70] A. Zhu and T. Brazil, "An adaptive Volterra predistorter for the linearization of RF high power amplifiers," in *Proc. MTT-S 2002*, Jun. 2002.

[71] A. Zhu, J. Pedro, and T. Cunha, "Pruning the Volterra series for behavioral modeling of power amplifiers using physical knowledge," *IEEE Trans. Microw. Theory Tech.*, vol. 55, pp. 813–821, May 2007.

# FBMC Implementation for a TVWS System

# 19

Dominique Noguet, Vincent Berg, Jean-Baptiste Doré, Dimitri Kténas

*CEA-Leti, Grenoble, France*

## CONTENTS

In this chapter, we discuss the implementation of an FBMC PHY for TVWS application. The baseline for this implementation is the recent IEEE 1900.7-2015 standard introduced in Chapter 2. To our knowledge, this is the first worldwide standard based on an FBMC air interface or, more generally speaking, on a nonorthogonal waveform. After a brief introduction, which highlights the benefits of FBMC in the context of TVWS, we describe the implementation using a frequency domain approach (namely Frequency Sampling FBMC, FS-FBMC). We show that this technique is very well suited for uncoordinated multiuser scenarios. We discuss the complexity of FS-FBMC implementation. Then, we analyze a flexible architecture that can manage FS-FBMC and OFDM modes with a discussion of the impact on the waveform on hardware constraints.

Orthogonal Waveforms and Filter Banks for Future Communication Systems. DOI: 10.1016/B978-0-12-810384-5.00019-0

## 19.1 DYNAMIC SPECTRUM ACCESS TO FRAGMENTED SPECTRUM CONTEXT

Several scenarios have been investigated for TVWS operation. In Chapter 2, the need for broadband systems is stressed, where the radio system guarantees few Mbps to few tens of Mbps for broadband access and WLAN use cases. Usually, such scenarios are covered by multicarrier systems using frequency multiplexes transmitted simultaneously over several subcarriers. The most classical approach is to use orthogonal carriers. With OFDM, the receiver can equalize each subcarrier independently. When the system is adequately specified, each subcarrier is processed as a narrow band signal under flat fading condition, despite the broadband nature of the overall multiplexed signal. This property leads to a simplified receiver architecture even under frequency-selective channel conditions. Because of these assets, OFDM has been the initial choice for almost all modern broadband wireless systems, such as Wi-Fi, WiMAX, 3GPP LTE, DVB-T, etc.

The major drawback of OFDM relies on its spectrum shape. Each rectangular shaped subcarrier in the time domain results in a $sinc(f)$ function in the frequency domain, where summation reveals high spectrum sidelobes (first lobe at $-13$ dB). Usually, this issue is solved by filtering sidelobes with analog SAW filters at the RF transmitter. Unfortunately, this results in nonagile radios since filtering is usually implemented as analog filters with limited flexibility. Other variants propose to smooth the transitions between the OFDM symbols by using digital windowing techniques [1]. Although this approach efficiently addresses the agility constraint, the roll-off of the filter remains limited unless the filter complexity is increased too significantly. Then, the guard interval has to be increased by the duration of the window at the cost of an important spectral efficiency reduction.

In the case of the TVWS, both frequency agility and sharp spectrum roll-off are expected. The agility requirement stems from the wide frequency span of channel opportunities across the UHF band (this depends on the country but typically 470 MHz to 790 MHz), and a sharp roll-off is requested to guarantee nonharmful interference to adjacent incumbents, which translates into ACLR specifications. Most regulators have set TVWS ACLR requirements 10 dB higher than in LTE (see Chapter 2). It was shown in [2] that guaranteeing such ACLR levels through digital filtering implies very complex filters. The complexity of the filter increases dramatically as the guard band is reduced since this increases the filter frequency steepness. This is illustrated in Fig. 19.1, where the complexity of filtered OFDM under TVWS spectrum mask is compared to the one of a 1024-FFT alone as a function of the used portion of the channel, in terms of the number of real multiplications. The calculation was achieved for an 8-MHz channel with 15-kHz subcarrier spacing and an equiripple filter (0.5 dB in band ripple) under 55-dB ACLR condition. It can be observed that the complexity of the filter designed to meet the ACLR specification is very significant. It varies as a function of the useful band for a given ACLR requirement. It is 10 times higher when the occupation reaches 85%.

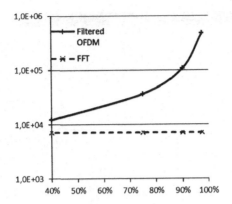

**FIGURE 19.1**

Complexity of filtered OFDM to meet 55-dB TVWS ACLR requirements.

The TV broadcast frequency planning translates into heavily fragmented TVWS. This means that the available channel map cannot be predicted. In some places, contiguous available channels are virtually inexistent. The situation in London is highlighted in [3], but similar fragmented available spectrum can be observed in many locations, even in rural areas. To enable broadband services under such conditions, the spectrum pooling approach is envisaged [4]. It consists of using the parallel nature of the multiplex to switch off the subcarriers in order to avoid interfering with a cochannel incumbent. Ideally, this technique should enable the creation of deep notches without any impact on the transmission on the active subcarriers, exploiting the independence of each subcarrier. In practice, as long as OFDM waveform is considered, similar smooth roll-off to the one described above is observed where notches are enforced. This is illustrated in Fig. 19.2. Unfortunately, the filter designed to steepen the edges of the spectrum does not have a positive impact on the notch [2,5]. Thus, on top of its large complexity, the digitally filtered OFDM does not have the flexibility to address TVWS fragmented spectrum properly.

FBMC is very helpful in this regard: it allows controlling the frequency response of each carrier by introducing a filter bank, centered on every active carrier and based on the same prototype response, which can be used to control adjacent leakage, and even to virtually null it.

Recently, FBMC-OQAM was adopted in the IEEE 1900.7-2015 standard [6]. The IEEE 1900.7-2015 standard specifies two different sizes for the prototype filter, $K = 3$ or $K = 4$, governing the level of protection of adjacent channels. Also, for the sake of flexibility, several modes are proposed (see Chapter 2, Table 2.2). They consider different number of carriers and two channelization modes (2 or 8 MHz). Thus, IEEE 1900.7-2015 can be used in any country authorizing TVWS operation and can cover medium to broadband channels with flexible bandwidth (2, 4, 6, 8, 10, ... MHz). Of course, the use of noncontiguous fragments is also possible.

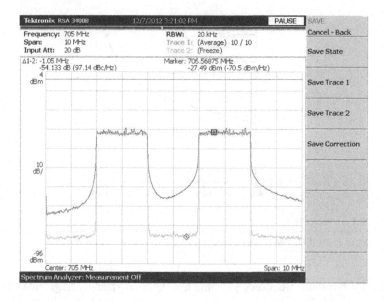

**FIGURE 19.2**

Shaping spectrum with FBMC (lower curve) vs. OFDM (upper curve).

The coded data are mapped to QPSK, 16QAM, or 64QAM modulation symbols, depending on the desired data rate. Symbols are then padded to make the transmitted burst length an integer multiple of multicarrier symbols. The generated block of QAM symbols is mapped to active carriers and modulated to an offset-QAM before being transformed into a time domain signal using FBMC waveform. A modulation and coding scheme (MCS) index is defined to describe the combination of the modulation and coding schemes that are used when transmitting data. Table 2.3 in Chapter 2 summarizes the possible MCS and the associated peak throughput (Mode 1K), where sampling rate is set to 15.36 Msps, and subcarrier spacing is set to 15 kHz as in LTE.

## 19.2 FS-FBMC PRINCIPLE

### 19.2.1 RECEIVER DESCRIPTION

In the case of FBMC, the choice of the prototype filter controls the localization in frequency of the generated signal and provides better adjacent channel leakage performance in comparison to OFDM. Offset Quadrature Amplitude Modulation (OQAM), combined with Nyquist constraints on the prototype filter, is used to guarantee orthogonality between adjacent symbols and adjacent carriers while providing maximum spectral efficiency [7].

Frequency sampling technique may be applied to design the prototype filter. The duration $L$ of the prototype filter is a multiple of the size of the FFT, $N$, so that $L = KN$, where $K$ is an integer and often referred to as the overlapping factor. FS-FBMC has been recently proposed in [8] and [9] as an alternative to polyphase network FBMC. This technique is inspired by the frequency sampling technique used to design the prototype filter. With this approach, the number of nonzero samples in the frequency response is given by $P = 2K - 1$. For $K = 4$, the frequency domain (FD) pulse response coefficients are equal to:

$$H_0 = 1; \quad H_1 = 0.971960 = H_{-1};$$
$$H_2 = \frac{\sqrt{2}}{2} = H_{-2}; \quad H_3 = \sqrt{1 - H_1^2} = H_{-3}. \tag{19.1}$$

FS-FBMC spreads the data over $P$ carriers by filtering the output of the OQAM process by the frequency domain pulse response of (19.1). The output is then processed through an inverse Fourier transform (IFFT) of size $K \times N$. OQAM precoding imposes that real and pure imaginary symbol values alternate on successive subcarrier frequencies and on successive transmitted symbols for any given subcarrier. This guarantees orthogonality between adjacent carriers since the coefficients of the prototype filter are real. The output of the IFFT is converted through a parallel-to-serial conversion and is accumulated with the following IFFT output data block stream delayed by $N/2$. This parallel-to-serial conversion is called overlap-and-sum. Once the transient period is over, $2K$ of the $KN$-IFFT output samples are added together at any given time.

At the receiver, the dual operation of the overlap-and-sum operation of the transmitter is a sliding window in the time domain (TD) at the receiver that selects $KN$ points every $N/2$ samples. FFT is then applied on every block of $KN$ selected points. One advantage of this architecture is that frequency domain time synchronization may be performed independently of the position of the FFT [9]. This is realized by combining timing synchronization with channel equalization. Another major benefit of FS-FBMC is that channel equalization may be limited to a one-tap complex-multiply operation while still sustaining significant channel impulse response delay spread. Fig. 19.3 illustrates the architecture of the FS-FBMC receiver.

The size of the FFT at the FS-FBMC receiver is $K$ times larger than the multicarrier symbol time duration. The signal at the output of the FFT is then oversampled by a factor $K$ compared to the carrier spacing. This property gives a significant advantage to FS-FBMC receiver when the channel is exhibiting large delay spread. Alternatively, it also suggests that the intercarrier spacing could be reduced. Of course, the classical PPN receiver scheme can also be applied, even where FS-FBMC is used at the transmitter [7] (Fig. 19.4B).

## 19.2.2 DETAILED DESCRIPTION OF THE FS-FBMC EQUALIZER

The proposed equalizer is placed directly after the FFT but before the FBMC prototype filtering. Assuming the channel delay spread $L$ small compared to the $KN$-point

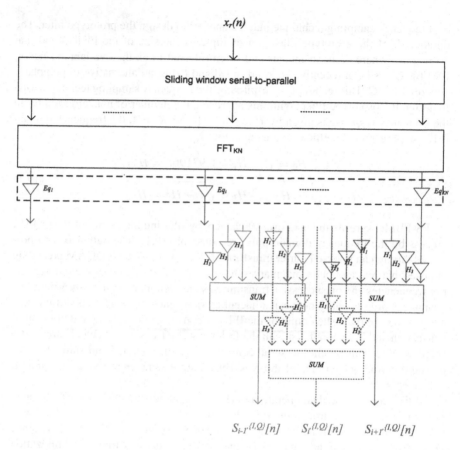

$$x_r(n)$$

**FIGURE 19.3**

FS-FBMC receiver principles.

FFT, the symbols at the output of the FFT may be written as follows:

$$R(k, p) = H(k)X(k, p) + Z(k, p) + Z_I(k, p), \qquad (19.2)$$

where $k$ is the subcarrier index, $p$ is the index of the FBMC symbol, $H(k)$ is the complex channel coefficients for subcarrier $k$, $Z(k, p)$ is the AWGN sample, and $Z_I(k, p)$ is the intersymbol interference (ISI) contribution. The channel is here again assumed static for the duration of the burst. When the matrix is diagonal, the ISI term can be omitted.

The goal of the equalizer is to recover $X(k, p)$ from the observation $R(k, p)$ and the estimated channel coefficient $H(k)$. Zero forcing (ZF) or Minimum Mean Square

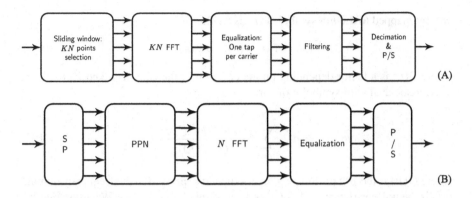

**FIGURE 19.4**

FBMC-OQAM receiver block diagram. (A) Using frequency spreading approach; (B) using classical PPN approach.

Error (MMSE) criteria are classically used and may be expressed as

$$\hat{X}(k, p) = \frac{H^*(k)}{Q(k)} R(k, p),\tag{19.3}$$

where $Q(k)$ is the optimized factor according to the ZF or MMSE criterion:

$$Q_{ZF}(k) = |H(k)|^2,\tag{19.4}$$

$$Q_{MMSE}(k) = |H(k)|^2 + \frac{\sigma_Z^2(k)}{\sigma_X^2(k)},\tag{19.5}$$

where $\sigma_X^2(k)$ is the expectation of the power of $X$ on subcarrier $k$, and $\sigma_Z^2$ is the expectation of the power of $Z$ on subcarrier $k$. In the following, the noise power is assumed to be constant over all the subcarriers.

Once the signal is equalized, the samples are filtered by the FBMC prototype filter before downsampling by a factor $K$. After payload extraction, OQAM inverse transform, LLR values are obtained. The LLR is defined as

$$LLR\big(b(m, n)\big) = \log \frac{P[b(m) = 1|\hat{Y}(n)]}{P[b(m) = 0|\hat{Y}(n)]}\tag{19.6}$$

$$= \log \frac{\sum_{\alpha \in \Omega_n^0} P[Y_n = \alpha|\hat{Y}(n)]}{\sum_{\alpha \in \Omega_n^1} P[Y_n = \alpha|\hat{Y}(n)]},$$

where $\Omega_n^0$ (resp. $\Omega_n^1$) is the set of symbols comprising a bit 0 (resp. 1) for bit $n$, $\hat{Y}(n)$ is the $n$th real symbol after filtering and inverse OQAM transform, and $b(m, n)$ is the

$m$th bit mapped to the $n$th symbol. $\hat{Y}(n)$ is written as

$$\hat{Y}(n) = \Gamma(n)Y(n) + Z_Y(n), \qquad (19.7)$$

where $\Gamma(n)$ is a factor depending on the criterion of the equalizer, and $Z_Y(n)$ is the noise associated with symbol $n$ of variance $\sigma^2_{Z_{Y(n)}}$:

$$\sigma^2_{Z_{Y(n)}} = \sigma^2_{Z_Y} \sum_{p=0}^{2K-1} \frac{G(p)^2 H(\mathbb{S}_n(p))^2}{Q(\mathbb{S}_n(p))^2}, \qquad (19.8)$$

where $G(p)$ is the $p$th prototype filter coefficient expressed in the frequency domain, $\sigma^2_{Z_Y}$ is the noise variance, and $\mathbb{S}_n(p)$ is a set of subcarrier indices. We group into this set the indices of subcarriers used for the computation of symbol $n$. The factor $\Gamma(n)$ may be expressed by

$$\Gamma(n) = \sum_{p=0}^{2K-1} \frac{G(p) H(\mathbb{S}_n(p))^2}{Q(\mathbb{S}_n(p))}. \qquad (19.9)$$

We assume that the conditional pdf of $\hat{Y}(n)$ is Gaussian and is expressed by

$$P\left[\hat{Y}(n)|Y(n) = \alpha\right] = \frac{1}{\sqrt{2\pi}\sigma_Z} \exp\left(-\frac{1}{2}\frac{|\hat{Y}(n) - \Gamma(n)\alpha|^2}{\sigma^2_Z}\right). \qquad (19.10)$$

By applying the Bayes rules, assuming equally distribution of symbols and using the max log approximation ($\log \sum_i \beta_i = \max_i \log \beta_i$), Eq. (19.6) may be rewritten as

$$LLR(b(m, n)) \propto \frac{1}{\sum_{p=0}^{2K-1} \frac{G(p)^2 H(\mathbb{S}_n(p))^2}{Q(\mathbb{S}_n(p))^2}}$$
$$\left( \max_{\alpha \in \Omega_m^0} \left(-|\hat{Y}(n) - \Gamma(n)\alpha|^2\right) \right.$$
$$\left. - \max_{\alpha \in \Omega_m^1} \left(-|\hat{Y}(n) - \Gamma(n)\alpha|^2\right) \right). \qquad (19.11)$$

Further complexity simplification of (19.11) may be considered with negligible performance loss such as the techniques described in [10].

## 19.2.3 IMPACT OF FFT MISALIGNMENT

When the receiver FFT is aligned to the transmitted FBMC symbol, the one-tap equalizer of the FS-FBMC recovers the distortion introduced by the channel. In this section, we analyze the offset introduced by applying the FFT at any given sampling instant rather than specifically aligning the samples to the start of the FBMC symbol.

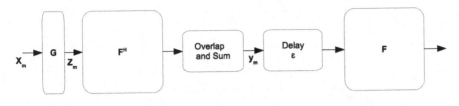

**FIGURE 19.5**

Block diagram of the FFT misalignment mathematical model at the FS-FBMC receiver.

The signal at the output of the transmitter $\mathbf{y_m}$ may be expressed as

$$\mathbf{y_m} = \mathbf{F}^H \mathbf{G}\mathbf{X_m} + \sum_{p=1}^{K-1} \mathbf{Q}_{\frac{pN}{2}} \mathbf{F}^H \mathbf{G}\mathbf{X_{m-p}} + \mathbf{Q}_{N-\frac{pN}{2}} \mathbf{F}^H \mathbf{G}\mathbf{X_{m+p}}, \tag{19.12}$$

where matrix $\mathbf{G}$ is the prototype filter matrix, and $\mathbf{Q_x}$ is defined by

$$\mathbf{Q_x} = \begin{bmatrix} \mathbf{0_{x \times N}} \\ \mathbf{I_{N-x}} \quad \mathbf{0_{N-x \times x}} \end{bmatrix}, \tag{19.13}$$

where $\mathbf{0_{x \times N}}$ is the zero matrix of size $x \times N$, and $\mathbf{I_x}$ is the identity matrix of size $x \times x$. Eq. (19.12) represents the sum of the $2K - 1$ filtered FBMC symbols that overlap over time. Multiplication by the matrix $\mathbf{Q_x}$ introduces $x$ sample delay between the block of samples.

If a synchronization mismatch of $\epsilon$ samples ($\epsilon > 0$) is introduced between the FFT of the transmitter and the FFT of the receiver, the signal at the output of the FFT of the receiver may be described as

$$\hat{\mathbf{Z}}_\mathbf{m} = \mathbf{F}\mathbf{Q}_\epsilon \mathbf{F}^H \mathbf{G}\mathbf{X_m} + \mathbf{F}\mathbf{Q}_\epsilon \mathbf{P_m} \tag{19.14}$$
$$+ \mathbf{F}\mathbf{Q}_{NK-\epsilon} \mathbf{F}^H \mathbf{G}\mathbf{X_m}.$$

Fig. 19.5 gives a representation of this mathematical model. The first term of (19.14) corresponds to the received part of the multicarrier symbol of interest, the second term to an interference term introduced by the overlapped symbols, and the last term from the next first symbol that leaks at the input of the $KN$-point FFT. The expression of $\mathbf{P_m}$ is defined as

$$\mathbf{P_m} = \sum_{p=1}^{K-1} \mathbf{Q}_{\frac{pN}{2}} \mathbf{F}^H \mathbf{G}\mathbf{X_{m-p}} + \mathbf{Q}_{N-\frac{pN}{2}} \mathbf{F}^H \mathbf{G}\mathbf{X_{m+p}}. \tag{19.15}$$

Power loss and amount of interference power generated by a misalignment of the receiver FFT are two metrics that can be analyzed. The power loss ratio $P_l(\epsilon)$ between

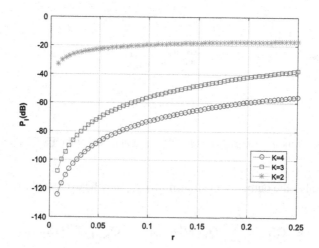

**FIGURE 19.6**

Power loss $P_l$ (dB) versus $r = \epsilon/N$ for various overlapping factors $K$.

a perfectly synchronized signal and a nonsynchronized signal has been defined by

$$P_l(\epsilon) = 1 - \frac{\text{trace}(\mathbf{FQ}_\epsilon \mathbf{F}^H \mathbf{GG}^H \mathbf{F}^H \mathbf{Q}_\epsilon{}^H \mathbf{F})}{\text{trace}(\mathbf{GG}^H)}. \tag{19.16}$$

The interference power ratio $I_l(\epsilon)$ is defined as

$$I_l(\epsilon) = \frac{\text{trace}(\mathbf{FQ}_{\text{NK}-\epsilon} \mathbf{F}^H \mathbf{GG}^H \mathbf{F}^H \mathbf{Q}_{\text{NK}-\epsilon}{}^H \mathbf{F})}{\text{trace}(\mathbf{FQ}_\epsilon \mathbf{F}^H \mathbf{GG}^H \mathbf{F}^H \mathbf{Q}_\epsilon{}^H \mathbf{F})}. \tag{19.17}$$

These two metrics have been evaluated considering the PHYDYAS prototype filter [7] for various values of the overlapping factor $K$ (the prototype filter for $K = 4$ is defined by (19.1)). Results are illustrated in Figs. 19.6 and 19.7. Assuming a coarse synchronization to the nearest most appropriate $KN$-point FFT, misalignment of the $KN$-FFT has been expressed in offset relative to $N$ using the variable $r = \epsilon/N$. Since a $KN$-point is processed every $N/2$ samples, the interval of interest for $r$ is within [0; 0.25].

For values of the overlapping ratio $K$ of 3 and above, the power loss due to the FFT position mismatch is below $-38$ dB. Only a negligible fraction of energy is lost. This is due to the shape of the time domain pulse created by the convolution of each carrier by the prototype filter. The amount of energy located at the tail of the pulse is therefore very small. For the interference ratio, Fig. 19.7 clearly indicates that the level of interference is negligible when $K = 4$. However, if $K$ is reduced to a value of 2, then the amount of interference generated by the nonsynchronized FFT becomes significant and may limit the performance of the receiver. It can be concluded that, for sufficiently large values of the overlapping ratio, signal-to-interference generated

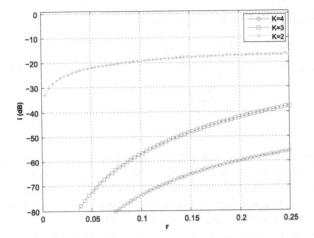

**FIGURE 19.7**

Interference to Signal Ratio $I_l$ (dB) versus $r = \epsilon/N$ for various overlapping factors $K$.

by a misalignment of the FFT is negligible when using an FS-FBMC receiver architecture. Furthermore, the performance limits introduced by the misalignment have been quantified.

The misalignment of the FFT should however be compensated before applying the matched filter. If the interference term is considered negligible (i.e., $K$ is large enough), it is possible to express the signal, $\hat{\mathbf{Z}}_\mathbf{m}$ at the output of the receiver FFT as follows:

$$\hat{\mathbf{Z}}_\mathbf{m} = \mathbf{F}\mathbf{Q}_\epsilon \mathbf{F}^H \mathbf{G}\mathbf{X}_\mathbf{m} + \mathbf{F}\mathbf{Q}_\epsilon \mathbf{P}_\mathbf{m}, \qquad (19.18)$$

$$\hat{Z}_m(k) = \Theta(0, \epsilon, KN) Z_m(k) e^{j2\pi \frac{k\epsilon}{KN}} \qquad (19.19)$$

$$+ \sum_{u=0, u \neq k}^{KN-1} \Theta(u - k, \epsilon, KN) Z_m(u) e^{j2\pi \frac{u\epsilon}{KN}}$$

$$+ \hat{P}_m(k),$$

where $\Theta(x, \epsilon, N_c)$ is defined by

$$\Theta(x, \epsilon, N_c) = \frac{1}{N_c} \frac{e^{j\pi \frac{x(N_c - \epsilon)}{N_c}}}{e^{j\pi \frac{x}{N_c}}} \frac{\sin\left(\frac{\pi}{N_c} x(N_c - \epsilon)\right)}{\sin\left(\frac{\pi}{N_c} x\right)}. \qquad (19.20)$$

The limit of $\Theta(x, \epsilon, KN)$ as $x$ tends to zero is thus equal to

$$\lim_{x \to 0} \Theta(x, \epsilon, KN) = \frac{NK - \epsilon}{NK}. \qquad (19.21)$$

Consequently, Eq. (19.19) can be written as

$$\hat{Z}_m(k) = \frac{NK - \epsilon}{NK} Z_m(k) e^{j2\pi \frac{k\epsilon}{KN}}$$

$$+ \sum_{u=0,u\neq k}^{KN-1} \Theta(u-k, \epsilon, KN) Z_m(u) e^{j2\pi \frac{u\epsilon}{KN}}$$

$$+ \hat{P}_m(k).$$

(19.22)

It clearly appears that the signal at the output of the FFT is simply modified by a phase ramp, which can be compensated before the filtering by the matched filter. The amplitude coefficient of the first term of (19.22) corresponds to the power loss evaluated in (19.16) and is depicted in Fig. 19.6. Interference introduced by the other terms is evaluated in (19.17) and depicted in Fig. 19.7.

In the multiuser scenario, provided that the start of transmission is detected, the same FFT may be used for users located in different frequency bands without particular constraints on the location of the FFT processing.

## 19.3 FS-FBMC-BASED IMPLEMENTATION AND COMPLEXITY EVALUATION

### 19.3.1 TRANSMITTER ARCHITECTURE

The tolerance on synchronization of the receiver and the reasonable complexity overhead of the single-user receiver were key motivations to further analyze the performance and complexity of the FS-FBMC receiver for transmission on fragmented spectrum, as depicted in the TVWS context. A flexible FBMC transmitter–receiver or transceiver has been implemented on the FPGA of the T-FLEX [11] platform. The transmitter architecture is composed of three main elements, forward error correction, data mapping, and modulation followed by a digital front-end (Fig. 19.8), and is compliant with the IEEE 1900.7-2015 standard. Forward error correction (FEC) is implemented around a convolutional encoder of constraint length $k = 7$ and rate $\frac{1}{2}$. The code may be punctured to support variable encoding rates. To keep a modular approach and map the code to the multicarrier modulation, the convolutional code is segmented by blocks of a few multicarrier symbols. The trellis is closed at the beginning and end of each FEC block.

The second module (Mapping and Modulation module) maps and modulates the encoded bits onto the multicarrier modulation. The coded data are mapped to a QPSK, 16QAM, or 64QAM. Symbols are then padded to complete the transmitted burst into an integer multiple of multicarrier symbols. A synchronization preamble is added to the burst structure. The preamble is used by the receiver to perform synchronization and channel estimation. The generated block of data and preamble symbols are mapped to the active carriers and modulated to an Offset-QAM before being inverse transformed into a time domain sequence. A polyphase network (PPN) filter

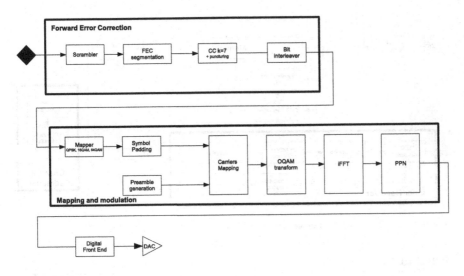

**FIGURE 19.8**

Block diagram of the FBMC transmitter.

structure completes the data stream and shapes the output of the IFFT over the duration of the prototype filter. A digital front-end completes the transmitter to allow data rate adaptation with the digital-to-analog (DAC) converter sampling rate. It consists of a set of filters and interpolators. The architecture is illustrated in Fig. 19.8.

An OFDM transmitter exhibits a very similar architecture except for the mapping and modulation module since in that case neither the OQAM transform nor the PPN structure is required. It should be noted that the OQAM modulation imposes that the iFFT is called twice as often as it would be for an OFDM transmitter. Indeed, the OQAM transformation effectively converts a block of complex data at the symbol rate into two blocks of real data at twice that rate (to keep the same overall bitrate). The carrier mapping module allows for a dynamic burst by burst selection of the active carriers.

## 19.3.2 RECEIVER ARCHITECTURE

As it was highlighted in Section 19.2.3, a flexible architecture for multiuser asynchronous reception on fragmented spectrum is able to exploit the advantages of FBMC if the signal is efficiently demodulated in the frequency domain without a priori knowledge of the FFT timing alignment (i.e., the location of the FFT block; this property is called asynchronous FFT). A receiver architecture based on this assumption is depicted in Fig. 19.9. An asynchronous FFT of size $KN$ is processed every blocks of $N/2$ samples generating $KN$ points, i.e., if $\mathbf{r_m}$ is the $m$th received vector, a $KN$-point FFT is computed for samples $k = (n + m \times N/2)$ with $n = 0, 1, \ldots, NK - 1$. These successive $KN$ points are stored in a memory unit.

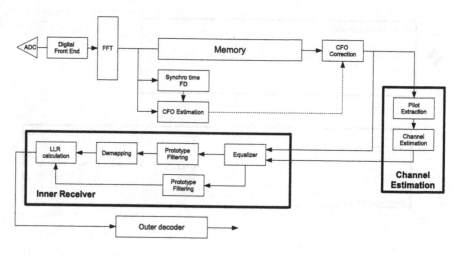

**FIGURE 19.9**

Block diagram of the FBMC receiver.

The detection of a start of burst is then achieved on the frequency domain (i.e., at the output of the FFT) using a priori information from the preamble. CFO is first estimated using the pilot subcarrier information of the preamble by computing the phase of the product between two consecutive FBMC symbols at the location of the pilot subcarriers. The propagation channel is assumed to be static for the duration of the burst. As described in [12], when large CFO correction is required, a first step in the estimation process consists of scanning the subcarriers around the pilot subcarrier locations to determine the subcarrier with the highest energy. A tracking algorithm of the CFO may complete the synchronization process when the duration of the burst is large and the accuracy of the preamble-based detection algorithm does not meet the required level [13]. CFO compensation is then performed in the frequency domain using a feed-forward approach.

The channel coefficients may be estimated on the pilot subcarriers of the preamble. In [14,15], a similar approach was already considered by introducing a phase term to correct the CFO. This technique is completed by an efficient algorithm that compensates intercarrier interference. The channel is then estimated on the pilot subcarriers before being interpolated on every active subcarrier. The use of a $KN$-point FFT makes the interpolation particularly specific to this receiver, and a description of the proposed algorithm is detailed in Section 19.2.3.

Once the channel is estimated on all the active subcarriers, a one-tap per subcarrier equalizer is applied before filtering by the FBMC prototype filter. Demapping and Log-Likelihood Ratio (LLR) computation complete the inner receiver architecture. A soft-input Forward Error Correction (FEC) decoder finally recovers the original message.

**Table 19.1** FBMC transmitter – FPGA hardware resource utilization

| Function | Resource utilization | | | |
|---|---|---|---|---|
| | **Slice Regs** | **LUTs** | **DSP48E1** | **RAM Blks** |
| Forward Error Correction | 319 | 325 | 1 | 2 |
| Mapping and Modulation | 10,981 | 7665 | 29 | 17 |
| Total FBMC Transmitter | 11,300 | 7990 | 30 | 19 |

**Table 19.2** FBMC receiver – FPGA hardware resource utilization

| Function | Resource utilization | | | |
|---|---|---|---|---|
| | **Slice Regs** | **LUTs** | **DSP48E1** | **RAM Blks** |
| FFT | 6615 | 4394 | 19 | 35 |
| Delay Line and Synchro | 7260 | 7605 | 38 | 71 |
| Channel Estimation | 13,915 | 9718 | 49 | 12 |
| Equalization and Demapping | 11,535 | 9433 | 38 | 7 |
| FEC Decoder | 2439 | 5493 | 1 | 8 |
| Control Logic | 13,206 | 13,453 | 10 | 38 |
| Total FBMC Receiver | 54,970 | 50,096 | 155 | 171 |

The asynchronous frequency domain processing of the receiver combined with the high stop-band attenuation of the FBMC prototype filter provides a receiver architecture that allows for multiuser asynchronous reception on different subbands. FFT and Memory Unit are common modules, while the remaining modules of the receiver should be duplicated as many times as the number of parallel asynchronous users the system may tolerate.

### 19.3.3 COMPLEXITY ANALYSIS

The transmitter and receiver architecture described in Sections 19.3.1 and 19.3.2 have been implemented and mapped to a Xilinx Kintex-7 FPGA platform. A complexity analysis has been performed for both transmitter and receiver. Complexity is given by the amount of resources that are used by the design to be implemented on the Kintex-7 FPGA. These are divided into Slice Registers, LUT, DSP48E1 cells, and Block RAMs. Slice Registers correspond to the amount of register cells, whereas LUT corresponds to the amount of combinatorial logic in the design. DSP48E1 cells are digital signal processing (DSP) cells dedicated to multiplication and accumulation (MAC) operations. FPGA architectures also require internal RAM storage for FIFO operations or random memory accesses and are evaluated in number of RAM Blocks of 18 kbits.

The amount of cells used by the transmitter is summarized in Table 19.1.

The amount of cells used by the FBMC receiver is summarized in Table 19.2.

The Xilinx Kintex-7 XC7K325T FPGA, which has been used to map, the design holds 407,600 Slice Regs, 326,800 LUTs, 840 DSP48E1 cells, and 890 equivalent

18 kbit blocks of RAM. Without any particular effort of design optimization, the receiver occupies less than 25% of a Xilinx Kintex-7 (XC7K325T) FPGA. Control Logic of the receiver is taking almost a quarter of the design area, this is because the receiver considered a flexible implementation of the design. It is worth pointing out that the FFT, which could have been identified as the most complex module of the receiver and only consumes around 10% of the actual receiver FPGA implementation. Clearly, the overhead of the FS-FBMC implementation has very little impact on the FFT size itself.

## 19.4 TOWARDS VLSI IMPLEMENTATION OF FS-FBMC
### 19.4.1 VLSI IMPLEMENTATION REQUIREMENTS

The architecture presented in Section 19.3 relies on a distributed memory for its design. For VLSI implementations such as an ASIC, the architecture should be reconsidered. Memory blocks are usually implemented using Static Random Access Memory (SRAM) in ASIC. A SRAM module is typically composed of the memory cells to store data, logic cells to interface the memory cells of the design (address decode, data drivers, etc.), and its associated power supply lines. Memory cells are usually heavily optimized in area and are generated using memory compilers to create a hard macrolayout for the SRAM block. This imposes to reduce the amount of RAM blocks used in the design in order to minimize the design area.

A memory centric architecture adapted to this constraint is depicted in Fig. 19.10. The proposed architecture considers dynamic support for both OFDM and FBMC to support both TVWS (IEEE 1900.7-2015 and IEEE 802.11af-2013 standards) and LTE applications.

A central memory unit dedicated to the physical layer (PHY) is at the core of the receiver. A set of coprocessor units are able to access this central memory through a high-speed PHY receiver data bus. These modules include a frequency domain synchronization coprocessor (Fig. 19.10), an FFT coprocessor (FFT and Active Carrier selection in Fig. 19.10), a DSP processor, an equalization/demapping coprocessor (equalization, filtering, and demapping coprocessor in Fig. 19.10), and an outer decoder processor. A control plane dedicated to transfer control information has been omitted on purpose in Fig. 19.10 to improve clarity of the figure. The information transiting through the control plane is of relatively low throughput.

The sampled signal received at the analog-to-digital converter is first conditioned by the digital front end to the appropriate sampling frequency into a baseband signal. A TD synchronization processor is at the output of the digital front end and determines the beginning of the burst when in OFDM mode. The module runs in parallel to the FFT module that can either be controlled by the TD synchronization module (when in OFDM mode) or be in the free running mode. Appropriate control sets the size of the FFT ($N$ when in OFDM mode or $KN$ when in FBMC mode). The FFT module is followed by an active carrier selection module that selects the active carriers and can write the result to the Memory unit through the PHY receiver bus. In the

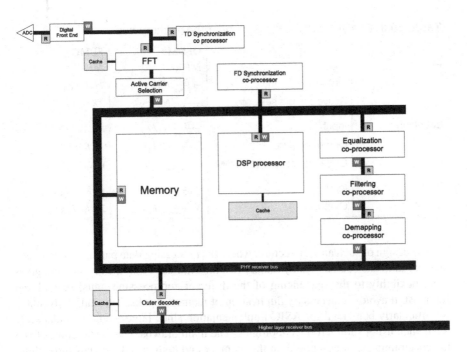

**FIGURE 19.10**

Memory centric FBMC and OFDM receiver.

case of the 8-MHz IEEE 1900.7-2015 mode, 504 carriers are typically selected out
of the 1024 points at the output of the FFT.

A FD synchronization coprocessor can then read the blocks of data samples at
the output of the FFT through the PHY receiver bus to either estimate CFO when in
OFDM mode or estimate CFO and detect start of burst when in FBMC mode. FD
synchronization output control signals are then shared through the control plane bus.

Equalization, demapping, and LLR computation are hard-wired functions assuming a data-flow architecture. Compared to classical OFDM processing, FBMC
includes an extra frequency domain filtering module. The module is therefore by-
passed when in OFDM mode. Once the demapping is done, LLR values are written
back to the shared memory for further processing by the Outer decoder. As mentioned in the previous section, demapping follows a very similar process between
FBMC and OFDM.

A dedicated digital signal processor (DSP) that can access to its dedicated cache
memory has been considered for processing operations such as deframing, pilot extraction, and channel estimation. This choice has been driven by the amount of control
that these operations required, which are therefore more adapted for implementation
by an embedded software processing unit. Finally, a dedicated outer module with its
internal (cache) memory completes the receiver. The output of the outer decoder is
connected to the higher-layer bus.

**Table 19.3** Analytic sample throughput at the memory bus

|  |  | OFDM | FBMC |
|---|---|---|---|
| **FFT** | Input | $N F_{CS}$ | $N F_{CS}$ |
|  | Output | $\frac{N}{N+N_{GI}} N F_{CS}$ | $2KN F_{CS}$ |
|  | Active carriers selection | $\frac{N}{N+N_{GI}} N_a F_{CS}$ | $2K N_a F_{CS}$ |
| **Equalizer** | Input | $\frac{N}{N+N_{GI}} N_a F_{CS}$ | $2K N_a F_{CS}$ |
|  | Output filtering | $\frac{N}{N+N_{GI}} N_a F_{CS}$ | $N_a F_{CS}$ |
|  | Output demapper ($2^m$-QAM) | $m N_a \frac{N}{N+N_{GI}} F_{CS}$ | $m N_a F_{CS}$ |
| **DSP** | Input | $\gamma \frac{N}{N+N_{GI}} N_a F_{CS}$ | $\gamma 2K N_a F_{CS}$ |
|  | Output | $N_a \frac{N}{N+N_{GI}} F_{CS}$ | $2K N_a F_{CS}$ |

The design of such an architecture, where the processing data path is centered on a memory unit, has been driven by mainly two motivations. First, the architecture gives more flexibility to the sequencing of the different coprocessing units; second and foremost, it avoids unnecessary duplication of memory banks. This latter advantage is particularly beneficial for ASIC implementation since large memory banks scale well when using submicron technologies. The main drawback however comes from the constraints that are imposed on the memory and high-speed data bus throughput since all the samples written to or read from the memory have to go through the same interface (i.e., the PHY receiver bus). The memory bus throughput has therefore been estimated in the following section to evaluate the relevance of the proposed architecture.

## 19.4.2 MEMORY BUS THROUGHPUT ESTIMATION

To estimate the constraints that have been put on the PHY receiver bus, an analysis of the throughput has been realized for this architecture. Throughput has been first evaluated analytically for the key modules of the receiver (FFT and the equalizer) for OFDM and for FBMC reception. The results have been summarized in Table 19.3 in samples per second. The following parameters have been introduced: $F_{CS}$ refers to the frequency spacing between the carrier of the multicarrier modulation, $N_a$ is the number of active carriers, $m$ is the modulation order, and $\gamma$ is the ratio of pilot carriers over all carriers. For this analysis, three different standards have been considered:

- IEEE 1900.7-2015 (1K mode) as the example for TVWS FBMC modulation considering 8-MHz channel;
- IEEE 802.11af-2013 (VHT MODE_1C) as the example for TVWS OFDM modulation considering 8-MHz channel;
- 3GPP LTE 10-MHz mode as a reference.

With the architecture of Fig. 19.10, a reconfigurable architecture able to switch between these standards could be envisaged. In the cases of IEEE 1900.7-2015 (1K mode) and IEEE 802.11af-2013, $\gamma$ is equal to 1.19% and 5.26%, respectively. It is equal to 4.76% in the case of LTE 10-MHz mode.

At the input of the FFT, the data throughput is analytically the same when the receiver receives OFDM or FBMC signals and is equal to the carrier spacing times the number of carriers in samples per second. Typically, for the worst case of IEEE 1900.7-2015, the input throughput at the FFT is equal to 15.36 Msamples/s. For OFDM, the FFT output average throughput is then divided by the ratio between $N$ and $N + N_{GI}$ as the guard interval is removed. Then active selection further reduces the throughput by $N_a/N$. The output of the FFT module is then used as an input to the equalizer, and LLR calculation increases the throughput by the modulation order $m$.

For FBMC, when FS-FBMC is considered, the FFT output throughput is however multiplied by $2K$ and therefore significantly increased. Similarly to OFDM, active carrier selection reduces the throughput by $N_a/N$. The output of the FFT module is also used as an input to the equalizer, where the throughput is divided by $2K$ once the prototype matched filtering is applied. Throughput is then increased as per OFDM by the modulation order $m$ after LLR calculation. The throughput of data processed through the DSP module essentially consists of the channel estimation and interpolation. The output of the FFT is read on the pilot tones only, and input throughput is therefore equal to the throughput of the FFT output (after active carriers selection) scaled by $\gamma$ for both OFDM and FBMC. Then channel state information is interpolated and output for every active carriers. Output throughput is thus equal to the equalizer output throughput of the FFT after active carrier selection.

A numerical application for the IEEE 1900.7-2015 1K mode (64-QAM), IEEE 802.11af-2013 VHT MODE_1C (256-QAM), and 3GPP LTE (64-QAM) has then been derived to evaluate the throughput of data at the bus and summarized in Table 19.4. The total $W$ is defined as the sum of the throughput of the output of the active carrier selection module, the output of the demapper, and the output of the DSP processor. Likewise, the total $R$ is the sum of the input of the equalizer processor, the input of the DSP processor, and the input of the outer decoder.

Since throughput also depends on the finite precision of the registers implemented in the receiver function, the following assumptions have been made: the input of the FFT is assumed to be a complex 12-bit input signal. Its output is assumed to be on 16 bits because of the FFT bit growth. Then, input to the equalizer includes both FFT output and Channel state information on 16-bit registers. Finally, LLR values are estimated to be sufficient on 6-bit registers. With these assumptions, for OFDM, the PHY receiver bus should sustain an overall write throughput of 840 Mb/s and read throughput of 853 Mb/s. Assuming a 32-bit (resp. 640-bit) transfer bus, this is equivalent to a data bus throughput of 26 Mw/s (resp. 13 Mw/s) for write operations and 26 Mw/s (resp. 13 Mb/s) for read operations (LTE case). This is relatively low when ASIC submicron implementations are considered.

However, for FBMC implementation, when similar quantization levels as the levels considered for OFDM reception are assumed, the architecture gives an overall aggregated throughput on the PHY receiver bus that is 5.9 times larger for write operations and for read operations. When a 32-bit (resp. 64-bit) data bus is considered,

**Table 19.4** Constraint on bus throughput for 10-MHz bandwidth

| | | Standard | | | | | Considering quantization | | |
|---|---|---|---|---|---|---|---|---|---|
| | | IEEE 1900,7 (FBMC, 64QAM, 8MHz) | IEEE 802,11af (OFDM, 256QAM, 8MHz) | LTE (OFDM, 64QAM, 10MHz) | | | IEEE 1900,7 (FBMC, 64QAM, 8MHz) | IEEE 802,11af (OFDM, 256QAM, 8MHz) | LTE (OFDM, 64QAM, 10MHz) |
| FFT | Input | 15.36 | 7.11 | 15.36 | Msamp/s | 2x12 | 368.64 | 170.64 | 368.64 |
| | Output | 122.88 | 5.69 | 14.35 | Msamp/s | 2x16 | 3932.16 | 182.08 | 459.23 |
| | Active Carrier Selection | 60.48 | 5.07 | 8.41 | Msamp/s | 2x16 | 1935.36 | 162.24 | 269.08 |
| Equalizer | Input | 60.48 | 5.07 | 8.41 | Msamp/s | 4x16 | 3870.72 | 324.48 | 538.16 |
| | Output filterin | 7.56 | 5.07 | 8.41 | Msamp/s | / | / | / | / |
| | Output | 45.36 | 40.53 | 50.45 | MLLR/s | 1x6 | 272.16 | 243.18 | 302.72 |
| DSP | Input | 0.72 | 0.27 | 0.4 | Msamp/s | 2x16 | 23.04 | 8.64 | 12.81 |
| | Output | 60.48 | 5.07 | 8.41 | Msamp/s | 2x16 | 1935.36 | 162.24 | 269.08 |
| Outer decoder | Input | 45.36 | 40.53 | 50.45 | MLLR/s | 1x6 | 272.16 | 243.18 | 302.72 |

| | | | | |
|---|---|---|---|---|
| | Total W | 4142.88 | 567.66 | 840.88 |
| | Total R | 4165.92 | 576.3 | 853.69 |
| | Total W+R | 8308.80 | 1143.96 | 1694.57 |
| 32 bit bus | Total W | 129.46 | 17.74 | 26.28 |
| | Total R | 130.18 | 18.01 | 26.68 |
| | Total W+R | 259.65 | 35.75 | 52.95 |
| 64 bit bus | Total W | 64.73 | 8.87 | 13.14 |
| | Total R | 65.09 | 9.00 | 13.34 |
| | Total W+R | 129.82 | 17.87 | 26.48 |

FBMC reception requires an aggregated throughput of 154 Mw/s (resp. 77 Mw/s) for read operations and 157 Mw/s (resp. 78 Mw/s) write operations. When considering a single-port RAM access, this imposes a frequency rate of read or write operation of 155 MHz when a 64-bit bus is considered. This requirement is well within the capabilities of submicron silicon technologies. Furthermore, throughput constraints are well balanced between write and read operations.

Then, albeit FBMC complexity overhead with FS-FBMC is relatively low (typically, 30% increase) [16], the constraint on the memory access is much higher (typically, 5 times higher). However, the values are far under the limits of current memory technology. Therefore, this difference is not a roadblock to VLSI implementation. An increase in power consumption is rather understood when the FBMC mode is used.

## 19.5 CONCLUDING REMARKS

Although OFDM is very well established and has a number of intrinsic benefits, nonorthogonal waveforms have gained recent interest. The excellent performance on ACLR and interference control and its flexibility in dynamic spectrum access context are the main asset of FBMC. Recent approaches using FS-FBMC are also showing an extra advantage since it manages unsynchronized users to share the same spectrum. These benefits are required in the context of the TVWS as was highlighted in this chapter. Similar requirements are now being considered in the context of

5G, where FBMC is considered as a potential contender. One of the main criticism against FBMC has been its implementation complexity. Recent research, discussed in this chapter, has shown that FS-FBMC-based implementation can maintain FBMC implementation complexity at reasonably limited increase, in particular, when demanding scenarios make FBMC advantages over OFDM desirable. On top of this limited complexity, this chapter has shown that a reconfigurable implementation of a multimode FBMC/OFDM transceiver can be envisaged with a very high level of hardware resource sharing between these modes.

# REFERENCES

[1] B. Farhang-Boroujeny, "OFDM versus filter bank multicarrier," *IEEE Signal Processing Magazine*, vol. 28, pp. 92–112, May 2011.

[2] D. Noguet, M. Gautier, and V. Berg, "Advances in opportunistic radio technologies for TVWS," *EURASIP Journal on Wireless Communications and Networking*, vol. 2011, no. 1, pp. 170–181, 2011. http://dx.doi.org/10.1186/1687-1499-2011-170.

[3] M. Nekovee, "A survey of cognitive radio access to TV white spaces," *International Journal of Digital Multimedia Broadcasting*, 2010, ID 236568.

[4] T. A. Weiss and F. K. Jondral, "Spectrum pooling: An innovative strategy for the enhancement of spectrum efficiency," *IEEE Communications Magazine*, vol. 42, no. 3, pp. S8–S14, 2004. http://dx.doi.org/10.1109/MCOM.2004.1273768.

[5] V. Berg, J.-B. Doré, and D. Noguet, "A flexible radio transmitter for TVWS based on FBMC," in *16th Euromicro Conference on Digital System Design*, Santander, Spain, Sep. 2013.

[6] *Radio Interface for White Space Dynamic Spectrum Access Radio Systems Supporting Fixed and Mobile Operation*, IEEE Std., 2015.

[7] B. M. Bellanger, *et al.*, "FBMC physical layer: A primer," Jun. 2010. [Online]. Available: http://www.ict-phydyas.org.

[8] M. Bellanger, "FS-FBMC: A flexible robust scheme for efficient multicarrier broadband wireless access," in *2012 IEEE Globecom Workshops*, Dec. 2012, pp. 192–196.

[9] J.-B. Doré, V. Berg, N. Cassiau, and D. Kténas, "FBMC receiver for multi-user asynchronous transmission on fragmented spectrum," *EURASIP Journal on Advances in Signal Processing*, vol. 2014, no. 1, pp. 41–60, 2014. http://dx.doi.org/10.1186/1687-6180-2014-41.

[10] F. Tosato and P. Bisaglia, "Simplified soft-output demapper for binary interleaved COFDM with application to HIPERLAN/2," in *IEEE International Conference on Communications, ICC 2002*, vol. 2, 2002, pp. 664–668.

[11] D. Noguet, V. Berg, X. Popon, M. Schuler, and M. Tessema, "T-FleX: A mobile SDR platform for TVWS flexible operation," in *EICE Technical Report SR2012-51, vol. 112, no. 240*, 2012, pp. 93–99.

[12] N. Cassiau, D. Kténas, and J.-B. Doré, "Time and frequency synchronization for downlink CoMP with FBMC," in *The Tenth International Symposium on Wireless Communication Systems 2013, ISWCS 2013*, 2013, pp. 46–50.

[13] P. Amini and B. Farhang-Boroujeny, "Packet format design and decision directed tracking methods for filter bank multicarrier systems," *EURASIP Journal on Advances in Signal*

*Processing*, vol. 2010, pp. 7:1–7:11, Jan. 2010. http://dx.doi.org/10.1155/2010/307983 [Online].

[14] T. Fusco, A. Petrella, and M. Tanda, "Data-aided symbol timing and CFO synchronization for filter bank multicarrier systems," *IEEE Transactions on Wireless Communications*, vol. 8, no. 5, pp. 2705–2715, 2009.

[15] T. Stitz, A. Viholainen, T. Ihalainen, and M. Renfors, "CFO estimation and correction in a WiMAX-like FBMC system," in *IEEE 10th Workshop on Signal Processing Advances in Wireless Communications, SPAWC '09*, 2009, pp. 633–637.

[16] V. Berg, J.-B. Doré, and D. Noguet, "A flexible radio transceiver for TVWS based on FBMC," *Microprocessors and Microsystems*, vol. 38, no. 8 (Part A), pp. 743–753, Nov. 2014.

**CHAPTER**

# Real-Time Implementation and Experimental Validation of a DL FBMC System

# 20

**Oriol Font-Bach\*, Nikolaos Bartzoudis\*, David Lopez Bueno\*, Xavier Mestre\*,
Philippe Mège[†], Laurent Martinod[†], Tor André Myrvoll[‡], Vidar Ringset[‡]**

*Centre Tecnològic de Telecomunicacions de Catalunya (CTTC/CERCA), Barcelona, Spain\**
*AIRBUS Defence and Space, Elancourt, France[†].*
*SINTEF Digital, Trondheim, Norway[‡]*

## CONTENTS

| | | |
|---|---|---|
| **20.1** | **Introduction** | 514 |
| **20.2** | **End-Use Scenario** | 514 |
| **20.3** | **Signal and System Specifications** | 516 |
| | 20.3.1 LTE CP-OFDM DL System | 516 |
| | 20.3.2 FBMC/OQAM DL System | 518 |
| **20.4** | **Baseband Design and Implementation** | 521 |
| | 20.4.1 Broadband FC-FBMC Transmitter: FPGA-Assisted SDR | 527 |
| |     *20.4.1.1 Single-Antenna Transmitter* | 527 |
| |     *20.4.1.2 Multiantenna Transmitter* | 532 |
| | 20.4.2 Broadband FC-FBMC Receiver: SDR Using the IRIS Framework | 533 |
| |     *20.4.2.1 Single-Antenna Receiver* | 533 |
| |     *20.4.2.2 Multiantenna Receiver* | 535 |
| **20.5** | **Real-Time Demonstrator Setup** | 536 |
| | 20.5.1 Transmitters | 537 |
| | 20.5.2 Propagation Channel | 537 |
| | 20.5.3 Receivers | 537 |
| | 20.5.4 Overall Configuration | 538 |
| **20.6** | **Experimental Validation** | 539 |
| | 20.6.1 PMR Terminal Performance | 540 |
| | 20.6.2 DL FC-FBMC Receiver Performance | 544 |
| **20.7** | **Concluding Remarks** | 544 |
| **Acknowledgments** | | 546 |
| **References** | | 546 |

Orthogonal Waveforms and Filter Banks for Future Communication Systems. DOI: 10.1016/B978-0-12-810384-5.00020-7
Copyright © 2017 Elsevier Ltd. All rights reserved.

**513**

## 20.1 INTRODUCTION

To assess the implementation feasibility and provide a practical experimental validation of the filterbank multicarrier (FBMC) scheme proposed in the EMPhAtiC project, a real-time hardware demonstrator was developed using two complementary software-defined radio (SDR) design methodologies. The spectral shape and spectral contention of the EMPhAtiC FBMC waveform makes it a prime candidate for a number of cases where agile wireless communications are required. Exploiting unused or underutilized fragmented spectrum in sub-6-GHz licensed bands without provoking interference to other inband or adjacent transmissions is the case of interest covered in this chapter. In particular, the focus is set on radio bands below 1 GHz featuring rich signal propagation characteristics. The introduction of new multicarrier waveforms and efficient cohabitation of the radio spectrum below 6 GHz are considered key enablers of fifth-generation (5G) wireless access communications.

On the implementation side, the goal of this chapter is showcase efficient digital design and SDR programming practices. It also aims to practically quantify the computational complexity cost of the defined FBMC scheme when compared with an equivalent 4G long-term evolution (LTE) waveform. The experimental validation verifies a subset of the analytical findings reported in other chapters of this book. In concrete, it underpins that efficient spectrum multitenancy of different radio access technologies (RAT) can be made feasible by exploiting the superior interference protection characteristics of the proposed FBMC waveform (i.e., when coexisting with primary transmissions in the same licensed band).

## 20.2 END-USE SCENARIO

Broadband access services are currently being introduced within the frequency allocation of professional mobile radio (PMR) networks used for public protection and disaster relief (PPDR) purposes. This is meant to satisfy emerging field operation requirements of fire, police, and rescue services, which depend on data centric applications such as situation awareness, video streaming, facial recognition, access to databases, and image/data retrieval, to mention just a few.

However, the shortage of available spectrum in specific sub-1-GHz frequency bands impedes to add broadband services. This is the reason why PMR operators and spectrum holders have already selected to exploit their underutilized licensed spectrum resources by developing multistandard base station (BS) platforms in the 400-MHz band that combine existing terrestrial trunked radio (TETRA) and TETRA for police (TETRAPOL) networks with broadband LTE technology. Likewise, PMR operators capitalized the inherent signal propagation benefits of the 400-MHz band with a minimum investment by introducing dedicated broadband coverage for PPDR purposes while offering roaming capability with commercial or private networks in other bands.

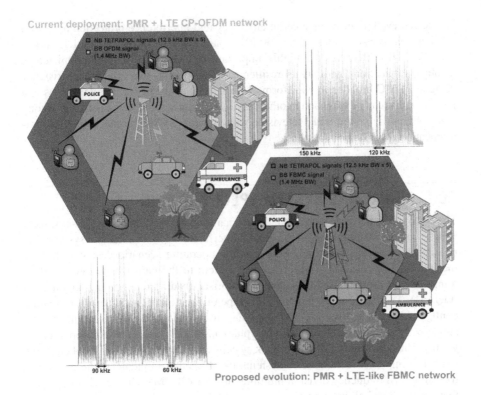

**FIGURE 20.1**

Target end use scenario: reuse of available fragmented PMR spectrum.

Despite the obtained benefits for PMR operators, such multistandard cellular systems usually imply the need of spectrum refarming and regrouping of PMR transmission to create the required contiguous unused spectrum for the LTE transmissions. This is a lengthy and cumbersome process involving different authorities and stakeholders. The alternative to spectrum refarming is to synthesize broadband LTE waveforms capable of occupying the spectral holes left by PMR transmissions. However, in such a fragmented spectrum scenario, the spectral shape of the LTE signals cannot optimally use the spectral holes left by sparse primary PMR transmissions (top right part of Fig. 20.1). A significant portion of spectrum between adjacent PMR and LTE transmissions (i.e., guard bands) must be sacrificed to guarantee interference immunity to PMR transmissions. Thus, to achieve better spectrum occupancy, the mentioned guard bands will have to be minimized or ideally suppressed. In addition, the economic delivery of PMR-based PPDR services requires to optimize the data rate as a function of the available frequency spectrum and as a function of the total number of deployed cells. In this chapter, we show that by replacing the physical layer (PHY) of LTE with the EMPhAtiC FBMC waveform significant benefits can be achieved in terms of spectral efficiency and interference protection to inband and adjacent transmissions at the 400-MHz band (as seen in the bottom left part of Fig. 20.1). Both features are highly important in mission critical communications or other applications with signal resilience requirements. The efficient coexistence of agile FBMC broadband waveforms with narrowband PMR communications is a paradigm that can be extended in other radio coexistence scenarios, involving, among many others, digital terrestrial television white spaces (TVWS) or machine-type communication (MTC) transmissions.

## 20.3 SIGNAL AND SYSTEM SPECIFICATIONS

The coexistence between a broadband and a narrowband system within the same band is based on the assumption that the mutual interference between the two systems can be kept at an acceptable level. Based on the operating scenario described before, the interference caused by the broadband system to the narrowband one could be minimized by using an FBMC scheme with offset quadrature amplitude modulation (OQAM). The inherent flexibility of this type of filter bank can be used to generate configurable waveforms that could simultaneously include filtered multitone (FMT), single-carrier, and FBMC (based on polyphase networks) signals with different bandwidths. The performance of a filterbank system with respect to out-of-band emission is greatly improved compared to systems like LTE that implements the cycle prefix orthogonal frequency division multiplexing (CP-OFDM) scheme. This important feature of FBMC allows one to create inband spectral holes in the broadband signal, which can be used to accommodate the narrowband PMR system. The width of each spectral hole depends on the air interface standard of the narrowband system. A TETRA signal with 25-kHz bandwidth requires three or four empty FBMC carriers, as shown in Figs. 20.2, 20.4, and 20.5. Since the channel spacing is 25 kHz in TETRA and 15 kHz in the FBMC system, there are three different positions of the TETRA transmission relative to the FBMC system. These are shown in Figs. 20.4 and 20.5. Similarly, a TETRAPOL signal with bandwidth 12.5 kHz requires only two or three empty FBMC carriers, as shown in Fig. 20.3.

Using CP-OFDM (LTE) instead of FBMC results in much higher interference at the narrowband system, as shown in Fig. 20.6. For optimum exploitation of the frequency band, it is necessary to allocate frequencies for the PMR systems in a manner that allows the broadband system to operate efficiently. It is therefore preferable but not mandatory to cluster PMR transmissions. By doing so the broadband FBMC system can allocate more groups of contiguous carriers (i.e., frequency channels), which results in an improved quality of the demodulated signal.

### 20.3.1 LTE CP-OFDM DL SYSTEM

The signal format of the LTE downlink frame is illustrated in Fig. 20.7. The duration of one frame is 10 ms. A frame is divided into ten subframes. Each subframe is

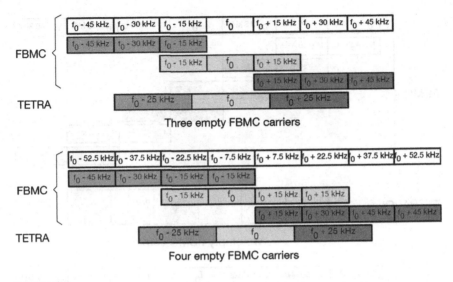

**FIGURE 20.2**

Coexistence FBMC and TETRA. The red, green, and blue colors (in the web version) indicate the empty carriers in the FBMC system when the corresponding color is used for TETRA. Upper part of figure: Three empty carriers in FBMC system. Lower part: Four empty carriers in the FBMC system.

divided in two slots. For the extended CP, each slot contains six OFDM symbols. For normal CP, there are seven OFDM symbols within a slot. The length of the CP is 32 samples for the extended and 10 samples for the normal one. The baseband sample frequency is 1.92 MHz when considering the lowest LTE bandwidth of 1.4 MHz. In the frequency direction, the frame is divided in chunks of twelve carriers with a total of 72 active carriers. The DC-carrier located in the middle of the spectrum is left unused (i.e., to mitigate the coupling with the local oscillator in the RF front-end of the transmitter). The Secondary and Primary synchronization signals (SSS and PSS, respectively) are located in symbols 4 and 5 in subframes 0 and 5. A resource block, which is the smallest quantity that can be allocated to the user, consists of twelve carriers in the frequency direction and one slot (six OFDM symbols) in the time direction. The duration of one resource block is 0.5 ms. The reference signals are located as indicated by the red squares (in the web version) in Fig. 20.7.

For the two-antenna MIMO configuration, the frame format is slightly modified as the number of reference signals is doubled. The signal format is shown in Fig. 20.8. For antenna port 0, the red squares (in the web version) are the reference signal symbols, whereas the dark blue ones are left empty. For antenna port 1, the allocation is opposite. The dark blue cells contain the reference signals, whereas the red ones are left empty. This facilitates the channel estimation at the receive side.

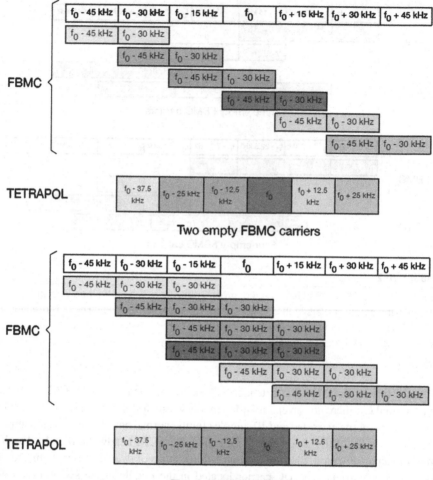

**FIGURE 20.3**

Coexistence FBMC and TETRAPOL. Upper part of figure: Two empty carriers in FBMC system. Lower part: Three empty carriers in the FBMC system.

## 20.3.2 FBMC/OQAM DL SYSTEM

FBMC transceivers are implemented with two classical ways, via polyphase networks (PN) or via fast convolution (FC) [1]. The FC-FBMC approach was followed in the present implementation. A short transform length of 8 FFT bins per subcarrier spacing (i.e., 16 points), combined with a long transform length of 1024 points, was selected to maintain a reasonable tradeoff between performance and run-time com-

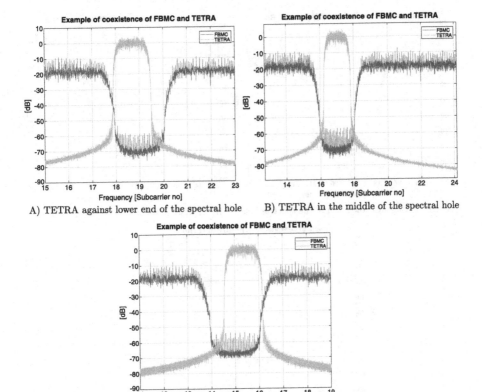

A) TETRA against lower end of the spectral hole    B) TETRA in the middle of the spectral hole

C) TETRA against the upper end of the spectral hole

**FIGURE 20.4**

Coexistence FBMC and TETRA. Three empty carriers in the FBMC system.

putational load. A nonoverlapping block length of 10 samples ensured good in-band and out-of-band interference levels.

To provide a fair comparison of the waveforms, the DC-FBMC DL frame structure maintained a high degree of similarity with the specifications of the CP-OFDM LTE (release 9) for the frequency division duplexing (FDD) mode. The absence of the CP results in a DL frame containing 150 FC-FBMC symbols, compared to the 120 symbols in the case of CP-OFDM (LTE). The content of the first subframe (after OQAM mapping) is shown in Fig. 20.9.

As in the case of LTE, the 10-ms DL FC-FBMC frame is divided into ten subframes, each consisting of 15 FC-FBMC symbols. Each carrier comprising an FC-FBMC symbol consists of two half-symbols, which is part of the OQAM mapping. The frequency axis is divided into blocks of 12 carriers. Other physical frequency allocations of active carriers can also be contemplated if the coexistence with narrowband TETRA signals needs to be optimized. To leave a 25-kHz slot unaffected by

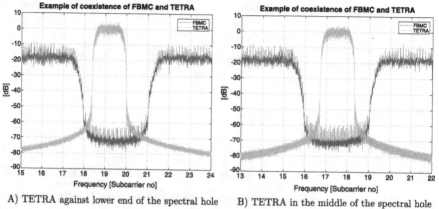

A) TETRA against lower end of the spectral hole     B) TETRA in the middle of the spectral hole

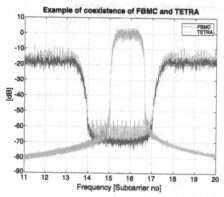

C) TETRA against the upper end of the spectral hole

**FIGURE 20.5**

Coexistence FBMC and TETRA. Four empty carriers in the FBMC system.

the FC-FBMC signal, three of four FC-FBMC carriers have to be turned off, which is significantly less than one resource block (12 carriers) in the frequency domain. The active carriers of the TETRA system should therefore be clustered in a manner that optimizes the broadband system capacity. The first three symbols in a frame contain the preamble. The structure 0P 00 P0 (i.e., inserting null half-symbols) ensures a certain isolation, i.e., reduction of intersymbol interference, to the neighboring symbols.

The pilot scheme is different from the LTE case, where there is no coupling between neighboring symbols, as long as the Doppler and channel multipath effects can be neglected. However, in FC-FBMC, the complex symbols are coupled both in time and frequency as shown in Fig. 20.10, where a real-valued signal in the red cell couples (in the web version) with the neighboring symbols indicated in green color. The complex channel estimate in the FC-FBMC receiver is based on the detection of the complex-valued received pilots. It is therefore necessary to eliminate

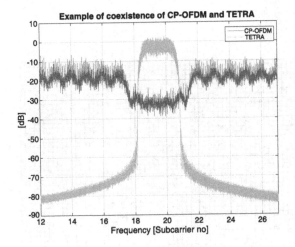

**FIGURE 20.6**

Coexistence CP-OFDM and TETRA. Four empty carriers in the OFDM system.

the influence of the surrounding data on the imaginary part of the received pilot. This was made feasible by assigning a value to the imaginary part (indicated by yellow color in Fig. 20.10). The value of this "auxiliary pilot" is calculated as the sum of the surrounding contributions (the green cells in Fig. 20.10) divided by the value in the darker green cell. The scheme shown below corresponds to the calculation of pilots in the third symbol within each subframe. For those in symbol 11, the coupling is the same, but the pilot is placed in this case in the first half OQAM symbol, whereas the auxiliary pilot in the second half. The physical mapping of the carriers is also shown in Fig. 20.9.

As far as the frame structure for the two-antenna MIMO FC-FBMC scheme is concerned, the pilots for antenna 0 are located in symbols 3 and 11, as in the SISO configuration. Auxiliary pilots are inserted in symbols 6 and 14 to produce zeros in the positions for the pilots on antenna 1. The configuration for antenna 1 is the reverse; the pilots are located in symbols 6 and 14, and the auxiliary pilots producing zeros are placed in symbols 3 and 11. The preamble is transmitted only on antenna 0. The precoding matrix used for spatial multiplexing is the following:

$$\mathbf{P} = \begin{bmatrix} 1/2 & 1/2 \\ 1/2 & -1/2 \end{bmatrix}.$$

## 20.4 BASEBAND DESIGN AND IMPLEMENTATION

SDR offers the fastest way to implement and test in realistic conditions DSP algorithms or entire baseband systems. In most of the cases, the underlying software code

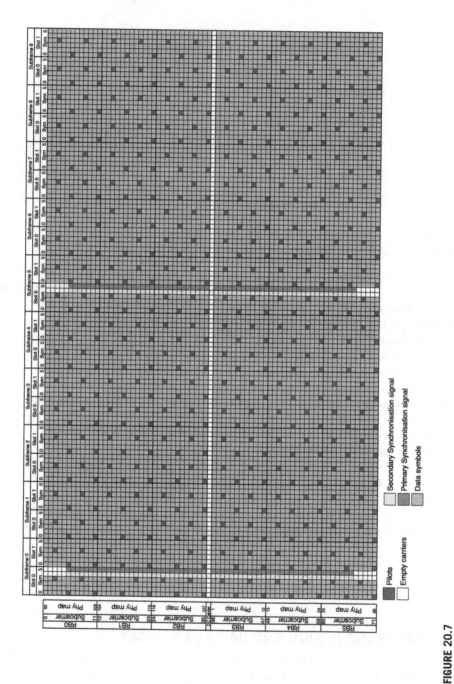

**FIGURE 20.7**

The CP-OFDM (LTE) DL signal frame structure for the SISO configuration.

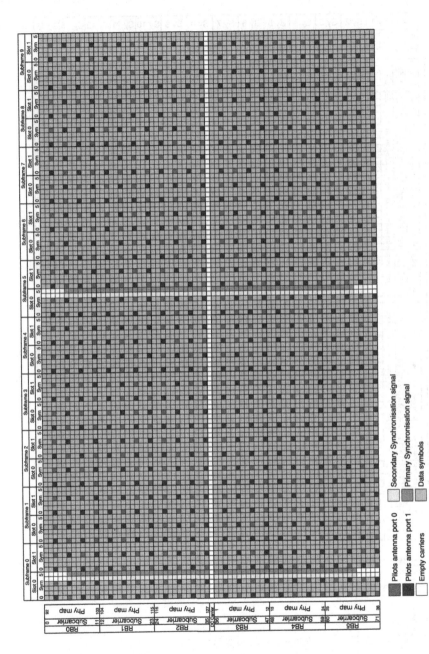

**FIGURE 20.8**

The CP-OFDM (LTE) DL signal frame structure for the MIMO configuration.

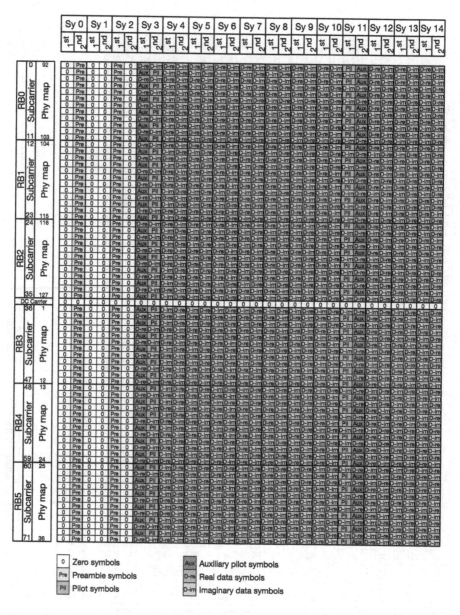

**FIGURE 20.9**

The FC-FBMC DL signal frame structure for the SISO configuration.

| Symbol no | Half symbol | Carrier N-1 | | Carrier N | | Carrier N+1 | |
|---|---|---|---|---|---|---|---|
| | | Re | Im | Re | Im | Re | Im |
| k-3 | 2 | -0.0055 | +0.0000 | 0.0000 | 0.0000 | -0.0055 | +0.0000 |
| k-2 | 1 | 0.0000 | -0.0046 | 0.0062 | 0.0000 | 0.0000 | +0.0046 |
| | 2 | 0.0080 | +0.0000 | 0.0000 | 0.0000 | 0.0080 | +0.0000 |
| k-1 | 1 | 0.0000 | +0.0433 | -0.0744 | 0.0000 | 0.0000 | -0.0433 |
| | 2 | -0.1194 | +0.0000 | 0.0000 | 0.0000 | -0.1194 | +0.0000 |
| k | 1 | 0.0000 | -0.2030 | 0.5675 | Auxiliary pilot | 0.0000 | +0.2030 |
| | 2 | 0.2398 | +0.0000 | 1.0000 | 0.0000 | 0.2398 | +0.0000 |
| k+1 | 1 | 0.0000 | +0.2030 | 0.5675 | 0.0000 | 0.0000 | -0.2030 |
| | 2 | -0.1194 | +0.0000 | 0.0000 | 0.0000 | -0.1194 | +0.0000 |
| k+2 | 1 | 0.0000 | -0.0434 | -0.0735 | 0.0000 | 0.0000 | +0.0434 |
| | 2 | 0.0082 | +0.0000 | 0.0000 | 0.0000 | 0.0082 | +0.0000 |
| k+3 | 1 | 0.0000 | +0.0044 | 0.0067 | 0.0000 | 0.0000 | -0.0044 |
| | 2 | -0.0044 | +0.0000 | 0.0000 | 0.0000 | -0.0044 | +0.0000 |

**FIGURE 20.10**

FC-FBMC filter contribution to neighboring symbols.

of SDR systems not only can serve for real-time testing and validation, but it can also be used as a classic software simulator. When the SDR development begins with the design of a high-level simulation model, the necessary modifications to transform this model to a proper real-time operating SDR system may typically require a moderate effort. One of the main reasons behind the success and adoption of SDR for academic and commercial purposes is the use of commodity, general-purpose, and inexpensive equipment, which can be combined with open source and freely available baseband processing components. Open-source SDR projects like the GNU radio offer a solid and rich SDR framework, which is constantly updated and extended by its own users and developers. Although there exist different ways to built an SDR system, a classical setup comprises (i) a computer with one or several general-purpose processors (GPPs) hosting the baseband functions and (ii) signal conversion and radio frequency (RF) transceiver hardware that is connected to the computer via a standard input/output (I/O) interface. The software code can be developed using different programming environments including schematic-entry design tools.

Despite the convenience and flexibility of software-only baseband implementations, their applicability to specific operating scenarios is bound to limitations related to processing capacity, energy consumption, or form factor. For instance, developing broadband real-time SDR systems is quite challenging due to the stringent processing requirements derived by the signal bandwidth, the baseband sample rate, or the algorithmic complexity of the underlying DSP functions. Although modern multicore GPPs feature remarkably high processing capacity and dedicated high-performance instruction sets, it is of key importance to analytically quantify the computational complexity of the baseband system before proceeding with the actual SDR implementation. On top of that, a GPP would not only host the SDR functionality, but it would execute in parallel a series of other tasks and processes using a non-real-time operating system (OS). Depending on the DSP algorithm intrinsics, a series of performance bottlenecks can impede the real-time baseband development of the SDR system. Such bottlenecks may include (i) the upper-bound processing capacity of the processor itself, (ii) the I/O communication interface between the computer and the RF front-end, (iii) the capacity of an internal bus (e.g., when different processing elements are being utilized), and (iv) the access of external memory components (e.g., when combining very low-latency and large-storage requirements).

Offloading the bit-intensive functions from the GPP to dedicated hardware co-processors is a popular solution to overcome the previously defined performance limitations. Field-programmable bit arrays (FPGA) have been widely used in SDR development to accelerate processing-demanding DSP algorithms of real-time broadband communication systems. Modern system-on-chip (SoC) devices leverage the benefits of different processing components and memory elements residing in the chip. This may include multicore embedded GPPs, hardware accelerators (including programmable logic area), on-chip ultrafast memory, and high-speed communication interfaces, enabling by this way the definition of custom coprocessing architectures. The processing capacity of such SoC devices can be exploited by following a hardware–software (HW/SW) codesign approach, which adequately partitions the baseband functions according to their processing load. Moreover, modern SoCs feature reduced form-factors with low-energy footprints, making them an attractive solution for a series of end-applications.

The development of the real-time DL FC-FBMC system presented in this chapter combined both classical and FPGA-assisted SDR implementation approaches. Fig. 20.11 shows the development methodology used in each case. A high-level simulation model was a common starting point for both design flows. The exchange of test vectors between the development teams at different stages of the coding process, ensured the functional correctness of the system prior to its experimental validation. The high-level models of the DL FC-FBMC transmitter and receiver have also served as a comparison basis in the cosimulations with a real-time implementation code. This procedure helped to optimize the hardware design and to minimize implementation losses. More details about the specific steps of this implementation methodology can be found in [2].

The translation of a high-level simulation model to a functional real-time implementation requires to account for a number of practical real-life programming issues and physical constraints. Table 20.1 provides a summary of some of these issues. The following subsections will cover in detail the design and implementation caveats of the DL FC-FBMC transmitter and receiver entities with the intention to provide a guideline on how to efficiently handle the SDR implementation of a real-time broadband wireless communication system according to the two SDR development methodologies.

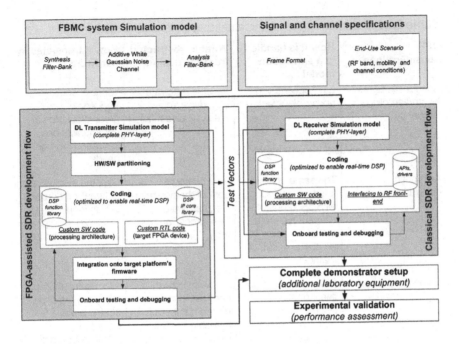

**FIGURE 20.11**

SDR-based development flow.

## 20.4.1 BROADBAND FC-FBMC TRANSMITTER: FPGA-ASSISTED SDR

To serve the needs of the spectrum cohabitation scenario featuring PMR transmissions together with inband LTE or FC-FBMC broadband transmissions, the baseband development of the real-time DL broadband transmitter had to include both the aforementioned technologies (LTE and FC-FBMC). This allowed to assess the implementation complexity and spectral agility of the two PHY-layer schemes in radio cohabitation use cases. The DL frame configuration of the two broadband signals could be configured to deactivate subcarriers with a granularity of 15 kHz in order to accommodate the primary narrowband PMR transmissions. Apart from the single-antenna scheme, the two baseband transmitters also featured a two-antenna scheme based on the open-loop spatial multiplexing configuration defined in the LTE standard. The baseband design targeted a Xilinx Zynq XC7Z045 FPGA-based SoC device, which has the native ability to be programmed using an HW/SW codesign approach. This FPGA-assisted SDR development environment offered both high computational capacity (e.g., required by the real-time FC-FBMC processing) and software flexibility (i.e., dynamic adaptation of the transmitted signal).

### 20.4.1.1 Single-Antenna Transmitter

A high-level block diagram of the broadband transmitter is shown in Fig. 20.12. To leverage the benefits of the SoC device, the bit-intensive baseband functions were

**Table 20.1** Common issues of implementing a simulation model

| DSP considerations | How it is handled in a simulation model | What is abstracted from the designer | How it translates in hardware |
| --- | --- | --- | --- |
| Changes in signal length/sampling frequency | E.g., upsampling *upsample(S,factor)* | Variable data-rate (i.e., dynamic memory allocation) | Variable working frequency, cross-clock domain communications, increased storage needs |
| Arithmetic calculations | The same operator/ function is used with operands of different type (e.g., real or complex numbers, arrays or matrices) | Real computational cost of the operation (i.e., hardware re-sources, latency, etc.) | E.g., the multiplication of two 8-bit complex values requires four 8-bit (real) multipliers and three 16-bit (real) adders |
| Numerical precision | (i) Double precision floating point (ii) Emulation of fixed-point using quantizers | Bit-width growth in contrast to the limited number of available hardware resources (i.e., operands need to be represented using $N$ bits) | – Increased storage requirements and control plane complexity – Truncations are required and may impact the system performance |
| Accessing a given subset of samples | A complete frame is stored and any (group of) sample(s) can be accessed using its index(es) | Latencies and storage requirements as result of accessing noncontiguous sample(s) (i.e., a new sample is available at each clock cycle) | E.g., an operation requires sample ($N$) and sample($N + k$): (i) buffering k ($M$-bit) samples (ii) latency of $k$ clock cycles |
| Signal impairments | (i) Ideal conditions (ii) Addition of synthetic impairments to (i) (e.g., static channel, noise) | Hardware-originated impairments that can lead to further implementation losses (e.g., nonidealities from the RF front-end) | Increased computational complexity in the DSP algorithms to estimate and compensate them |

implemented in the programmable logic (PL) area of device, whereas the generation of the dynamically reconfigurable frame was implemented in the processing system (PS) of the device. In the following, the design of the FC-FBMC transmitter will be presented in more detail.

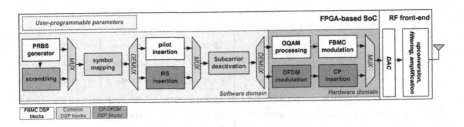

**FIGURE 20.12**

Block diagram of the single-antenna transmitter.

The first task carried out by the transmitter is the generation of the frame in the frequency domain; each FC-FBMC symbol can be seen as a matrix where QAM data are allocated. High flexibility is required in this initial DSP stage in order to be able to temporarily deactivate subcarriers in those frequencies where PMR transmissions already exist. The storage capacity is also elevated considering that each frame comprises 19,200 samples. Hence, the frame generation was implemented as a software process in the PS side of the Zynq FPGA-based SoC, whereas an adequately dimensioned embedded memory structure acted as the interface to the PL -FPGA-area.

To minimize the utilization of embedded RAM blocks in the FPGA implementation, the software-based frame generation process had to be optimized according to the real-time processing requirements, whereas a new sample had to be provided at each clock cycle to the large iFFT, and a number of prior buffering stages and added latencies had to be accounted due to the internal operation of the synthesis filterbank.

First, the OQAM processing applies an upsampling by a factor of two, returning a signal with interleaved I/Q components at the double baseband sample rate of 3.84 Msamples/s. Additionally, due to the overlap and save operation, only ten out of sixteen OQAM samples processed by each short-FFT are new; the remaining six samples are utilized in the previous FFT processing stage. Hence, if we think of the input of the large iFFT as a matrix, with 128 columns, where each corresponds to the input of a short-FFT, the FC-FBMC processing of a complete frame (i.e., 19,200 samples) requires thirty large iFFT iterations (Fig. 20.13). Under these circumstances, if the doubled baseband sample rate is maintained, then it would be required 0.53 ms to gather the required samples to feed each large iFFT (i.e., 128 short-FFTs×16 samples at 3.84 Msamples/s rate). Nevertheless, when taking into account the overlapping factor, the required samples will be available in only 0.33 ms (i.e., 128 short-FFTs×10 new samples at 3.84 Msamples/s rate). During the remaining time, 768 received extra samples have to be used in the following iFFT processing stage. This accumulative sample overflow greatly increases the buffering and processing requirements (e.g., two short-FFTs processing in parallel). On top of that, the outputs of the short-FFTs need to be combined between them in a nonsequential manner. The latter means that only the half amount of samples are eventually forwarded to the large iFFT. Finally,

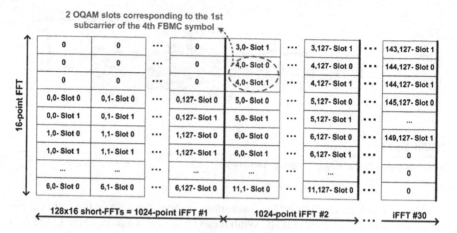

**FIGURE 20.13**

Organization of the samples in the FC-FBMC processing chain.

only a subset of the output samples of the iFFT processing block need to be forwarded to the DAC device, which must be fed with an uninterrupted sequential data flow (operating at 1.92 Msamples/s). These variable data-path storage requirements, depicted in Fig. 20.14, increase the complexity of the control and memory planes. An efficient baseband design can greatly reduce both of them. The starting point toward this end was to develop a software process running at the PS that generated the overlapped 16-sample structures before the OQAM processing. The latter is the first DSP block mapped to the PL domain.

An optimized custom digital design employing advanced register transfer level (RTL) design techniques was vital at the PL-FPGA-side to satisfy the stringent computational requirements of the FC-FBMC processing while minimizing the utilization of logic and memory resources. This was particularly important when considering that the FPGA-based SoC hosted both the FC-FBMC and CP-OFDM LTE baseband implementations. Fig. 20.15 shows the RTL design of the single-antenna FC-FBMC transmitter.

As already mentioned before, the overlapped QAM structures are fed to the OQAM processing stage at a rate of 3.84 Msamples/s, which in turn is doubled at 7.68 Msamples/s. This time division multiplexing enabled the use of a single instance of a 16-point FFT engine and a single 1024-point iFFT engine to implement the overall FC-FBMC scheme (i.e., both were based on a resource-optimized Xilinx IP core). A further reduction of the processing resources was achieved by simplifying and optimizing the arithmetic calculations of other DSP stages (e.g., the phase rotation of the OQAM samples implemented by the short-FFT BIN creation stage, which ensures phase continuity between short blocks).

The RTL design had to address a number of challenges, including the crossing of different clock domains and an efficient handling of a variable number of sample

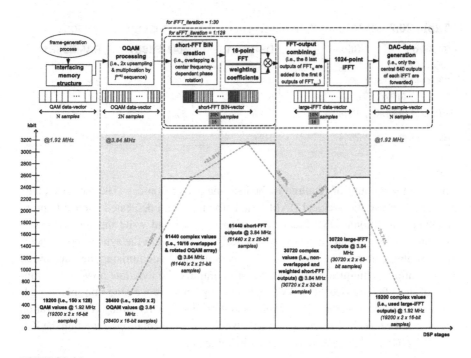

**FIGURE 20.14**

Variable data-path requirements of the proposed FC-FBMC scheme.

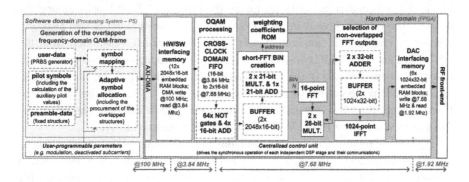

**FIGURE 20.15**

RTL design of the single-antenna FC-FBMC transmitter.

groups. On top of that, the number of bits used by each sample was also changing between processing blocks, as a result of the utilization of fixed-point logic to implement the arithmetic operations. Hence, a latency-aware memory plane was designed to serve the flexible intermediate storage needs of the pipelined DSP architecture and

**Table 20.2** FPGA utilization metrics of the transmitter implementations

| System | Slices | DSP48E1 | RAMB18E1 | RAMB36E1 |
|---|---|---|---|---|
| SISO LTE | 6% | 2% | 1% | 1% |
| MIMO LTE | 9% | 4% | 1% | 2% |
| SISO FC-FBMC | 6% | 5% | 1% | 4% |
| MIMO FC-FBMC | 13% | 11% | 2% | 10% |
| All four Txs | 26% | 18% | 4% | 16% |

optimize the amount of required embedded memory resources. The memory plane features adaptive hierarchical RAM-based structures and dedicated control logic to implement additional DSP functions during the read and write memory accesses. A centralized baseband control unit was designed to govern the synchronous operation and intercommunication of all the processing blocks comprising the transmitter. Clock-gating techniques were employed to minimize the dynamic power consumption of the FPGA-based FC-FBMC scheme (e.g., both processing and memory elements could be disabled when their operation was not required). The utilization metrics of the real-time single-antenna transmitter implementations, targeting the Xilinx ZC7Z045 SoC device, can be found in Table 20.2.

### 20.4.1.2 *Multiantenna Transmitter*

The modular and hierarchical structure of the RTL design of the single-antenna FC-FBMC DL transmitter not only facilitated the sharing of DSP blocks among the different transmitters hosted in the FPGA-based SoC device, but it was also crucial for accelerating the implementation of the two-antenna transmitter. Both the CP-OFDM and FC-FBMC two-antenna transmitters used the open-loop spatial multiplexing scheme defined in the LTE standard specifications (rel. 9). The baseband sample rate was the double of the single-antenna system. Considering the utilized frame format, the MIMO transmitter could reach a peak data-rate slightly above 714 kbits/s when using 16-QAM and having all its subcarriers active.

Fig. 20.16 provides a high-level block diagram of different DSP stages comprising the multiantenna transmitter. As it can be seen, the MIMO FC-FBMC design is basically an extension of the previously described SISO system and presents the same HW/SW block partition. Apart from replicating most of the DSP blocks in each transmit antenna (e.g., two instances of a 16-point FFT engine and two 1024-point iFFT engine instances were required), a precoding stage applied a fixed-point multiplication of each incoming sample with a predefined $2 \times 2$ precoding matrix. The centralized control unit was also extended to accommodate the storage and latency requirements of the MIMO scheme. The FPGA utilization metrics of the real-time MIMO transmitter implementations on the Xilinx ZC7Z045 SoC device are also shown in Table 20.2.

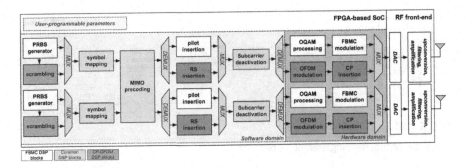

**FIGURE 20.16**

Block diagram of the two-antenna transmitter.

## 20.4.2 BROADBAND FC-FBMC RECEIVER: SDR USING THE IRIS FRAMEWORK

The *implementing radio in software* (IRIS) SDR framework was used to develop the FC-FBMC receiver. IRIS was developed as part of the FP7 CREW Project [3]. It facilitates an SDR implementation based on the data-flow paradigm, featuring components, and controllers as its main building blocks. Two main classes of components are available, the PHY and the stack. Whereas the stack components are geared toward networking and enable a two-way communication, PHY components are used to implement DSP blocks with well-defined input and outputs. As a result, the PHY-layer of the FC-FBMC receiver is represented as a data-flow structure, where the different stages are defined as PHY components and conform an acyclic graph. The latter features an unidirectional flow of information relying on intrastage memory buffers. On top of that, components are organized in engines that live on a single execution thread. This facilitates the balancing of the processing load between the different cores of the host GPP. Likewise, the resulting structure eases the resource-access control since no expensive synchronization solutions are required to enforce mutual exclusion when communicating within a PHY engine (e.g., semaphores or locking solutions). Controllers live outside the engines and communicate with the components through events, which are registered by the addressee and can contain data payloads. Furthermore, they enable the reconfiguration of the components at run-time. Each DSP block (i.e., component) is independently implemented in C++. An extensible markup language (XML) description file is then used to specify their organization in engines, interconnections, and required parameterization values.

### 20.4.2.1 *Single-Antenna Receiver*

A high-level block diagram of the single-antenna DL FC-FBMC receiver is shown in Fig. 20.17. The implementation includes two initial processing blocks. The first builds the required interfacing code to control and configure the *universal software*

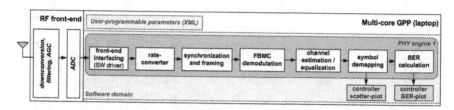

**FIGURE 20.17**

Block diagram of the single-antenna receiver.

*radio peripheral* (USRP™) N210 device (e.g., sample rate, carrier frequency), which acts as the RF transceiver. The USRP device returns a series of 32-bit floating-point complex baseband samples to the host PC. The second processing block applies a sample rate conversion before delivering the I/Q samples to the SDR FC-FBMC receiver, due to specific technical features of the utilized USRP hardware. In more detail, the sampling rate of the ADC needs to be $\frac{100}{N}$ Msamples/s, with $N$ being a multiple of two integers. Taking into account this restriction, the closest sampling rate to our target rate of 1.92 Msamples/s is 3.125 Msamples/s (i.e., $\frac{100}{32}$). As a result, a fractional rate-conversion filter had to be included, with an interpolation/decimation ratio of $\frac{384}{625}$. An efficient polyphase structure was utilized considering the large number of required filter coefficients. As a result, eight additions and eight multiplications are required per sample, which are efficiently implemented through the streaming single instruction multiple data (SIMD) extension (SSE) for Intel® processors (i.e., the i7-3720QM multicore GPP hosted the SDR implementation).

To implement the remaining PHY components, the software code was carefully written to minimize the messaging between them. This was achieved by adequately adjusting the granularity of the modular structures and by optimizing latencies and computational resources. An illustrative example of the latter is provided by the joint synchronization and framing stage shown in Fig. 20.18. Both frame detection (i.e., selection of the sample set to be forwarded to the FC-FBMC demodulation stage) and a coarse estimation of the CFO were obtained by performing a sliding correlation between the incoming samples and the known preamble values. To implement a coherent detector, the real and imaginary components of each sample needed to be handled separately, which resulted in the implementation of four concurrent real correlations (i.e., the preamble includes two FC-FBMC symbols). As a result, a single 1024-point FFT and four equivalent iFFT instances were required. This is equivalent to a computational complexity (i.e., the number of operations per sample) of $\frac{N \log N + C \cdot N}{N-L}$, where $N$ is the FFT length (i.e., 1024), $L$ is the overlap factor (i.e., 384), and $C$ includes operations such as linear time searches and absolute value computations. The fully optimized FFTW3 software library [4] provided a resource-efficient implementation of the required (i)FFT processing blocks.

Once the preamble is detected, the symbols comprising the payload (i.e., user-data and pilots) are forwarded to the FC-FBMC demodulation stage. The fast-convolution

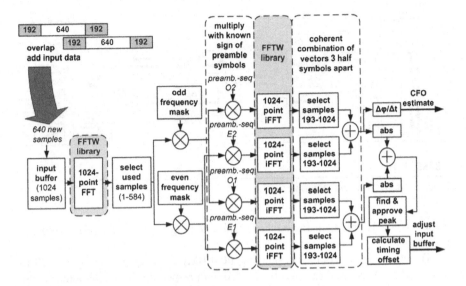

**FIGURE 20.18**

Joint synchronization and framing processing structure.

analysis filterbank is implemented using a 1024-point FFT, a series of matched filters in the frequency domain, and a 16-point iFFT. The computational complexity depends on the ratio between the FFT lengths and the overlap factor, resulting in a number of operations per sample in the order of $\frac{N \log N + 2N + KM \log M}{K}$, where $N$ is the large FFT length (i.e., 1024), $M$ the number of points of the short iFFT (i.e., 16), $L$ is the overlap factor (i.e., 384), and $K$ is the number of channels (i.e., 128). This results in 160 arithmetic operations per each real symbol.

In an attempt to keep the computational requirements within a confined processing budget, a linear interpolation was used to implement a simple pilot-based channel estimation algorithm. Consequently, a single-tap zero-forcing equalizer was used to estimate the received symbols, resulting in a linear computational complexity for the number of processed symbols. Similarly, a look-up table was utilized to implement a lightweight symbol demapping stage. Finally, the demodulated symbols were used to calculate the BER and plot the received constellation.

The SDR implementation of the receiver runs on a modern computer fitted as already mentioned with an i7-3720QM multicore GPP clocked at 2.60 GHz. The utilization of the processor reached up to 60% of a single core, whereas the memory footprint was negligible (i.e., tens of megabytes).

### 20.4.2.2 Multiantenna Receiver

As in the case of the transmitter, the SDR implementation of the MIMO FC-FBMC receiver largely reused the code developed for its single-antenna counterpart. Hence,

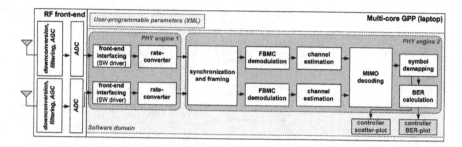

**FIGURE 20.19**

Block diagram of the two-antenna receiver.

the text below is only focusing on those differentiating factors of the resulting implementation.

The block diagram of the two-antenna FC-FBMC receiver is shown in Fig. 20.19. The number of received samples is doubled with respect to the single-antenna case. The interleaved received samples for two antennas needed to be separated before starting to be processed by the receiver. The synchronization was realized with independent correlations computed for each antenna sample-stream (i.e., doubling the number of implemented arithmetic operations with respect to the SISO receiver). The frame detection was thus triggered by the first peak occurrence in any of the two branches. The channel estimation and equalization stages quadruplicated the computational complexity with respect to the single-antenna case. In more detail, the $2 \times 2$ MIMO configuration required the estimation of the channel response between each transmit and receive antenna (i.e., solving four linear equations for each received symbol). As a result, the MIMO receiver requires more than twice the resources when compared to the previously described SISO SDR receiver implementation. As a result, the DSP computation had to be distributed across several GPP cores.

Given the FC-FBMC system specifications (i.e., small signal bandwidth, overlap and save filterbank scheme, and simple DSP algorithmic) and the observed computational load of the full software implementation, it is made clear that a FC-FBMC receiver with more stringent signal specifications or improved baseband signal processing algorithms will be difficult, if not unfeasible, to implement with the classic SDR approach. As it has been already mentioned before, the bit-intensive SDR programming challenges could be tackled by employing FPGA-based SoCs in a hardware-accelerated SDR implementation flow.

## 20.5 REAL-TIME DEMONSTRATOR SETUP

Two different hardware setups have been used for the real-time demonstrator. The first helped to conduct a measurement campaign whose goal was to quantify the level of interference received by the TETRAPOL terminal (i.e., primary communication system) when coexisting either with an LTE or FC-FBMC DL transmission in the same band (i.e., recreating the scenario seen in Fig. 20.1 with close to real-life emulated mobility conditions). The second hardware setup involved the same secondary transmissions (either LTE or FC-FBMC), but this time, the goal was to assess the perceived quality of the device-to-device voice communication using two TETRAPOL terminals configured with the direct mode operation (DMO). A detailed list of the boards, devices, and equipment used for the measurement campaign is given in the following. The two setups are briefly described in the final subsection.

## 20.5.1 TRANSMITTERS

- A vector signal generator (Agilent E4438C ESG) emulated a real-time PMR transmitter by cyclically reproducing the known TETRAPOL I/Q frame-sequence (loaded as a .MAT file) and providing the RF signal centered at 382.53 MHz.
- The Xilinx ZC706 board hosted both the DL FC-FBMC and LTE real-time baseband implementations. The Xilinx board carried the AD-FMCOMMS3 RF transceiver from Analog Devices, which provided the FC-FBMC or LTE RF signal centered at 382.88 GHz.

## 20.5.2 PROPAGATION CHANNEL

- The RF signals were fed to the Elektrobit Propsim C8 real-time channel emulator, which recreated realistic channel propagation conditions and applied mobility effects. The signals at the output of the channel emulator were driven through a series of RF step attenuators to control the power ratio between the two coexisting transmissions. The power-adjusted RF signals were then combined, and a replica of the combined RF signals fed the FC-FBMC and PMR receivers, respectively.

## 20.5.3 RECEIVERS

- A TETRAPOL terminal (M9620SG2) tuned in the frequency channel of interest and set in test mode was connected through a custom cable to a computer, where a software application calculated the average bit error rate (BER) observed at the terminal for a given number of frames.
- The FC-FBMC DL signal reception was made feasible using an Ettus Research USRP N210 device, which applied RF down-conversion, analog-to-digital signal conversion, and digital down conversion. The baseband complex samples were forwarded to a computer through a gigabit Ethernet (GigE) cable connection. There the samples were retrieved and forwarded to the FC-FBMC receiver SDR implementation running in the host processor, where the frame-based BER metrics were calculated and stored.

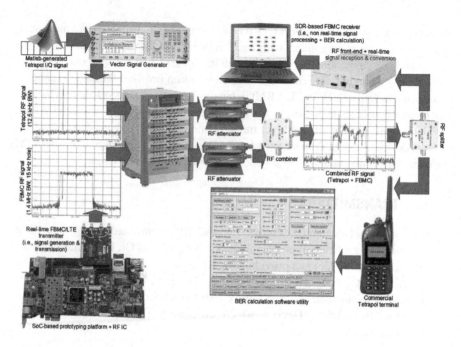

**FIGURE 20.20**

The demonstrator setup configured to assess the performance of the primary TETRAPOL communication when coexisting at the same band with different broadband waveforms.

## 20.5.4 OVERALL CONFIGURATION

The complex hardware setup used for the measurement campaign is seen in Fig. 20.20. This configuration is used for the quantitative assessment of the TETRAPOL terminal performance under different broadband signal coexistence conditions (e.g., number of disactivated subcarriers, power relation between the signals, and mobile channels). It can also be used to quantify the performance of the broadband FC-FBMC system. The broadband transmitter DL frame configuration (either LTE or FC-FBMC) can be flexibly tuned through a dedicated programming interface. The latter, apart from allowing one to set parameters like the modulation scheme or which of the two transmitters is currently activated, also allows one to disactivate subcarriers at the broadband signal in order to accommodate existing in-band transmissions. This means that the transmitter, apart from being able to serve a fixed frequency planning (a priori knowledge of the frequencies where narrowband transmission exist), can be employed for opportunistic use of the available spectrum. In that case, spectrum sensing must be added to the system. A second minimalist variation of this setup is seen in Fig. 20.21. This helped us to get a tangible qualitative assessment of the superior waveform characteristics featured in FC-FBMC when compared with LTE under the same waveform coexistence conditions.

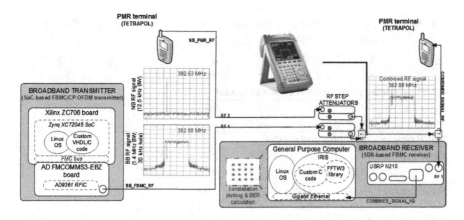

**FIGURE 20.21**

The demonstrator setup configured to assess the perceived voice quality when two TETRAPOL terminals communicate in the same band where the FC-FBMC and LTE broadband waveforms coexist.

## 20.6 EXPERIMENTAL VALIDATION

The development of a real-time demonstrator enabled a credible validation and performance assessment of the proposed scheme under realistic operating conditions. The DL FC-FBMC system was validated using an incremental testing approach. The first baseline hardware setup included a cabled baseband-to-baseband connection omitting signal impairments. Once the functionality of the system was verified, cabled RF transceivers were added to the hardware setup. Finally, the system was tested under different channel propagation conditions, which were made feasible either through over-the-air transmissions or by using the EB C8 Propsim real-time multichannel emulator.

The experimental validation of the real-time DL FC-FBMC system is based on a comprehensive measurement campaign that was conducted using the previously described hardware setup. In more detail, there was evaluated the impact of a secondary DL broadband transmission on the performance of a mission-critical TETRAPOL system operating at the 400-MHz band. Four major parameters were modified between the different test cases: the broadband signal type (i.e., LTE or FC-FBMC), the subcarrier power ratio between the coexisting transmissions (i.e., the level of interference to the PMR system), the size of the spectral hole used in the secondary signal to accommodate the TETRAPOL transmission (i.e., the number of deactivated subcarriers), and the channel conditions (i.e., channel type and mobility). The BER of the TETRAPOL and FC-FBMC system was calculated by averaging 10,000 frames for each of the numerous configuration cases.

**Table 20.3** Relative delay and average power of the channel taps. The taps of each channel follow a Rayleigh distribution

| Tap | 1 | 2 | 3 | 4 | 5 | 6 | 7 |
| --- | --- | --- | --- | --- | --- | --- | --- |
|  | 8 | 9 | 10 | 11 | 12 | | |
| *3GPP extended pedestrian A channel model* | | | | | | | |
| Delay (ns) | 0 | 30 | 70 | 90 | 110 | 190 | 410 |
| Power (dBm) | 0.0 | −1.0 | −2.0 | −3.0 | −8.0 | −17.2 | −20.8 |
| *3GPP typical urban channel model* | | | | | | | |
| Delay (ns) | 0 | 100 | 300 | 500 | 800 | 1100 | 1300 |
|  | 1700 | 2300 | 3100 | 3200 | 5000 | | |
| Power (dBm) | −4.0 | −3.0 | 0.0 | −2.6 | −3.0 | −5.0 | −7.0 |
|  | −5.0 | −6.5 | −8.6 | −11.0 | −10.0 | | |
| *ITU vehicular A channel model* | | | | | | | |
| Delay (ns) | 0 | 310 | 710 | 1090 | 1730 | 2510 | – |
| Power (dBm) | 0 | −1.0 | −9.0 | −10.0 | −15.0 | −20.0 | – |

Three different channel models were selected for the experimental validation of the system (Table 20.3). Namely, the 3GPP extended pedestrian A (EPA, 3 km/h), typical urban (TU, 50 km/h), and the ITU vehicular A (VA, 50 km/h).

## 20.6.1 PMR TERMINAL PERFORMANCE

The impact of the hole-size left by the broadband system on the performance of the TETRAPOL system was validated utilizing a VA channel. For each of the different considered spectral hole sizes, the hardware setup was configured to obtain a similar BER performance at the TETRAPOL terminal (i.e., a value around 0.4%, which represents a high-quality voice communication) by adjusting the subcarrier power-ratio of the two coexisting transmissions. Likewise, there was evaluated the effect of the spectral hole length on the performance of the PMR system as a function of its received signal power (top of Fig. 20.22). The results show that by increasing the guard-band size the measured interference at the PMR system was reduced up to 10 dB when using the FC-FBMC waveform and negligible effects when the secondary broadband transmission was the LTE one. The benefits of using the FC-FBMC DL system were maintained under fixed received signal power at the TETRAPOL terminal (i.e., −98 dBm) and for a wide range of interference conditions (bottom of Fig. 20.22). Furthermore, when the received signal power, guard-band size, and resulting BER at the TETRAPOL terminal were maintained, the secondary DL FC-FBMC transmission was capable of using a subcarrier power level up to 30 dB higher than that of the PMR system when compared to equivalent setup using a DL LTE transmission. A useful output of the experimental validation was that a two-subcarrier-wide spectral hole offers the optimum trade-off between spectral efficiency and signal coexistence of broadband FC-FBMC and TETRAPOL transmissions.

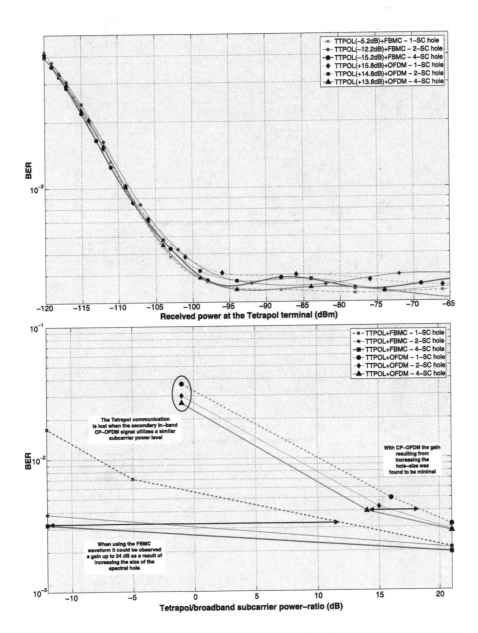

**FIGURE 20.22**

Performance of the PMR terminal in coexistence with broadband transmissions providing spectral holes of different sizes and subcarrier power-level ratios.

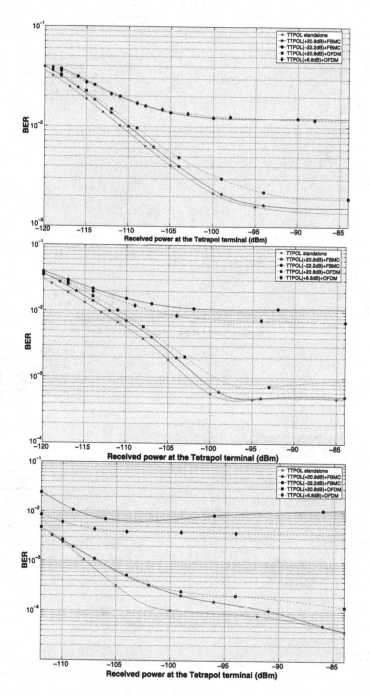

**FIGURE 20.23**

Performance of the PMR terminal evaluated in relation to the interference received from the in-band broadband transmissions for different received signal powers.

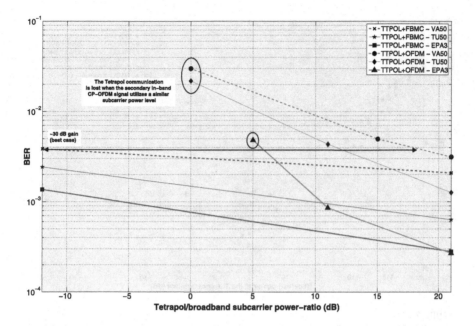

**FIGURE 20.24**

Performance of the PMR terminal evaluated in relation to the interference received from the in-band broadband transmissions for a fixed received power of −98 dBm.

The interference-protection benefits of the FC-FBMC waveform were then evaluated under different mobile channels for a wide range of power ratios between the broadband and narrowband transmissions. These started from a primary PMR signal that was 21 dB above the broadband one, up to a case where the TETRAPOL signal was almost 22 dB below the secondary transmission. As it can be seen in Fig. 20.23, similar results were observed for the VA (top), TU (middle), and EPA (bottom) channels. A gain up to 30 dB with respect to the LTE transmitter was once again observed when considering a fixed received signal power at the TETRAPOL terminal for all considered channel models (Fig. 20.24). It is important to note that the TETRAPOL terminal stopped working (i.e., could not synchronize) whenever the received power of the PMR signal was close (i.e., less than 5 dB above) to that of the LTE transmitter, whereas a good voice quality was still attained when switching to the FC-FBMC waveform (in fact, even for higher transmitted powers).

Finally, there was also evaluated the effect of utilizing a multiantenna scheme in the broadband systems. For both MIMO FC-FBMC and LTE transmissions, a 30-kHz spectral hole was provided. Nearly identical results were obtained with those of the single-antenna transmitter under a VA channel (Fig. 20.25). The MIMO FC-FBMC waveform provides an interference protection margin of up to 29 dB compared to its LTE counterpart when measuring the performance of the TETRAPOL terminal.

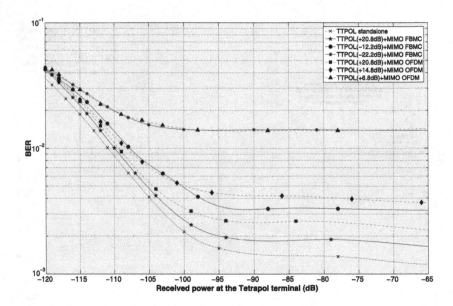

**FIGURE 20.25**

Performance of the PMR terminal evaluated in relation to the interference received from the in-band MIMO broadband transmissions.

## 20.6.2 DL FC-FBMC RECEIVER PERFORMANCE

The correct operation of the real-time DL FC-FBMC baseband implementation was validated by obtaining a basic set of BER metrics at the receiver (Fig. 20.26). A fixed 30-kHz guard band accommodated the TETRAPOL transmission. Two indicative configurations were considered, one where the subcarrier power-level ratio was favorable for the FC-FBMC system and an opposite one. No clear variation of the FC-FBMC receiver performance was observed when the subcarrier power level of the primary system was below that of the broadband one. The comparison reference was the case where there was no TETRAPOL transmission at the assigned band. Even when the system operates on the contrary situation, with a TETRAPOL signal power nearly 21 dB above the FC-FBMC one, the observed performance-loss was found to be minimal. In fact, it was only noticeable under the EPA channel, the most sensible one to variations in the power ratios among the coexisting signals during the PMR performance analysis. It is important to notice that the broadband transmitters do not include channel coding (hence, the figures show the raw BER).

## 20.7 CONCLUDING REMARKS

FBMC modulations is an efficient PHY-layer scheme with inherent ability to operate under spectral coexistence scenarios. Contrary to CP-OFDM waveforms like

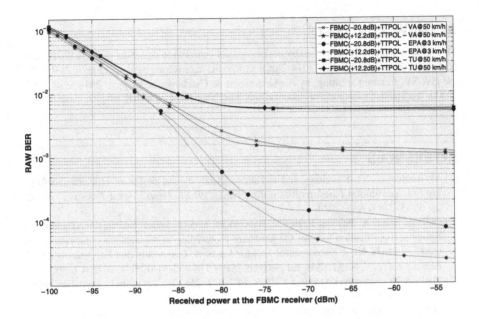

**FIGURE 20.26**

Performance observed at the broadband FC-FBMC receiver under different operating conditions.

LTE, FC-FBMC signals feature a near optimal spectral containment, providing at the same time excellent interference protection to adjacent and in-band transmissions. Such characteristics can be exploited in underutilized frequency bands below 1 GHz for PPDR or other end-applications. The implementation feasibility of spectrally agile FC-FBMC signals has been verified using two different SDR development approaches. The prime waveform characteristics of FC-FBMC have been evaluated and experimentally quantified targeting a real-life waveform coexistence scenario including PMR communications. The SDR implementations demonstrated that FPGAs should be employed when the baseband computational requirements are high (i.e., assisted SDR design approach). The interference protection that FC-FBMC waveforms are able to offer to coexisting waveforms can be directly mapped to other spectrum cohabitation cases that share similar functional and operational conditions. An experimental evaluation of such scenarios would have to analyze the FC-FBMC waveform particulars in a case-by-case basis (e.g., ideal size of the spectral holes and power relation of the coexisting transmissions). Although the experimental work focused on a single primary waveform spectrum cohabitation case, the interference protection benefits are encouraging the adoption of FC-FBMC waveforms in those bands where primary critical communication systems operate.

## ACKNOWLEDGMENTS

This work was partially supported by the European Commission under the projects EMPhAtiC (GA 318362) and H2020-ICT-2014-2 project Flex5Gware (Grant agreement no. 671563) for the HW-SW function splitting; by the Generalitat de Catalunya under grant 2014 SGR 1551; and by the Spanish Government under project TEC2014-58341-C4-4-R.

## REFERENCES

[1] M. Renfors, J. Yli-Kaakinen, and F. Harris, "Analysis and design of efficient and flexible fast-convolution based multirate filter banks," *IEEE Trans. Signal Process.*, vol. 62, pp. 3768–3783, Aug. 2014.

[2] O. Font-Bach, A. Pascual-Iserte, N. Bartzoudis, and D. López Bueno, "MATLAB as a design and verification tool for the hardware prototyping of wireless communication systems," in *MATLAB, A Fundamental Tool for Scientific Computing and Engineering Applications, vol. 2*, V. N. Katsikis, Ed., InTech. ISBN 978-953-51-0751-4, 2012 (Ch. 9).

[3] "IRIS SDR framework (Project CREW FP7-ICT 258301)." [Online]. Available: http://www.crew-project.eu/iris.

[4] "FFTW Fourier transform library." [Online]. Available: http://www.fftw.org.

# List of Acronyms and Math Symbols

| | |
|---|---|
| **3GPP** | 3rd Generation Partnership Project |
| **4G** | 4th Generation |
| **5G** | 5th Generation |
| **ADC** | Analog to Digital Converter |
| **ADSL** | Asymmetric Digital Subscriber Line |
| **ACK** | ACKnowledgment |
| **ACLR** | Adjacent Channel Leakage Ratio |
| **ACPR** | Adjacent Channel Power Ratio |
| **AFB** | Analysis FilterBank |
| **APCO** | Association of Public safety Communications Officials |
| **ASIC** | Application-Specific Integrated Circuit |
| **ASK** | Amplitude-Shift Keying |
| **ATA** | Autonomous Timing Advance |
| **AWGN** | Additive White Gaussian Noise |
| **B2B** | Back-to-Back |
| **BDFM** | Biorthogonal Frequency-Division Multiplexing |
| **BEP** | Bit Error Probability |
| **BER** | Bit Error Rate |
| **BFDM-OQAM** | Biorthogonal Frequency-Division Multiplexing with Offset-QAM subcarrier modulation |
| **BLAST** | Bell Labs Layered Space Time [code] |
| **BLEP** | Block Error Probability |
| **BLER** | Block Error Rate |
| **BLT** | Balian–Low Theorem |
| **BLUE** | Best Linear Unbiased Estimator |
| **BPSK** | Binary Phase-Shift Keying |
| **BS** | Base Station |
| **CAS** | Carrier Assignment Scheme |
| **CB-FMT** | Cyclic Block-Filtered MultiTone |
| **CC** | Circular (or Cyclic) Convolution |
| **CCDF** | Complementary Cumulative Distribution Function |
| **CCE** | Cross-Correlation Estimator |
| **CCI** | CoChannel Interference |
| **CD** | Chromatic Dispersion |
| **CDF** | Cumulative Distribution Function |
| **CDM** | Code-Division Multiplexing |
| **CDMA** | Code-Division Multiple Access |
| **CEPT** | European Conference of Postal and Telecommunications administrations |
| **CFE** | Closed-Form Estimator |
| **CFO** | Carrier Frequency Offset |
| **CFR** | Channel Frequency Response |
| **CG** | Conjugate Gradient |
| **CIR** | Channel Impulse Response |
| **CMA** | Constant Modulus Algorithm |

| | |
|---|---|
| **CMFB** | Cosine-Modulated FilterBank |
| **CMOS** | Complementary Metal-Oxide-Semiconductor |
| **CMT** | Cosine-modulated MultiTone |
| **CNA** | Constant Norm Algorithm |
| **COFDM** | Coded Orthogonal Frequency-Division Multiplexing |
| **CoMP** | Coordinated MultiPoint |
| **COQAM** | Circular Offset Quadrature Amplitude Modulation |
| **CQI** | Channel Quality Indicator |
| **CP** | Cyclic Prefix |
| **CP-OFDM** | Cyclic Prefix Orthogonal Frequency-Division Multiplexing |
| **CR** | Cognitive Radio |
| **CRC** | Cyclic Redundancy Check |
| **CRLB** | Cramér–Rao Lower Bound |
| **CRS** | Cell-specific Reference Signal |
| **CS** | Control Symbol |
| **CSI** | Channel State Information |
| **CSMA-CA** | Carrier Sense Multiple Access with Collision Avoidance |
| **CSP** | Conjugate Symmetry Property |
| **CW** | Continuous Wave |
| **D2D** | Device-to-Device |
| **DAB** | Digital Audio Broadcasting |
| **DAC** | Digital to Analog Converter |
| **DADP** | Dual Asynchronous Distributed Pricing |
| **DC** | zero-frequency [initially, Direct Current] |
| **DFB** | Distributed FeedBack |
| **DFE** | Decision Feedback Equalizer |
| **DFT** | Discrete Fourier Transform |
| **DFT-FB** | DFT FilterBank |
| **DL** | DownLink |
| **DMO** | Direct Mode of Operation |
| **DMT** | Discrete MultiTone |
| **DPD** | Digital PreDistortion |
| **D-PRACH** | Data Physical Random Access CHannel |
| **DSATUR** | Degree SATURation |
| **DSL** | Digital Subscriber Line |
| **DSO** | Digital Switch Over |
| **DSP** | Digital Signal Processing |
| **DTFT** | Discrete-Time Fourier Transform |
| **DTR** | Dispersive Tone Reservation |
| **DTT** | Digital Terrestrial Television |
| **DTWE** | Discrete Time Wilson Expansion |
| **DVB-T** | Digital Video Broadcasting-Terrestrial |
| **DWMT** | Discrete Wavelet MultiTone |
| **DZT** | Discrete Zak Transform |
| **ECL** | External Cavity Laser |
| **ECMA** | European Computer Manufacturers Association |
| **EESM** | Exponential ESM |
| **EGF** | Extended Gaussian Function |

| | |
|---|---|
| **E-IAM-C** | Extended IAM-C |
| **EIRP** | Equivalent Isotropically Radiated Power |
| **EM** | Expectation Maximization |
| **EMFB** | Exponentially Modulated FilterBank |
| **EOM** | Electro-Optic Modulator |
| **ESM** | Effective SINR Mapping |
| **EU** | EUropean |
| **ETU** | Extended Typical Urban [channel model] |
| **EVA** | Extended Vehicular-A [channel model] |
| **EVM** | Error Vector Magnitude |
| **FB** | FilterBank |
| **FBMA** | FilterBank Multiple Access |
| **FBMC** | FilterBank MultiCarrier |
| **FBMC/COQAM** | FilterBank MultiCarrier with Circular Offset Quadrature Amplitude Modulation |
| **FBMC/OQAM** | FilterBank MultiCarrier with Offset-QAM subcarrier modulation |
| **FBMC-QAM** | FilterBank MultiCarrier with QAM subcarrier modulation |
| **FB-SC** | FilterBank Single-Carrier |
| **FC** | Fast-Convolution |
| **FC-FB** | Fast-Convolution FilterBank |
| **FC-OFDM** | Flexible Configured OFDM |
| **FD** | Frequency Domain |
| **FDD** | Frequency-division duplexing |
| **FDM** | Frequency-Division Multiplexing |
| **FEC** | Forward Error Correction |
| **FF** | Fast Fading |
| **FFT** | Fast Fourier Transform |
| **FIFO** | First-In, First-Out |
| **FIR** | Finite Impulse Response |
| **FLO** | Frequency-Limited Orthogonal |
| **FMT** | Filtered MultiTone |
| **F-OFDM** | Filtered OFDM |
| **FP7** | Framework Programme 7 |
| **FPGA** | Field Programmable Gate Array |
| **FS** | Frequency Sampling |
| **FS-FBMC** | Frequency-Spreading FilterBank MultiCarrier |
| **F-T** | Frequency-Time |
| **FWM** | Four-Wave Mixing |
| **GFDM** | Generalized Frequency-Division Multiplexing |
| **GI** | Guard Interval |
| **GigE** | Gigabit Ethernet |
| **GLDB** | GeoLocation DataBase |
| **GMSK** | Gaussian Minimum-Shift Keying |
| **GPP** | General Purpose Processors |
| **GSM** | Global System for Mobile communications |
| **HD-FDD** | Half Duplex Frequency-Division Duplexing |
| **HOS** | High-Order Statistics |

| | |
|---|---|
| HP | Help Pilot |
| HPA | High Power Amplifier |
| HT | Hilly Terrain [channel model] |
| HW | HardWare |
| IAM | Interference Approximation Method |
| IBO | Input Back-Off |
| IC | Interference Cancellation |
| ICI | Inter-Carrier Interference |
| ICM | Interference Cancellation Method |
| ICT | Information and Communications Technology |
| IDFT | Inverse Discrete Fourier Transform |
| IDMA | Interleave Division Multiple Access |
| IEEE | Institute of Electrical and Electronics Engineers |
| IFFT | Inverse Fast Fourier Transform |
| IIC | Iterative Interference Cancellation |
| i.i.d. | independent and identically distributed |
| IIM-CBF | Intrinsic Interference Mitigating Coordinated Beamforming |
| IIR | Infinite Impulse Response |
| ILA | Indirect Learning Architecture |
| IMDD | Intensity Modulation Direct Detection |
| IoT | Internet of Things |
| IOTA | Isotropic Orthogonal Transform Algorithm |
| I/Q | In-phase/Quadrature [complex data signal components] |
| IQM | In-phase/Quadrature Modulation [complex data signal components] |
| IRIS | Implementing Radio In Software |
| ISI | Inter-Symbol Interference |
| ITU | International Telecommunication Union |
| ITU-R | International Telecommunication Union Radiocommunication sector |
| LDPC | Low-Density Parity-Check [code] |
| LLR | Log-Likelihood Ratio |
| LMMSE | Linear Minimum Mean-Squared Error |
| LMS | Least Mean Squares |
| LO | Local Oscillator |
| LPSV | Linear Periodically Shift Variant |
| LS | Least Squares |
| LTE | Long-Term Evolution |
| LTE-A | Long-Term Evolution-Advanced |
| LTI | Linear Time-Invariant |
| LUT | Look Up Table |
| M2M | Machine-to-Machine |
| MAC | Medium Access Control |
| MAC | Multiplication and AcCumulation [DSP operations] |
| MBB | Mobile BroadBand |
| MBMS | Multimedia Broadcast Multicast Service |
| MBSFN | Multimedia Broadcast Single Frequency Network |
| MC | MultiCarrier |
| MCM | MultiCarrier Modulation |

| | |
|---|---|
| **MCS** | Modulation and Coding Scheme |
| **MDFT** | Modified Discrete Fourier Transform |
| **MI** | Mutual Information |
| **MIMO** | Multiple-Input-Multiple-Output |
| **MISO** | Multiple-Input-Single-Output |
| **ML** | Maximum Likelihood |
| **MLE** | Maximum Likelihood Estimator |
| **MLP** | MultiLayer Perceptron |
| **MLSE** | Maximum Likelihood Sequence Estimation |
| **MLT** | Modulated Lapped Transform |
| **MMA** | MultiModulus Algorithm |
| **MMIB** | Mean MI per coded Bit |
| **MMSE** | Minimum Mean-Squared Error |
| **MP** | Memory Polynomial |
| **MRC** | Maximum Ratio Combining |
| **MS** | Mobile Station |
| **MSE** | Mean-Squared Error |
| **MTC** | Machine Type Communication |
| **MSK** | Minimum-Shift Keying |
| **MU** | MultiUser |
| **MUI** | Multi-User Interference |
| **MUSIC** | MUltiple SIgnal Classification [algorithm] |
| **MZM** | Mach–Zehnder Modulator |
| **NACK** | No ACKnowledgment |
| **NATO** | North Atlantic Treaty Organization |
| **NC-OFDM** | NonContiguous Orthogonal Frequency-Division Multiplexing |
| **NLMS** | Normalized Least Mean Squares |
| **NLSE** | NonLinear Schrödinger Equation |
| **NG** | Natural Gradient |
| **NMSE** | Normalized Mean-Squared Error |
| **NN** | Neural Network |
| **NPR** | Near Perfect Reconstruction |
| **NPSTC** | National Public Safety Telecommunications Council |
| **NTIA** | National Telecommunications and Information Administration |
| **N-WDM** | Nyquist Wavelength-Division Multiplexing |
| **OBE** | Out-of-Band Emission |
| **OFDM** | Orthogonal Frequency-Division Multiplexing |
| **OFDMA** | Orthogonal Frequency-Division Multiple Access |
| **OFDM/OQAM** | Orthogonal Frequency-Division Multiplexing with Offset-QAM subcarrier modulation |
| **OFDP** | Orthogonal Finite Duration Pulse |
| **OMA** | Open Mobile Alliance |
| **OMC** | Orthogonal Multiple Carrier |
| **OOB** | Out Of Band |
| **OQAM** | Offset Quadrature Amplitude Modulation |
| **O-QAM** | Orthogonally multiplexed QAM |
| **OQPSK** | Offset Quadrature Phase-Shift Keying |
| **OSIC** | Ordered Successive Interference Cancellation |

| | |
|---|---|
| **PaIC** | Partial Interference Cancellation |
| **PAM** | Pulse Amplitude Modulation |
| **PAPR** | Peak-to-Average Power Ratio |
| **PBS** | Polarization Beam Splitter |
| **PCI** | Perfect Channel Information |
| **PDU** | Protocol Data Unit |
| **PDM** | Polarization Division Multiplexing |
| **Ped-A** | Pedestrian-A [channel model] |
| **PF** | Proportional Fair |
| **PFF** | Prototype Filter Function |
| **PHY** | PHYsical layer |
| **PL** | Programmable Logic |
| **PLC** | Power Line Communications |
| **PLL** | Phased-Locked Loop |
| **PMR** | Professional (or Private) Mobile Radio |
| **PN** | Pseudo Noise |
| **PoC** | Push over Cellular |
| **POCS** | Projection Onto Convex Sets [algorithm] |
| **P-OFDM** | Pulse-shaped OFDM |
| **POP** | Pair Of Pilots |
| **PPDR** | Public Protection and Disaster Relief |
| **PPN** | PolyPhase Network |
| **PR** | Perfect Reconstruction |
| **PRACH** | Physical Random Access Channel |
| **PRB** | Physical Resource Block |
| **PRT** | Peak Reserved Tone |
| **PS** | Polarization Scrambler |
| **PSD** | Power Spectral Density |
| **PSK** | Phase-Shift Keying |
| **PSN** | Public Safety Networks |
| **PSS** | Primary Synchronization Signals |
| **PSWF** | Prolate Spheroidal Wave Function |
| **PTT** | Push-To-Talk |
| **PUSCH** | Physical Uplink Shared CHannel |
| **QAM** | Quadrature Amplitude Modulation |
| **QAM-FBMC** | Quadrature Amplitude Modulation-FilterBank MultiCarrier |
| **QCQP** | Quadratically Constrained Quadratic Program |
| **QR** | Quasi-Rectilinear |
| **R** | Rectilinear |
| **RACH** | Random Access CHannel |
| **RAM** | Random Access Memory |
| **RAN** | Radio Access Network |
| **RB** | Resource Block |
| **RB-F-OFDM** | Resource Block Filtered-OFDM |
| **RC** | Raised Cosine |
| **RE** | Resource Element |
| **RF** | Radio Frequency |

| | |
|---|---|
| **RLS** | Recursive Least Squares |
| **RMS** | Root Mean-Squared [error] |
| **RMSE** | Root Mean-Squared Error |
| **RR** | Round Robin |
| **RRC** | square Root Raised Cosine |
| **RRM** | Radio Resource Management |
| **RTL** | Register Transfer Level |
| **RX** | Receiver |
| **SAIC** | Single-Antenna Interference Cancellation |
| **SAS** | Subcarriers Assignment Scheme |
| **SAW** | Surface Acoustic Wave |
| **SC** | Single-Carrier |
| **SC-FDMA** | Single-Carrier Frequency-Division Multiple Access |
| **SDR** | Signal-to-Distortion power Ratio |
| **SDR** | Software Defined Radio |
| **SE** | Spectral Efficiency |
| **SEL** | Soft Envelope Limiter |
| **SER** | Symbol Error Rate |
| **SFB** | Synthesis FilterBank |
| **SFBC** | Space Frequency Block Code |
| **SIC** | Successive Interference Cancellation |
| **SIMO** | Single-Input-Multiple-Output |
| **SINDR** | Signal-to-Interference-plus-Noise-and-Distortion Ratio |
| **SINR** | Signal-to-Interference-plus-Noise Ratio |
| **SIR** | Signal-to-Interference Ratio |
| **SISO** | Single-Input-Single-Output |
| **SLNR** | Signal-to-Leakage-plus-Noise Ratio |
| **SLR** | Signal-to-Leakage Ratio |
| **SM** | Spatial Multiplexing |
| **SMFB** | Sine-Modulated FilterBank |
| **SMT** | Staggered-modulated MultiTone |
| **SNDR** | Signal-to-Noise-and-Distortion Ratio |
| **SNR** | Signal-to-Noise Ratio |
| **SoC** | System on Chip |
| **SO** | Second-Order |
| **SOI** | Signal-Of-Interest |
| **SOP** | State Of Polarization |
| **SOS** | Second-Order Statistics |
| **SPM** | Self-Phase Modulation |
| **SRAM** | Static Random Access Memory |
| **SSMF** | Standard Single Mode Fiber |
| **SSPA** | Solid-State Power Amplifier |
| **SSS** | Secondary Synchronization Signals |
| **STBC** | Space-Time Block Code |
| **STC** | Space-Time Coding |
| **STFT** | Short-Time Fourier Transform |
| **STO** | Symbol Timing Offset |

| | |
|---|---|
| **STTC** | Space-Time Trellis Coding |
| **SVD** | Singular Value Decomposition |
| **SW** | SoftWare |
| **TB** | Transport Block |
| **TBS** | Transport Block Size |
| **TD** | Time Domain |
| **TDD** | Time-Division Duplex |
| **TDL** | Tap Delay Line |
| **TDM** | Time-Division Multiplexing |
| **TETRA** | Terrestrial Trunked Radio |
| **TETRAPOL** | Terrestrial Trunked Radio for POLice |
| **T-F** | Time-Frequency |
| **TFL** | Time-Frequency Localization |
| **TGF** | Tight Gabor Frame |
| **THP** | Tomlinson–Harashima Precoder |
| **TLO** | Time-Limited Orthogonal |
| **TMO** | Trunked Mode Operation |
| **TMUX** | TransMUltipleXer |
| **TR** | Tone Reservation |
| **TS-IC** | Two-Stage Interference Cancellation |
| **TS-OSIC** | Two-Stage Ordered Successive Interference Cancellation |
| **TTI** | Time Transmit Interval |
| **TU** | Typical Urban [channel model] |
| **TV** | Television |
| **TVWS** | Television White Space |
| **TWTA** | Traveling-Wave Tube Amplifiers |
| **TX** | Transmitter |
| **UE** | User Equipment |
| **UFMC** | Universal Filtered MultiCarrier |
| **UF-OFDM** | Universal Filtered OFDM |
| **UL** | UpLink |
| **USRP** | Universal Software Radio Peripheral |
| **VDSL** | Very-high-bit-rate DSL |
| **Veh-A** | Vehicular-A [channel model] |
| **Veh-B** | Vehicular-B [channel model] |
| **WCP** | Windowed Cyclic Prefix |
| **WCP-OFDM** | Weighted Cyclic Prefix OFDM |
| **WDM** | Wavelength-Division Multiplexing |
| **WiMAX** | Worldwide interoperability for Microwave Access |
| **WISP** | Wireless Internet Service Providers |
| **WL** | Widely Linear |
| **WLP** | Widely Linear Processing |
| **WOLA** | Weighted OverLap and Add |
| **WRAN** | Wireless Regional Area Network |
| **WSD** | White Space Device |
| **WSSUS** | Wide-Sense Stationary Uncorrelated Scattering |
| **XML** | eXtensible Markup Language |
| **XPM** | Cross Phase Modulation |

| | |
|---|---|
| **ZF** | Zero Forcing |
| **ZT** | Zak Transform |
| | |
| **N** | set of (positive) natural numbers |
| **R** | set of real numbers |
| **C** | set of complex numbers |
| **diag(x)** | diagonal matrix with the vector **x** on its diagonal |
| **diag(X)** | column vector of the diagonal entries of the matrix **X** |
| **tr**($\cdot$) | trace of a matrix |
| $(\cdot)^{\mathrm{T}}$ | transposition |
| $(\cdot)^{\mathrm{H}}$ | Hermitian transposition |
| $(\cdot)^{*}$ | complex conjugation |
| $\otimes$ | Kronecker product |
| $\oslash$ | elementwise division |
| $\star$ | linear convolution |
| $\circledast$ | circular convolution |
| $\mathcal{N}(\cdot, \cdot)$ | normal distribution |
| $\mathcal{CN}(\cdot, \cdot)$ | complex normal distribution |
| **Pr** | probability |
| $\mathbb{E}\{\cdot\}$ | expected value |
| $\lfloor x \rfloor$ | largest integer not larger than $x$ |
| $\lceil x \rceil$ | smallest integer not smaller than $x$ |
| $((\cdot))_{M}$ | modulo $M$ |
| **j** | imaginary unit, $\mathrm{j} = \sqrt{-1}$ |
| **e** | basis of the natural logarithm |
| **Re**$\{\cdot\}$ or $(\cdot)^{\mathrm{R}}$ | real part of a complex quantity |
| **Im**$\{\cdot\}$ or $(\cdot)^{\mathrm{I}}$ | imaginary part of a complex quantity |
| $|\cdot|$ | absolute value or modulus of a (complex) number |
| $\angle\{\cdot\}$, **arg**$\{\cdot\}$ | argument (angle) of a complex number; takes values in $[0, 2\pi)$ |
| $\|\cdot\|$ | norm |

# Index

## Symbols

4G, 55, 150
5G, 149, 158, 452, 514

## A

Active constellation extension (ACE), 473
Alamouti coding, 148, 408, 416
Analysis channelizer, 108, 124
Analysis filter bank (AFB), 82, 101, 106,
    110, 167, 200, 222, 260, 286, 300,
    306, 318, 329, 343, 361, 378, 396
AWGN channel, 236

## B

Balian–Low Theorem (BLT), 138
Beamformers, 348
Bell Labs layered space-time (BLAST)
    coding, 377, 400
Biorthogonal frequency division
    multiplexing (BFDM), 20
Blind equalization, 327
Blind estimation, 290, 381
Blind synchronization, 241
Broadband PPDR, 57

## C

Carrier frequency offset (CFO), 77, 236, 242
Channel
    equalization, 20, 183, 244, 311, 323, 385,
        469, 495
    estimation, 150, 168, 262, 271, 291, 316,
        325, 379, 400, 429, 442, 502, 517,
        536
    matrix, 22, 385, 416, 424, 435
Channel frequency response (CFR), 113,
    182, 200, 232, 260, 342, 356, 370,
    377, 431
Channel impulse response (CIR), 100, 259,
    307, 345, 368, 381, 432, 444, 468
Channel models
    extended typical urban (ETU), 211, 289,
        321, 391
    extended vehicular-A (EVA), 387
    Vehicular-A (Veh-A), 326, 411

Channel-aware allocator, 455
Channelization, 42, 94, 98, 108, 121, 158,
    170, 182, 245
Chromatic dispersion (CD), 75, 82
Circular (or cyclic) convolution (CC), 146
Closed-form estimator (CFE), 236
Co-channel interference (CCI), 311
Coded orthogonal frequency-division
    multiplexing (COFDM), 134
Coexistence, 38, 46, 61, 516
Cognitive radio, 20, 36, 45
Conjugate symmetry property (CSP), 231
Constant modulus algorithm (CMA), 76, 327
Constant norm algorithm (CNA), 327
Coordinated beamforming schemes, 352,
    370, 427
Coverage, 56
Cross-correlation estimator (CCE), 236
Cyclic prefix orthogonal frequency-division
    multiplexing (CP-OFDM), 11, 61,
    67, 100, 143, 233, 262, 303, 356,
    379, 408, 422, 516

## D

Decimation, 93, 109, 140
Decision feedback equalizer (DFE), 308,
    367, 400
Decoding matrices, 352, 354, 425
Device-to-device (D2D) communication, 66,
    447
DFT FilterBank (DFT-FB), 89, 101
Digital audio broadcasting (DAB), 134
Digital predistortion (DPD), 478
Direct mode of operation (DMO), 65, 537
Discrete Fourier transform (DFT), 99
Discrete multitone (DMT), 133
Discrete wavelet multitone (DWMT), 140
Discrete Zak transform (DZT), 164
Dispersive selected mapping (DSLM), 474
Dispersive tone reservation (DTR), 476
Downsampling, 106, 111, 186, 281, 396, 497
Dual asynchronous distributed pricing
    (DADP) algorithm, 449
Duplex separation, 63